We are becoming used to the fact that the climate is changing due to human influences, but in our lifetimes most of us will be much more affected by decadal climate variability. This book provides a comprehensive review, covering DCV in the Pacific, Atlantic and Indian Oceans as well as in tropical cyclones and extratropical winds. Highlighted throughout the book are potential links between DCV and solar variability, a fascinating topic that has engaged our minds for centuries. Written by an expert with more than 30 years' experience, this book should be an invaluable resource for students and researchers interested in how our climate will evolve over the coming decades.

Doug Smith
Decadal Climate Prediction Leader, Meteorological Office Hadley Centre, Exeter, UK

Provides a comprehensive account of decadal time-scale variations in the global and regional climate, that are of societal relevance. There is a detailed discussion on solar influences on the Earth's climate variability, along with supporting examples of natural DCV and their interactions with human-induced climate change. This book will be useful for policy makers also.

Krishnan Raghavan
Senior Scientist, Indian Institute of Tropical Meteorology, Pune, India

This book is a tour de force by the author who has spent his career studying decadal climate variability. He brings new insights to the vast scope of this topic, providing clearly understandable descriptions of the various aspects. This compelling treatment of a highly relevant area of climate science will clarify these complex topics for readers who have varying familiarities with science. The book can be read as not only a general overview of decadal climate variability, but also as a primer on the scientific details of such variability and predictions. Additionally, it serves as a guide to the impacts that arise as a consequence of climate variability on the timescales of seasons to decades.

Gerald Meehl
Senior Scientist, National Center for Atmospheric Research, Boulder, Colorado, USA

Vikram Mehta has been actively involved in decadal climate variability (DCV) research for decades. This book is an eyewitness report about the fascinating science of DCV.

Mojib Latif
GEOMAR Helmholtz Centre for Ocean Research, Kiel, Germany

Natural Decadal Climate Variability

Natural Decadal Climate Variability

Phenomena, Mechanisms, and Predictability

Vikram M. Mehta

CRC Press is an imprint of the
Taylor & Francis Group, an informa business

First edition published 2020
by CRC Press
6000 Broken Sound Parkway NW, Suite 300, Boca Raton, FL 33487-2742

and by CRC Press
2 Park Square, Milton Park, Abingdon, Oxon, OX14 4RN

CRC Press is an imprint of Taylor & Francis Group, LLC

ISBN: 978-1-138-48414-6 (hbk)
ISBN: 978-1-351-05290-0 (ebk)

Typeset in Times
by codeMantra

Contents

Foreword

"It was the best of times, it was the worst of times; it was the age of wisdom, it was the age of foolishness; it was the epoch of belief, it was the epoch of incredulity; it was the season of light, it was the season of darkness; it was the spring of hope, it was the winter of despair; we had everything before us, we had nothing before us; we were all going direct to Heaven, we were all going direct the other way – in short, the period was so far like the present period, that some of its noisiest authorities insisted on its being received, for good or for evil, in the superlative degree of comparison only". – Charles Dickens, *A Tale of Two Cities*

Charles Dickens (1812–1870 Current Era (CE)) might be describing the status of climate research and its societal applications in the early 21st century CE. All components of the climate system have never been observed more or better than at this time at the end of the second decade of the century – from *in situ* sensors in the atmosphere measuring winds, temperatures, and atmospheric constituents; on land measuring soil moisture and temperature, and snow and ice extent and thickness; and in oceans measuring currents, temperature, and salinity down to thousands of meters depth; to sensors on satellites in Earth's orbit indirectly estimating atmospheric, land and vegetation, ocean, and ice and snow properties by measuring emitted and reflected radiation at microwave to visible wavelengths. We know more about physics, chemistry, and biology of the Earth System now than at any previous time in the Earth's history. But, nations quibble about sharing the data from the Earth observing systems because they fear encroachments on their national security and economic interests. Computers capable of mind-boggling processing power and mathematical-computer models of the Earth System have made moderately skillful predictions of climate phenomena such as the El Niño–La Niña events possible several seasons in advance, and efforts are now underway to predict climate and its impacts several years to several decades in advance. But, efforts to apply these predictions for societal benefits are stymied by a general lack of applicable research for climate adaptation in various societal sectors and the availability of meager resources for implementation of predictions and associated adaptation options. Then, there is the central climate issue of our times; namely, potential global climate change during the 21st century CE and beyond due to human activities on the Earth in the last several hundred years. The climate modeling community is well on the way to developing the necessary Earth System Models and associated data processing and analysis techniques to foretell climate one or more decades in advance, including potential anthropogenic climate changes (ACCs) in the future, but further progress is mired in real or imaginary fears over economic impacts of recommended actions to slow or reverse possible global climate changes, and in the greed of vested interests to maintain the status quo. We live in the Age of Information, and the general public is inundated with detailed, often conflicting, information about the ever-changing climate and confused, even disinterested, by the public and often acrimonious debate among climate scientists. So-called "climate change deniers" prey on this public confusion and ignorance to spread disinformation and paralyze actions.

In the meantime, societal costs due to floods, droughts, tropical storms, heat waves, and other weather/climate events are rising sharply. The United Nations Office for Disaster Risk Reduction reported[1] in 2018 CE that 91% of all disasters between 1998 and 2017 CE were due to such weather/climate events, which cost US $2,245 billion. During this period, over 500,000 people lost their lives; and well over one billion people were injured, rendered homeless, displaced, or in need of emergency assistance due to such events. Much of this burden of weather/climate disasters falls on developing countries, especially on the poor in such countries. Can we do something to reduce such losses and human suffering? Can we ensure water, food, energy, and health securities for all by adapting to adverse weather/climate events and benefiting from them when possible? We can if we put the understanding, and modeling and prediction capabilities which have been built up over the last 100 years to work for us. We can if we develop adaptation options and if we can predict with usable skill the worldwide climate a decade or more in advance. But, most importantly, we can if we know why extreme weather/climate events occur, if we can attribute such events to natural climate variability or ACC, and if we can predict them and develop adaptation options for various affected societal sectors. To mention one example of the usefulness of decadal climate variability (DCV) information, an estimate of the monetary value of DCV information to the agriculture sector in the Missouri River Basin (MRB) – the largest river basin in the USA and a major "bread basket" of the world – is that the correct prediction of important DCV phenomena one year in advance can be worth approximately US $80 million annually; even the correct prediction of just the phase of the DCV phenomena one year in advance can realize a sizeable fraction of this monetary value. These monetary benefits would be derived from adaptation of the MRB agriculture sector to the predicted DCV information (Fernandez et al., 2016; Mehta, 2017). A recent study has extended these estimates to the entire US agriculture enterprise and found that the benefits of a perfect DCV phase forecast one year in advance can be worth US $1.1 billion (Rhodes and McCarl, 2020). Extrapolated to the worldwide agriculture enterprise, monetary benefits can easily be many billions of US dollars annually.

The word "climate" today is synonymous with ACC. This is a fallacy that can be attributed to the climate science community's and the news media's understandable desire to focus the world's attention on ACC. In fact, however, the changing climate is due to both natural climate variability, including DCV, and ACC. In attributing observed climate changes to causes, it is very important to distinguish between DCV and ACC because of interactions between the two. Societal impacts associated with DCV, especially on precipitation and temperature, dry and wet hydrologic epochs, river flows, agriculture, fisheries, inland water-borne transportation, and other sectors are well documented, as are their socio-economic and political consequences catalyzing rises and falls of civilizations in the past, in my previous book (*Natural Decadal Climate Variability: Societal Impacts*; CRC Press; 2017). The purpose of the present book is to describe major DCV phenomena, their hypothesized causes

[1] www.unisdr.org/files/61119_credeconomiclosses.pdf

and mechanisms, and estimates of potential decadal climate predictability and the gathering momentum on actual decadal climate prediction. Important developments in the fledgling sub-field of interactions between DCV and ACC are described to highlight the importance of understanding and skillful prediction of DCV in the attribution and mitigation of ACC, and also in adapting to the changing climate.

When I started working in this field in late 1980s CE, the long history of DCV research was confined to statistical associations between climate variables and sunspot and lunar nodal cycles, and sporadic attempts to use these associations for climate and impacts prediction. When I undertook to model DCV as a purely internally generated coupled ocean–atmosphere phenomenon in my Ph.D. research, even some very well-known scientists on my supervisory committee at the Florida State University (FSU) were skeptical about the wisdom of wading into what was then a very uncharted field. In the 30 years since then, the field has made rapid and substantial strides in description of DCV phenomena, understanding their mechanisms with state-of-the-art climate models, and experimental – even tentatively operational – efforts to make skillful decadal climate and impacts predictions. I have been very fortunate to witness these astounding developments and contribute to them. So, this book is focused on presenting a history of DCV research in this Modern Era while providing a historical context going back over two centuries. A part of this history is about a renewed effort to understand and predict potential impacts of solar activity cycles on the Earth's climate, with a much greater emphasis on the physics and chemistry of the atmosphere–ocean system. This renewed effort is now attempting to understand how solar activity cycles can influence climate in the presence of internally generated DCV. The encouraging results, described in this book, of this effort bring vividly to life the words of L. Howard – a very well-known applied mathematician specializing in fluid dynamics and geophysical flows, first at the Massachusetts Institute of Technology and then at FSU later in his career – that the possibility of interactions between internally generated DCV and the centuries-long beliefs about solar and lunar cycle influences on climate would make for a very rich symphony of climate music. Results of research in the Modern Era bear witness to Howard's belief and point to more exciting developments waiting in the future. Unfortunately, Howard is not with us to see his words coming true. We must also thank the countless number of scientists and non-scientists whose firm belief in climate variability forced by solar activity cycles motivated them to analyze whatever meager (by today's standards) data were available and endure ridicule for their beliefs. They were the pioneers of DCV research.

Scientists pride themselves on their objectivity, but no one is infallible and some subjectivity in research and its description is inevitable, especially when there are competing narratives about phenomena and their mechanisms. So, I urge readers to use this book only as a starting point in their own journeys in climate research. The emphasis in this book is on steering clear of mathematical and statistical equations and arcane details of analysis techniques while summarizing essential knowledge and presenting as objective a view as possible. Important references to published papers and books are cited extensively to point readers to the details not described in this book. The intention is to provide an introduction to DCV and its predictability to graduate students, faculty members and other teachers and researchers, and anyone who is interested in learning about a very important component of the puzzle of the changing climate. This book also puts forward ideas about further avenues of research and gaps in the research done so far. Very importantly, it is a companion book to the vast collection of books, research articles, monographs, and articles in the news media about ACC. Lastly, the *Societal Impacts* book and this book – although both are focused on natural DCV and its impacts – are not in any way meant to refute or deny or cast aspersions on published research on ACC and its potential impacts.

In the Foreword to my *Societal Impacts* book, I described the motivation for my journey in the field of science research and applications over the last 40 years. There are several people to whom I owe immense gratitude for making the journey-within-the-journey towards this book possible. During my work in National Aeronautics and Space Administration (NASA) – Goddard Space Flight Center (GSFC) from 1990 to 2002 CE after finishing Ph.D., I had the good fortune to have W.K.M. Lau as a mentor who supported my independent development of research on DCV. I also benefited from nearly 10 years of research interactions with M. Suarez and P. Schopf in NASA-GSFC, who were busy unraveling mysteries of El Niño–Southern Oscillation as a coupled ocean–atmosphere phenomenon as the 1990s CE dawned. A fortunate introduction to T. Delworth (National Oceanic and Atmospheric Administration (NOAA) – Geophysical Fluid Dynamics Laboratory, USA) during a visit to Russia in 1992 CE on a US delegation led to a very fruitful research partnership; I also gratefully acknowledge my discussion with him about decadal variability of the North Atlantic Oscillation for this book. I thank Eric Lindstrom of the NASA – Physical Oceanography Program for his constant encouragement and support of DCV research over the last 20 years. I also thank M. Alexander (NOAA – Earth System Research Laboratory (ESRL), USA), M. Collins (University of Exeter, UK), C. Deser (National Center for Atmospheric Research (NCAR), USA), C. Folland (U.K. Meteorological Office-Hadley Centre (UKMO-HC), UK), J. Hurrell (NCAR), M. Latif (Universität Kiel, Germany), Z. Liu (Ohio State University, USA), G. Meehl (NCAR, USA), M. Newman (NOAA-ESRL, USA), S. Power (Bureau of Meteorology Research Centre, Australia), B. Rajagopalan (University of Colorado, USA), A. Scaife (UKMO-HC, UK), N. Schneider (University of Hawaii, USA), D. Smith (UKMO-HC, UK), R. Sutton (University of Reading, UK), W. White (Scripps Institution of Oceanography, USA), and many others who have been fellow travelers on this journey of DCV research from the early 1990s CE and whose research has stimulated my own work.

This book would not have happened without the limitless patience of the publishers CRC Press, since writing this book became a long-term project due to my other, numerous and more urgent, commitments. I am utterly grateful to my Editor, Irma Britton of CRC Press, for patiently and gently coaxing the *Societal Impacts* book and this book out of me. I thank Saranya P.N. and the team at codeMantra for managing copy-editing and proof stages of the book very patiently and cleanly, and Carly Cassano of Taylor & Francis for managing the production process very smoothly and responsively. Katherin Mendoza prepared graphics for the original research that is presented in this book. Katherin also acted as this book's "internal" Editor. Like the *Societal Impacts* book, this book would never have been completed without Katherin. My wife, Amita Mehta, a NASA scientist specializing in remote sensing of the Earth–atmosphere system, has supported my research journey with enthusiasm and stoicism; this journey would not have been possible without her support. Amita also patiently read through early drafts of all chapters of this book and gave constructive suggestions which have improved this book substantially. Last but not the least, my daughter, Unnati Mehta, encouraged me for many years to finish the two books and I am very happy to present her the second book.

This book is dedicated to my mother Nilprabha Mehta, a career school teacher specializing in languages, who taught me the virtues of patience, perseverance, and correct grammar. I hope that my writings have reflected these virtues.

Vikram M. Mehta
2 April 2020

Author

Dr. Vikram Mehta received his M.Sc. degree in Physics in 1977; and Post-graduate Diploma in Space Sciences and Their Applications in 1979, both from Gujarat University, Ahmedabad, India. He was an Indian Space Research Organization (ISRO) Research Fellow at the Space Applications Centre, ISRO, Ahmedabad, India, from 1979 to 1982, working on Microwave Remote Sensing. He then studied Upper Atmosphere Physics at the University of Saskatchewan, Canada, from 1982 to 1984, receiving a Post-Graduate Diploma in Physics. A strong interest in more applications-oriented scientific research led him to the Department of Meteorology at the Florida State University, Tallahassee, Florida, USA, where he received his M.S. degree in 1986 and Ph.D. degree in Meteorology in 1990. Vikram was then a research scientist for over 12 years in NASA – Goddard Space Flight Center, further specializing in identification of ocean–atmosphere interactions that give rise to natural decadal climate variability (DCV). Strongly motivated to use climate science for societal benefits, Vikram founded the Center for Research on the Changing Earth System (CRCES), a non-profit, scientific research organization in Columbia, Maryland, USA, in 2002, with initial funding from NASA. In addition to funding from NASA, CRCES has also received financial support from the National Oceanic and Atmospheric Administration, the National Science Foundation, the U.S. Department of Agriculture – National Institute of Food and Agriculture, and the U.S. Army Corps of Engineers – Institute for Water Resources for research and other activities on DCV and its societal impacts. Currently, Vikram is the President and Executive Director of CRCES.

Under his leadership, CRCES has studied the physics of DCV; its predictability; and DCV impacts on water resources, crop yields, and inland water-borne transportation in the Missouri and Mississippi River Basins in the USA. CRCES has also estimated the monetary value of DCV information to the agriculture sector in the Missouri River Basin and has developed options for the agricultural sector in the Missouri River Basin to adapt to DCV. Vikram's major research interests are understanding and predicting DCV; assessment of DCV impacts on water resources, agriculture, public health, and the economy; and climate and public policy. Vikram has published over 150 research papers and conference/workshop presentations. CRC Press published Vikram's book in 2017, titled *Natural Decadal Climate Variability: Societal Impacts*, about worldwide DCV impacts on water, crop production, fisheries, inland water-borne transportation, and hydro-electricity generation; and socio-economic and political consequences of impacts on these societal sectors. Vikram was one of the founders of the Indian Centre for Climate and Societal Impacts Research (ICCSIR) in Ahmedabad, India, and led its development from 2008 to 2013. For his pioneering research and leadership in the worldwide climate community, the Non-Resident Indian Welfare Society of India presented Vikram the Mahatma Gandhi Medal in a ceremony in the House of Lords in the U.K. Parliament in October 2012 and the Hind Ratna (Jewel of India) Award in New Delhi in January 2015. Among his major hobbies are piloting single-engine airplanes, reading, and listening to music.

1 Prologue

1.1 PREPARING FOR THE JOURNEY

Cycles of light and water, and adaptations to them, are at the heart of the human story on the Earth. Early humans observed the diurnal (daily) solar, the monthly lunar, the annual solar, and other astronomical cycles. They also observed that amount and duration of daylight, precipitation (rain, snow), and temperature also undergo some of these cycles. River flows and lake levels, vegetation growth and decay, and edible fruits and vegetation also undergo some of these cycles. Eventually, humans learned to plant and harvest crops, use and store water, and plan journeys by land or water according to these cycles. Rituals and religious practices such as annual/semi-annual harvests, festivals to celebrate the appearance of the Sun over a particular location, sacrifices to propitiate Gods, and ceremonies marking important passages of life, to name a few, were linked to the light and water cycles. Astronomers in the Sindhu (Indus) Valley Civilization between 2,000 and 3,000 years Before Current Era (BCE) developed rules to track cycles of the Sun and the Moon for ritual purposes (Subbarayappa, 1989). Babylonian astronomers around 1,000 BCE also recognized that astronomical phenomena are periodic and applied mathematics to predict their motions (Pingree, 1998). In the 3rd century BCE, as a precursor to modern statistical prediction, Babylonian astronomers compiled observations of individual planet's cycles, and favorable and unfavorable events which occurred at specific times during the cycles. These data bases were then used to predict auspicious times for events. These astronomers also built mathematical models of planetary cycles and used them to predict phases of a planet's cycle (Pingree, 1998). It is important to mention that except for the Sun's daily and annual cycles, the other astronomical cycles were used as predictors of political/societal events without any scientific hypothesis or understanding about their possible connections. But, the idea that nature's cycles govern human and other life on the Earth is rooted in the human experience for millennia and has contributed very substantially to human evolution. Perhaps this idea about nature's cycles also subconsciously underlies naming of climate phenomena as oscillations – for example, the Southern Oscillation, the North Atlantic Oscillation, the Pacific Decadal Oscillation (PDO), and the Atlantic Multidecadal Oscillation.

Other cycles noticed more recently in the last 200–300 years are the approximately 11 (Schwabe, 1844)- and 22 (Hale, 1908; Hale and Nicholson, 1925)-year cycles of spots on the visible disk of the Sun, and the 18.6-year cycle of nodes of the Moon's orbit around the Earth (Bradley, 1728). Continuing with the tradition of associations of astronomical cycles with water and food, the Sunspot cycles and the lunar nodal cycle were also associated with approximately cyclical variations in dryness–wetness and food production, and this scientific interest has continued into the 21st century (see, for example, Herschel (1801), Carrington (1863), Poey (1873), Jevons (1879), Proctor (1880), Poynting (1884), Chambers (1886), Currie (1974, 1976, 1981,

1984, 1993), Currie and Fairbridge (1985), Currie and O'Brien (1988), Currie et al. (1993), O'Brien and Currie (1993), Cook et al. (1997), Cerveny and Shaffer (2001), Yndestad (1999, 2003, 2006), Garnett et al. (2006), Pustil'nik and Yom Din (2004a, b, 2009, 2013), McKinnell and Crawford (2007), Love (2013), and others). Due to the one to two decade periods of the Sunspot and lunar nodal cycles, the climate variability associated with them became known as decadal climate variability (DCV).

That only external forcings can generate DCV was the paradigm which, due to insufficient understanding of ocean and atmosphere dynamics, was based on the belief that any climate variations must be forced externally. This paradigm started to shift with the Bjerknes (1969) hypothesis that interactive coupling between atmospheric and oceanic dynamics may be able to generate climate variations even in the absence of variations in external forcing. In this new paradigm of internally generated climate variability, sea surface temperatures (SSTs) (and related air–sea heat fluxes) and surface wind stress were the agents which coupled oceanic and atmospheric dynamics and thermodynamics. It was – and still is – generally believed that the atmosphere cannot generate variability at timescales longer than one or two years, at the most, due to its relatively low thermal and mechanical inertia, whereas the much larger inertias of the ocean and its slow to very slow waves and circulations may be able to generate interannual to multidecadal and longer timescale variability on their own. Theoretically, however, nonlinear dynamics of atmospheric flow can also generate interannual to multidecadal variability as we will see later in this book.

In the late 1980s Current Era (CE), following the development of a theory of the El Niño (EN)–La Niña (LN) events as an internally generated, coupled ocean–atmosphere phenomenon, tentative efforts began to model DCV also as an internally generated, coupled ocean–atmosphere phenomenon (see, for example, Mehta (1991, 1992), Latif and Barnett (1994), Chang et al. (1997), Mehta and Delworth (1995), Delworth and Mehta (1998)). In parallel, the end of the Cold War in the early 1990s CE enabled public releases of long-term ocean temperature and salinity data that the United States (US), British, Soviet, and other navies held in their archives. These and other data collected by merchant ships since the 1850s CE soon became available to researchers for analyses. Several poorly quantified or hitherto unknown phenomena emerged from these analyses such as decadal variability of the tropical Atlantic SST gradient (TAG for brevity), the PDO/the Interdecadal Pacific Oscillation (IPO), and decadal variability of frequency and intensity of the EN-LN phenomenon (see, for example, Deser and Blackmon (1993), Mehta and Delworth (1995), Balmaseda et al. (1995), Zhang et al. (1997), Rajagopalan et al. (1998), Mehta (1998), Power et al. (1999)). Then, analyses of archived data and their assimilation in models of the atmosphere and the oceans made possible discoveries of decadal–multidecadal variability in long known atmospheric phenomena such as the North Pacific Oscillation, the North Atlantic Oscillation, and the Southern Oscillation; and in the tropical warm pools (TWPs) (Wang and Mehta, 2008), adding further richness to the symphony of music generated by DCV phenomena. What are these DCV phenomena and what are their causes? What do we know about mechanisms of DCV? Do solar cycles play any role in DCV? Are evolutions of DCV phenomena predictable? These questions are addressed in this book. Worldwide societal impacts of DCV phenomena, and the need for their understanding and prediction were extensively and intensively described in Mehta (2017), so they are not included in this book.

In the climate science literature, there are three words used to describe non-steady climate – cycles, variations/variability, and change. What do these words convey? The *Merriam-Webster Dictionary* defines these three words as follows.[1] A cycle, as described in the previous paragraphs, can convey at least two meanings – (1) an interval of time during which a sequence or a recurring succession of events or phenomena is completed; and (2) one complete performance of a vibration, electric oscillation, current alternation, or other periodic process. When the ancient civilizations were observing the Sun's daily and annual motions in the sky and their associations with light and water, they referred to the phenomena as cycles in the first meaning. The same meaning was conveyed when cyclical motions of planets and stars were observed and associated with favorable or unfavorable events. Even today, farmers refer to dry and wet sequences (with drought and flood as extremes) as cycles. Neither the ancient civilizations nor today's farmers meant strictly periodic oscillations with constant amplitude as in the second meaning. The definition of variability is the ability to vary, and variation is the act or process of varying. A change is defined in at least two ways – (1) to make a shift from one to another; and (2) to undergo a transformation, transition, or modification. In the context of climate, variability and change have specific meanings. Climate variability is defined as precipitation and temperature undergoing cyclic or nearly cyclic (in all its meanings) variations, and climate change is defined as precipitation and temperature undergoing a uni-directional change; that is, they only increase or decrease. Over time, variables in addition to precipitation and temperature have also been included in the definition of climate.

Climate impacts or at least influences many aspects of human life: water–air–land-–vegetation, commonly referred to as the environment; agriculture; health; transportation; natural resources such as forests and fisheries; and birds, animals, and terrestrial and marine ecosystems. Impacts/influences of DCV on some of these societal sectors are described in detail by Mehta (2017). Climate impacts/influences affect quality of life and economies from local to global scales to varying extents. Let us take the example of one of the most important substances for life on the Earth, namely, freshwater. Freshwater is the lifeblood of civilizations. Ancient civilizations developed near major rivers such as the Nile, the Euphrates, the Tigris, the Sindhu (Indus), the Ganga (Ganges), and the Yangtze. Spurts in the evolution of civilizations on their banks were related to extremes of flow in these and other important rivers driven by climate variability and change. The inter-relationships among availability of freshwater, the rise of civilizations, and societal well-being or vulnerability depend heavily on population and population density, internal cohesion, infrastructure, and the resilience deliberately or inadvertently built into the society. The vulnerability also depends on socio-economic-political-religious dynamics of the society. Strains due to these dynamics, sometimes with geopolitical consequences, can be lessened or exacerbated by climate variability and changes. Rising populations and growing urbanization can magnify societal vulnerabilities as can land use–land cover changes.

[1] www.merriam-webster.com/dictionary.

Thus, climate variability, especially at decadal to centennial timescales, and change are very important factors in the rise and fall of civilizations (Mehta, 2017).

After this brief introduction, natural climate variability and anthropogenic climate changes are introduced in Sections 1.2 and 1.3, respectively. Then, in preparation for subsequent chapters, major Earth System components important in natural climate variability and anthropogenic climate changes, and the annual cycles of their major attributes are introduced in Section 1.4. Next, important caveats of this book are described in Section 1.5. Finally, the plan of the book is outlined in Section 1.6.

1.2 NATURAL CLIMATE VARIABILITY

Electromagnetic radiation from the Sun is the primary driver of the Earth's climate system. Due to the tilt of the Earth's rotation axis and the Earth's orbit around the Sun, the amount of solar radiation reaching the Earth's surface is a function of latitude and time. At any time in the year, there is a gradient of solar radiation reaching the Earth's surface. The annual cycle of the Sun's apparent north–south movement across the Earth's surface defines seasons. In this apparent north–south movement, the Sun's most northern position is at 23.5°N latitude (the Tropic of Cancer) on approximately June 21; then, it starts to move southward and reaches the Equator on approximately September 21 and its most southern position at 23.5°S latitude (the Tropic of Capricorn) on approximately December 21. Then, the apparent position starts to move northward again reaching the Equator on approximately March 21 and the most northern position again on approximately June 21. These four astronomical events are known in the Northern Hemisphere (NH) as summer solstice, autumnal equinox, winter solstice, and vernal equinox, respectively. The naming of these events is the opposite in the Southern hemisphere (SH). Annual cycles of weather and climate generally follow the Sun, but the Earth System's response is delayed due to thermal inertia of oceans, snow-ice, and land; and due to heat transports by winds and ocean currents. Incidentally, the latitude belt between the Tropic of Cancer and the Tropic of Capricorn is usually referred to as the tropics or the tropical belt.

Climate is traditionally defined as 30-year averages of precipitation[2] and air temperature (Baede, 2015) at any location on the surface of the Earth. This definition has evolved to include averages over shorter periods down to three months; and to include variables such as near-surface winds and humidity, sea-level pressure (SLP), soil moisture, and SST. Weather is the day-to-day or week-to-week variability of precipitation, temperature, clouds, winds, and storminess. In other words, weather is the instantaneous realization of climate. Climate and weather at any location depend on latitude, longitude, altitude, land cover, and proximity to oceans/seas and other large water bodies.

There are several external and internal causes of climate variability and changes. These causes and the resulting climate variability are introduced in this section. External causes are variations/changes in solar emissions; variations in lunar gravitational force on the Earth – due to the Moon's orbit around the Earth and the

[2] In meteorology, precipitation includes all liquid and solid forms of water falling from the atmosphere, such as rain, snow, sleet, hail, ice, and other forms.

precession of the Moon's orbit around the Earth with an 18.6-year period; and effects of the Sun's magnetic field variations and their interactions with solar and galactic cosmic rays. Major internal causes of climate variations and changes are ocean–atmosphere interactions; land– atmosphere interactions; ice–ocean–atmosphere interactions; nonlinear dynamical interactions within the atmosphere and the oceans; volcanic aerosols such as sulfur dioxide, silicates, hydrochloric acid, and ash; and natural aerosols such as sand and sea salt.

Climate variations have a very broad frequency spectrum ranging from one cycle in two years to one cycle in hundreds of thousands of years or even longer. Natural interannual – or, year-to-year – climate variability, such as EN-LN events, and their predictability have attracted much attention from researchers and the general public since the second quarter of the 20th century CE. As mentioned in the introduction to this chapter, natural DCV research began in mid-19th century CE, but the pace, understanding, and excitement have increased in the last 15–20 years because of the availability of historical, archived observations; and increasing realism of climate simulated by complex Earth System Models (ESMs). Moreover, characteristics of natural interannual climate variability and its predictability vary at decadal timescales. Natural DCV can also confound the detection, quantification, and prediction of anthropogenic climate change. Last but perhaps the most important, the entire spectrum of natural climate variability and anthropogenic climate changes must be predicted in order to predict climate evolution over multiyear to multidecadal periods. The importance of DCV research is being increasingly recognized in the contemporary era, including by international research programs such as the World Climate Research Program. Thus, DCV research in the last 15–20 years is driven by the twin motivations of intellectual curiosity and the need to assess and predict societal impacts. The motivation for this book is provided by the need to synthesize knowledge of empirically identified phenomena, mechanisms of these phenomena, and fledgling efforts to predict the future evolution of these phenomena.

1.3 ANTHROPOGENIC CLIMATE CHANGES

There are also anthropogenic, or originating in human activities, causes of climate variability and change. Anthropogenic variations and changes in atmospheric constituents such as carbon dioxide, methane, sulfur dioxide, and combustion products such as ash and soot can cause climate variations and changes. Effects of carbon dioxide and methane, so-called "greenhouse gases", are well known.[3] These gases in the atmosphere absorb heat emitted by the Earth's surface and re-radiate it towards the surface which increases surface temperature. Warmer SSTs resulting from this greenhouse effect contribute towards increased evaporation. This increased water vapor in the atmosphere also absorbs heat from the surface, re-radiates it back towards the surface, and thus further warms the surface. This positive feedback to the original heating due to increased anthropogenic greenhouse gases increases the total heating several times compared to the direct effects.[4] Sulfur dioxide and its

[3] https://climate.nasa.gov/causes/.

[4] https://www.ncdc.noaa.gov/monitoring-references/faq/greenhouse-gases.php?section=watervapor.

compounds are more reflective due to their lighter colors and reflect more of the Sun's visible radiation, which reduces the amount of light reaching the Earth's surface. This reduction in the incoming solar radiation cools the Earth–atmosphere system.[5] Soot is darker in color and so absorbs more of the Sun's visible radiation, warming the air and surfaces on which it is deposited.[6] Thus, substantial variations and changes in these and other emissions from industrial processes, internal combustion engines, and other sources can influence/impact climate.

Land use–land cover changes such as deforestation, afforestation, urbanization, and agriculture can also cause climate variations and changes. Changes in vegetation cover change the reflectivity of the surface and change the amounts of the Sun's visible radiation reflected from and absorbed by the surface. Vegetation cover changes also influence transfers of water vapor, heat, and momentum between the surface and the atmosphere as outlined in Section 1.4.5. Increases in the built environment – such as buildings, and roads and other paved surfaces – also influence surface–atmosphere heat transfer processes such that more heat is retained by the built environment creating the so-called urban heat island effect.[7] Increased numbers of vehicles with internal combustion engines and associated air pollution, storm water flooding, and changes in land cover are some of the other climatic consequences of urbanization. All these anthropogenic effects of the Earth's climate have been observed or simulated with process models as well as with global ESMs.

The study of interactions between natural climate variability and anthropogenic climate changes is in its infancy. The current knowledge about the interactions is described in Chapter 8.

1.4 MAJOR EARTH SYSTEM COMPONENTS

Seasons have been defined traditionally by NH mid-latitude meteorologists as three-month averaging periods which reflect seasons in NH mid-latitudes according to the Sun–Earth orbital and rotational movements described earlier. These three-month periods are December–January–February (DJF; winter), March–April–May (MAM; spring), June–July–August (JJA; summer), and September–October–November (SON; autumn or fall). Seasons in SH, of course, follow the opposite cycle to that in NH. Seasons in monsoonal countries are defined according to rainy periods; for example, the Indian summer monsoon season is from June to September and the winter monsoon season is from December to March, the East Asian monsoon season is from May to August, the Australian monsoon season is from September to February, the West African monsoon season is from June to October, and the North American monsoon season is from late June to September. The NH, mid-latitude definition of seasons is generally used in this book, and other definitions are used as appropriate.

To set the stage for describing roles of the atmosphere, the oceans, the land–vegetation system, and ice and snow in DCV phenomena, fundamental attributes

[5] http://volcanology.geol.ucsb.edu/gas.htm.

[6] https://blogs.ei.columbia.edu/2016/03/22/the-damaging-effects-of-black-carbon/.

[7] https://scied.ucar.edu/longcontent/urban-heat-islands.

of these Earth System components, including the average annual cycle of some important variables of each component, are described here.

1.4.1 The Atmosphere

The Earth's atmosphere is a gaseous fluid composed of 78% nitrogen by volume, 21% oxygen, 0.9% argon, 0.04% carbon dioxide, and other gases. Water vapor can contribute 0.001% to 5% of the atmosphere's volume. The total mass of the atmosphere is approximately 5.15×10^{18} kg. The vertical structure of the atmosphere is mainly due to the force of gravity on constituent gases, with approximately 50% of the atmospheric mass below 5.6 kms, 90% below 16 kms, and 99.9999% below 100 kms. The weight of the atmospheric mass above any point on the Earth's surface is the pressure exerted by the atmosphere above the point, and depends on weather and elevation. The reference standard pressure at sea level is 101,325 Pa (or, in the popular units used by meteorologists, 1,013.25 millibars or hectoPascals (hPa)). The estimated density of air at sea level is 1.2 kg m^{-3}. The specific heat capacity of air at 300 K (26.85°C) temperature is 1,005 J kg^{-1} K^{-1}. Motions and heat transfer processes, including waves, in the atmosphere are governed by laws of fluid dynamics and thermodynamics. As mentioned in Section 1.2, substantial changes in carbon dioxide and water vapor contents of the atmosphere can make very important impacts on climate even though their relative volumes in the atmosphere are very small.

Two principal layers of the atmosphere which interact with the oceans and other Earth System components are the troposphere and the stratosphere. Depending on latitude, the troposphere and the stratosphere extend from the surface to approximately 12 kms and from 12 to 50 kms, respectively. Very near the Earth's surface is the planetary boundary layer (PBL) which, although within the troposphere, has distinct properties and plays a very important role in atmospheric interactions with the underlying surface. Properties and roles of the PBL are described in Section 1.4.5. As we will see in later chapters, the troposphere and the stratosphere play very important roles in DCV phenomena. The lower troposphere plays a very important role in interactions with the underlying oceans and land. Therefore, annual cycles of lower tropospheric winds, SLP, precipitation, and surface air temperature (SAT) are described here.

Global SLP and wind maps for each of the twelve calendar months are shown in Figure 1.1. The data are from the National Center for Environmental Prediction (NCEP) – National Center for Atmospheric Research (NCAR) atmospheric reanalysis (Kalnay et al., 1996) at 2.5° longitude – 2.5° latitude resolution from January 1951 to December 2019. Figure 1.1 shows SLP and 850 hPa (atmospheric pressure level at approximately 1.5 km above the Earth's surface) winds. The winds at 850 hPa are shown because this pressure level is reasonably representative of water vapor transport in the atmosphere which drives precipitation, and near enough to land and ocean surfaces to indicate winds at the surface. Figure 1.1 shows that there are high-pressure centers in subtropical latitudes and low-pressure centers in subpolar latitudes, and that there is a pronounced annual cycle of SLP. In the NH, high-pressure centers are strongest and largest from March to August–September, and low-pressure centers are strongest and largest from November to February. In the SH, high-pressure

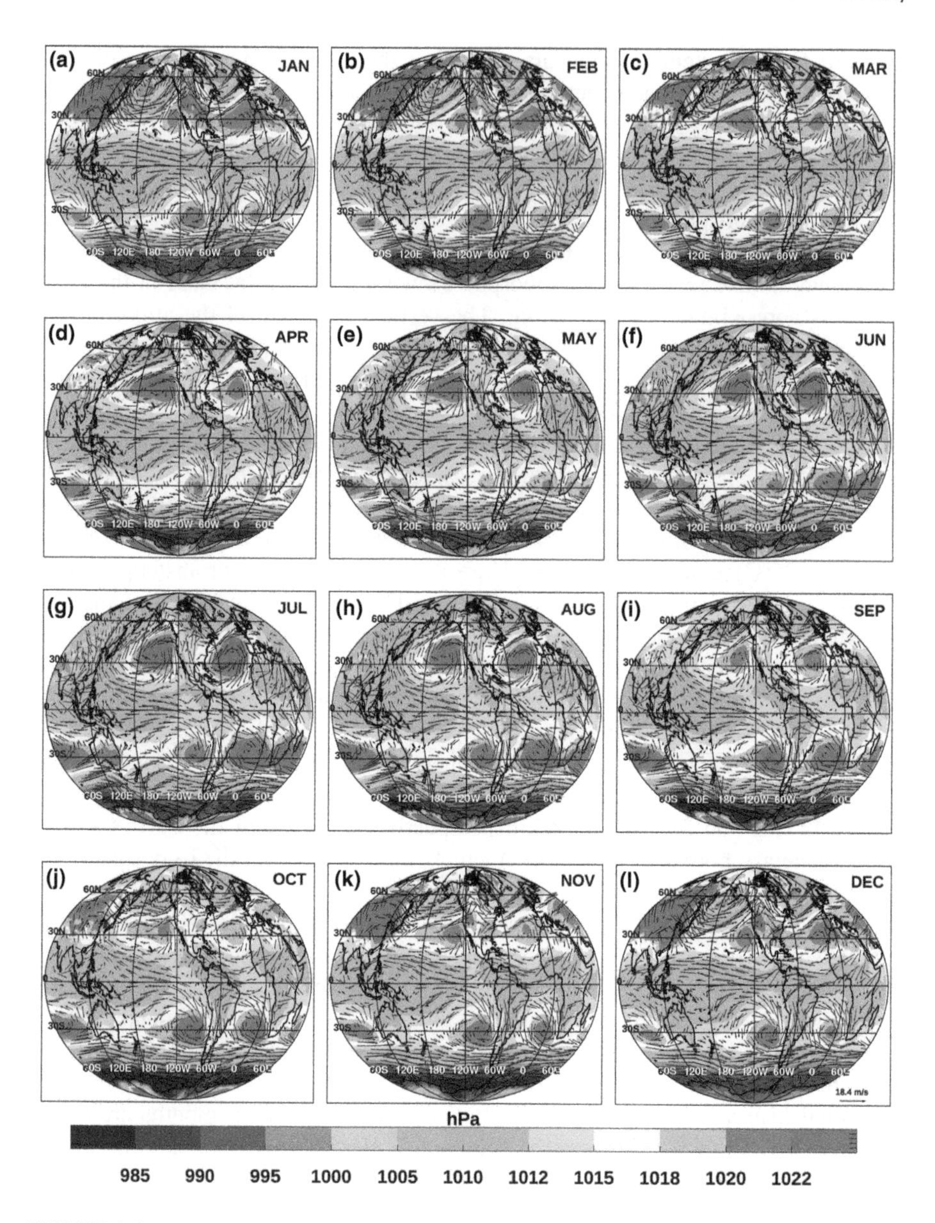

FIGURE 1.1 Monthly average SLP and 850-hPa winds from 1951 to 2019 CE.

centers are strongest and largest from June to November, and low-pressure centers have approximately the same strength and size during the entire year. The 850 hPa winds around high-pressure centers are clockwise in NH and counter-clockwise in SH due to the Earth's rotation around its axis. These winds from the two Hemispheres meet near the Equator such that the winds around the tropical belt are generally easterly, or from east to west. The 850 hPa winds around subpolar low-pressure centers are generally counter-clockwise in NH and clockwise in SH. Consistently with these SLP centers, 850 hPa winds in the mid-latitudes are westerly, or from west to east. As is apparent in Figure 1.1, the wind patterns also have a pronounced annual cycle. In monsoonal regions, winds change direction twice a year and so easterly (or northeasterly) winds change to westerly (or southwesterly) winds during the summer monsoon season. For example, winds over monsoonal southern India are from the southwest or the west beginning in May–June and are from the north or the northwest beginning in October–November. In another example of changes in monsoonal flows, winds over southeast Asia are from the south or the southwest in April–May and are from the east beginning in October–November. Monthly averages of precipitation and SAT are shown in Figures 1.2 and 1.3, respectively. The precipitation data are from the Global Precipitation Climatology Centre (Schneider et al., 2011) at 1° longitude – 1° latitude resolution from January 1951 to December 2019. Figure 1.2 shows that maximum precipitation is in the SH or near the Equator in the December to March period (SH summer and early autumn), moves northward to the Equator in April–May, then to north of the Equator and NH tropics–subtropics in June–July–August–September (NH late spring, summer, and early autumn), finally advancing southward in October–November. This annual migration causes two rainy seasons in some regions, especially near the Equator; and pronounced monsoonal precipitation regions in west Africa, south and southeast Asia, and northern Australia as Figure 1.2 shows. There is seasonal variability of precipitation in middle and higher latitudes on all continents. There are also substantial east–west gradients in precipitation on all continents. Low-level winds (Figure 1.1) from oceans which bring water vapor that precipitates over land are generally consistent with major precipitation regions in Figure 1.2.

Figure 1.3 shows SAT over oceans and land from the previously cited NCEP-NCAR reanalysis. The warmest SATs are near the Equator and the coldest are near the poles due to the distribution of incoming solar radiation. Also, in the Northern and Southern Hemispheres, the warmest SATs are in local summer and the coldest are in local winter. As described in Section 1.4, the apparent position of the Sun moves southward from the Equator on September 21, is above 23.5°S on December 21, and returns to the Equator on March 21, so the SH receives the maximum daylight during this period and the NH receives the minimum daylight during this period. These differences are apparent in Figure 1.3. SATs in the SH are warmest in the October to March period, and those in the NH are coolest in the same period. Then, as the apparent position of the Sun moves northward, warm SATs follow. At NH tropical latitudes, SATs are warmest in April–May–June. Due to different specific heat capacities of oceans and land, a lag between SATs over oceans and land is also apparent. An interesting feature in Figure 1.3 is that the Tibetan Plateau – with an average elevation of 4,500 m above mean sea level – is cold generally throughout the year, with the coldest SATs in the December to March period. As the land south of the Himalaya Mountains warms up in April–May–June, the approximately north–south SAT gradient between the Indian sub-continent and the Tibetan Plateau becomes very large; this is also reflected in the SLP gradient (Figure 1.1) which plays a very important role in the southwest (or summer) monsoon in the Indian sub-continent.

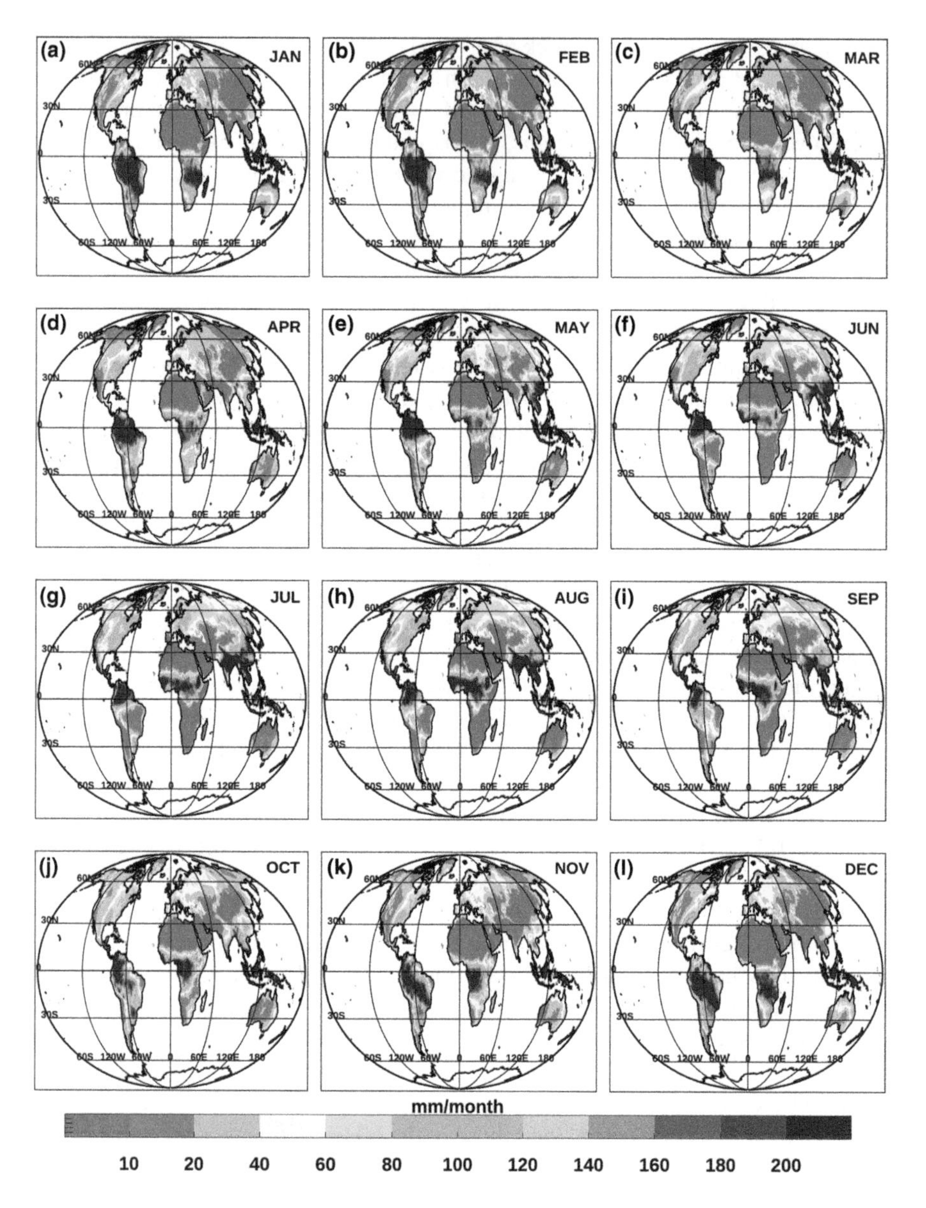

FIGURE 1.2 Average monthly precipitation from 1951 to 2019 CE.

1.4.2 The Oceans

The oceans are also a fluid, composed principally of water and salt. They cover 71% of the Earth's surface and contain 97% of the total water on the Earth. Average salinity of sea water is 3.5% by volume. The density of surface water ranges from 1,020 to 1,029 kg m^{-3}, and the specific heat at constant pressure, a function of temperature and salinity, is approximately 4,000 J kg^{-1} K^{-1}. Motions and heat transfer processes

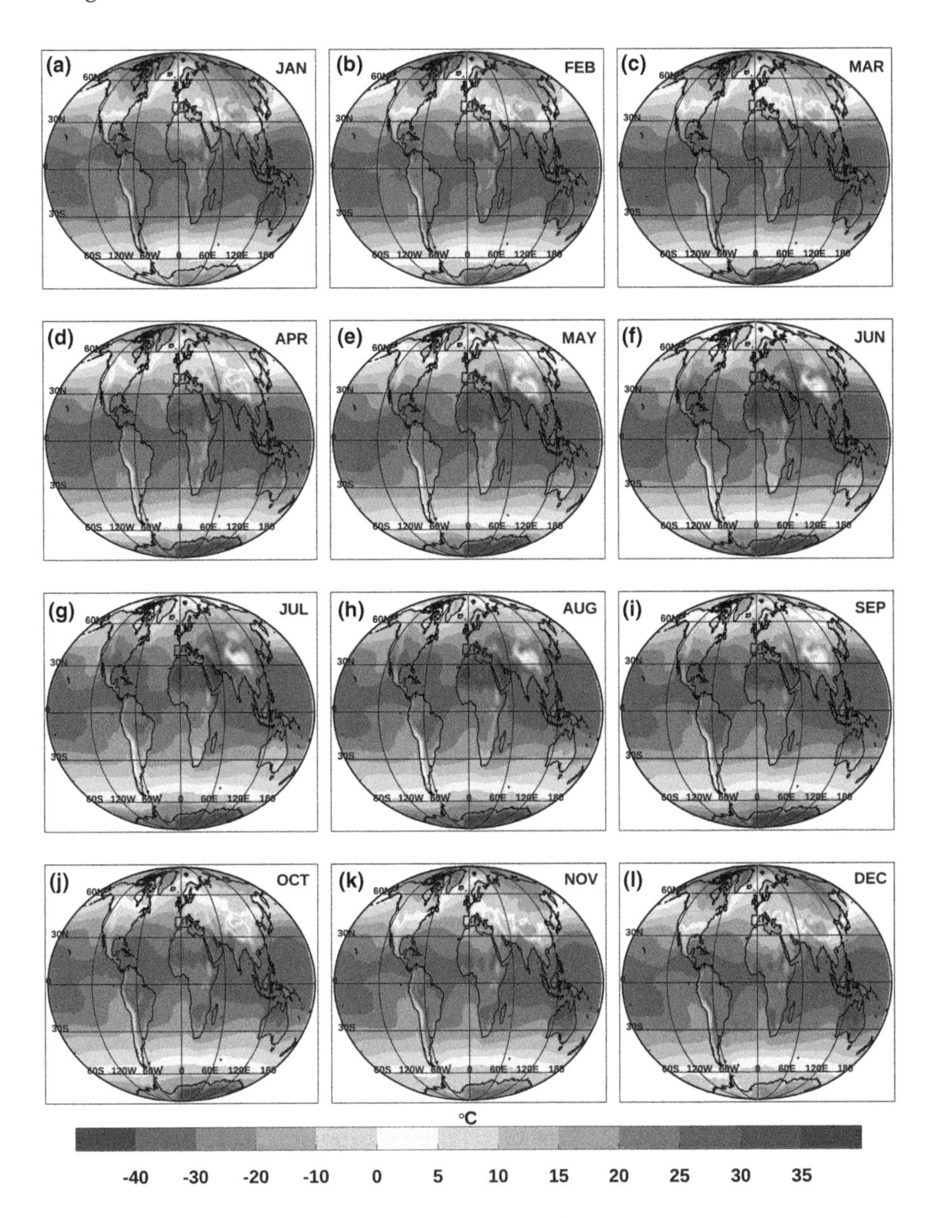

FIGURE 1.3 Average monthly SAT from 1951 to 2019 CE.

in the oceans, including waves, are governed by laws of fluid dynamics and thermodynamics. The very large differences in densities and specific heat capacities of the atmosphere and the oceans imply that the upper 2 m of the world's oceans can store as much heat as the entire atmosphere. These very large differences in densities and specific heat capacities of the atmosphere and the oceans are also the reasons for very large differences in typical speeds of atmospheric winds and ocean currents and in typical speeds of waves in the atmosphere and the oceans. The much slower speeds of ocean currents and waves provide the slow timescales of DCV phenomena as we will see in later chapters. The very large thermal inertia of the oceans due to a very large heat capacity also plays a very important role in DCV phenomena.

The heat content of the oceans from the surface to 200 m depth plays a very important role in natural climate variability due to heat exchanges with the atmosphere. Because of long records of SST measurements, going back to mid-19th century CE in some regions, SST has been used as an indicator of upper ocean heat content in climate research. Based on SST data from Huang et al. (2017) at 2° longitude – 2° latitude resolution from January 1951 to December 2019, average monthly SST patterns are shown in Figure 1.4. The warmest SSTs are at tropical latitudes, and the coldest are at mid and high latitudes. As is evident, warm SSTs generally follow the annual cycle of solar radiation with a time lag of a few weeks due to thermal inertia of sea water. A comparison of Figures 1.3 and 1.4 shows that, over the oceans, SATs are generally similar to SSTs in patterns and magnitudes.

The TWPs contain some of the warmest sea water in the world. The TWPs are characterized by SSTs persistently warmer than 28°C. According to this definition, the principal TWPs are located in the eastern Indian Ocean, the western Pacific Ocean, the western Atlantic Ocean, the Caribbean Sea, and the Gulf of Mexico. There are also small TWPs in the Red Sea and the eastern Pacific Ocean near the Central America coast. As shown in Figure 1.4, the eastern Indian and western Pacific Ocean TWPs are very large; they cover approximately 6 million kms^2 and 12 million kms^2 average surface areas, respectively. Since the atmosphere's capacity to hold water vapor is an exponential function of temperature, the intensity of deep atmospheric convection can be very sensitive to changes in SST over the TWP regions. Even small changes in TWP SST can cause large changes in atmospheric convection, which, in turn, can dramatically alter atmospheric circulations locally and significantly modify global-scale atmospheric wave activity and atmospheric heating.

The 850 hPa winds are generally representative of winds down to the ocean surface and, therefore, the 850 hPa winds can be approximated as ocean surface winds. Winds at the ocean surface exert a force, known as wind stress, on the surface water. This wind stress drives currents which constitute the so-called wind-driven ocean circulation. Figure 1.5 shows major, annual-average currents in the world oceans. A comparison of Figure 1.5 with Figure 1.1 shows that the streamlines of currents, shown by black lines and arrows, in the former figure are generally similar to wind vectors in the latter figure. Figure 1.5 also shows temperatures of major currents (red and blue arrows). Poleward flowing currents are generally warm and equatorward flowing currents are generally cold. The clockwise and counterclockwise flowing currents, known as gyres, play very important roles in DCV phenomena because their interactions with the atmosphere can generate anomalously warm/cold water which is then transported by the gyres at their slow speeds over ocean basin scales; and because of their interactions with slow-moving, ocean basin scale waves.

Due to the rotation of the Earth around its axis, a pseudo-force known as the Coriolis force turns winds and currents towards the right in the NH and towards the left in the SH. Figure 1.5 shows that the two southward-flowing currents in the NH – the California Current and the Canary Current – turn away from the coasts

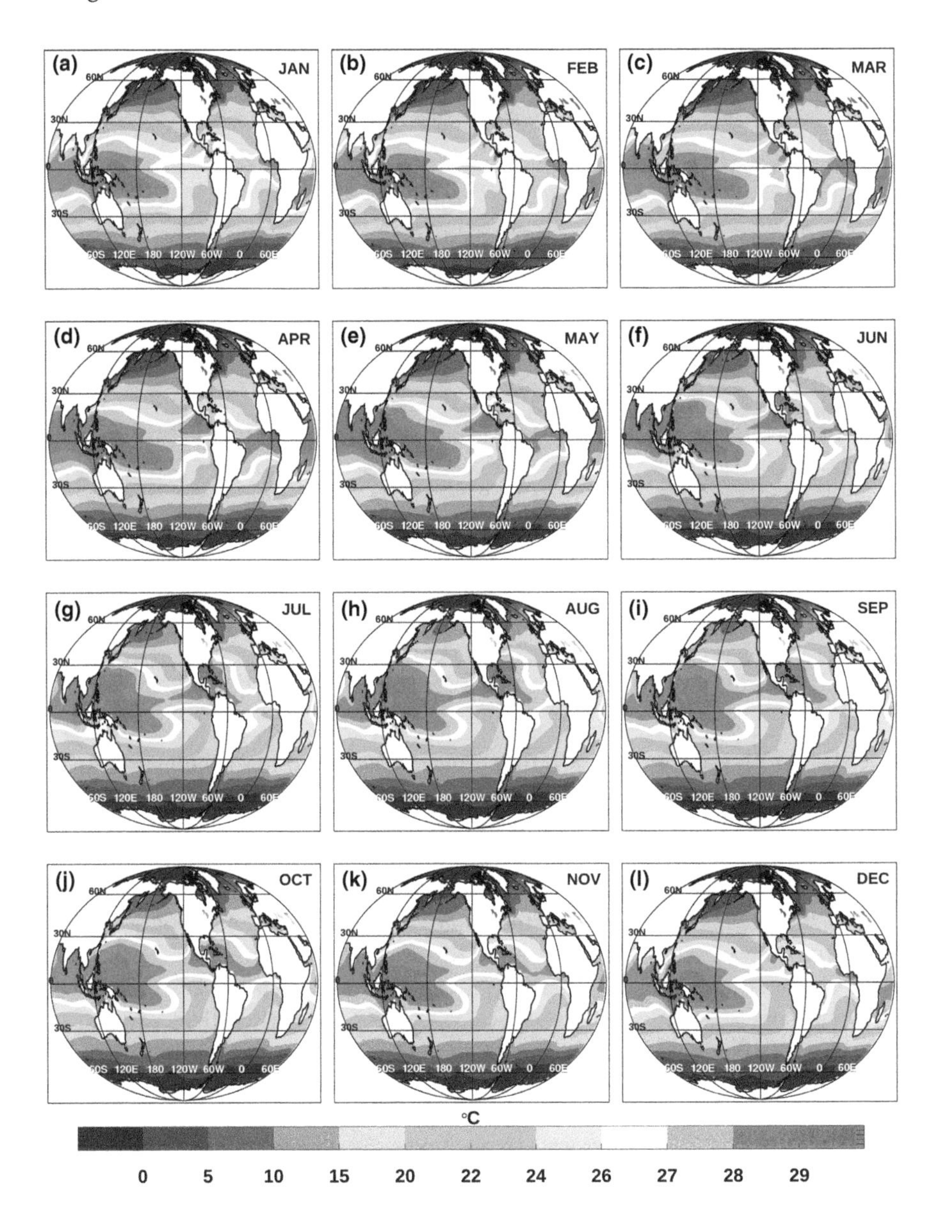

FIGURE 1.4 Average monthly SST from 1951 to 2019 CE.

of North America and North Africa, respectively; in a similar way, northward-flowing currents in the NH – the Kuroshio and the Gulf Stream – also turn away from the coasts of Asia and North America, respectively, as a consequence of the Coriolis force. Currents in the SH turn to the left as Figure 1.5 shows for the Peru (or Humboldt) Current; and for currents off the east and west coasts of Australia, and west coast of southern Africa. Associated with this water moving away from the coasts is water that rises from below the surface which is colder; this process

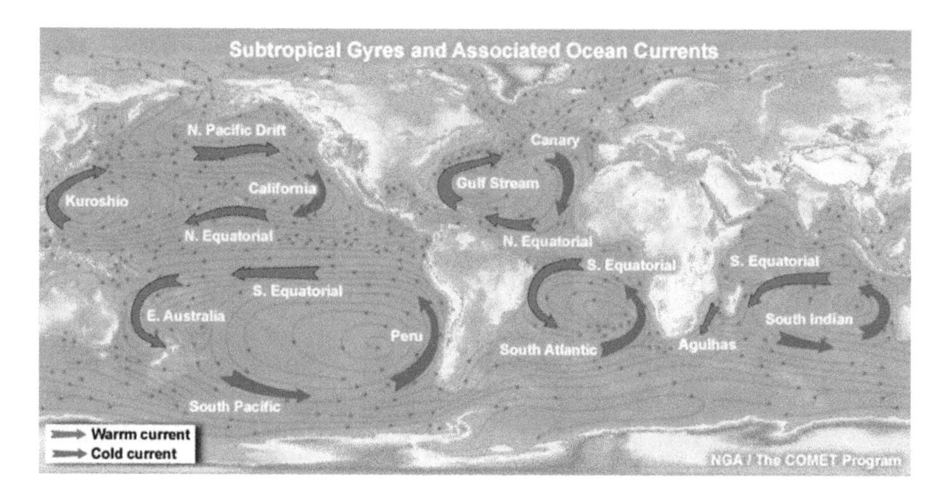

FIGURE 1.5 Global upper ocean circulations and subtropical gyres; red marks the warm ocean currents, and blue marks the cool ocean currents. The source of this material is the COMET® Website at http://meted.ucar.edu/ of the University Corporation for Atmospheric Research (UCAR), sponsored in part through cooperative agreement(s) with the National Oceanic and Atmospheric Administration (NOAA), US Department of Commerce (DOC). (©1997–2017 University Corporation for Atmospheric Research. All Rights Reserved.)

which brings up water from below the surface along coastlines is known as coastal upwelling. Similarly, in the SH, the Humboldt or Peru Current along the southern and central South American coast, the South Atlantic Current along the southern Africa Coast, and the South Indian Current along the western Australian coast bring up deeper, colder water. In contrast, east to west wind stress on the western sides of the oceans "piles up" warmer water along the east coasts of continents, giving rise to warm currents such as the Kuroshio in the Pacific, the Gulf Stream in the North Atlantic, and the currents along the east coasts of Australia, southern South America, and southern Africa. As mentioned earlier, the Coriolis force turns currents flowing from east to west along the Equator to the right in the NH and to the left in the SH. This current divergence at the surface results in water rising from below the surface. This process is known as equatorial upwelling, which usually brings up colder water from below the surface. Figure 1.4 shows that there are large-scale SST gradients in the north–south and east–west directions. Therefore, if currents are from warmer to colder SST, they transport warmer water and if currents are from colder to warmer SST, they transport colder water. This heat transport process changes SSTs and sub-surface ocean temperatures as do the coastal and equatorial upwelling processes. As we will see in some of the subsequent chapters, these heat transport and upwelling processes are very important in DCV phenomena.

As in the atmosphere, ocean water also has a well-mixed layer due to penetrative solar radiation, stirring by winds, and mixing by waves. Properties of water in this layer are relatively well mixed. At the base of this layer is a strong vertical gradient of temperature known as the thermocline which separates the upper well-mixed and (relatively) warm layer from the deeper and colder ocean. Near the Equator, the stress on the ocean surface due to easterly winds "piles up" water on the western sides of the Pacific and Atlantic Oceans as mentioned earlier. In the Indian Ocean, the net stress due to stronger and longer duration southwesterly monsoon winds and less strong easterly winds results in piling up of water towards Indonesia and other islands of the Maritime Continent. These actions result in deeper thermoclines in equatorial western Pacific and western Atlantic, and shallower thermoclines on the eastern sides of these Oceans. In the Indian Ocean, the wind stress actions cause the thermocline to be deeper on the eastern side than on the western side. The TWPs in western Pacific, western Atlantic, and eastern Indian Oceans (Figure 1.4) are partial manifestations of the deeper thermoclines in those regions. Equatorial thermoclines in the three Oceans are shown in Figure 1.6 which shows longitude-depth cross sections of ocean temperatures from western Indian Ocean (east coast of Africa) eastward to eastern Atlantic Ocean (west coast of Africa). The 20°C line, generally identified as the equatorial thermocline, is marked in Figure 1.6 and shows the thermocline tilting downwards from east to west in the Pacific and Atlantic Oceans. A stronger vertical temperature gradient in the eastern Indian Ocean compared to that in the western Indian Ocean is also evident. Incidentally, the presence of warm pools in western Pacific Ocean and western Atlantic Ocean, and in eastern Indian Ocean is also evident from the presence of the 28°C isotherm in Figure 1.6. As we will see in

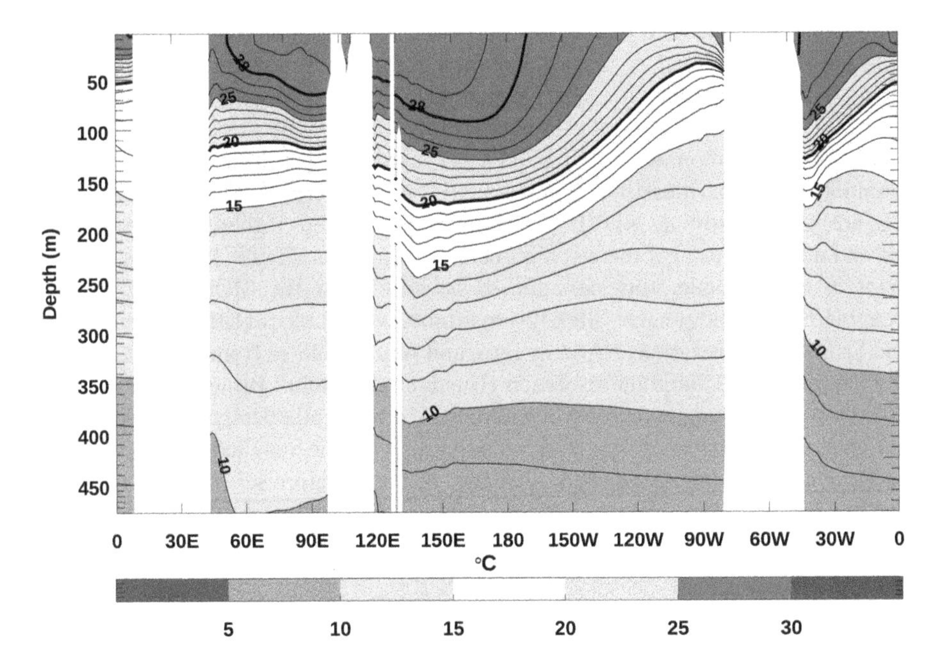

FIGURE 1.6 Longitude-depth (m) profile of annual-average ocean temperatures (°C) at the Equator in the Indian, Pacific, and Atlantic Oceans. The 28°C and 20°C isotherms (thick black lines) indicate the SST generally used to define tropical warm pools and the temperature generally used to define the equatorial thermocline, respectively. Data are from the Estimating the Circulation and Climate of the Ocean (ECCO) project from 1993 to 2017 CE.

some later chapters, wind stress, thermocline depth, and SSTs in the eastern Pacific constitute a positive feedback system very important in interannual to multidecadal climate variability. We will also see that the deep thermocline and warm ocean temperatures in the eastern Indian and western Pacific Oceans play a very important role in interannual to decadal climate variability.

There are other very important attributes of ocean properties such as salinity and density, and worldwide circulations such as the thermohaline circulation (also known as "the global conveyor belt") which are integral to decadal-multidecadal variability phenomena. They will be described in appropriate chapters as necessary.

1.4.3 The Land–Vegetation–Snow-Ice System

Land – continents and islands – covers 29% of the Earth's surface. Ten percent of the land area is covered in glacial ice. Agricultural land covered 37.4% of the world's land area in 2015.[8] Out of this total, permanent pasture land was 67.3%, arable land was 29.3%, and permanent cropland was 3.4%. Here, agricultural land is defined as land used for producing food for humans, permanent pasture land is defined as natural or artificial grass land and shrub land for grazing livestock, arable land is defined as land for growing crops which have to be re-planted every year, and permanent cropland is defined as land where vineyards, orchards, and trees for nuts are grown but not re-planted every year. Annual cycles of solar radiation, precipitation, and temperatures drive annual cycles of natural vegetation on the Earth and also of artificial vegetation such as crops where there is rain- and snow-fed agriculture. One global indicator of vegetation is the Normalized Difference Vegetation Index (NDVI) which indicates the photosynthesis capacity of vegetation. Vegetation reflects visible and infrared radiation differently from other land covers (such as sand). These differential reflections, measured by instruments on Earth-orbiting satellites since 1979 CE, are used to estimate NDVI. The Index is close to zero if there is no green matter, and it is close to +1 if there is high-density vegetation. Figure 1.7 shows monthly maps of NDVI made from radiation measurements by the MODerate resolution Imaging Spectroradiometer (MODIS) instruments on the US National Aeronautics and Space Administration (NASA)'s Terra and Aqua satellites from 2000 to 2019 CE at 1° longitude – 1° latitude resolution (Huete et al., 2002).[9] Brownish colors show little or no vegetation, and greenish colors show photosynthetically active vegetation in Figure 1.7. The annual cycle of vegetation following the Sun, and precipitation and SAT cycles in Figures 1.2 and 1.3, is clearly visible in Figure 1.7.

Land surface also plays a very important role in natural climate variability. Specific heat capacity, soil moisture and type, albedo, and roughness are some of the important attributes which are integral to the response of land surface to climate variability as well as in generating climate variability. Specific heat capacity of land surface ranges from 800 to 2,000 J kg^{-1} K^{-1}, depending on constituents of the surface.

[8] FAOSTAT data on land use (www.fao.org/faostat/en/#data/EL). Retrieved on August 4, 2018.

[9] GES DISC Northern Eurasian Earth Science Partnership Initiative Project (2006), MODIS/Terra Monthly Vegetation Indices Global 1x1 degree V005, Greenbelt, MD, Goddard Earth Sciences Data and Information Services Center (GES DISC), Accessed: (3/30/2019), https://disc.gsfc.nasa.gov/data collection/MODVI_005.html.

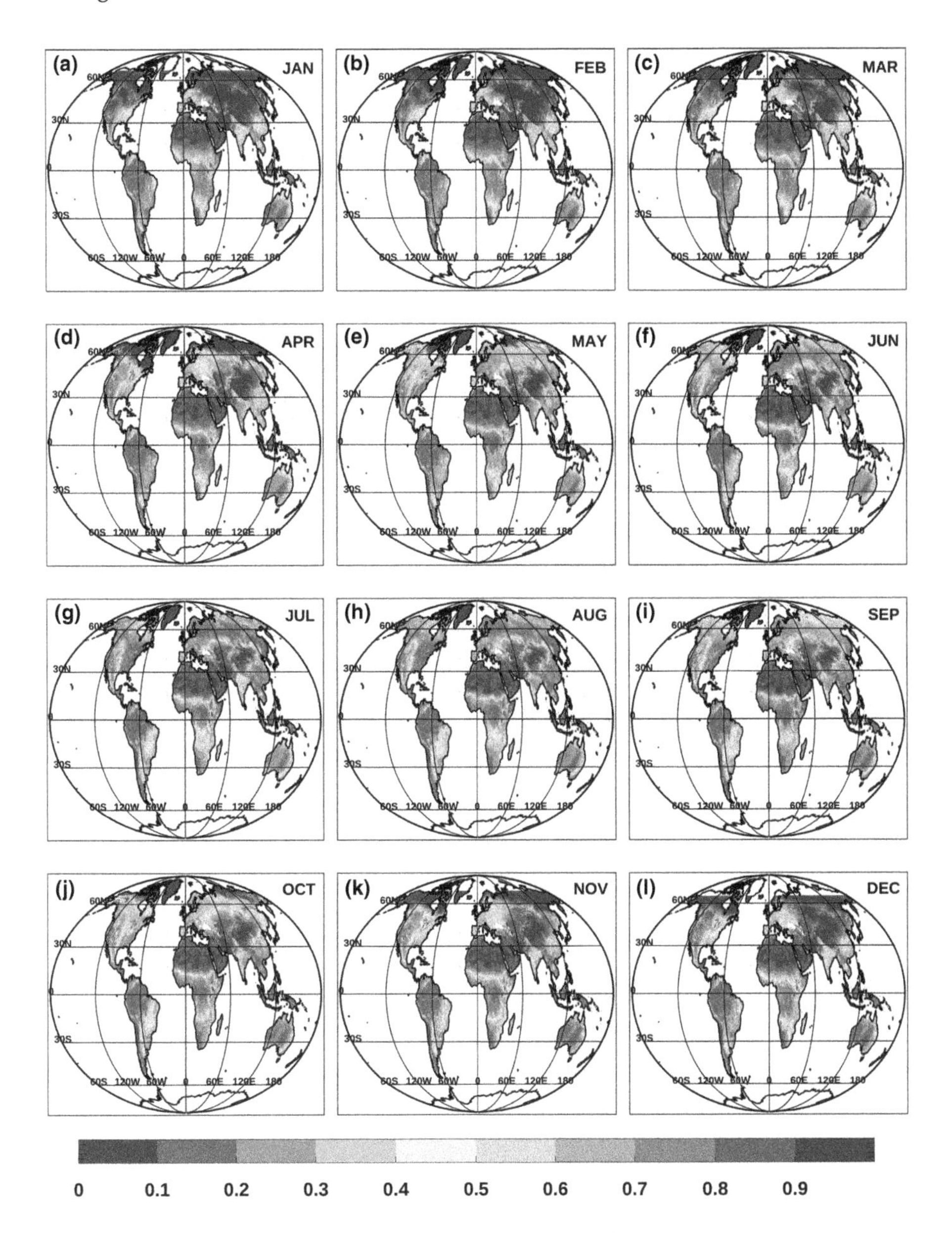

FIGURE 1.7 Monthly average NDVI from 2000 to 2019 CE.

Liquid water content increases the heat capacity of dry land. Therefore, the heat capacity of dry or vegetated land is at the lower end of the range and that of wet sand or ice is at the upper end of the range. Due to this range of heat capacities, dry land heats up much faster than wet or vegetated land for the same amount of heat energy reaching the surface, introducing a timescale which is dependent on the amount of wetness or vegetation. The evapotranspiration (ET) process evaporates water from surface and deeper soil, and by transpiration through vegetation. The timescale of ET depends on the amount of soil moisture and also the depth from which the moisture is drawn up and evaporated. Since precipitation, temperature, surface winds, and ET undergo annual cycles, soil moisture also undergoes an annual cycle. Deeper soil layers store the moisture for longer times than surface soil. Soil depths from the surface to 200 cm are in the root zone of vegetation. Figure 1.8 shows monthly average maps of soil moisture from 1951 to 2019 CE, integrated from 10 cm to 40 cm depth from NASA's Global Land Data Assimilation System at 1° longitude – 1° latitude resolution (Rodell et al., 2004).[10] Integrated soil moisture maps from the surface to 10 cm, from 40 cm to 100 cm, and from 100 cm to 200 cm depths show generally similar horizontal patterns for each month as Figure 1.8. The per-cm soil moisture content from the surface to 200 cm appears to be generally similar in most regions, indicating that one month is a sufficiently long time for soil moisture in the top 200 cm to equilibrate with precipitation, ET, and run-off. As a comparison between Figures 1.2 and 1.8 shows, regions of higher rainfall and greater root zone soil moisture are generally co-located in all months as are regions of lower rainfall and less soil moisture. It is also interesting to note the consistency between root zone soil moisture and NDVI (Figure 1.7) in the tropics in all months and at higher latitudes in warm season. The latter observation points to the importance of temperature in the life cycle of green vegetation.

Less than 1% of the water on the Earth is freshwater, with 97% to 98% of it in rivers, lakes, and other water bodies on the surface; and in ground water reservoirs. Out of all the water on the Earth, 2% to 3% is in glaciers and ice caps. Precipitation, evaporation, topography, gravity, and soil properties determine the movement and storage of water at any location on and under the Earth's surface. We saw the annual cycle of total precipitation in Section 1.4.1. A substantial fraction of precipitation in mid- and high latitudes during winter is in the form of snow. Figure 1.9 shows the average annual cycle of snow cover in the NH from 1972 to 2018 CE at 25× 25 km resolution (Helfrich et al., 2007).[11] As is clearly evident, the maximum snow cover is from November to March in northern North America, northern Europe, and northern Asia; and the minimum snow cover is in July–August. Southward expansion of snow cover from the Arctic region begins in September, reaches its southernmost extent in January–February, and then recedes northward till August. Greenland is an exception in that it is nearly 100% snow covered throughout the year. It is interesting to note that there are some snow cover and glacial ice throughout the year in the appropriately named Himalayas (abode of the snow), the northern Rocky Mountains, and the Alps.

---

[10] Beaudoing, Hiroko and M. Rodell, NASA/GSFC/HSL (2015), GLDAS Noah Land Surface Model L4 3 hourly 1.0 x 1.0 degree V2.0, Greenbelt, Maryland, USA, Goddard Earth Sciences Data and Information Services Center (GES DISC), Accessed: (*March 2020*), 10.5067/L0JGCNVBNRAX.

[11] Brodzik, M. J. and R. Armstrong. 2013. *Northern Hemisphere EASE-Grid 2.0 Weekly Snow Cover and Sea Ice Extent,* Version *4*. (Snow Data). Boulder, CO. NASA National Snow and Ice Data Center Distributed Active Archive Center. doi: https://doi.org/10.5067/P7O0HGJLYUQU. (March 2020).

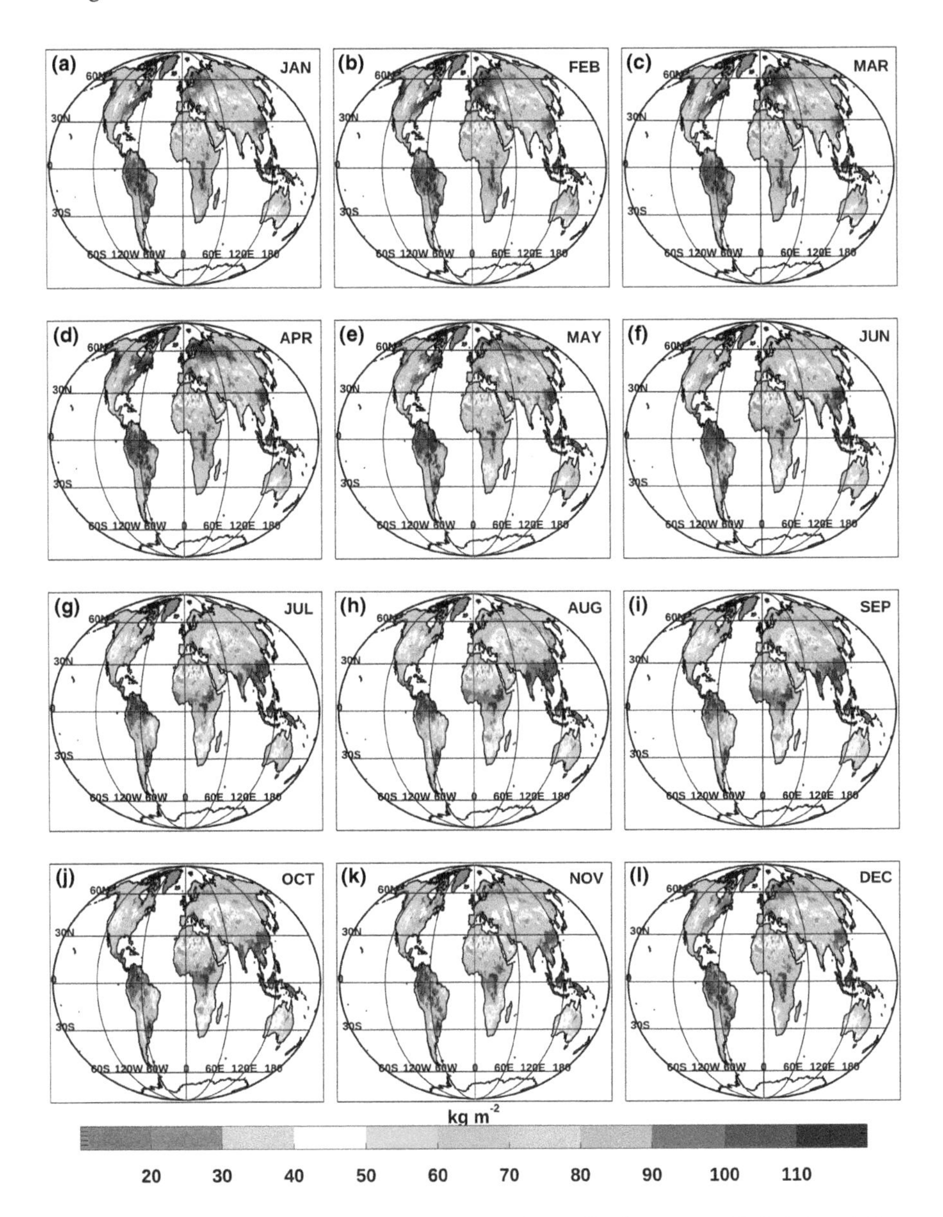

FIGURE 1.8 Monthly average soil moisture (kg m^{-2}) from the Global Land Data Assimilation System, averaged from 10 to 40 cm depth from 1951 to 2019 CE.

1.4.4 Sea Ice

Sea ice forms when sea water freezes. After formation, sea ice is under the influences of air and water temperatures, winds, and ocean currents. Therefore, sea ice which is not anchored to land can move according to prevailing winds and currents, and can melt/re-form according to air and water temperatures. The importance of sea ice in

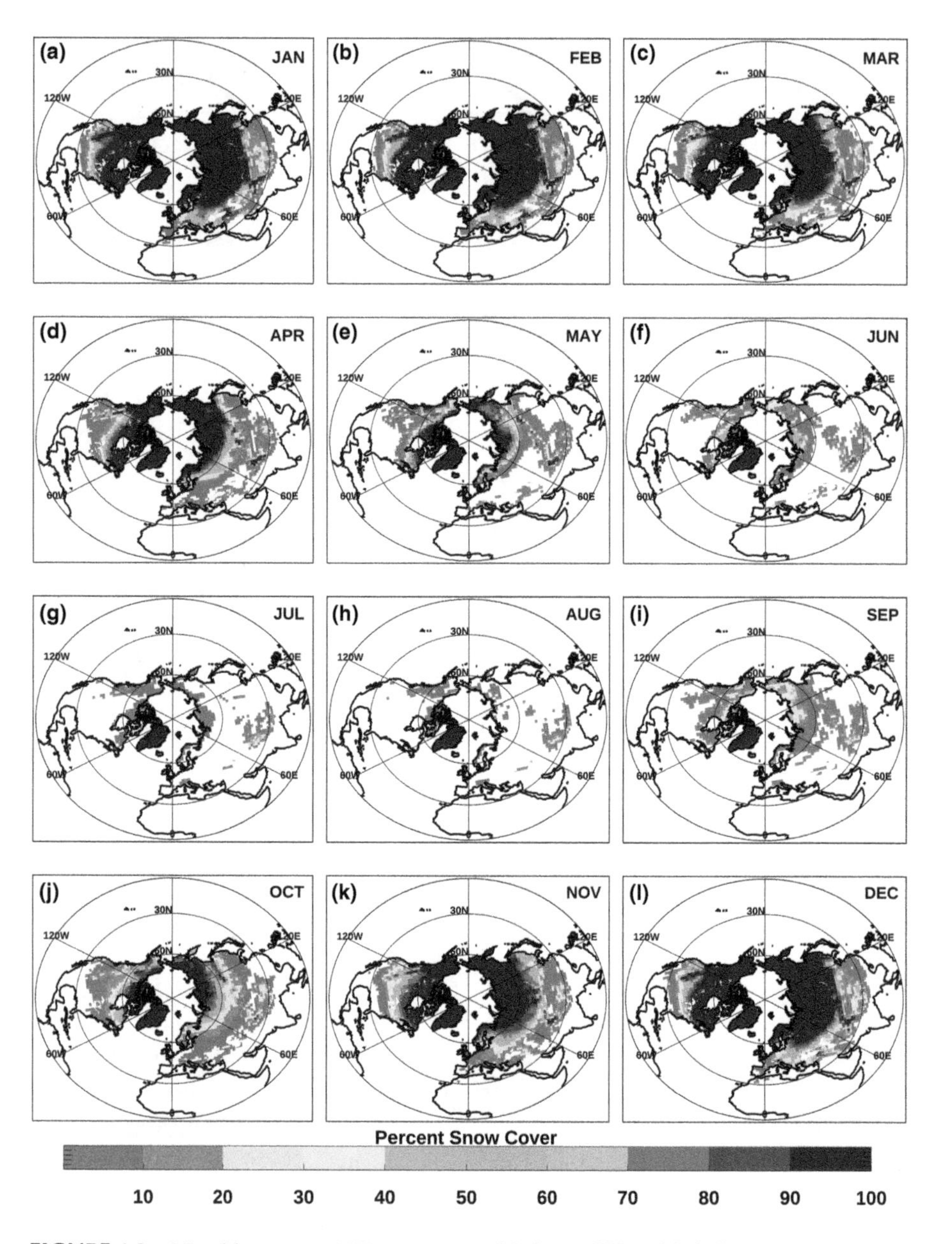

FIGURE 1.9 Monthly average NH snow cover (%) from 1972 to 2018 CE.

the DCV story is because sea ice changes the albedo of the underlying ocean, modulates ocean–atmosphere heat and water vapor transfers, adds freshwater to oceans when and where transported sea ice melts, and the ice formation (melt) increases (reduces) salinity of remaining sea water. These processes can change atmospheric response to oceanic heating and ocean circulations driven by horizontal and/or vertical differences in density of sea water (so-called thermohaline circulations), respectively, which, in turn, introduce multiyear and longer timescales of variability to the climate system.

Sea ice covers approximately 7.3% of the Earth's surface and approximately 11.8% of the ocean surface (Weeks, 2010). In both Northern and Southern Hemispheres, sea ice extent is maximum in local winter and minimum in local summer. Sea ice thickness estimates, based on submarine data, suggest variable thickness ranging up to 3 to 4 m in the Arctic.[12] Due to actions of annual cycles of solar radiation, SAT, SST, and winds, sea ice cover also undergoes an annual cycle. Figure 1.10a shows monthly, NH sea ice cover averaged from 1979 to 2018 CE at 25 × 25 km resolution (Comiso et al., 2017).[13] There is some sea ice cover as far south as the southern tip of the James Bay in Canada and northern coast of Asia as early in winter as November when the Arctic Ocean is completely covered with sea ice. Minimum NH sea ice cover is in August–September and maximum is from January to April. SH sea ice cover (Figure 1.10b) has the opposite phase, with the minimum ice cover from January to March and the maximum in August–September–October.

1.4.5 Planetary Boundary Layer and Surface–Atmosphere Interactions

When a fluid flows in contact with a surface, it forms a layer in which velocities vary from zero at the bounding surface to maximum at the top of the layer due to the viscosity of the fluid and friction with the surface. The same process results in the formation of an atmospheric boundary layer, more generally known as the PBL, when air flows in contact with the Earth's surface. Due to the velocity shear, the PBL is turbulent over its entire depth, the turbulence mixes air properties horizontally and vertically in the PBL, and the friction with the surface dissipates a substantial amount of momentum of the air flow and generates an almost logarithmic vertical profile of air speed in the PBL. This turbulent transfer of momentum downwards, and water vapor and heat upwards connects the troposphere and the underlying oceans and land, and is crucially important in natural climate variability due to ocean–atmosphere and land–atmosphere interactions. The PBL varies in depth from a few hundred meters during night time to 2 to 3 kms during day time. PBL depth is also dependent on topography and surface roughness. There can be several sub-layers within the PBL, each defined by properties of air within it. The most important sub-layer for atmosphere–surface interactions and climate is the surface layer which is approximately 10 m deep in calm conditions. In this layer, vertical changes in temperature, water vapor, and wind velocity are very substantial.

Now, we will see the PBL's role in surface–atmosphere interactions, starting with surface energy balance. As described in the beginning of Section 1.1, solar radiation

---

[12] National Snow and Ice Data Center (nsidc.org/cryosphere/sotc/sea_ice.html).

[13] Comiso, J. C. 2017. *Bootstrap Sea Ice Concentrations from Nimbus-7 SMMR and DMSP SSM/I-SSMIS, Version 3.* [North and South polar regions]. Boulder, CO. NASA National Snow and Ice Data Center Distributed Active Archive Center. doi: https://doi.org/10.5067/7Q8HCCWS4I0R. [March 2020].

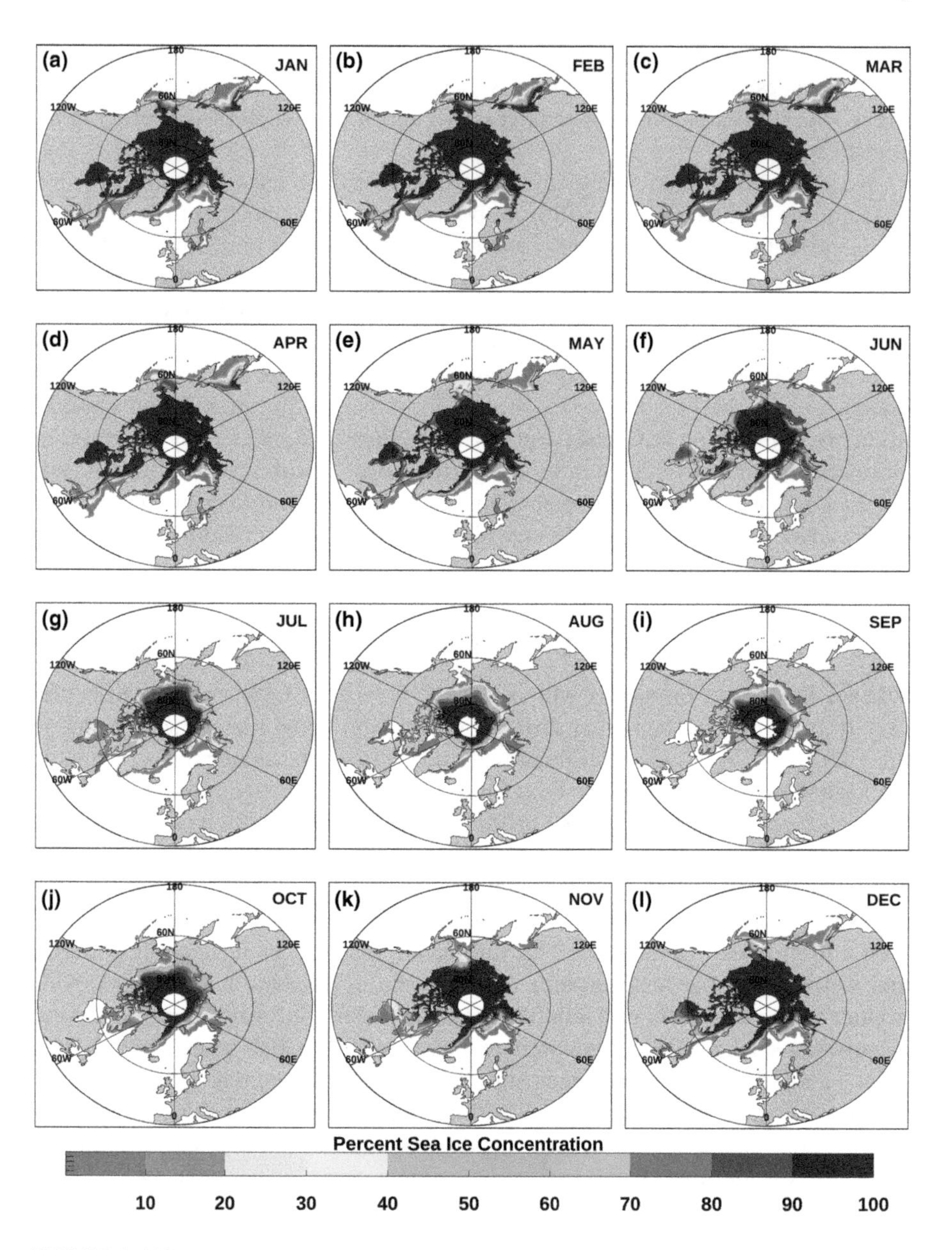

FIGURE 1.10A Monthly average NH ice cover (%) from 1979 to 2018 CE.

is the primary energy source for the Earth System. Figure 1.11 shows how this primary energy is distributed among several components of surface energy balance on land and ocean. The sum of net downward shortwave radiation and net downward longwave radiation provides the energy (net radiation Rn) to drive evaporation and sensible heat transfer from the surface. The evaporated water contains latent heat of evaporation which is released when the water vapor condenses and precipitates

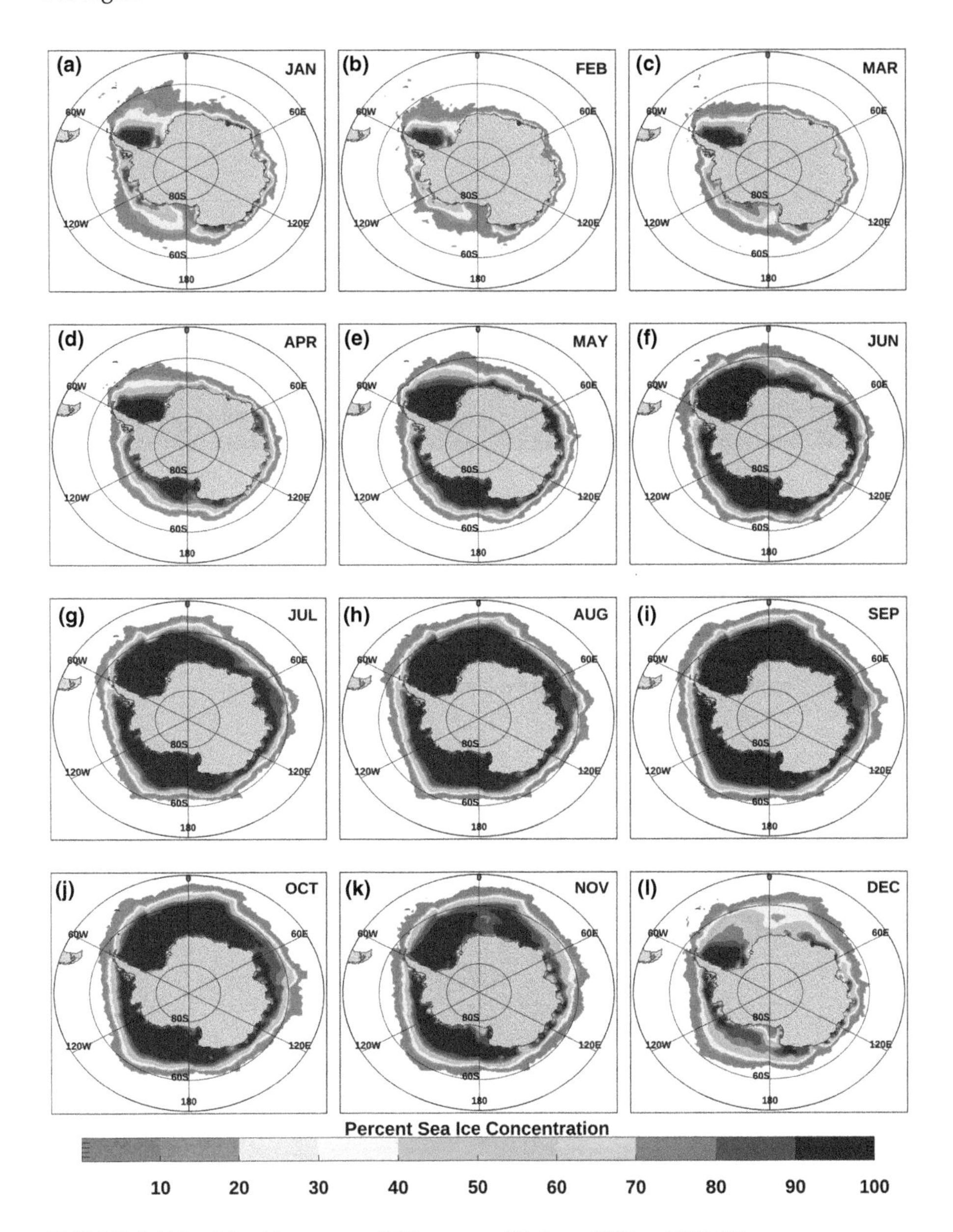

FIGURE 1.10B Monthly average SH ice cover (%) from 1979 to 2018 CE.

as liquid or solid water. Some of the net radiative energy (Rn) is also transferred to sub-surface soil or ocean (G). In the oceans, currents transport heat horizontally (Δf). Changes in surface energy balance directly affect vertical heat transport and the vertical structure of the PBL. For example, it was mentioned in Section 1.4.3 that the amount of soil moisture can influence the timescale of ET. Important processes in land–atmosphere energy and water interactions are shown in Figure 1.12.

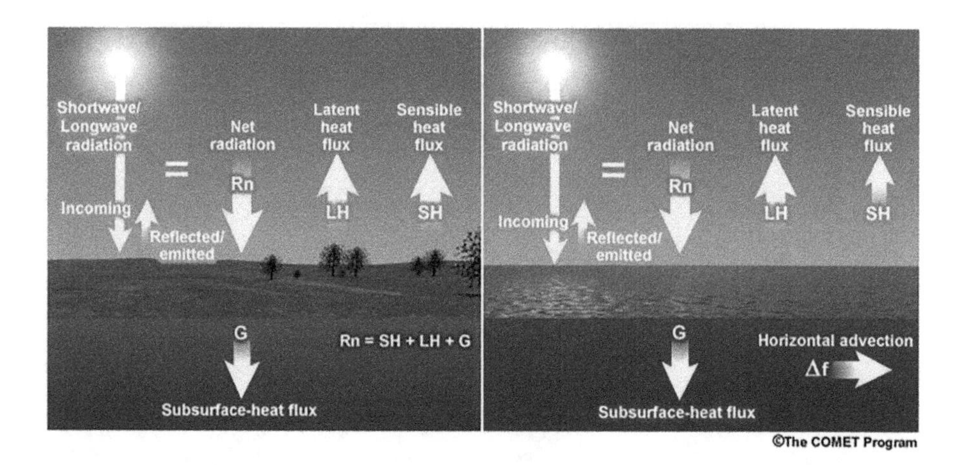

FIGURE 1.11 The Earth's energy balance on land (a) and ocean (b). Rn, LH, SH, G, and Δf denote net radiation, latent heat, sensible heat, heat transfer between surface and sub-surface layers, and horizontal heat flux, respectively. The source of this material is the COMET® Website at http://meted.ucar.edu/ of the University Corporation for Atmospheric Research (UCAR), sponsored in part through cooperative agreement(s) with the National Oceanic and Atmospheric Administration (NOAA), US Department of Commerce (DOC). (©1997–2017 University Corporation for Atmospheric Research. All Rights Reserved.)

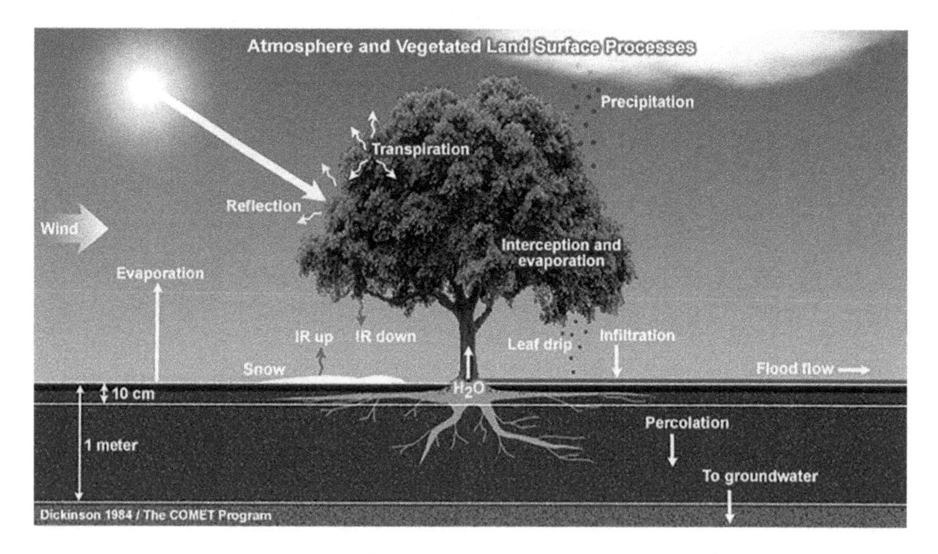

FIGURE 1.12 Energy- and water-related land–atmosphere interaction processes. The source of this material is the COMET® Website at http://meted.ucar.edu/ of the University Corporation for Atmospheric Research (UCAR), sponsored in part through cooperative agreement(s) with the National Oceanic and Atmospheric Administration (NOAA), US Department of Commerce (DOC). (©1997–2017 University Corporation for Atmospheric Research. All Rights Reserved.)

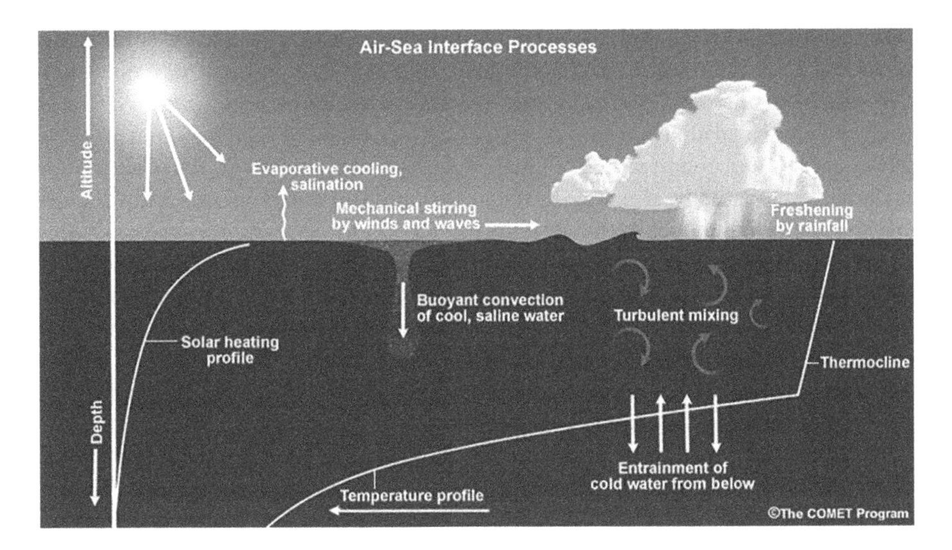

FIGURE 1.13 Energy- and water-related ocean–atmosphere interaction processes. The source of this material is the COMET® Website at http://meted.ucar.edu/ of the University Corporation for Atmospheric Research (UCAR), sponsored in part through cooperative agreement(s) with the National Oceanic and Atmospheric Administration (NOAA), US Department of Commerce (DOC). (©1997–2017 University Corporation for Atmospheric Research. All Rights Reserved.)

Precipitation, directly or dripping from vegetation, can flow along the surface; or can infiltrate into the ground, percolate, and become groundwater. Water is transpired from below the land surface by roots of vegetation, which then evaporates from vegetation surfaces. The two processes are combined as ET. Water also evaporates from the land surface. ET and water run-off on/under the surface leave from the surface, and precipitation arrives to the surface, with the net difference between them known as freshwater flux. As mentioned earlier and shown in Figure 1.11, the energy to evaporate water at the surface is provided by net radiation.

There are more basic and complex processes involved in ocean–atmosphere interactions as Figure 1.13 shows. Unlike land–atmosphere interactions, the primary incoming energy - solar radiation - penetrates tens or hundreds of meters into the ocean, heating not only the surface but also deeper water. Further complexity is due to the ocean being a fluid which flows horizontally and vertically transporting heat. Such flows are driven by winds as well as by water density gradients created by changes in salinity and temperature caused by evaporation, precipitation, ice melt, river flows, sensible heat, and Rn. Winds and density gradients stir the upper ocean, mixing water properties much deeper than vertical mixing on land.

1.5 CAVEATS

While empirical data have the credibility of objectively derived truth, to the extent that it is possible to do so, these data can only be used in deriving statistical and case studies of what (may have) happened in the past, especially spatial and temporal patterns of DCV. Since physics of DCV phenomena involve primarily oceanic and coupled ocean–atmosphere processes, and since global empirical data on important variables other than SSTs are collected systematically only since the 1950s CE, results of empirical analyses can only be used to speculate or even hypothesize about physics of DCV phenomena and compare empirical characteristics of a phenomenon with those simulated by a model. As data about atmospheric processes and variables are collected systematically for at least the last 60 to 70 years, the understanding of how DCV phenomena may be influencing atmospheric circulations and hydrometeorology on land is more certain. One also has to be mindful of the possibility that statistical results may be due to chance. To the extent possible, empirical analyses results are linked to simulated DCV phenomena in this book. But, despite such apparent linkages, there is no guarantee that inferred or implied mechanisms of DCV phenomena are not partially or entirely due to other causes.

As a final note on comparisons between empirical and model-simulated phenomena, it must be stated that the DCV phenomena are non-stationary in time. That is, their statistics such as average (mean), standard deviation, and higher moments are not constant with respect to time. Also, DCV phenomena are not strictly cyclic or periodic. In addition, there is a substantial amount of climate variability at shorter and longer timescales, and there is also background noise, especially at smaller regional scales. Therefore, large numbers of samples of each climate phenomenon are required to determine its characteristics to a high degree of certainty. Although 60 years of observed data contain a substantial and significant amount of information, they are not sufficient to determine characteristics of DCV with a high degree of certainty; for example, there are, at the most, only six cycles of decadal variability – as defined earlier in this chapter – in a 60-year long time series. Therefore, comparisons between results of analyses of such time series with results of simulations described in this book may not be completely conclusive. In the absence of much longer time series of worldwide empirical data on important quantities, however, we must use consistency and plausibility to verify/support results and draw conclusions. There are very long – thousands of years in some regions – time series of paleoclimate proxy data such as tree rings and lake sediments from which information about precipitation, temperature, and dryness/wetness is derived (Mehta, 2017). While such long time series can yield long-term perspectives on a region's climate, there are also problems associated with such data as described where appropriate in this book. Despite all caution, nature, while not capricious, may sometimes lead us astray!

Finally, some caveats about decadal predictability are necessary. The area of climate science associated with decadal climate predictability is at a nascent stage (see, for example, Meehl et al. (2009), Mehta et al. (2011)). Although the collective scientific community experience of seasonal to interannual climate prediction appears to be shortening the learning period about decadal climate prediction, we may be many years or even decades away from useful and skillful decadal climate predictions (see Meehl et al. (2014) and Mehta et al. (2018) for reviews). But, what appeared to be impossible even a decade ago seems to be possible in some near or not-very-distant future. Therefore, the decadal predictability research described in this book should be looked upon as a partially filled glass rather than a largely empty glass. Also, decadal hindcast and forecast data from many ESMs may become available from another Coupled Model Intercomparison Project (CMIP6) in the 2020–2021 CE timeframe, which may quantitatively or even qualitatively change the results and conclusions derived hitherto. But, that is inevitable in a rapidly moving area of scientific research. The description in this book, however, will still serve current and future students and researchers in familiarizing themselves with the history and fundamentals of decadal climate prediction.

1.6 PLAN OF THE BOOK

This book is organized in three themes: phenomena, mechanisms, and predictability. Except for Chapter 10 (Predictability), a phenomenon and its hypothesized mechanisms are described in the same chapter for easier comprehension and understanding. Therefore, after this Prologue, observed DCV in association between solar influences is described in Chapter 2; the PDO and the IPO are described in Chapter 3; and decadal variability of the mysterious tropical Atlantic dipole is described in Chapter 4. The Pacific and Atlantic decadal variabilities are followed by decadal variability of the tropical Indian Ocean in Chapter 5. Then, decadal variability of the North Atlantic Oscillation/Northern Annular Mode and that of the Southern Annular Mode are described in Chapters 6 and 7, respectively. The state-of-the-science from the nascent field of interactions between natural DCV and anthropogenic climate change is described in Chapter 8. Moving to decadal variability in higher frequency phenomena, variability in frequency and intensity of tropical cyclones is described in Chapter 9, followed by decadal climate predictability and prediction research in Chapter 10. Chapter 11 is the Epilogue, which summarizes highlights of the book, outlines outstanding problems in DCV understanding and prediction, recommends a path forward to make further progress, and makes closing remarks.

Although abbreviations are defined the first time they are used in each chapter, a table of abbreviations and their definitions is given in Appendix 1 for easy reference. All cited references are listed at the end of the book.

2 Solar Influences and the Earth's Climate Variability

2.1 INTRODUCTION

Life on the Earth cannot exist without solar energy. This central fact was known to humans since pre-historic times, reiterated every day by the rising of the Sun, and every year by the Sun's apparent north–south motion and the resultant seasonality of weather and climate. Due to the indispensable dependence on solar energy, solar eclipses created panic and many cultures devised various rituals to scare away or propitiate the demons "swallowing" the Sun. Therefore, it is not surprising that apparent blemishes or dark spots on the Sun's visible disk, perhaps viewed in reflected images or through cloud matter before the invention of the telescope, were believed to reduce the Sun's energy reaching the Earth which, in turn, was thought to affect weather and climate. This line of thinking gave rise to the search for and attribution of the solar connection to weather and climate variability, which has continued in spurts for over two millennia as we saw very briefly in Chapter 1.

According to anecdotal evidence, records of Sunspot observations in China go back to at least 28 Before Current Era (BCE),[1] perhaps to even 800 BCE.[2] The earliest drawing of Sunspots was apparently by John of Worcester's Chronicle in 1128 Current Era (CE)[2]; Chinese and Korean astronomers may also have witnessed these Sunspots in 1128–1129 CE[2]. In 1611 CE, four Europeans, almost simultaneously but independently, observed Sunspots through telescopes. These were Galileo Galilee of Italy, Johannes Fabricius of Holland, Christoph(er) Scheiner of Germany, and Thomas Herriot of England. Galileo was the first to hypothesize that Sunspots were objects on the Sun's surface.[3] Since then, systematic observations of Sunspots; reconstructions of solar irradiance based on proxy data; statistical associations between these indicators of solar activity and climate variables; ground-based and satellite-based observations and measurements of the Sun's radiative, particulate, and electromagnetic field emissions; and simulation experiments with a variety of climate models have continued to shed further light on the Sun–climate relationship and have led to the development of physics- and chemistry-based hypotheses about mechanisms of this relationship. From statistical associations among sparse data to the developing understanding of a broad range of physical mechanisms has been an exciting journey delving into one of the oldest enigmas humans have faced. This chapter tells the story of this continuing journey. During this journey, several comprehensive literature reviews have been published on the Sun–climate relationship. They are cited in this chapter, and the reader is referred to the cited literature

[1] http://galileo.rice.edu/sci/observations/sunspots.html
[2] http://chandra.harvard.edu/edu/formal/icecore/The_Historical_Sunspot_Record.pdf
[3] http://chandra.harvard.edu/edu/formal/icecore/The_Historical_Sunspot_Record.pdf

for comparative merits and credibility of published research on the subject of the Sun–climate relationship. Only the published research on decadal climate variability (DCV) and solar influences is reviewed in this chapter to provide highlights of the oldest and very intriguing search for causes and mechanisms of DCV. Recent advances in simulating and understanding how various types of solar emissions can influence the Earth's climate have revived the credibility of over two centuries of statistical studies; therefore, this history is reviewed first, followed by more recent observational and modeling studies. The purpose of this chapter is to provide an overview of scientific disciplines involved in understanding the role of solar influences in DCV. Readers should refer to cited publications for specialized details. Lastly, there are also many observational studies of associations between the 18.6-year lunar nodal cycle and climate variability on the Earth. These studies are not included in this book, and readers are referred to published literature on this subject (see, for example, Currie (1993), O'Brien and Currie (1993), Currie and Vines (1996), Cook et al. (1997), Cerveny and Balling (1999), Chen (1999), Yndestad (1999, 2003, 2006), Camuffo (2001), Cerveny and Shaffer (2001), and McKinnell and Crawford (2007))

After this brief introduction, the Sun's structure and composition, and its emissions are described in Section 2.2. Associations between Sunspots and solar emissions, and the Earth's paleoclimate variability and instrument-observed climate variability are described in Sections 2.3 and 2.4, respectively. Various hypothesized mechanisms and simulations with a variety of climate models to understand solar influences on the Earth's climate at decadal timescales are then reviewed and discussed in Section 2.5. This chapter ends with a summary and recommendations for future research in Section 2.6.

2.2 THE SUN AND ITS EMISSIONS

2.2.1 THE SUN

The Sun is an average star with an approximately 1.4 million km diameter, which is as long as the combined diameters of 110 Earths. It rotates on its axis with an approximately 27 to 30-day period, with the equatorial region rotating faster than the polar regions. The Sun is composed of 70% hydrogen; 28% helium; and heavier elements, such as carbon, nitrogen, oxygen, neon, magnesium, silicon, and iron.[4] The matter in the Sun is in a state called plasma, an ionized gas. The Sun–Earth distance varies from approximately 146 to 152 million km over one year.

The Sun's energy originates from nuclear reactions in the Sun's core. The solar core is approximately 350,000 km in diameter – 25% of the total diameter of the Sun. Nuclear fusion reactions at a temperature of 15 million °C convert hydrogen to helium in the solar core and convert the mass difference between them to energy. This energy then passes through five layers (Figure 2.1) before it is radiated into space. These layers are known as the radiative zone, the convective zone, the photosphere, the chromosphere, and the corona. The radiative zone begins at the edge of the core and extends to 70% of the Sun's diameter or 980,000 km from the center

[4] http://coolcosmos.ipac.caltech.edu/ask/4-What-is-the-Sun-made-of-

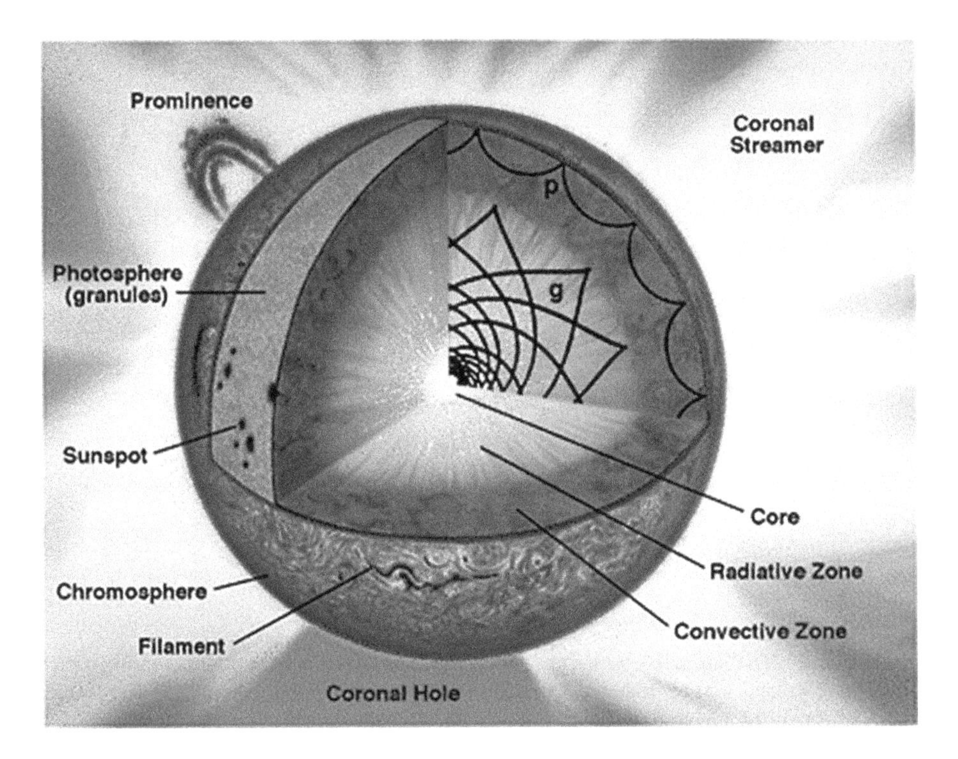

FIGURE 2.1 Layers and major features of the Sun. (Courtesy: The Solar and Heliospheric Observatory (SOHO) project at National Aeronautics and Space Administration (NASA). http://sohowww.nascom.nasa.gov/explore/img/mdigraphic.gif.)

of the Sun. The energy generated in the core is transmitted through the radiative zone in the form of radiation scattered by largely stationary gas. Due to the density of gas in the radiative zone and the number of times a photon of energy is scattered within the radiative zone, it is estimated that the time taken for a photon to go through the radiative zone is 1 million years. Fluid motions begin at the top of the radiative zone in the interface with the convective zone. It is believed that the Sun's magnetic field is generated by a dynamo in this interface.[5] The convective zone is approximately 200,000 km thick and extends to the Sun's visible surface. Lower temperatures (2 million °C) in the convective zone than those in the solar core make heavier elements less ionized (fewer electrons are stripped out of the ions), allowing them to retain more heat, to create a temperature gradient within the layer, and to make the fluid more unstable and result in convective motion. The convective motion of the fluid carries the energy from the core to the surface. As the fluid rises, it cools to such an extent that the temperature at the top of the convective layer, in the photosphere, is "only" 5,700°C! The photosphere is the Sun's visible surface and is 100 km thick. The chromosphere surrounds the photosphere. Temperature of the gas in the chromosphere increases to 20,000°C from the photosphere's 5,700°C. The hydrogen

[5] solarscience.msfc.nasa.gov/interior.shtml

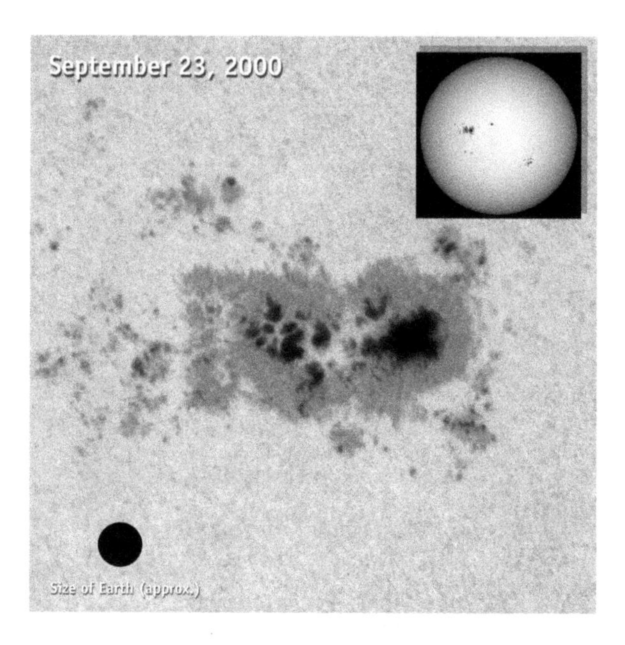

FIGURE 2.2 The Sun's disk with some sunspot groups as big as Jupiter (upper right), with a close-up of sunspots bigger than the Earth shown at lower left. (Courtesy: The SOHO project at NASA. https://sohowww.nascom.nasa.gov/gallery/images/large/sunspot00_prev.jpg.)

gas in the chromosphere emits a reddish light at this temperature, giving the layer its name (chromo – color). The corona is the Sun's outer atmosphere surrounding the chromosphere and extending past the Earth to the outer regions of the solar system.

Sunspots, flares, faculae, and prominences are major features apparent on the photosphere.[6] Sunspots are apparent dark spots on the Sun's visible disk (Figure 2.2). They typically occur in pairs or groups and are restricted in their position on the Sun between approximately 35°N and 35°S solar latitudes. Sunspot sizes vary between 4,000 and 30,000 km, and can be even larger than the Earth as shown in Figure 2.2. The magnetic field in a Sunspot is concentrated, which temporarily inhibits transport of hot plasma from the interior of the Sun, so the area of the Sunspot is cooler and darker than the surrounding matter. The inner part of the Sunspot, the umbra, is darkest and coolest, whereas the outer part, the penumbra, is less dark. The most distinctive feature of Sunspots is that their number varies over several timescales. Bands of Sunspots first form at solar mid-latitudes, widen, and then move towards the equator as each Sunspot cycle progresses. Figure 2.3a shows observed Sunspot numbers from 1749 to 2004 CE and predicted Sunspot numbers from August 2004 to December 2007. Figure 2.3b shows real Morlet coefficients of the wavelet spectrum[7] of the Sunspot number time series in Figure 2.3a in the time-oscillation timescale space (Weng, 2005). Very strong cycles at approximately 11 years, the Schwabe cycles

[6] scied.ucar.edu/sun-features-regions

[7] The wavelet transform decomposes a time series into amplitude evolution of its component oscillations.

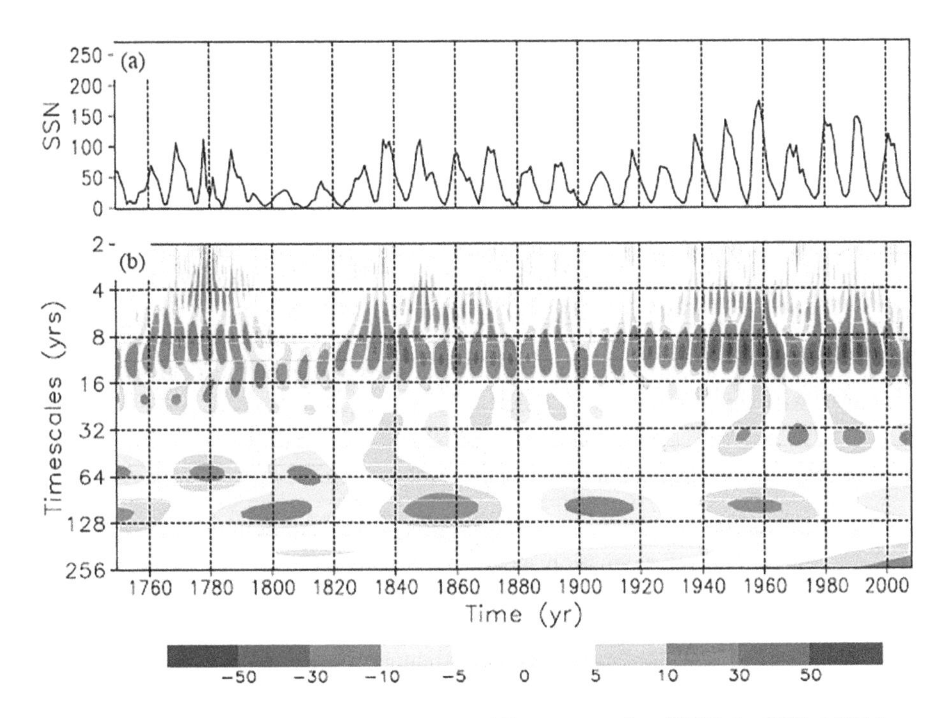

FIGURE 2.3 (a) The time series of the annual Sunspot number (SSN) for 1749–2007 (predicted Sunspot number after August 2004) and (b) its real Morlet wavelet coefficients for timescales between 2 and 256 years. (From Weng, 2005.)

(1843), are prominent in Figure 2.3a and b. The period of the Schwabe cycle (or, the cycle length), however, is not constant and varies between 9 and 14 years. The number of Sunspots in each cycle is also not constant, and there are periods when very small numbers of Sunspots have been observed as from the 1790s to 1830s CE and in the first decade of the 21st century CE in Figure 2.3. Besides the Schwabe cycle, Figure 2.3b also shows that variability of Sunspot numbers in Figure 2.3a can range from 2–3 years to approximately 180 years, with specific periods evident within the wavelet spectrum. At the beginning of each Schwabe cycle, new spots and groups of spots appear at high solar latitudes, and appear to move towards the solar equator as the cycle progresses until reaching within 5° solar latitude near the end of the cycle. The magnetic field polarity of a new high-latitude group is opposite to that of old, low-latitude groups. In conjunction with the Sunspot cycle, the magnetic poles of the Sun reverse their polarity on the same time scale; thus, the magnetic field of the Sun has an approximately 22-year cycle, the so-called Hale cycle (Hale and Nicholson, 1925). In one Hale cycle, Sunspot groups with the same magnetic field polarity appear at high solar latitudes approximately every 22 years or two Schwabe cycles. The numbers of Sunspots, flares, and other solar storms go through a maximum and a minimum during every Schwabe cycle. Relatively high solar activity is known as the solar maximum, and relatively low activity is known as the solar minimum. These observations of the Schwabe and Hale cycles stimulated statistical analyses of associations between solar cycles and climate variability on the Earth as we will see later.

Another visible feature in the photosphere is a solar flare – a sudden, rapid, and intense variation in the Sun's apparent brightness. A solar flare occurs when accumulated magnetic energy in the photosphere is suddenly released. During a flare, radiation is emitted across almost the entire electromagnetic spectrum. The amount of energy released in a typical flare is equivalent to millions of 100-megaton bombs exploding simultaneously or ten million times greater than the energy released from a volcanic eruption. A facula is smaller than a Sunspot, but is often more numerous and brighter than its surroundings. Faculae are associated with strong magnetic fields. A prominence is a large bright "loopy" feature extending outward from the Sun (Figure 2.1). Stable prominences may persist in the corona for several months. A typical prominence extends over many thousands of kilometers, and the mass contained within a prominence is typically 10^{14} kg.

2.2.2 The Sun's Emissions

The Sun radiates electromagnetic energy at almost all wavelengths from radio waves to gamma rays. The solar emission spectrum from the ultraviolet to infrared wavelengths is shown in Figure 2.4. The total energy emitted from the Sun is approximately 63 million W m^{-2}. Most of the electromagnetic energy emitted from the Sun is at visible wavelengths centered at 500 nanometers (nm); the Sun also emits substantial amounts of energy at ultraviolet and infrared wavelengths; and small amounts of energy at radio, microwave, X-ray, and gamma ray wavelengths. An indicator of the emitted ultraviolet energy is the core-to-wing ratio of the solar magnesium (Mg) ii spectral line. An indicator of the solar emissions at radio frequencies is the 10.7 cm wavelength radiation.

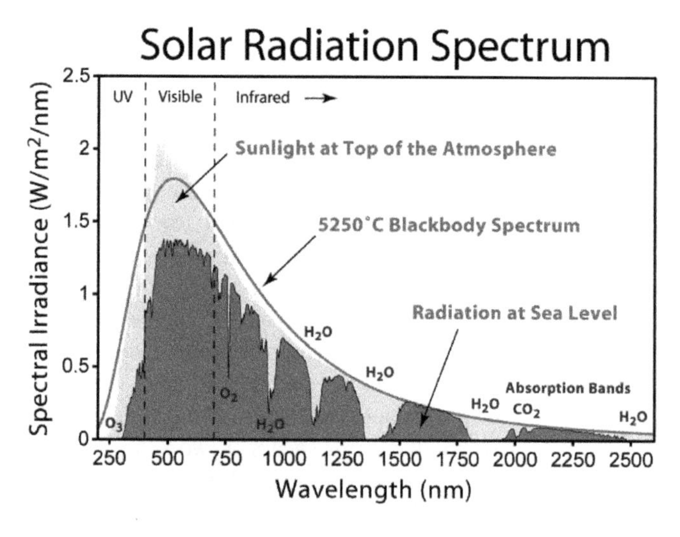

FIGURE 2.4 The spectrum of solar radiation from 200 to 2750 nm at the top of the atmosphere (light gray) and at sea level (dark gray). The blackbody radiation spectrum of the body at 5,250°C is also shown for comparison. (Courtesy: Wikimedia.)

The amount of solar electromagnetic energy incident on top of the Earth's atmosphere is approximately 1,367 W m^{-2}. The atmosphere reflects and scatters almost 60% of the incident solar radiation, and the remaining approximately 40% passes through the atmosphere to the Earth's surface as Figure 2.4 shows. Gamma rays, X-rays, and ultraviolet radiation with wavelengths shorter than 200 nm are selectively absorbed in the atmosphere by oxygen and nitrogen, heating the upper atmosphere. Most of the solar ultraviolet radiation with a range of wavelengths from 200 to 300 nm is absorbed by ozone in the stratosphere. Infrared solar radiation with wavelengths greater than 700 nm is partially absorbed by carbon dioxide, ozone, and water vapor in the atmosphere.

In addition to electromagnetic radiation and magnetic field, the Sun also emits electrons and protons, the so-called "solar wind", including massive clouds of charged particles and magnetic field known as coronal mass ejections. Many phenomena on the Earth such as geomagnetic storms and the aurora borealis and australis – the Northern and Southern Lights, respectively – are directly caused by the solar wind. The solar magnetic field is frozen into the solar wind and is "pulled out" of the Sun by the latter as it travels through the solar system. The Earth's magnetic field is influenced by charged particles and by the solar magnetic field embedded in the solar wind.

Charged particles from outside the solar system, the so-called galactic cosmic rays (GCRs), also impinge upon the Earth's atmosphere. GCRs are highly energetic particles originating in the Milky Way galaxy and consist mostly of hydrogen. These particles travel through the galaxy and are stripped of their electrons, making them fully ionized. As a result, they are accelerated to nearly the speed of light by the galactic magnetic field. The solar and Earth's magnetic fields modulate the GCR flux near the Earth, and the latter field deflects them away from the equatorial region towards the polar regions. When a solar wind cloud travels past the Earth, the magnetic fields embedded among the solar wind deflect incoming GCR particles, thereby decreasing the number of GCR particles impinging upon the Earth's atmosphere.

These various solar features and emissions are inter-related, and undergo cycles as shown in Figure 2.5 for the 1975 to 2010 CE period. Beginning at the top (Figure 2.5a) with images of the Sun at maximum and minimum activities, subsequent figures show Sunspot numbers (Figure 2.5b), 10.7 cm radio flux (Figure 2.5c), the Mg ii line core-to-wing ratio (Figure 2.5d), open solar magnetic flux (Figure 2.5e), GCR particle count (Figure 2.5f), satellite-measured total solar irradiance (Figure 2.5g), and one of the indicators of the geomagnetic field the Ap flux (Figure 2.5h). As Figure 2.5 shows, the solar emissions undergo the Schwabe cycle, with the GCR count varying in opposite phase to other indicators of solar activity as explained earlier. It is interesting to note that the total solar irradiance, and ultraviolet and radio fluxes are maximum almost at the same time as the Sunspot number. This is because the darkening of the solar disk caused by spots at the solar maximum is more than compensated by brightening of the disk caused by solar faculae. Mechanisms of solar dynamics causing the Schwabe and Hale cycles are beyond the scope of this book.

As we will see later in this chapter, solar radiative emissions, solar magnetic field, and solar and GCR particles have been hypothesized as "agents of influence" in

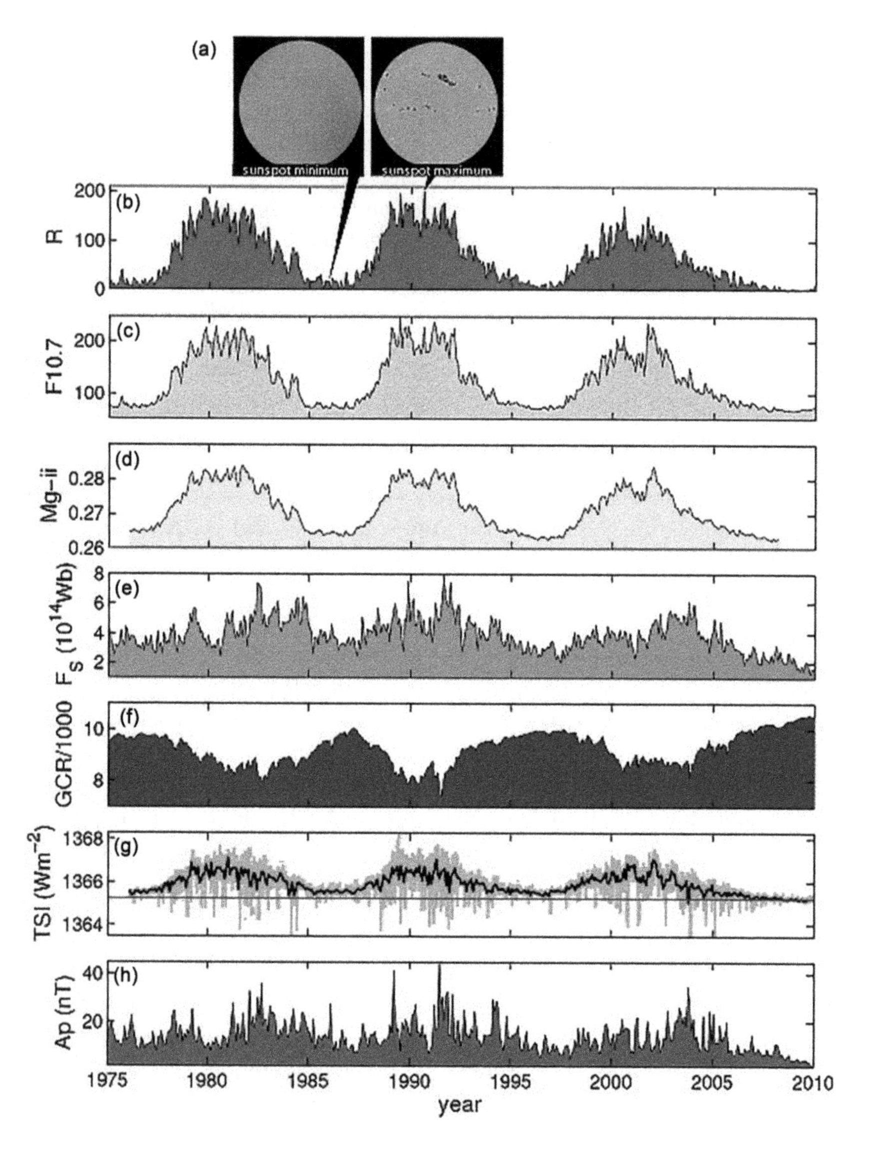

FIGURE 2.5 (a) Images of the Sun at sunspot minimum and sunspot maximum. Observed variations of (b) the sunspot number R (a dimensionless weighted mean from a global network of solar observatories, given by $R = 10N + n$, where N is the number of sunspot groups on the visible solar disk and n is the number of individual sunspots); (c) the 10.7 cm solar radio flux, F10.7 (in W m^{-2}Hz^{-1}, measured at Ottawa, Canada); (d) the Mg ii line (280 nm) core-to-wing ratio (a measure of the amplitude of the chromospheric Mg II ion emission, which on time scales up to the solar cycle length has been found to be correlated with solar UV irradiance at 150–400 nm); (e) the open solar flux FS derived from the observed radial component of interplanetary field near Earth; (f) the GCR counts per minute recorded by the neutron monitor at McMurdo, Antarctica; (g) the PMOD composite of TSI observations; and (h) the geomagnetic Ap index. All data are monthly means except the light gray line in (g), which shows daily TSI values. (After Gray et al., 2010.)

the Earth's climate variability. We begin by reviewing tales told by ancient climate observers – paleoclimate proxy data – about possible solar influences at decadal timescales.

2.3 SOLAR CYCLES AND PALEOCLIMATE VARIABILITY

Instrument-based records of precipitation and temperature in most regions of the world are only approximately 150 years long, except for a few regions such as Central England and some parts of China where such records extend back several hundred years. Therefore, climate scientists take recourse to what are known as paleoclimate proxy indicators to analyze climate variability over the last several centuries or millennia. In this context, paleo refers to ancient geologic past, and proxy refers to a parameter or entity which represents an environmental variable or entity which was not observed or estimated directly in the past. Some of the better-known paleoclimate proxies are tree rings, oxygen and carbon isotopes in various types of materials, tropical corals, and sediments in lakes and oceans. These proxies are sensitive to precipitation, temperature, and/or salinity, and are used to infer environmental conditions. These proxies, however, are sensitive not only to environmental variables, but also to other conditions such as density of forests, diseases, and wildfires, to name a few. Therefore, paleoclimate variables or entities inferred from proxy indicators are subject to misinterpretation and must be verified in other ways. Despite confounding influences, the proxies do yield useful information about past climate and indicate evidence of worldwide climate variability and changes in the near and distant past. Therefore, the presence of DCV, and its associations with the Schwabe and Hale cycles in paleoclimate proxies are reviewed here briefly and a few specific examples are mentioned. For each type of paleoclimate proxy, we will first see briefly why it may yield useful climate information and some major problems in interpreting the proxy directly in terms of climate.

2.3.1 TREE RINGS

Let us start with tree rings as climate observers; a detailed description is given in Mehta (2017). Annual tree growth appears as rings in the cross section of a tree. Changes in the thickness of tree rings over time indicate changes in length of, or water availability during, the growing season, making certain types of trees climate observers. A tree ring consists of a layer of lighter color that grows in the spring and a layer of darker color that grows in the summer. At locations where tree growth is limited by water availability, trees will produce wider rings during wet and cool years than during hot and dry years. Drought or a severe winter can cause narrower rings. If the rings have the same width over the entire tree's cross section, it implies that climate was the same during the entire period of tree growth. By counting the rings and measuring their widths from the center of the tree trunk outwards, the age and health of the tree and the growing season of each year can be determined with high accuracy. Although it is difficult to completely remove confounding effects mentioned earlier and isolate climate-related effects alone, ring widths can be translated into quantitative information about durations and lengths of each type of hydrologic epoch the tree experienced. The quantitative information can be calibrated in terms of precipitation, temperature, drought index, or other climate indicators, thereby enabling a time history of such epochs to be formed over the lifetime of a tree. Certain species of trees are more sensitive to their environment than others and so only the sensitive species can be used to assemble proxy climate records. After all these problems are overcome, techniques such as carbon dating must be used to accurately relate tree rings to time, so that a time series of ring widths, and dry/wet epochs – the so-called dendrochronology – can be assembled. Braving these and other difficulties, many climatologists have assembled multicentury or even millennial records of tree-ring-based climate reconstructions in many regions of the world. In this chapter and some later chapters, we will see tree-ring data used to infer paleoclimate variability.

Almost all published research on Sunspot cycle associations with tree-ring data from various regions of the world consists of correlation coefficients, spectrum, cross-spectrum, and other statistical analyses between Sunspot numbers and tree-ring-derived quantities. Many studies used spectrum analysis techniques on tree-ring data, found spectral peaks approximately at 11- and 22-year periods, and attributed such spectral peaks to solar cycles. Some researchers also associated solar cycle length with various attributes of tree rings. Some representative results are reviewed here.

2.3.1.1 Analyses of Tree Rings

Perhaps the oldest solar cycle attribution of tree-ring variability goes back 290.6 million years in the Permian, the last period in the Paleozoic Era. Luthardt and Rößler (2017) found 10.6-year periodicity in 79-year-long tree-ring width data from an early Permian fossil forest in southeast Germany, which was in the tropical region of the Pangaea (or Pangaia) supercontinent surrounded by the Panthalassa superocean during the Permian Period.[8] A series of volcanic eruptions covered the forest ecosystem in hot ash and lava, preserving the forest in petrified form. Since the ring width period is very close to the 11-year solar cycle period in the 79-year sequence of tree rings from partially broken or upright trees, Luthardt and Rößler attributed the cyclic ring width variability to solar influence on monsoonal paleoclimate of the region via impacts of solar cycle modulation of GCRs on cloud formation and precipitation. This discovery implies that DCV, which influenced tree-ring formation, existed 290 million years ago and also that the 11-year Schwabe cycle may have been active over such a long geologic timescale. The hypothesized GCR-cloud formation mechanism of solar influence on climate is discussed later in this chapter.

Many researchers who analyzed tree-ring variability in the last several centuries also alluded/attributed 11- and 22-year cycles in tree-ring-based inferences about climate variability to the solar influence. Representative studies which analyzed 100-year or longer tree-ring time series in various parts of the world are Blasing and Duvick (1984), reconstructed precipitation in Iowa and Illinois, United States of America (USA), 1680–1980 CE; Murphy (1990, 1991), tree rings, Australia, 1028–1975 CE; Raspopov et al. (2001, 2004), tree rings, Russia, 1629–1974 CE; Rigozo

[8] www.britannica.com/science/Permian-Period

et al. (2004), tree rings, southern Brazil, 1797–1997 CE; and Rigozo et al. (2007), tree rings, Chile, 1587–1994 CE.

2.3.1.2 Analyses of Tree Rings and Sunspot Numbers

A variety of time series and spectrum analysis techniques have been used to study possible associations between tree-ring variability and Sunspot numbers. Many of these studies have also shown non-stationarity of tree ring–Sunspot relationships, perhaps implying a lack of physical causality or modulation of the relationship by longer-term solar or other variability. For example, wavelet spectrum and cross-spectrum analysis techniques were used by He et al. (2007) to analyze variability in a multicentury (1657–2004 CE) tree-ring width chronology, developed from 58 samples taken from pine trees in Shenyang, China. The analyses included annual tree-ring growth, Sunspot numbers, and several climate indices. The annual tree-ring growth showed a significant, but non-stationary, association with Sunspot numbers in the 5 to 8-, 10 to 12-, and 20 to 30-year periods. He et al. (2007) also found that the 11-year Sunspot cycle had high amplitudes during the 1720–1800 CE, 1830–1870 CE, and 1940–1990 CE periods, and that most interannual El Niño and La Niña events occurred in low Sunspot number years; these two types of events refer to warmer than average and colder than average sea surface temperatures in eastern equatorial Pacific, respectively.

A proxy record from the Mt. Logan region in Canada was used by Sinclair et al. (1993) to infer a solar cycle–tree rings–hare population connection. They used the information that dark marks in the rings of white spruce less than 50 years old in Yukon, Canada, are correlated with the number of stems scratched by snowshoe hares, and the frequency of these scratch marks is positively correlated with the density of hares in this region. They also used the information that the frequency of marks in trees germinating between 1751 and 1983 CE is positively correlated with the hare fur records of the Hudson Bay Company. It was found that both tree marks and hare numbers are correlated with Sunspot numbers, and there is a 10-year periodicity in the correlations. Phase analysis showed that tree marks and Sunspot numbers have periods of nearly constant phase difference during 1751 to 1787, 1838 to 1870, and 1948 to the early 1990s CE, and these periods coincide with those of high Sunspot maxima. The nearly constant phase relations between the annual net snow accumulation on Mt. Logan; and tree mark ratios, hare fur records before approximately 1895 CE, and Sunspot numbers during periods of maxima in the Sunspot cycles suggest a solar cycle–climate–hare population–tree mark association.

Zhou and Butler (1998) determined correlation coefficients between tree-ring width and the Sunspot cycle length for 69 tree-ring data sets from the USA, Canada, Chile, Argentina, Belgium, France, Germany, Poland, Finland, Turkey, China, Australia, and New Zealand, with each data set at least 594 years long. The analyses implied that wider tree rings, perhaps due to more optimum growth conditions, were associated with shorter Sunspot cycles. It was also found that climate impacts of solar cycle length variations accumulated over several decades and the degree of accumulation was dependent on the elevation and geographical location of the trees, with a distinct difference in characteristics between trees above and below 2,500 m elevation.

N.R. Rigozo and co-authors have analyzed numerous tree-ring width time series in Brazil and Chile. Rigozo et al. (2004) obtained a tree-ring width time series from 1797 to 1997 CE using samples of *Araucaria angustifolia* trees from Santa Catarina State in southern Brazil. Using maximum entropy and iterative regression spectral analysis techniques, they found several periodicities of ring widths associated with solar activity variations. A cross-correlation analysis between Sunspot number and ring width time series showed a lag of zero year, implying that the trees responded to solar activity variations within one year. Rigozo et al. (2007) searched for solar cycle signals in approximately 400-year long ring width time series from two locations, one at 1,000 m elevation and one near the Pacific Ocean, in Chile. Spectral and wavelet analysis techniques showed evidence of the Schwabe and Hale cycles, along with longer period cycles. It was also found that the two cycles were present intermittently in Sunspot and tree-ring data. The cross-wavelet signal around the Schwabe cycle period was more significant during periods of high Sunspot activity than during periods of low Sunspot activity. Based on these results, Rigozo et al. speculated that the tree-ring response may be stronger at the Pacific Ocean coast due to the possible heating of the Pacific Ocean water following an increase of solar radiation during high Sunspot activity. Later in this chapter, we will see hypotheses about how the Pacific Ocean temperatures can be influenced by variability of solar emissions and model simulations to test these hypotheses.

As this brief review shows, numerous studies of tree-ring chronologies from many parts of the world going back to almost 300 million years show nearly cyclic variability at decadal timescales in independent analyses as well as in joint analyses with Sunspot number time series. At the least, these studies show the presence of DCV which influenced tree-ring growth regardless of whether the DCV was caused by solar cycles.

2.3.2 Oxygen and Carbon Isotopes

Compositions of water, marine sediments, ice cores, and fossils can also tell stories about past climate. For example, oxygen nuclei in water contain eight protons, but can contain eight or more neutrons. Nuclei with the same number of protons but different number of neutrons have different atomic masses but the same chemical properties, and are known as isotopes. The oxygen isotope with eight protons and eight neutrons, ^{16}O, is the most common, with the isotope with eight protons and ten neutrons, ^{18}O, the next in abundance. Due to different evaporation and condensation rates of these two isotopes, their relative amount expressed as ratio (a measure of it expressed as $\delta^{18}O$) can contain signatures of past climate. The oxygen isotope ratio of water in glaciers and ice sheets also contains information about past climate. The oxygen isotope ratio in shells of marine plants, corals, and animals is also influenced by ambient temperature, so it also contains signatures of past climate. Thus, the oxygen isotope ratio can be a very informative proxy of past climate, not only in liquid and frozen water but also in stalagmites formed by dripping water in caves. Another example is isotopes of carbon with six protons; and six, seven, and eight neutrons. These isotopes are known as ^{12}C, ^{13}C, and ^{14}C, respectively. ^{12}C is by far the most abundant, followed by the other two. A measure of the ratio of

^{13}C and ^{12}C abundances, expressed as $\delta^{13}C$, can contain information about photosynthetic activity dependent on ambient light level and, therefore, about climate. Incidentally, ^{14}C is highly radioactive and eventually decays to a stable isotope. This property of ^{14}C and a measure of the ratio of ^{14}C and ^{12}C is used to determine the age of organic matter in a process known as carbon dating. We will now see results of analyses of oxygen and carbon isotopes to infer DCV and its association with solar cycles.

In the Caribbean Sea–tropical North Atlantic Ocean region, Black et al. (2004) analyzed oxygen isotope records with almost-annual resolution from fossils of two species of planktic foraminifera[9] from the Cariaco Basin over the Caribbean and tropical North Atlantic spanning the last 2000 years. The oxygen isotope is influenced by sea surface temperature (SST) and salinity variations related to precipitation variations in the InterTropical Convergence Zone (ITCZ) region. A large-scale pattern of high correlations between foraminiferal $\delta^{18}O$ and SSTs was found over the period of instrumental-proxy data overlap, but the correlations weakened when extended back in time and instrumental SST records became discontinuous. Black et al. offered two plausible explanations for a long-term trend in one species' (the Globigerinoides ruber) $\delta^{18}O$ record. First, the increase in $\delta^{18}O$ may indicate that tropical summer and autumn SSTs have cooled by as much as 2°C over the last 2000 years, possibly as a result of a long-term increase in upwelling intensity, or comparisons to other studies of the ITCZ and regional evaporation and precipitation variability suggest that much of the $\delta^{18}O$ record may have been influenced by decadal-to-centennial variations in the average position of the ITCZ. Similarities between the second species' (the Globigerinoides bulloides) $\delta^{18}O$ record and the 11-year Schwabe cycle suggest that solar variability may be important in influencing the freshwater content and thermohaline circulations in the Caribbean–tropical North Atlantic Ocean.

In an example of the usefulness of oxygen isotope ratio as a proxy climate variable on land, Paulsen et al. (2003) found oxygen and carbon isotopic variations, with a time resolution of 1 to 3 years in a 1270-year record, in a stalagmite from the Buddha Cave in central China. In addition to changes corresponding to the Medieval Warm Period (900–1300 CE), the Little Ice Age (1300–1870 CE), and the 20th century CE warming, periodic variations at 33, 22, 11, 9.6, and 7.2 years were also found, indicating variations at Schwabe and Hale cycle periods.

Castagnoli et al. (2002a, b) analyzed a highly precise record of carbon isotope $\delta^{13}C$ in the planktonic Globigerinoides ruber extracted from the GT90/3 shallow water, Ionian Sea core for the period 590–1979 CE, with a resolution of 3.87 years. It is believed that $\delta^{13}C$ variations in symbiotic foraminifera mainly record the effects of productivity and of photosynthetic activity, varying with the ambient light level. Therefore, information about sea surface illumination at the time of the planktonic foraminifera growth was inferred from this proxy time series. Using Monte Carlo singular spectrum analysis, a highly significant 11-year signal was found in phase with the Schwabe cycle.

[9] Planktic foraminifera are single-celled living organisms which float at various water depths.

As in the analyses of tree rings, this brief review of oxygen and carbon isotopes analyses in ocean water and in stalagmites on land shows the existence of DCV regardless of whether it was caused by variability of solar emissions.

2.3.3 Lake Sediments

Sand, silt, and organic material can be transported by wind, water, or ice flows and deposited in lakes, rivers, and oceans. Particles flowing or settling into a lake during different years can have different sizes and/or composition. As a result, annual layers of sediments, known as varves in Swedish, can form and preserve a record of climatic conditions whose chronology can be established by carbon dating and whose composition can be established by chemical analyses. Various statistical analysis techniques can then be used to associate variability of sediments and their composition with astronomical, climatic, and other phenomena. For example, Damnati and Taieb (1995) found that sediment cores from Lake Magadi, Kenya, dating back to 12,000 and 10,000 years ago, showed alternating light and dark sediments. Spectral analysis of the thickness variation of sediment layers showed numerous peaks at 6.45, 8.23, 8.79, 9.44, 10, 13.82, 18.94, 24.18, and 30 years, leading to speculation by the researchers that the spectral peaks around 10 to 20 years may be associated with the Schwabe and Hale cycles. In another example, in an analysis of laminated sediments of Loch Ness, Scotland, from 1321 to 1963 CE, Cooper et al. (2000) found that lamination thickness was highly correlated with the number of weeks of ice off Iceland, the North Atlantic Oscillation index, and with Sunspot numbers. A spectral analysis of the thickness time series showed the presence of the Schwabe and Hale cycles. Milana and Lopez (1998) analyzed time series of varves and their spectra to search for cyclic signals in western Argentina. A peak around 12 years was apparent in most of the spectra, and the researchers associated it with the 11-year Schwabe cycle. A secondary peak near 24–26 years was associated with the 22-year Hale cycle. A record of Schwabe cycle-like signals was found by Munoz et al. (2002) in varves from the Pliocene epoch (approximately 5.3 to 2.6 million years BCE) in the Villarroya Pliocene Basin (La Rioja, northern Spain). These examples illustrate that lake sediment variability at approximately 10 to 20-year periods is ubiquitous and goes back millions of years. As in the case of proxy climate variability derived from tree rings, researchers' instinctive inclination has been to associate the 10 to 20-year sediment variability with solar cycles.

In Section 2.3, we got a flavor of what ancient climate observers – tree rings, oxygen and carbon isotopes, and lake sediments – recorded and learned these recorded stories revealed by indefatigable researchers. First and foremost, these stories show beyond any doubt that these observers have undergone DCV – in some cases, hundreds of millions of years ago. They also show the possibility that the Schwabe and Hale cycles may have been the drivers of DCV, even though the researchers who revealed this may not have a physical mechanism in mind to establish a cause-and-effect relationship. In the next section, we will hear what stories a plethora of instruments – the modern climate observers – in the last 150+ years are telling us about DCV and its possible association with the solar cycles.

2.4 SOLAR CYCLES AND INSTRUMENT-OBSERVED CLIMATE VARIABILITY

Initiated by individuals to observe weather in North America in the mid-17th century CE, groups of volunteer observers formed in the USA beginning in 1776 CE. The United States (US) Surgeon General James Tilton ordered Army posts in 1814 CE to begin meteorological observations, giving rise to the first observing network in North America,[10] which was formalized by the Smithsonian Institution in 1849 CE. The US National Weather Service was established in 1870 CE. In Europe, Fernando II de Medici – the Grand Duke of Tuscany from 1621 to 1670 CE – established a meteorological observations network in ten cities, among them Florence, the home of the Medicis; the collected data were sent to Florence at regular intervals (Bradley and Jones, 1992). A few stations in Central England have measured surface air temperature since 1659 CE, but the more reliable record is since 1772 CE. The United Kingdom (UK) Government established the UK Meteorological Office in 1854 CE. The British East India Company established several meteorological observatories in India beginning in 1785 CE and expanded the network beginning in the first half of the 19th century CE. The India Meteorological Department was established in 1875 CE as a branch of the colonial government.[11] In late 19th and early 20th centuries CE, national meteorological services were established in many countries in Europe and Asia, although meteorological observations were started in some countries much earlier. Therefore, reliable instrument-measured observations of precipitation and temperature are available in many countries for well over 100 years.

The recorded history of oceanographic observations goes back to at least the 15th century CE. Since then, there are records of shipboard measurements of various oceanographic phenomena and of meteorological phenomena observed at sea. The Challenger expedition in 1872–1876 CE was probably the first, organized, around-the-world cruise that made extensive oceanographic observations. Regular measurements of SSTs, however, did not begin till the 1850s CE. In the beginning, most SST observations were made at regular times along shipping routes from Europe to Africa, Asia, and North America. Geographical coverage even in these regions was generally sparse and uneven. The two 20th century CE World Wars interrupted regular measurements, and the opening of the Suez (1869 CE) and Panama (1914 CE) Canals permanently changed shipping routes and, therefore, geographical coverage of oceanographic measurements. In the Pacific Ocean, sparsity of shipping routes and the World Wars made possible regular observations only after the 1940s CE. Changes in SST measurement instruments and methods also complicated quantitative use of the available data. Woodruff et al. (1987), Bottomley et al. (1990), Parker (1994), Parker et al. (1995a), and Folland and Parker (1995) have described the geographical distribution and density of oceanographic and land surface observations, and the problems associated with changes in measurement instruments and methods.

[10] http://celebrating200years.noaa.gov/foundations/weather_obs/welcome.html#earlyyear

[11] http://www.imd.gov.in/pages/about_history.php

The availability of these data since the 17th and 18th centuries CE spawned a veritable "cottage industry" to look for Sun–climate relationships. The availability of electric and geomagnetic field measurements, the 10.7 cm radio flux, the GCR flux, and the solar charged particle flux in recent times has resulted in the expansion of possible Sun–climate connection mechanisms from Sunspots and solar irradiance to these other solar agents of influence. Despite confounding influences, analyses of instrument-measured climate data to search for solar connections have yielded useful climate variability information and indicate evidence of worldwide climate variability in the last two to three centuries. Therefore, associations between DCV as seen in instrument-measured climate data, and cyclic variability in solar emissions and influences are described here. Important, although somewhat subjectively selected, associations are described to convey the ubiquity of the quest for solar–climate connections at decadal timescales. These associations were found with correlation, spectrum and cross-spectrum, and other time series analysis techniques. These techniques, however, analyze linear associations between solar and geophysical variables, and, as argued by Weng (2005), the Earth's climate system is nonlinear and forced responses of the system to solar variability can occur at different timescales compared to the forcing timescales. Therefore, linear analyses of possible associations may not always tell the true story and may even tell a misleading story. This possibility should be kept in mind.

2.4.1 Associations of Surface Climate Variations with Solar Cycles

Many researchers analyzed rainfall, temperature, and other weather and climate observations on land. When they found quasi-oscillatory features in such data at 10 to 20 year periods, they usually attributed such features to the Schwabe and Hale cycles.

Robert Currie and co-researchers have rigorously analyzed precipitation, surface pressure, and temperature observations in North America. Currie (1974) analyzed the late 19th and 20th century CE, annual-average surface air temperatures at 78 North American stations using the Maximum Entropy Method of spectral estimation and found the average oscillation period at 10.6 ± 0.3 years with an average amplitude of 0.1°C. In a later study, Currie (1979) used monthly-average temperatures at 100 stations, surface pressure at 73 stations, and rainfall at 80 stations in North America, typically from 1881 to 1960 CE. The 10 to 11-year signal in temperature was present only northward of 35°N and eastward of 105°W, with 0.29°C ± 0.15°C area-average amplitude. Currie and O'Brien (1988) also analyzed annual-average precipitation data from 1835 to 1984 CE in the northeastern USA and found 10 to 11-year period oscillation with approximately 1 cm annual-average amplitude. In a subsequent study, Currie and O'Brien (1990) found oscillations with an average period of 10.6±0.5 years in 70% of 288 rain-gauge records in the states of California, Colorado, Kansas, Missouri, North Carolina, Ohio, Oregon, Washington, and West Virginia in the USA. These researchers attributed the nearly 11-year oscillation in temperature and rainfall to the Schwabe cycle. Their finding spectral peaks at approximately 11-year period in several climate variables and at many locations in North America in time series a century or longer lent credence to the existence of

DCV. The amplitudes of these oscillations in rainfall and temperature, however, appeared to be too large to be the result of direct forcing by solar irradiance variability alone. In another example of a tantalizing hint about Sun–climate relationship, Clegg and Wigley (1984)'s spectral analysis of 510 years (1470–1979 CE) of wetness/dryness index for seven locations in the Beijing region, China, suggested that there were spectral peaks at approximately 10- and 20-year periods. These spectral peaks, however, were either not statistically significant in data from all locations or such peaks were unstable during the over 500-year-long record.

In other geophysical observations, Currie (1976) found a 10 to 11-year signal, with typical 1 cm amplitude, in monthly-average sea level measured by tide gauge records in 33 Northern European harbors from 1890 to 1964 CE. Currie (1980) also found such a signal in a measure of the Earth's rotation rate, with 0.16 ms amplitude of the length-of-day changes. Although Currie attributed both these signals to the Schwabe cycle, the 11-year length-of-day cycle lagged behind the Schwabe cycle by 3.4 years.

2.4.2 Sunspot Numbers

Nineteenth and 20th century CE research on impacts of solar variability on climate expanded the search for Sunspot cycle signatures in precipitation, temperature, atmospheric pressure, winds, tropical and mid-latitude storms, lake levels, tide levels, and river floods. Poey (1873) compiled lists of tropical cyclones from 1700 to 1855 CE in the Atlantic and Indian Oceans from historical newspaper accounts, ships' log books, and weather records. By comparing his lists with Sunspot number records, he suggested that variations in numbers of tropical cyclones in the North Atlantic Ocean and the southern Indian Ocean were associated with Sunspot variations. Subsequently, other researchers have continued this line of inquiry to the present and have used Schwabe cycle extrapolations to predict numbers of hurricanes in the Atlantic region. This research is described further in Chapter 11.

2.4.2.1 Rainfall and Temperature

Using observations made in the mid-19th century to the early 20th century CE and using the superposed epoch analysis method, Clayton (1923) and Shaw (1928) found that the correlation between Sunspot numbers and rainfall was positive near the Equator and at latitudes greater than 40°. There was more rainfall during years of Sunspot maxima compared to years of Sunspot minima. In the mid-latitudes, the superposed epoch analysis revealed less rainfall during years of Sunspot maxima compared to years of Sunspot minima.

As mentioned earlier, marine air and sea surface temperatures have been measured globally since the mid-19th century CE. Several interesting insights relating climate variability to solar variability have come out of analyses of these records. Newell et al. (1989) found a prominent 22-year period oscillation in global and hemispheric marine temperatures from 1856 to 1986 CE and suggested that this oscillation may be related to the 22-year solar Hale cycle. Reid (1991) found a significant and high correlation between globally averaged SST from 1860 to 1980 CE and the envelope of the Schwabe cycle over the same period. Running averages of the Sunspot number and SST time series from Reid (1991) are shown in Figure 2.6.

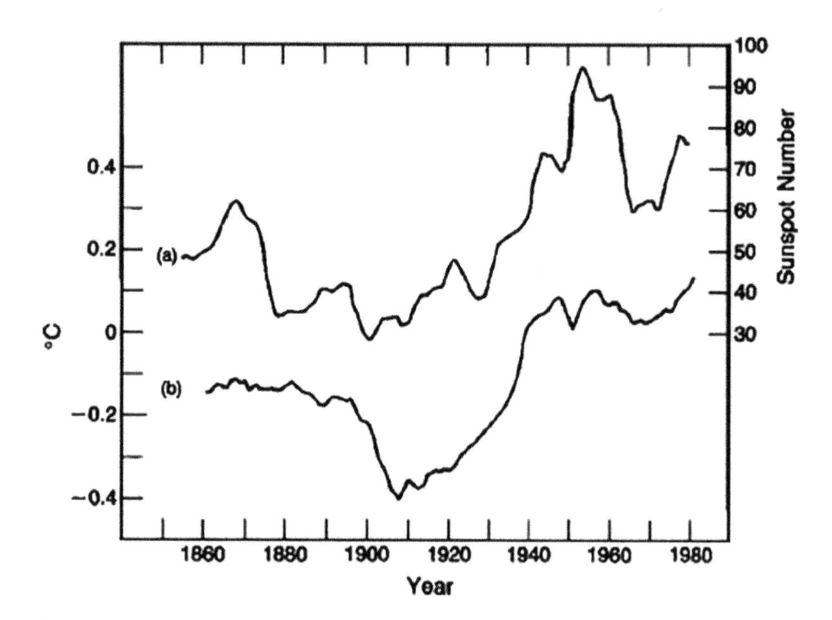

FIGURE 2.6 Eleven-year running means of the Zürich sunspot number (curve a) and global-average SST anomalies (curve b). (From Reid, 1991.)

As described later, this correlation could be explained by variations in the Sun's total irradiance in phase with the envelope of the Schwabe cycle.

Many studies analyzed co-variability between solar cycles and climate variables in Asia. Based on 88 years of data, Bhalme and Mooley (1981) found that the cross-spectrum of the time series of the Flood Area Index of India and Sunspot numbers has a significant peak around the 22-year period. They found that the two variables were nearly in phase. A harmonic dial analysis showed that large-scale floods occurred consistently in India during the maxima of the 22-year Hale cycle. Currie (1984) estimated the spectrum of the Bhalme and Mooley Flood Area Index and also found spectral peaks at approximately 20- and 9.9-year periods. However, Currie associated the 20-year spectral peak with the 18.6-year lunar nodal cycle. Hiremath and Mandi (2004) and Hiremath (2006) used 130 years' (1871–2000 CE) data to quantify correlations between solar cycles and rainfall in India. They found significant positive correlations between Sunspot cycles and monsoon rainfall in March–April–May and June–July–August–September periods. It was also found that the rainfall data show 22-year periodicity, similar to that found in Sunspot number and solar irradiance data, and in the Bhalme and Mooley Flood Area Index.

Azad et al. (2010), Narasimha and Bhattacharyya (2010), Joshi and Pandey (2011), van Loon and Meehl (2012), and Roy and Collins (2015) have analyzed dominant timescales of variability in the Indian monsoon rainfall. Spectral peaks at 9 to 13-year periods have been found in some of these studies, which were associated with the Schwabe cycle. van Loon and Meehl (2012)'s compositing of rainfall around maxima in the Schwabe cycle also found increased rainfall at these times. Roy and Collins (2015), however, argued that this compositing technique may misidentify El

Niño signal in rainfall as a solar cycle signal. As if the solar variability–climate relationship was not complex enough, Narasimha and Bhattacharyya (2010) found that SSTs in the eastern equatorial Pacific Ocean (the Niño region) were cooler during maxima in the Schwabe cycle and warmer during minima in the Schwabe cycle. They argued that the Indian monsoon rainfall may be influenced directly by solar variability as well as indirectly by the solar variability's influence on the Niño region SSTs, an argument supported by Roy and Collins (2015) as well. Incidentally, Mehta and Lau (1997) found that multidecadal variability of the Indian monsoon rainfall was positively correlated with solar irradiance variability, and the Niño region SSTs were negatively correlated with solar irradiance variability. Thus, the solar variability – Niño SSTs – Indian monsoon rainfall relationships were empirically found to be similar at decadal and multidecadal timescales in the Narasimha and Bhattacharyya (2010) and Mehta and Lau (1997) studies, respectively. The hypothesized physics of these relationships are described later in Section 2.5. Elsewhere in Asia, Juan et al. (2005) found that annual precipitation in Beijing area during 1749–2001 CE was correlated with Sunspot cycles.

As mentioned earlier, temperatures have been measured in Central England since mid-17th century CE. Njau (2003) found that variations in these temperatures since 1660 CE, and rainfall in England and Wales since 1726 CE can be associated with a 80 to 90-year Sunspot cycle (possibly the Gleissberg cycle), the Hale cycle, and the Schwabe cycle, with the 80 to 90-year cycle causing changes in these associations. Moving further south to Africa, Seleshi et al. (1994) found a statistical link between the Schwabe cycle and Addis Ababa, Ethiopia, rainfall during 1900–1990 CE and, reminiscent of Sir William Herschel's effort to predict commodity prices in London (Herschel, 1801), used it to develop a statistical rainfall forecasting scheme with Sunspot numbers as the predictor.

2.4.2.2 Lake Level

In another climate-related variable, Stager et al. (2005, 2007) showed that water level in Lake Victoria, East Africa, rose during every peak of the Schwabe cycle since the late 19th century CE. This correlation was statistically significant but unstable, with high or moderate positive correlation coefficients during the first third and the last third of the 20th century CE, and moderate negative correlation coefficient during the middle third of the 20th century CE. They also showed that lake level peaks associated with the Schwabe cycle were associated with above-average rainfall approximately one year before Sunspot maxima. A similar variability of lake levels also occurred in at least five other East African lakes, suggesting a regional coherence of this phenomenon. In a result reminiscent of the Lake Victoria study, Lyatkher (2000) and Shermatov et al. (2004) found a strong correlation between Sunspot cycles and water level in the Caspian Sea straddling Europe and Asia.

2.4.3 Solar Cycle Length

It has also been suggested that Sunspot cycle length (SCL) of both the Schwabe and Hale cycles plays an important role in solar influence on climate. Friis-Christensen and Lassen (1991) and Friis-Christensen (1993) suggested a direct influence of solar variability on global climate via the SCL, which is known to vary with solar activity such that shorter SCL implies high activity and longer SCL implies low activity. As Friis-Christensen and Lassen (1991) found, variations in SCL match variations of the instrumental record of air temperature over Northern Hemisphere land from 1860 to 1990 CE. Lassen and Friis-Christensen (1995) later confirmed the correlation between SCL, and reconstructed and instrumental records of Northern Hemisphere land air temperature going back to the second half of the 16th century CE. Schlesinger and Ramankutty (1992) and Berndtsson et al. (2001) also confirmed that SCL variations have significantly contributed to observed global temperature variations and Northern Hemisphere air temperature variations, respectively.

Weng (2005) further analyzed relationships among low-frequency components of Sunspot numbers and SCL; intensity of the Schwabe cycle; and decadal and longer timescale variations in global-average and North Pacific-average SST, and the Niño3.4 SST index. These variables are shown in Figure 2.7 from Weng (2005). As mentioned in Section 2.2.1, the SCL varies from approximately 9 to 14 years and leads low-frequency Sunspot numbers and its intensity by a few years as seen in Figure 2.7. It is also apparent that there is an approximate similarity between the

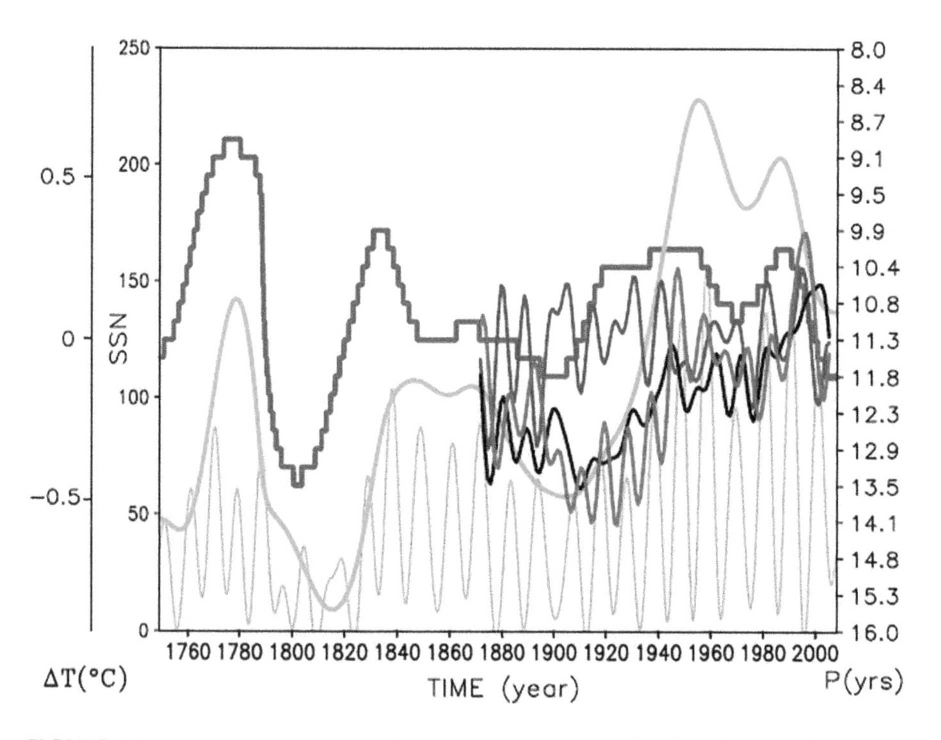

FIGURE 2.7 The low-pass-filtered (Sunspot number; orange) for 1749–2007 CE (predicted after August 2004), (SCL; blue) and intensity of the Schwabe cycle (green, range: 20–80); and low-pass filtered global-average SST (black), North Pacific-average SST (red) and Niño3.4 SST index (purple) for the winters of 1870–1871 to 2003–2004. To make the variation tendency of the low-pass-filtered Niño3.4 SST comparable with other time series, its magnitude was reduced by a factor of 3. (Adapted from Weng, 2005.)

Schwabe cycle characteristics, and global-average and North Pacific-average SSTs at multidecadal and longer timescales. All these variables reached a relative maximum from approximately 1940 to the late 1950s CE, a relative minimum from approximately 1960 to the early 1980s CE, and a relative maximum from approximately mid-1980s to 2000 CE. Following Friis-Christensen and Lassen (1991), Lassen and Friis-Christensen (1995), Berndtsson et al. (2001), and others, Weng (2005) also interpreted this similarity as a causal relationship between the envelope of the Schwabe cycle, its length, and its intensity; and climate variables such as SSTs averaged over global oceans and the North Pacific Ocean. Phase relationships between the Schwabe cycles and decadal "cycles" in the average SSTs, however, are not constant which would decrease the overall correlation between them. Also, the Niño3.4 SST index shows a different multidecadal evolution compared to the global-average and North Pacific-average SSTs. As we will see in Chapter 3, the North Pacific SST anomalies are a part of the Pacific Decadal Oscillation, and this apparent SCL association is also reflected in that oscillation. At the regional scale, Butler (1994) found that there was a strong correlation between SCL and the average temperature at Armagh Observatory in Northern Ireland from 1844 to 1992 CE in conformity with a similar result for Northern Hemisphere temperature by Friis-Christensen and Lassen (1991) mentioned earlier. Wilson (1998) also analyzed temperature records of the Armagh Observatory from 1844 to 1992 CE and found that SCL of the Hale cycles was highly and inversely correlated with average temperatures. The studies reviewed here show at least a statistical association between SCL and global-average and regional-average SSTs and surface air temperature, providing a strong motivation to understand the physics of such relationships.

2.4.4 Galactic Cosmic Rays, Solar Radio Flux, and Electromagnetic Fields

Another manifestation of climate variability is in cloudiness. There are several hypotheses outlining how GCR flux can influence cloudiness and thereby influence climate at solar cycle timescales. But, it is very difficult to observe cloudiness quantitatively, globally, and continuously in time for decades. Also, clouds can occur at many levels throughout the troposphere. If GCRs are influencing cloudiness, the physics depend critically on the level of cloudiness in the troposphere. Surface-based observations of cloudiness are limited by observer presence, both on land and on oceans. Also, surface-based naked eye or photographic observations see only the lowest level of clouds and cannot see higher levels. Satellite remote sensing-based cloudiness estimates, until relatively recently, relied on visible and infrared radiation in day time and infrared radiation in night time. Neither of these wave lengths provide unambiguous information about vertical distribution of clouds. Moreover, the temperature of the underlying surface can strongly influence infrared radiation-based cloudiness estimates. Despite these limitations and braving confounding influences, as in the case of paleoclimate proxy data, several researchers have used global cloudiness data compiled by the International Satellite Cloud Climatology Project (ISCCP) from remote sensing estimates made by several satellites and surface-based cloudiness data to search for solar variability–climate connections. It must be emphasized that the quality of the ISCCP cloudiness data, especially inter-satellite calibration after 1994 CE, has been questioned by several authors (see, for example, Kernthaler et al., 1999; Jorgensen and Hansen, 2000; Sun and Bradley, 2002, 2004; Marsh and Svensmark, 2003, 2004; Kristjánsson et al., 2002; Gray et al., 2010). Nevertheless, Svensmark and Friis-Christensen (1997) and Svensmark (1998) found a strong correlation between GCR flux and global-average cloudiness, with a higher correlation at high latitudes as would be expected due to the Earth's magnetic field deflecting electrically-charged GCR particles towards the poles. Following up on this initial result, Marsh and Svensmark (2003) conducted empirical orthogonal function analysis of worldwide, low cloud amounts and cloud-top pressure of low clouds from July 1983 to September 2001, including the suspect calibration period around September 1994 to January 1995 CE, from the ISSCP-D2 data set. They found that the most dominant empirical patterns of both cloud amount and cloud-top pressure of low clouds were highly correlated with GCR flux up to the calibration gap and then diverged. The GCR-low cloud signal was strongest in the regions where satellites had an unobstructed view of low clouds. They also found that the second most dominant patterns were highly correlated with the Niño3 index of El Niño–La Niña events up to the calibration gap period. Subsequently, Marsh and Svensmark (2004) suggested an adjustment of the ISSCP cloudiness data after the gap period which resulted in an improved correlation between low cloud properties and GCR flux after the calibration gap also. This is very much an ongoing scientific controversy as evidenced by criticisms of the adjustment by, among others, Sun and Bradley (2004) and Gray et al. (2010). It has also been argued by these latter two studies that surface-based cloudiness observations do not show associations with GCR flux over most of the globe. Another intriguing aspect of the Marsh and Svensmark (2003, 2004) studies is that highest correlations between low cloud properties and GCR flux are in tropics and subtropics rather than at higher latitudes as found by Svensmark (1998). So, it appears that, like many other scientific controversies, the GCR-clouds controversy will have to be resolved over time by better observed data and simulations with models based on fundamental principles of science.

In a somewhat related study, Soon et al. (2000) found the temperature anomaly of the lower troposphere, inferred from satellite-borne Microwave Sounding Unit radiometers, to be inversely correlated with the area of the Sun covered by coronal holes from January 1979 to April 1998. The coronal hole (Figure 2.1) area is a physical proxy for both the global-scale, 22-year geometrical and shorter-term, dynamical components of the GCR modulation, as well as the particle emission of the Sun. It was concluded that variable fluxes, either of solar charged particles or GCRs modulated by the solar wind or both, may influence tropospheric temperature on timescale of months to years.

At the regional scale, two studies have analyzed climate variability in the USA and Mexico with respect to hypotheses based on possible effects of GCR flux and solar magnetic field variations. Balling and Cerveny (2003) analyzed cloudiness, temperature, and atmospheric moisture data in the USA; and Sunspot number and GCR flux data to test the hypothesis that an increased GCR flux can increase cloudiness by providing an increased number of cloud condensation nuclei. They analyzed twice-daily dew-point temperatures at the surface and at four levels in the atmosphere from 850 to 300 hPa at 55 radiosonde stations for the 1957–1996 CE period; daily maximum and minimum temperatures at over 1000 stations; monthly cloud amount data from ISCCP in 2.5° longitude–2.5° latitude boxes from 1984 to 1996 CE; and Sunspot data as a proxy for GCR flux (Sunspot numbers and GCR flux are inversely related as described earlier in Section 2.2.2). They also analyzed cloud cover data and actual, ground-based GCR flux measurements over the 1953–2002 CE period. It was found that periods with low Sunspot numbers, i.e., periods with high GCR flux, were associated with significantly higher dew-point depressions; a higher diurnal temperature range; and less cloud cover. However, correlation coefficients were found to be low, even though highly significant. While the high significance levels suggest robust relationships among solar activity variability and cloudiness, the relatively low correlation coefficients suggest that solar activity variability may account for only a small amount of observed cloudiness variability over the USA in the Balling and Cerveny study. Chaudhari et al. (2015) found that the Indian monsoon rainfall and daily minimum temperature decreased, and daily maximum temperature increased when the Schwabe cycle-related GCR flux increased during 1953–2005 CE. This empirical observation may be physically consistent with the solar cycle–Indian monsoon rainfall studies described in Section 2.4.2 which found an increased rainfall around the times of Sunspot cycle maxima because the Sunspot and GCR cycles have opposite phases as described in Section 2.2.2.

Among other solar influence–regional climate variability studies, Mendoza et al. (2001) and Maravilla et al. (2004) found associations between temperature variability in central and northern Mexico, respectively, and geomagnetic activity over an approximately 70-year period from the 1920s to 1990s CE.

2.4.5 Associations of Atmospheric Variations with Solar Cycles

After the foregoing brief review of observed associations between surface climate variations and solar emissions, we now see associations between atmospheric variations in observed data and solar cycles. We begin this "ascent" in the atmosphere from the Earth's surface and progress to the stratosphere.

Solar cycle associations have been found in the atmosphere over the Pacific and Atlantic Oceans. In the North Pacific region, Christoforou and Hameed (1997) found that extremes in solar variability, as indicated by Sunspot numbers, correlated highly with the locations of the low surface pressure center near the Aleutian Islands and the high surface pressure center near the Hawaiian Islands. The Aleutian center migrated eastward during minimum solar activity, and the Hawaiian center migrated southward. Variations in surface pressures at the centers were also associated with solar variability. The strengths and locations of these two surface pressure centers affect intensities and tracks of extratropical storms in the North Pacific; therefore, it is possible that solar variability can affect climate over the North Pacific and North America by affecting the Aleutian and Hawaiian surface pressure centers.

Statistical associations between solar activity variability and weather variability over the North Atlantic Ocean have also been found. In an exploratory study, Boberg and Lundstedt (2002) found that decadal variations in the electric field of

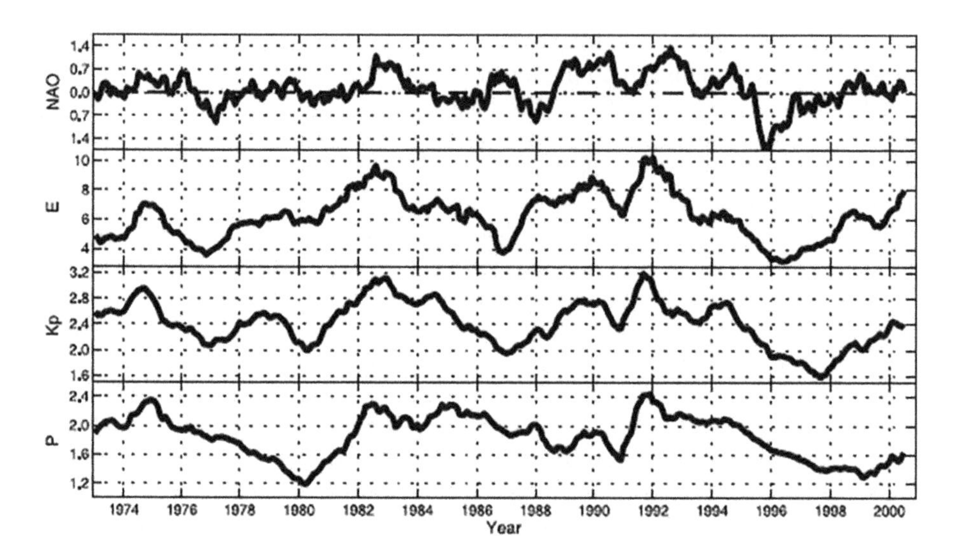

FIGURE 2.8 Time series of 12-month moving averages of the NAO index compared to variations in the solar wind and the geomagnetic activity. E is the electric field strength of the solar wind in units of 10^{-4} Vm^{-1}, and P is the solar wind pressure in units of $10^{6}m^{-1}s^{-2}$. Solar cycle minima occurred in May 1976, August 1986, and October 1996. (From Boberg and Lundstedt, 2002.)

the solar wind, presumably associated with the Schwabe cycle, are associated with decadal variations in the North Atlantic Oscillation (NAO), which is an oscillation in atmospheric mass – and, therefore, pressure – between subpolar and subtropical latitudes in the North Atlantic region. Figure 2.8 from Boberg and Lundstedt (2002) shows 12-month moving average NAO index, electric field strength of the solar wind, magnetospheric K_p index, and the dynamic pressure of the solar wind from 1973 to 2000 CE. Correlation coefficients between the NAO index, and the electric field and the K_p index were moderately high and with high statistical significance; the solar wind dynamic pressure correlation with the NAO index was low. Kodera (2002) provided further details of the possible Schwabe cycle–NAO relationship when he found that the NAO pattern is hemispheric except for the Pacific sector and the associated zonal wind pattern extends from the surface to the stratosphere in 17 winters of the maximum phase of the cycle indicated by the 10.7 cm radio flux from 1958–1959 to 1996–1997 CE. The correlation pattern is restricted to the eastern North Atlantic and the associated zonal wind pattern extends only to the troposphere in 22 winters of the minimum phase of the cycle. Kodera (2003) confirmed results of this study with 100 years of observed NAO data. It is possible that changes in the polar vortex at the time of solar cycle maxima are important for stratosphere–troposphere interactions as indicated also by other studies and this may be one of the ways maxima in solar ultraviolet (UV) radiation may be influencing the troposphere and surface climate (Section 2.5.2). Unlike the Boberg and Lundstedt (2002) study, however, Kodera (2003) found no correlation between maxima in the solar cycle and the NAO. It is possible that a relatively small overlap between the NAO time series analyzed in the two studies and non-stationarity of the NAO may be responsible for this difference in conclusions.

In another indicator of the North Atlantic atmospheric weather variability, Veretenenko et al. (2005) suggested possible links between long-term variations in cyclonic activity in the mid-latitude North Atlantic region, and solar activity and GCR flux variations on decadal to century time scales. They compared long-term variations of surface pressure in the North Atlantic from 1874 to 1995 CE with indices of solar and geomagnetic activity and GCR flux characterized by the concentration of the cosmogenic isotope^{10}Be. An approximately 80-year period (modulation in the amplitude of the Schwabe cycle) was found in surface pressure variations in the 45ºN to 65ºN latitude band during periods of intensive cyclogenesis, generally from Northern Hemisphere autumn to spring. They found that a long-term increase of surface pressure in this region coincided with a secular rise of solar/geomagnetic activity, accompanied by a decrease in GCR flux. Long-term decreases of surface pressure were found during periods of low (or decreasing) amplitudes of the Schwabe cycle. Similar features were also found in the spectral characteristics of geomagnetic activity indices, GCR flux, and surface pressure at middle latitudes in the North Atlantic region on the quasi-decadal time scale. In this study, effects of solar activity and GCR variations on surface pressure were more noticeable in the North Atlantic intensive cyclogenesis region near the eastern coast of North America. It is possible that the three studies found North Atlantic manifestations of solar cycle influences on the Northern Hemisphere polar vortex from the stratosphere, with solar UV and possibly solar wind particles and electromagnetic fields as the main agents of influence. We will see more of this possibility later in this chapter.

In the tropical Atlantic region, Lim et al. (2006) found that decadal components of relative humidity of the surface air and Sunspot numbers time series were correlated with a lag of two years. At the decadal timescale, the observed relative humidity change was dominated by specific humidity change over the tropical North Atlantic and by air temperature change over the tropical South Atlantic. Lim et al. (2006) suggested that this observed association between the Schwabe cycle and relative humidity variability may provide a positive feedback via clouds and water vapor to the direct climate forcing by solar irradiance variations. We will see in Chapter 4 that tropical Atlantic SSTs indeed undergo variability with a period of 12–13 years, which is close to the Schwabe cycle period.

Next, we move up from the Earth's surface into the troposphere.

2.4.5.1 The Troposphere

In one of the first analyses of tropospheric data with respect to solar cycles and reminiscent of the solar cycle–NAO association, Thresher (2002) found that the Antarctic circumpolar vortex in the 35°S–60°S latitude belt since the early 1940s CE exhibits decadal variability at its equatorward margin. Winds, rainfall, and sea-level pressure, with a period of 9 to 13 years, were found to be correlated with the Schwabe cycle. This variability was in phase around the Southern Hemisphere as shown by analyses of sea-level pressure and rainfall in Australia, South Africa, and South America. Soon thereafter, Gleisner and Thejll (2003)'s and Gleisner et al. (2005)'s analyses of the National Centers for Environmental Prediction (NCEP) – National Center for

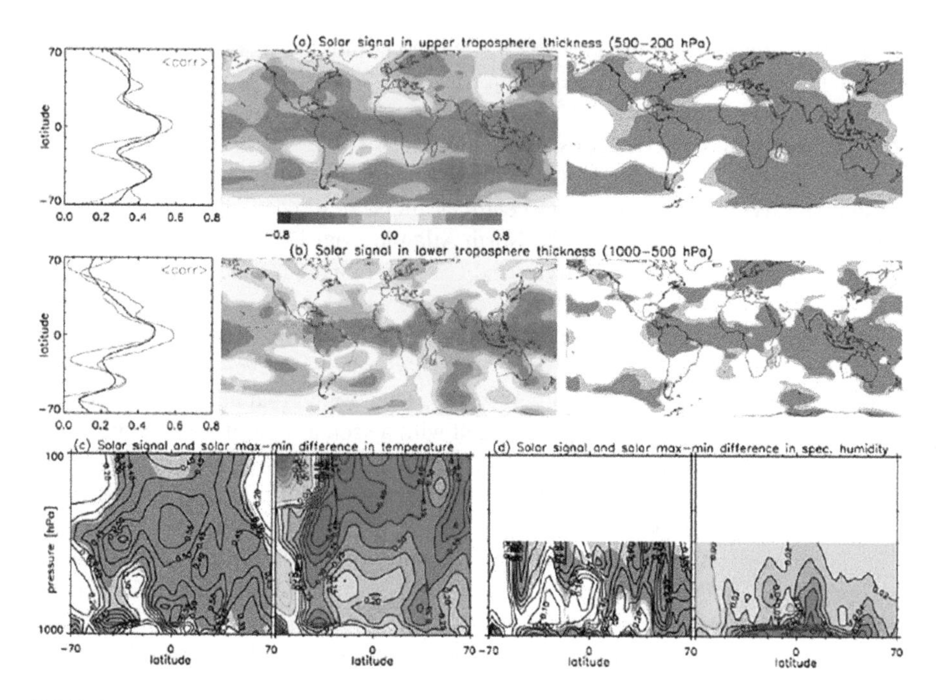

FIGURE 2.9 Correlation coefficients between 10.7 cm solar radio flux and (a) upper tropospheric and (b) lower tropospheric thicknesses from the NCEP-NCAR reanalysis. The corresponding 90% and 95% significance levels are shown on the right in gray shading, and zonal averages of correlation coefficients are shown on the left; globally and in 60°W–10°E and 140°E–150°W. The panels at the bottom show correlations between 10.7 cm flux and (c) zonally averaged temperature, and (d) zonally averaged specific humidity. Gray shadings denote the 90% and 95% significance levels in (c) and (d). Color figures (c) and (d) show the mean temperature and humidity differences between solar cycle maximum (annual 10.7 cm flux > 150) and minimum (annual 10.7 cm flux < 80). (From Gleisner and Thejll, 2003.)

Atmospheric Research (NCAR) (Kalnay et al., 1996) reanalysis[12] data found statistically significant variations in tropospheric temperatures, geopotential heights, water vapor distribution, and global-scale circulations in phase with the Schwabe cycle, as expressed in the 10.7 cm solar radio flux, from 1958 to 2001 CE. These results show atmospheric response to solar variability in the low- and mid-latitude troposphere, with the heating and near-surface moistening being maximum during maxima in the Schwabe cycles (Figure 2.9). Modulations of the Hadley, Walker, and Ferrel Circulations by the Schwabe cycle were also found. It is interesting to note in Gleisner and Thejll (2003) that in the upper troposphere between 500 and 200 hPa pressure levels, the strongest Schwabe cycle signals are near the Equator from the Amazon region in South America to Africa (perhaps the upper troposphere

[12] Surface, radiosonde, satellite, and other observations of winds, temperatures, surface pressure, and humidity assimilated into general circulation models of the atmosphere without Schwabe and Hale cycle variations in solar emissions.

signature of the near-surface signal found by Lim et al. (2006)), and in the western Pacific–Maritime Continent region. We will revisit these interesting observations in Chapters 4 and 5, respectively. The poleward shift of the Ferrel Circulation associated with the Schwabe cycle in the NCEP-NCAR reanalysis data is consistent with Bronniman et al. (2006)'s similar finding in historical, Northern Hemisphere, upper troposphere data since 1922 CE.

Continuing with the story of the Schwabe cycle association with tropospheric temperatures and circulations, analyses of the NCEP-NCAR reanalysis data have also shown that tropospheric jets in the mid-latitudes are weaker and more poleward in Schwabe cycle maxima years (Haigh et al., 2005; Haigh and Blackburn, 2006). Even though vertical velocity estimates from observed data are noisy and are prone to large errors, several analyses have shown Schwabe cycle associations with Hadley, Walker, and Ferrel Circulations as mentioned earlier. Labitzke and van Loon (1995) using station radiosonde data, and van Loon et al. (2004, 2007) using NCEP-NCAR reanalysis data found that the Hadley Circulation was stronger in Schwabe cycle maxima years. However, Haigh (2003) and Haigh et al. (2005) also analyzed NCEP-NCAR reanalysis data and found a weaker and broader Hadley Circulation in Schwabe cycle maxima years. Errors in estimating vertical velocities using observed data, alluded to earlier, may be responsible for the apparent discrepancy. In agreement with the Gleisner and Thejll (2003) observation of Schwabe cycle association with Walker Circulation variability, van Loon et al. (2007), Meehl et al. (2008), and Lee et al. (2009) found that the Walker Circulation was stronger in Schwabe cycle maxima years. Reminiscent of typical La Niña-like conditions, a below-average SST anomaly in the eastern equatorial Pacific, a poleward-shifted ITCZ, and a poleward-shifted South Pacific Convergence Zone was associated with the stronger Walker Circulation. Interestingly, the below-average SST anomaly was followed after two years by an above-average SST anomaly, perhaps due to a coupled ocean–atmosphere response to the initial, solar maximum forced SST anomaly (Meehl et al., 2008; White and Liu, 2008a). The studies reviewed here show conclusively the presence of the Schwabe cycle signal in tropospheric temperatures, winds, near-surface humidities, and SSTs.

2.4.5.2 The Stratosphere

We now move further up in the atmosphere and review major findings on Schwabe cycle associations with variability of temperatures, winds, and ozone in the stratosphere. In this upward journey, we will begin with the pioneering body of research by Karin Labitzke and Harry van Loon. As more and better observed data about the atmosphere began to become available, Labitzke (1987) embarked on perhaps the first physics-based quest to analyze if/how the almost two centuries long trail of statistical associations between solar cycles and the Earth's climate variability actually led to understanding of possible physics – in other words, if the statistics were just chance associations or they were actually underpinned by physics. An atmospheric phenomenon, the Quasi-Biennial Oscillation (QBO), provided a vital insight into how the Schwabe cycle influenced polar stratospheric temperature depending on the phase – westerly or easterly – of the zonal winds in the equatorial stratosphere which defines the QBO phase. Labitzke (1987) found that the smaller the number of Sunspots, the lower the stratospheric temperature in the QBO-West phase in Northern Hemisphere winter; no such relationship was found in the QBO-East phase. In a follow-up study, Labitzke and van Loon (1988) found that three Schwabe cycles from 1956 to 1987 CE were positively correlated with high-latitude, stratospheric temperatures and negatively correlated with stratospheric temperatures in mid- and low-latitudes in the QBO-West phase. They found that such relationships existed in the troposphere as well. The correlations changed sign during the QBO-East phase and occurrences (or lack thereof) of mid-winter warmings were also associated with the QBO phase during Schwabe cycle minima. These two studies complemented the long trail of statistics of surface climate associations with Sunspot cycles and contributed towards placing Sun–climate research on a more physics-based foundation. Subsequent to these early studies, Labitzke and co-authors, and others continued to further clarify the role of the QBO in Schwabe cycle influences on the stratosphere and speculated about implications of these influences on stratospheric and climate dynamics (see the summary in Gray et al. (2010) and references therein). On a personal note, the almost two-century-long published literature on the Sun–surface climate studies and the pioneering research by Labitzke (1987) and Labitzke and van Loon (1988) motivated me to begin my journey of exploration of DCV and its societal impacts in the late 1980s CE, a journey which has continued over the last 30 years.

Let us digress for a moment and see how variability of solar emissions can influence the stratosphere. The so-called ozone layer exists in the atmosphere from a height of 15 to 35 km, and peaks at approximately 22–25 km height. As we saw in Chapter 1 (Section 1.3.1), the stratosphere spans the height range from approximately 12 to 50 km depending on latitude. Thus, the bulk of the ozone layer is in the stratosphere. Ozone absorbs solar UV radiation and plays a very important role in radiative heating of the stratosphere. This property of ozone makes it the "agent of change" in the Sun–stratosphere relationship via its direct effect on radiative balance and temperature, and its indirect effect on stratospheric circulation (Gray et al., 2010). The absorption of solar UV radiation influences photolysis of molecular oxygen which, in turn, influences ozone production rate at low latitudes in the middle to upper stratosphere (Haigh, 1994). Stratospheric ozone concentration can also be influenced by energetic charged particles, whose flux is modulated by solar variability and whose concentrations at polar latitudes can play an important role in destroying ozone (Randall et al., 2007). Thus, two types of solar emissions can influence the production and destruction of stratospheric ozone. It is estimated that the average solar UV emission varies approximately 6% between a successive maximum and minimum of the Schwabe cycle (Lean, 1991; Gray et al., 2010). This variation can result in approximately 2%–4% variation in ozone concentrations in the tropical, upper stratosphere and mid-latitude, middle and lower stratosphere (Soukharev and Hood, 2006; Gray et al., 2010). Heating changes caused by horizontal and vertical changes in ozone concentrations can drive changes in stratospheric circulations and can also influence tropospheric circulations and temperatures. Therefore, understanding responses of ozone concentrations to solar UV and charged particle fluxes is crucial for understanding the "top-down" solar influences on climate as described in Section 2.5.

In describing observations of solar cycle influences on the stratosphere, reanalysis of atmospheric data is quite useful as we saw for the troposphere earlier. Figure 2.10

shows annual-average and zonal-average temperature differences between maxima and minima of the Schwabe cycles from 1,000 hPa (the Earth's surface) to 0.1 hPa (approximately 65 km height) (Gray et al., 2010). Statistically significant temperature differences of 1 to 2 K in the tropical and 3 to 4 K in the high-latitude stratosphere are apparent in Figure 2.10a. These temperature changes are approximately at the levels in the stratosphere where the 2% to 4% changes in ozone concentrations occur,

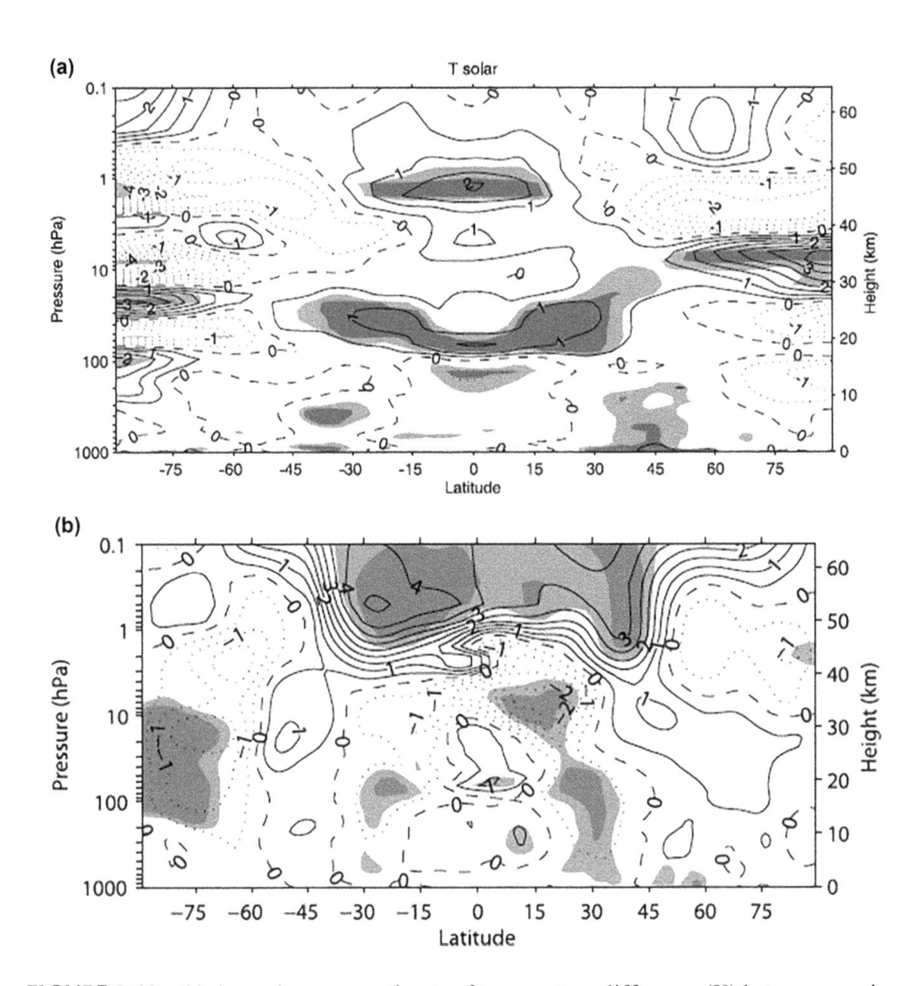

FIGURE 2.10 (a) Annual-average estimate of temperature difference (K) between maxima and minima of the Schwabe cycle derived from a multiple regression analysis of the European Centre for Medium Range Weather Forecasts (ECMWF) Reanalysis (ERA-40) data set. Dark and light shaded areas denote statistical significance at the 1% and 5% levels, respectively. (b) Annual-average differences in zonally averaged zonal wind (m s^{-1}) between maxima and minima of the Schwabe cycle derived from a multiple regression analysis of the ERA-40 data set. Dark- and light-shaded areas denote statistical significance at the 1% and 5% levels, respectively. Contour values are 0, ±0.5, ±1, ±2, and ±3 m s^{-1} and a contour interval of 2 m s^{-1} thereafter. Solid (dotted) contours denote positive (negative) values, and the dashed line is zero. (From Gray et al., 2010.)

which is consistent with solar influences on rates of ozone production and destruction mentioned earlier. The corresponding zonal wind differences can be 4 ms^{-1} and stronger in the tropical–subtropical upper stratosphere, with smaller but significant differences almost down to the surface (Figure 2.10b). A small but significant zonal wind signal is also apparent at polar latitudes. As described by Gray et al. (2010), ozone feedback and stratospheric dynamics also make substantial contributions to these apparent influences of the Schwabe cycle. An alternate method to estimate temperature changes associated with extrema in the Schwabe cycle is based on remote sensing of infrared radiation in a column from the surface to the atmosphere, and calibration of the radiation in terms of vertical temperature profile through the atmosphere. Analyzing temperature estimates derived with this method from the TIROS[13] Operational Vertical Sounder infrared radiometers, Scaife et al. (2000) and Randel et al. (2009) found somewhat smaller and vertically broader stratospheric temperature differences between Schwabe cycle extrema. Although this retrieval method does not use a dynamical model of the atmosphere and data assimilation techniques, it has coarser vertical resolution and problems in the temperature retrieval technique (Gray et al., 2010; Lee and Smith, 2003; Smith and Matthes, 2008).

In this entire Section 2.4, we saw that numerous indicators of surface and atmospheric climate – such as precipitation, surface air temperature, dryness–wetness, flood area index, sea level, lake level, and atmospheric pressure; atmospheric winds, temperature, and surface air humidity; and stratospheric ozone concentration – have undergone DCV in the past which has been associated via a variety of analysis techniques with indicators of solar variability – such as Sunspot number, 10.7 cm radio flux, UV flux, solar irradiance, GCR flux, and the electric field of the solar wind. To any unbiased observer, it is obvious that there is sufficient *prima facie* evidence which addresses the question "What?" to warrant detailed investigations to address the question "How?" with detailed hypotheses and with models based on physics and chemistry of the Earth–atmosphere system. Some of the studies reviewed here, and other studies have tried to develop insights into possible physics of how the observed DCV may be related to Schwabe and Hale cycles. In the next section, we will review the hypothesized mechanisms and attempts to test them with a variety of models.

2.5 HYPOTHESIZED MECHANISMS OF SOLAR INFLUENCES ON CLIMATE

After reviewing paleoclimate proxy and instrument-measured data indicating associations between solar emissions and their modulation of GCRs and geomagnetic field, and climate variability on the Earth, we will now see hypothesized mechanisms with the focus on DCV. Potential influences of each type of solar emission/modulation are described separately, followed by a synthesis of how more than one mechanism can operate together. One hypothesis about nonlinear, resonant excitation of interannual to decadal climate variability by near-cyclic variations in solar emissions is also described even though the hypothesis is not specific to any particular type of emission.

[13] Television InfraRed Observation Satellite.

2.5.1 Visible Radiation

We saw in Section 2.2.2 that visible radiation from the Sun spans the 350–725 nm range in wavelengths (Figure 2.4). As we also saw, the first hypothesis about solar influence on the Earth's climate involved visible radiation as the driver of the influence. In this first hypothesis about DCV, modulation of the visible radiation reaching the Earth's surface by the Schwabe and Hale cycles would modulate the heating of the Earth's surface and the atmosphere, and thereby influence climate. Elementary estimates based on the Earth's energy balance, however, showed that the global thermodynamic response to the observed variations in incident, visible solar radiation would be too small compared to the purported effects (Gray et al., 2010). As research on the El Niño–Southern Oscillation (ENSO) and other patterns of DCV provided insights into the physics of the coupled ocean–atmosphere system, the original hypothesis was modified. This modified hypothesis is known as the "bottom-up" hypothesis in which coupled ocean–atmosphere dynamics in some regions of the world and associated cloud feedbacks can generate a much larger response to observed variations in visible radiation (Cubasch et al., 1997, 2006; Meehl et al., 2003, 2008, 2009; Meehl and Arblaster, 2009). Specifically, varying visible radiation would be absorbed in relatively cloud-free subtropical oceans. In the solar maximum phase, increased radiation would increase evaporation which would then increase water vapor transport by trade winds into tropical precipitation regions from subtropical regions. Increased precipitation due to the increased water vapor convergence would increase latent heat release in the tropical atmosphere, which would strengthen trade winds and equatorial ocean upwelling, resulting in cooler SSTs. The stronger trade winds would increase subsidence in subtropical latitudes, further reducing clouds and increasing absorption of visible radiation by the subtropical ocean and, consequently, evaporation. Thus, in the bottom-up hypothesis, positive feedbacks via ocean–atmosphere dynamics and clouds are crucial to amplify responses to original visible radiation changes.

2.5.2 Ultraviolet Radiation

Ultraviolet radiation from the Sun spans the wavelength range 200–350 nm as shown in Figure 2.4 and varies between a Schwabe cycle maximum and minimum by a few percent of the average UV radiation, a much larger variation than that in the visible part of the solar spectrum. Since the solar UV radiation is absorbed by ozone in the stratosphere, the hypothesis about how variations in UV radiation can influence the Earth's climate is known as the "top-down" hypothesis (Haigh, 1996; Balachandran et al., 1999; Shindell et al., 1999; Kodera and Kuroda, 2002; Haigh, 2003; Haigh et al., 2005; Matthes et al., 2006; Meehl et al., 2009). This still-evolving hypothesis also involves feedbacks via further ozone formation and heating, and stratosphere–troposphere interactions. In this hypothesis, UV radiation emitted from the Sun increases at the time of maximum solar activity in the Schwabe cycle. More UV radiation in the stratosphere increases ozone due to atmospheric chemistry, which, in turn, increases UV radiation absorption. This positive feedback warms the stratosphere and creates a latitudinal temperature gradient. There is further stratospheric warming due to increased wave motions which amplifies the original warming. This modified stratospheric circulation, in turn, modifies tropical tropospheric circulation causing an enhancement and poleward expansion of the Hadley and Walker Circulations and tropical precipitation maxima (Haigh et al., 2005; Meehl et al., 2009; Gray et al., 2010; and references therein). In this way, solar influence can permeate from the top of stratosphere down to the surface of the Earth.

2.5.3 Combined Effects of Visible and Ultraviolet Radiations

Building on previous efforts to simulate effects of top-down and bottom-up mechanisms individually, Meehl et al. (2009) showed with numerical models of the troposphere–stratosphere system and the coupled, dynamical atmosphere–ocean system that the top-down and bottom-up mechanisms can act together during maxima in the 11-year solar cycle to strengthen off-equatorial, tropical precipitation maxima in the Pacific; cool eastern equatorial Pacific SSTs; and reduce low-latitude, especially subtropical, clouds to amplify the solar forcing at the surface and in the stratosphere. White and Liu (2008a) also used a global coupled ocean–atmosphere model to simulate effects of the bottom-up mechanism with an estimated radiative forcing due to the top-down mechanism added to the solar irradiance cycle at the surface. They found that the tropical Pacific climate responded to the thus-constructed Schwabe radiation cycle with a one- to three-year delay and generated an SST response somewhat similar to the observed SST variability at the Schwabe cycle timescale. The delayed-action oscillator mechanism invoked to explain the ENSO phenomenon (see, for example, Schopf and Suarez (1988)) also appeared to work in the White and Liu study. But, in their study, westward-propagating oceanic Rossby waves were excited farther away from the Equator than for the interannual ENSO; these waves and eastward-propagating coupled ocean–atmosphere waves at the Equator together generated a resonantly excited response to the Schwabe cycle-like radiative forcing. It is quite likely that the coupled ocean–atmosphere feedbacks operating in the Meehl et al. study and the resonant excitation of oceanic and coupled ocean–atmosphere waves together may be operating in the actual climate system.

The simulated responses in both studies cited here, however, while having some resemblance to observed associations among solar cycle maxima, SSTs, and rainfall in the tropical–subtropical Pacific, are still considerably weaker than the observed associations; both studies allude to model deficiencies as possible causes of the discrepancies. Perhaps as models are improved, details of stratospheric chemistry are explicitly and more accurately specified in the models, stratosphere–troposphere interactions become more realistic, and interactions between moist atmospheric dynamics and thermodynamics and the upper ocean become quantitatively more representative, simulated responses to solar cycles in visible and UV radiations might come closer to the observed associations.

2.5.4 Galactic Cosmic Rays

As described in Section 2.2.2, the GCR flux reaching the Earth is modulated by the solar wind and, therefore, undergoes solar cycles as shown in Figure 2.5f. Solar wind cycles have the same phase as solar irradiance cycles and, since an increase in solar wind flux decreases GCR flux, solar irradiance and GCR cycles have opposite phases. GCRs can generate ions in the entire atmosphere down to the surface by collisions with air molecules, which can also be expected to undergo cycles. Consequently, processes in the atmosphere which are caused by GCR-associated ion production can also be expected to undergo cycles. These processes are (Gray et al., 2010) (1) current flow in the global atmospheric circuit, causing charging of aerosol particles and cloud-edge water droplets, and (2) nucleation of ultrafine condensation nuclei (UCN) from trace gases such as sulfuric acid and water which can cause cloud droplet formation. The details of these processes are explained by Gray et al. (2010), so only a brief overview is provided here.

There is an electric current in the global atmosphere due to a conducting Earth's surface, finite air conductivity, and a conductive lower ionosphere (Wilson, 1906; Burke and Few, 1978; Harrison and Ingram, 2005; Bennett and Harrison, 2008). Solar wind-induced changes in GCR ionization in the atmosphere can modulate this global current down to the Earth's surface (Markson and Muir, 1980; Harrison and Usoskin, 2010), which can cause local droplet and aerosol charging at lower and upper boundaries of stratiform clouds. Thus, according to this hypothesis, solar cycle modulations of global electric current can modulate cloud formation via microphysical processes in which electric charge plays a very important role.

A more direct effect of GCR flux on cloud formation is hypothesized as follows: at the solar minimum time, GCR flux is increased because of decreased solar wind flux, which increases ion production in the atmosphere and increases sulfate aerosols, which then provide increased cloud condensation nuclei for cloud droplet formation. In this process, however, the altitude(s) at which these nuclei are increased has an important bearing on resulting temperature changes at the Earth's surface. An enhancement of low-altitude clouds increases reflected shortwave radiation from the Sun, leading to cooling at the surface. On the other hand, an enhancement of high-altitude clouds increases longwave radiation absorption and re-radiation, leading to warming at the surface. A testing of this hypothesis by a climate model showed that GCR-induced cloud condensation nuclei changes are two orders of magnitude smaller than observed changes in solar cycle-associated cloud properties (Pierce and Adams, 2009). As the foregoing shows, modeling studies so far are able to reproduce hypothesized mechanisms qualitatively, but a substantial amount of work is still needed to reproduce the observed changes quantitatively.

2.5.5 Nonlinear Resonant Response to Solar Cycles

The climate system is a highly nonlinear dynamical system. It can generate internal variability in response to constant, external forcing such as solar emissions, and it can also respond nonlinearly to varying external forcing. The primary time-varying forcings of solar origin are at diurnal and annual periods, and there are also secondary time-varying solar forcings at the Schwabe and Hale cycle periods as well as at longer periods. In addition, there are lunar tidal forcings at various timescales including the 18.6-year period. The possibility that there are nonlinear, resonant responses to time-varying external forcings has been examined with models of various complexities. Results of two sets of numerical model experiments to test nonlinear responses are briefly described here.

Weng (2005) conducted numerical experiments using a simple nonlinear system and found that the intensity of the annual cycle of solar forcings, modulated by the Schwabe cycle, can cause the generation of various timescales in regional SST variability. It was found that even a small change in the amount of visible solar radiation reaching the top of the Earth's atmosphere by 0.04% on the Schwabe cycle timescale can result in a regime change in the SST variability. The results show that some of the energy of internally generated variability of the model climate system can be transferred to the forced variability at multiple timescales due to nonlinear resonance. Thus, the response spectrum can be much richer than the forcing spectrum. This result with a simple nonlinear model suggests that observed climate variability at interannual to decadal and longer timescales can also be generated by nonlinear responses of the climate system just as by the linear mechanisms described in Sections 2.5.1 to 2.5.4.

Another study by White and Liu (2008b) simulated the generation of higher harmonics of the Schwabe cycle in a global coupled ocean–atmosphere model. This model can generate El Niño–La Niña like variability due to internal instabilities or driven by noise in the model with steady external forcing. It was found that El Niño-like interannual variability in the tropical Pacific climate system can be generated in the model by a nonlinear cascade of energy into 3rd and 5th harmonics which produces 3.6-year and 2.2-year cycles aligned with the nominal Schwabe cycle frequency of 11 years used to force the model. In these model experiments, 3 Wm^{-2} amplitude, based on estimates of effective surface heating after including positive feedbacks as by van Loon et al. (2007) and Meehl et al. (2008), was used. As in White and Liu (2008a)'s simulation of the response of the tropical Pacific ocean–atmosphere system to the 11-year solar cycle forcing, described in Section 2.5.3, the delayed-action oscillator mechanism appears to be at work in generating the harmonic response, with the period of the response proportional to the latitude of the oceanic Rossby wave generation. They also found that this nonlinear energy cascade generates non-commutable pairs of El Niño–La Niña events during the ascending part of the 11-year cycle and La Niña–El Niño events during the descending part of the 11-year cycle, both in one-century-long experiment with the model and in 105-year-long SST observations. This pair of pairs repeats every 11-year cycle. These solar cycle-forced El Niño–La Niña events explain over 50% of the interannual variance of the observed Niño3 SST anomalies. It must be emphasized that such a strong, nonlinear response of the global coupled ocean–atmosphere model occurs due to the approximately ten times larger amplitude (3 Wm^{-2} rather than 0.3 Wm^{-2} direct irradiance forcing at the top of the tropical atmosphere as estimated by Lean et al. (2005)) of the solar cycle forcing used in these experiments. Such a large amplitude is based on experiments with another model as described in Section 2.5.3, so the results reported by White and Liu (2008a, b) should be considered suggestive rather than definitive.

2.6 SUMMARY AND CONCLUSIONS

The Sun–climate relationship is an old enigma which has defied understanding for over 200 years. This relationship, especially as it pertains to DCV, is the subject of this chapter. After reviewing fundamental physics and observations of the Sun and its emissions, empirical relationships between ancient climate observers – so-called proxy climate indicators such as tree rings, oxygen and carbon isotopes, and lake sediments in many parts of the world – and indicators of solar variability at the Schwabe (approximately 11 years) and Hale (approximately 22 years) cycle periods were reviewed. Some of the proxy climate indicators tell tales about climate as far back as 12,000 years ago and, in one case, tree-ring data from approximately 290 million years ago as recounted in this chapter. Then, the Sun–climate relationship as seen in modern climate observations mainly since the mid-1800s CE – such as precipitation, temperature, lake levels, winds, atmospheric pressure, and cloudiness in many parts of the world – was reviewed. Analyses of direct instrument measurements (such as by rain gauges), instrument measurements assimilated in atmospheric models, and satellite remote sensing-based ozone and cloudiness estimates were reviewed.

As is well known, indicators of solar variability – Sunspot number, total solar and UV irradiance, 10.7 cm radio flux, and geomagnetic field – show the Schwabe and other cycles. The GCR flux coming into the solar system from the Milky Way galaxy is maximum at the time of solar activity minimum, so the GCR flux and solar variability indicators have cycles with opposite phases. All climate proxy data reviewed here show DCV in all geographic regions from where the data originated and the data also show statistical associations with cycles of solar variability, solar cycle length, or the GCR flux. While some of these associations may be due to chance or some other agent of influence on the climate system, later corroboration of the associations with instrument-measured climate data and with complex models of the climate system vindicate the belief of researchers over the last several centuries that solar variability influenced the Earth's climate variability.

After this tour d'horizon of the tales told by various empirical data, hypothesized mechanisms of the decadal Sun–climate relationship and efforts to test these mechanisms with models were reviewed. From the beginning of the instrument era and especially in the last 30 years or so since the advent of reasonably accurate measurements or estimates of solar emissions and since long time series of model-assimilated data started to become available, the development of the Sun–climate variability field has picked up pace. Measurements of UV radiation and visible solar irradiance have shown that the difference between UV radiation impinging upon the Earth at Schwabe cycle maxima and minima is approximately 6% of the average compared to the much smaller (approximately 0.1%) corresponding difference in visible radiation. These and other measurements/estimates, such as of stratospheric ozone, provided the motivation to hypothesize how the UV radiation cycle, in conjunction with chemical and dynamical feedback processes, may be driving a substantial response in the stratosphere which may also influence thermodynamics and dynamics of the troposphere and thereby influence surface climate in the so-called top-down process. Similarly, an evolving understanding over the last 40 years of the tropical–subtropical, coupled ocean–atmosphere system provided the motivation to hypothesize how even a small (percentage-wise) change in visible solar radiation can generate a disproportionately large linear as well as nonlinear responses due to positive feedbacks between trade winds, SSTs, and precipitation processes in the so-called bottom-up process. The hypothesized top-down and bottom-up processes are already being tested in state-of-the-art climate models which are showing that the two processes working together can generate climate system responses which are substantially similar, although not of the same magnitude, as the proxy and instrument-measured climate variability associated with the Schwabe cycle. Also, these climate model experiments are showing that feedback processes in the tropical–subtropical Pacific Ocean region can generate stronger trade winds, above-average precipitation in the tropical precipitation regions, and below-average SST anomalies in the eastern equatorial Pacific Ocean during maxima of the Schwabe cycle. These simulated responses to the Schwabe cycle are qualitatively similar to the responses found in empirical data analyses by Meehl et al. (2009) and Narasimha and Bhattacharyya (2010), and for multidecadal solar irradiance variability by Mehta and Lau (1997). The hypothesized mechanisms described in this chapter are summarized in Table 2.1.

The research reviewed in this chapter spans a period of well over a century. During this period, paleoclimate proxy data such as tree rings, oxygen and carbon isotopes, and lake sediments; instrument-measured land–ocean–atmosphere data; instrument-measured data assimilated in atmosphere and ocean models; estimated total and spectral solar irradiance data from paleoclimate proxies and satellite-based instruments; Sunspot numbers; 10.7 cm radio flux; solar and terrestrial magnetic fields; ground-based measurements of GCR flux; and other types of data were used in disparate analyses and statistical associations with solar variability were estimated. To date, there is not one published effort to collect all available data of all types and rigorously analyze them with multiple techniques to bring out internally consistent pictures of how these myriad indicators/variables of the Earth–atmosphere system

TABLE 2.1
Solar Cycle Influences and Hypothetical Climate Responses

External Forcing	Processes	Primary Response Variables
Solar visible radiation	*Bottom-up*: Absorption of short-wave radiation by the subtropical ocean, subtropical evaporation, water vapor transport, rainfall and latent heat release in tropical regions, equatorial ocean upwelling	Upper-ocean temperature, winds, clouds, tropical ocean circulations, and temperatures
Solar ultraviolet radiation	*Top-down*: Absorption by stratospheric ozone and latitude-dependent heating, formation of ozone, stratospheric wave dynamics, stratosphere–troposphere interactions, tropical rainfall	Stratospheric ozone, temperatures, winds, and waves; tropospheric circulations
Solar cycle modulation of Galactic cosmic rays (GCRs)	Modulation of current flow in global electric circuit Formation of cluster ions and cloud condensation nuclei Formation of ultrafine condensation nuclei (UCN) from trace gas vapors	Clouds

are inter-related at the Schwabe and Hale cycle timescales, and what appear to be the roles of the solar cycles (maxima, minima, and all phases in between) in DCV. Such an effort, despite requiring numerous researchers with complementary expertise, would definitely yield a treasure trove of useful information which can be used to formulate comprehensive hypotheses about how the Earth–atmosphere system might be responding to all types of solar emissions. Outcomes from such an analysis effort can also be used to plan and test simulations with Earth System Models with ever-increasing complexities.

Instrument-based atmosphere–ocean–land observations, including satellite-based observations, are being conducted routinely by national and many international observing systems, and the data are being routinely assimilated in model systems with ever-increasing complexities and sophistication. It is vitally important that these measurements and assimilation activities be continued. It is equally important that permanent surface-based and satellite-based efforts to measure total and spectral solar irradiance, 10.7 cm radio flux, GCR flux, and solar and terrestrial magnetic fields be instituted, along with activities to cross-calibrate such measurements across different types of observing systems.

There are several outstanding problems in understanding the observed DCV and the increasingly likely role of solar cycles in DCV. From the point of view of understanding the observed DCV in the tropical Atlantic and Pacific regions, although several possible, purely internal mechanisms have been proposed, there is no conclusive evidence that any of them works in nature as we will see later in the book. A better understanding, perhaps including a role of the Schwabe and Hale cycles, may contribute to not only more accurate attribution of observed climate variability and changes, and better testing of climate models used for regional prediction but also to skillful decadal climate prediction.

From the point of view of the evolving hypotheses/mechanisms, it is important that these mechanisms be tested by the expanding pool of observed atmospheric and oceanic data sets, so that these hypotheses/mechanisms can be developed further to serve as a benchmark for testing climate models. It is also important to further explore the roles of feedbacks via the hydrologic cycle and ocean–atmosphere interactions, including ocean heat transports.

It cannot be argued that the observed data and climate models are perfect, and that some clever analyses will test and further develop the bottom-up and top-down hypotheses/mechanisms. Although the coverage of a wide variety of atmospheric and oceanic observations is increasing, and they are being assimilated in increasingly more sophisticated models, observing systems have varied in quantity and quality of observations in the past and these variations influence assimilated data products and introduce spurious climate variations. Also, as alluded to above, the climate system can generate DCV due to purely internal processes and can confound the detection of solar-forced signals; the ENSO phenomenon is one example, and impacts of volcanic eruptions is another. In spite of these (and other) outstanding problems, it is essential that we continue to address the scientific problem of solar-forced climate variability with available observations and models.

The field of Sun–climate relationship has come a long way from the early attempts more than two centuries ago, a journey sustained during difficult times by known and unknown researchers who faced ridicule for their belief that solar variability influenced climate on the Earth. In the meantime, the field of internally generated climate variability developed as increasing amounts and types of observed data led to insights into physical processes and applications of fluid dynamics principles led to development of first independent and then coupled ocean–atmosphere models. It will be a grand challenge for climate researchers now to understand how the externally forced and internally generated climate variabilities co-exist or interact because, after all, they are both working in the same climate system. It is very important, however, that this co-existence or interaction is understood and their impacts predicted because DCV generated by both types make very substantial impacts on worldwide water and food securities, and on other societal sectors (Mehta, 2017). After having come this far, we owe this to the early pioneers and to future generations.

3 Slowly Oscillates the Pacific

3.1 INTRODUCTION

In a land that had experienced multiyear droughts and famines for centuries, and suffered tragic deaths of millions of its people, the decades of 1860s and 1870s Current Era (CE) were particularly bad for India (Mehta, 2017). First, there was a drought from 1865 to 1874 CE, which devastated eastern and northwest India and killed 4 to 5 million people due to the failure of monsoon rains. Then, another shorter but equally ferocious drought from 1876 to 1878 CE laid waste to western and southern India, and killed more than 5 million people. The British colonial government was besieged by these calamities and decided to set up a government department to find ways to predict monsoon rains well in advance, so that mitigating measures could be applied. The India Meteorological Department (IMD) was thus founded in September 1875 CE. For the next two decades, the department searched for precursor relationships between the Indian monsoon rainfall; and atmospheric pressure (generally known as sea-level pressure or SLP) measurements from stations in the Indian Ocean region, Australia, and South Africa; Himalayan snowfall estimates in pre-monsoon winter; and Sunspot numbers. After moderate success till the late 1890s CE, forecasts based on the statistical relationships began to fail, and the government came under renewed criticism after failing to forecast and mitigate effects of droughts and famines from 1896 to 1900 CE. As per "desperate times call for desperate measures",[1] the IMD appointed Gilbert Walker (later Sir Gilbert) – an alumnus of Trinity College, Cambridge, and a mathematician, *but without any background in meteorology or climatology* – as its Director General in January 1904 CE (Allan, 2017).

Sir Gilbert immediately started expanding the geographical scope of searching for precursors to essentially the entire world wherever there were meteorological stations. He and his associates soon found some interesting statistical associations between the Indian monsoon rainfall and SLP variations in the eastern tropical Pacific–northern Australia region, the Alaska–Hawaii region, and the Iceland–Iberian Peninsula region. In this process, at least equally importantly, Sir Gilbert and his team of IMD researchers found three very important climate phenomena serendipitously. They named the eastern tropical Pacific-northern Australia phenomenon the Southern Oscillation (SO), the Alaska–Hawaii phenomenon the North Pacific Oscillation (NPO), and the Iceland–Iberian Peninsula phenomenon the North Atlantic Oscillation (Walker and Bliss, 1932). In all three, there was (is) an oscillation in SLP, which implies an oscillation in atmospheric mass between two centers of action in each phenomenon, around a central or average value. This was perhaps

[1] An expression originally attributed to the Greek physician Hippocrates.

the first instance of a climate phenomenon named an oscillation, but it would not be the last as we will see in this and subsequent chapters. Incidentally, these three oscillations occupy very important places in the pantheon of climate phenomena to this day. The SO and the NPO are characters in the story of the Pacific Decadal Oscillation (PDO) and the Interdecadal Pacific Oscillation (IPO) as we will see in this chapter.

Halfway around the world from India, the United States (US) Weather Bureau forecasters found in 1916 CE that a higher than average SLP over Alaska was associated with a more southerly track of winter storms, more rain in the USA, and a likelihood of colder than average temperatures east of the Rocky Mountains. Before Sir Gilbert took on the responsibilities of the Director General of the IMD, he had visited government meteorological organizations in the United Kingdom (UK), the USA, France, and Germany in 1902–1903 CE to familiarize himself with such organizations' work since he had no previous background or experience in weather and climate research (Allan, 2017). Due to his excellent background in mathematics and perhaps due to his fresh view of meteorology, Sir Gilbert connected the newly discovered NPO with the subseasonal variability in storm track over the North Pacific and weather over the USA about which he had learned in the US Weather Bureau (Walker and Bliss, 1932; Linkin and Nigam, 2008). The subsequent development of the weekly to monthly timescale NPO is comprehensively summarized by Linkin and Nigam (2008), so we will go directly to the discovery of a decadal NPO-like SLP oscillation, which is pertinent to the main story of this chapter.

Krishnamurti et al. (1986) were studying stationary and traveling components of the SO, using a near-global SLP data set from 1961 to 1976 CE, and found that zonally-averaged SLP perturbations, filtered on the interannual SO timescales, traveled from the South Polar region to the North Polar region during one decade followed by a reversal of the direction of travel during the next decade. Krishnamurti et al. (1986) also found such decadal timescale, north–south travel of interannual SLP anomalies in a Northern Hemisphere only data set from 1900 to 1980 CE. Although this study was little noticed by the climate research community, in hindsight it can be pointed to as one of the first studies which showed some type of decadal climate variability (DCV) without invoking solar cycles in the renaissance of DCV research as the 1980s CE were coming to an end. Then, as so often happens in scientific research, it was not till mid-1990s CE that this story came to life again.

As in the case of the serendipitous discovery of three oscillations by Sir Gilbert Walker and his team, the PDO and the IPO phenomena were discovered while scientists were puzzled by inexplicable variations in two quantities of societal importance. In the case of the PDO, it was the salmon catches along the coast of the Pacific Northwest in the USA and along the Alaskan coastline. In the case of the IPO, it was wheat production in Australia. Variations in these two quantities motivated scientists to search for their causes in natural phenomena. Following up on several previous studies of the relationship between catches of a variety of fish types – especially salmon – in the Pacific fishing regions near Alaska and Northwest USA and sea surface temperatures (SSTs) in the North Pacific Ocean region, Mantua et al. (1997) found a pattern of decadal–multidecadal variations that spanned the entire North Pacific Ocean such that there were SST anomalies of one sign (positive or negative) in the tropical–subtropical North Pacific and of the opposite sign in the mid-latitude North Pacific. They named it the PDO.[2] Almost at the same time when the PDO pattern was associated with variations in fish catch in the North Pacific Ocean, Power et al. (1999) were trying to understand why the relationship between interannual SO variations, and Australian climate – including rainfall, surface air temperature, and Murray River streamflow – and Australian wheat production varied at decadal and longer timescales. It was found that an SST pattern spanning the Pacific Ocean modulated the relationship between the SO and Australian climate at decadal and longer timescales.[3] Following Folland et al. (1998), Power et al. (1999) named this SST pattern the IPO. This chapter describes the over 20 years' journey to define not only the PDO and IPO patterns, but also other decadal and longer timescale variability patterns of the Pacific climate system, and understand their physics.

This journey is cluttered with at least three terms alluding to timescales of variability – decadal (10 to 20 years),[4] interdecadal (10 to 100 years), and multidecadal (20 to 100 years). As explained in Chapter 2, the term "decadal climate variability" was first used to describe variability associated with the solar Schwabe (approximately 11 years) and Hale (approximately 22 years) cycles, and, later, to describe any climate variability at those timescales. For the sake of reducing the semantic clutter, only decadal and multidecadal are used here. Somewhat similar semantic clutter exists on naming geographical extents of patterns, determined by analysis domains. Therefore, the timescale-geographical extent combinations are separated into tropical Pacific decadal variability and pan-Pacific decadal–multidecadal variability. These two geographical and timescale separations are necessary to formulate testable hypotheses and verify them with models as we will see in Section 3.4. Also, given the sparseness of instrument-measured data in the Pacific ocean–atmosphere system before the 1950s CE, the available data may not be sufficient even to characterize tropical Pacific decadal variability (as defined here) with some semblance of representativeness. The pre-1950s CE data are grossly inadequate to characterize pan-Pacific multidecadal variability. Some analyses of time-limited data to study so-called climate shifts over approximately a decade are also referred to in the published literature as decadal variability; such studies are not separately included in this chapter, but they are mentioned within the overall context of variability.

After this brief introduction, spatial and temporal patterns identified in instrument-measured data, and indications of past Pacific decadal–multidecadal variability found in paleoclimate proxy data are described in Sections 3.2 and 3.3, respectively. Then, hypothesized mechanisms, including effects of low-latitude volcanic eruptions, and their simulations by a variety of models are described in Section 3.4. Finally, the foregoing is summarized and conclusions are presented in Section 3.5.

---

[2] Associations between DCV phenomena, and fish and crustacean catches in the Pacific, Atlantic, and Indian Oceans are described in detail in Mehta (2017).

[3] Associations between DCV phenomena and worldwide agricultural production are described in detail in Mehta (2017).

[4] In the 1990s CE, the lower period was reduced to eight years as applications of filters to separate decadal and longer timescale variability from interannual ENSO variability became widespread.

3.2 SIGNATURES IN INSTRUMENT-MEASURED DATA

A brief history of the evolution of networks of scientific instruments for atmosphere, land, and ocean observations is described in Chapter 2 (Section 2.4). As mentioned there, instrument-measured ocean data were collected systematically since the 1850s CE. Century-long and longer time series of SST, SLP, and surface air temperature over the oceans became publicly available after the end of the Cold War in the early 1990s CE. Upper-air data (wind, temperature, humidity) are available since the 1940s and 1950s CE over most of the world. These atmosphere and ocean data have been assimilated in global general circulation models to produce dynamically and thermodynamically balanced, global atmosphere and ocean data. In this section, we will see what these data tell us about decadal and longer timescale variability in the Pacific Ocean and its associated atmospheric patterns.

Section 3.2 is divided into two sub-Sections. The pan-Pacific phenomena generally referred to as the PDO and the IPO are described in Section 3.2.1. Then, decadal ocean–atmosphere variability in the tropical Pacific region is described in Section 3.2.2.

3.2.1 Pan-Pacific Decadal–Multidecadal Variability

We began this chapter with an introduction to serendipitous discoveries of SLP oscillations at interannual and longer timescales mainly in the Pacific region. We first see the observational evidence of pan-Pacific decadal–multidecadal climate variability, starting with the discovery of the PDO and the IPO.

As century-long and longer time series of instrument-measured and gridded SST data started becoming available in the early to mid-1990s CE, the journey of exploration began. One of the first analyses of Pacific SST data was by Zhang et al. (1997) who applied a low-pass filter to century-long time series of gridded SST data to remove oscillation periods shorter than six years and then applied the empirical orthogonal function (EOF)–principal component (PC) analysis to the filtered Pacific SST data from 30°S to 60°N. The dominant SST EOF pattern to emerge from this analysis had one sign of variability in the tropics–subtropics and opposite sign in mid-latitudes. Also, this pattern was much broader in the meridional direction than the El Niño SST pattern, with maximum amplitude in central and western equatorial Pacific between 160°W and 180° longitudes, unlike the El Niño pattern's maximum amplitude in eastern equatorial Pacific. The corresponding PC time series showed variability at decadal–multidecadal timescales with a warming trend superimposed on it. Zhang et al. (1997) called this EOF-PC combination "ENSO-like interdecadal variability" – a reference to the interannual El Niño-Southern Oscillation (ENSO) phenomenon.

Almost simultaneously with the Zhang et al. (1997) study, and following up on several previous studies (see Mehta (2017) and references therein) of the relationship between catches of a variety of fish types – especially salmon – in the Pacific fishing regions near Alaska and Northwest USA, and SSTs in the Pacific Ocean region, Mantua et al. (1997) found an EOF-PC combination of decadal–multidecadal variations which they named the PDO. The spatial pattern of the PDO SST anomalies in the positive phase and associated 850 hPa wind anomalies from the National Centers for Environmental Prediction (NCEP)-National Center for Atmospheric Research (NCAR) reanalysis data in four three-month average periods – December–January–February (DJF), March–April–May (MAM), June–July–August (JJA), and September–October–November (SON) – and the PDO index time series are shown in Figure 3.1a and b, respectively. In Figure 3.1a, in the positive or warm PDO phase, the SSTs in the tropical–subtropical Pacific and along the North American coast are warmer than average, and those in the mid-latitude central and western Pacific are cooler than average. In the negative or cool PDO phase, the SSTs in the tropical–subtropical Pacific and along the North American coast are cooler than average, and those in the mid-latitude central and western Pacific are warmer than average. Wind anomalies at 850 hPa converge over the warmer surface water in the tropics, and there is an anomalous cyclonic circulation over the negative or cold SST anomalies in the North Pacific region. Seasonal differences in the SST and wind patterns are noticeable with the DJF patterns having the largest anomalies and the JJA patterns the smallest. Figure 3.1b shows that the three-month average PDO index from January 1900 to November 2015 CE has undergone variability at interannual

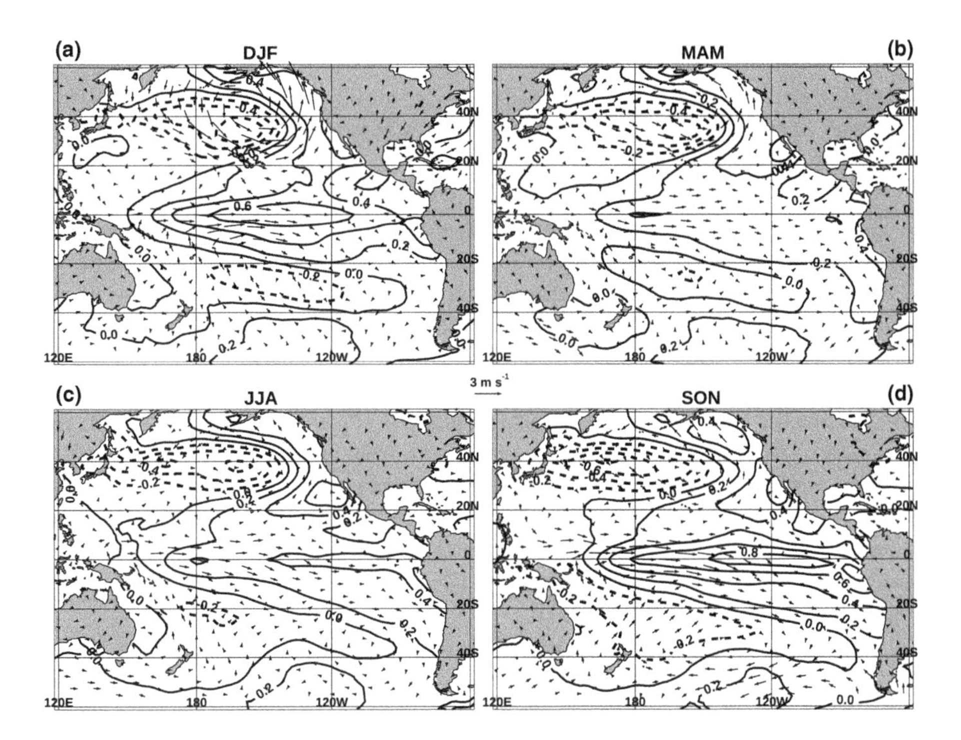

FIGURE 3.1A Regression coefficient (°C/unit index) of the PDO index regressed on seasonal-average SST anomalies; and regression coefficient (vectors) between the PDO index, and eastward and northward winds. (a) DJF, (b) MAM, (c) JJA, and (d) SON. Scale vector in the center of the figure shows regression coefficient scale of wind anomalies in m s^{-1}/unit index. Contour interval of regression coefficient of SST anomalies is 0.2. SST and wind data from December 1950 to November 2018 CE were used.

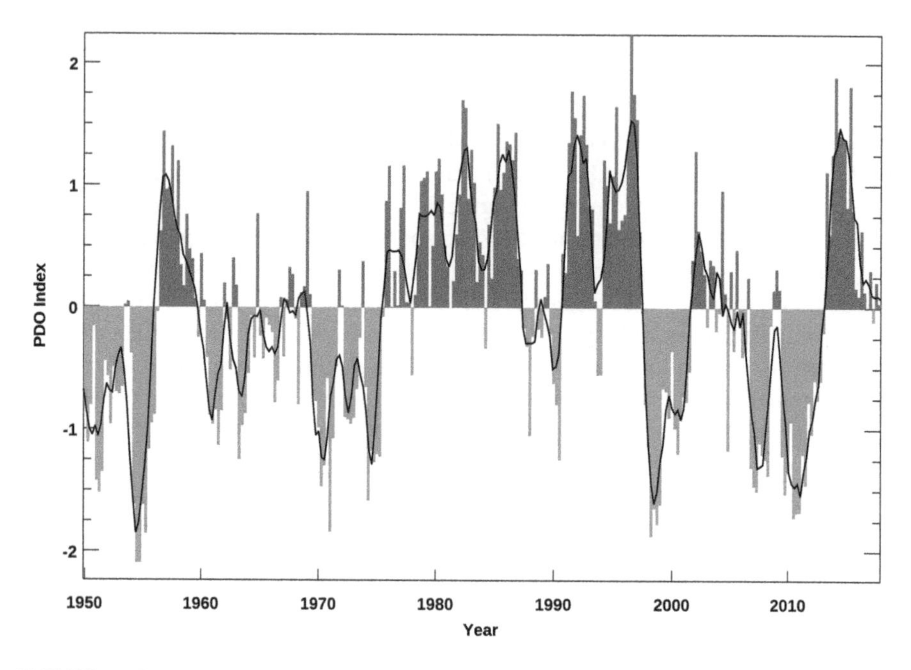

FIGURE 3.1B The three-month average (DJF, MAM, JJA, SON) PDO Index (bars) from December 1950 to November 2018 CE and its four-season running-average smoothed version (black line).

to multidecadal periods, but the smoothed index clearly shows the prominence of decadal – multidecadal variability.

Based on indications of a pan-Pacific SST variability pattern in analyses of low-pass-filtered, global SST data by Folland et al. (1998), Power et al. (1999) found that an SST pattern in the Pacific Ocean, oscillating at decadal and longer timescales, modulated the relationship between the SO and Australian climate, and named this SST pattern the IPO. In its original form (Folland et al., 1998; Power et al., 1999), the IPO pattern (Figure 3.2) spans the Pacific Ocean from approximately 60°S to 60°N latitudes, and is generally similar to the PDO pattern north of the Equator. While the PDO index time series is usually derived without any filtering of the SST data and shows variations at timescales of a few months to a few decades, the IPO index time series reflects the low-pass filtering applied to SST data with only timescale longer than 13 years present in the time series.

After the PDO and the IPO were discovered, discussions soon started within the climate science community about the physical reality of these SST patterns, their similarities and differences, and possible mechanisms underlying these so-called oscillations. At about the same time, evidence of decadal variability of ENSO attributes and their predictability (Chapter 6) also started emerging, further complicating the interpretation and understanding of all three. How can there be so many climate variability patterns at similar timescales in one (Pacific) ocean–atmosphere system? Are there different mechanisms causing them? Which patterns make substantial worldwide impacts on societal sectors such as water resources and agriculture?

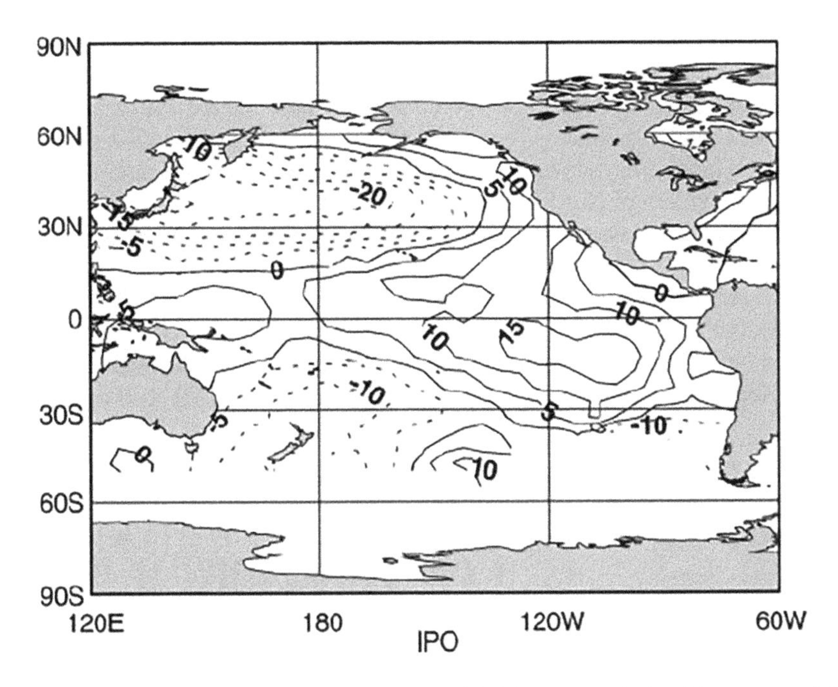

FIGURE 3.2 The positive or warm phase of the IPO SST pattern derived from an empirical orthogonal function analysis of low-pass-filtered (>13 years), near-global SST data from 1911 to 1995 CE. (Adapted from Power et al., 1999.)

Among major handicaps in addressing these questions are sparseness – to the point of virtual non-existence – of even ocean surface observations in the Pacific region before the end of the Second World War in 1945 CE; very different density of observations in time and space in the North and South Pacific Oceans even after 1945 CE (much higher in North Pacific compared to South Pacific); and very spotty subsurface data anywhere in the Pacific before the 1980s CE. Therefore, the parable about blind men trying to understand what an elephant is by running their hands on its body was (and, to a large extent, still is) applicable to decadal–multidecadal variability in the pan-Pacific region. Undaunted by these handicaps, the climate variability literature for next 20 years was rich with data analyses, model simulations, and theoretical studies arguing and counter-arguing about the origin and mechanisms of these SST patterns. Later, the identification of a purported Southern Hemisphere counterpart to the PDO, the South Pacific Decadal Oscillation (SPDO; Shakun and Shaman, 2009), added further confusion to the already-crowded field of Pacific climate variability phenomena. We will see the proposed mechanisms and simulated pan-Pacific variability patterns in Section 3.4, but now let us see if we can bring clarity to this discussion about the physical reality and identification of these SST patterns before we try to understand the mechanisms.

Let us focus first on the pan-Pacific SST patterns which oscillate at decadal–multidecadal timescales. Folland et al. (2002) showed that the conventionally defined North Pacific PDO and pan-Pacific IPO indices were highly correlated.

Henley (2017) compared the PDO, the IPO, the SPDO, and Niño3.4 SST patterns and indices, using Extended Reconstructed Sea Surface Temperature (ERSST) version 4 data (Huang et al., 2015). He also compared low-pass-filtered (>13 years) components of the respective indices. These SST patterns and their respective, non-filtered and filtered, index time series are shown in Figure 3.3. In this figure, the PDO is the first EOF pattern of North Pacific northward of 20°N, and the SPDO is the first EOF pattern of South Pacific southward of 20°S. The IPO index is based on the difference between SST anomalies averaged over the central equatorial Pacific (10°S and 10°N) and the average of SST anomalies in the Northwest (25°N and 45°N) and Southwest (50°S and 15°S) Pacific regions (Henley et al., 2015). Thus, this index is not based on an EOF analysis of pan-Pacific SST data, unlike the original IPO pattern and index, and the PDO and SPDO patterns and indices. All spatial patterns in

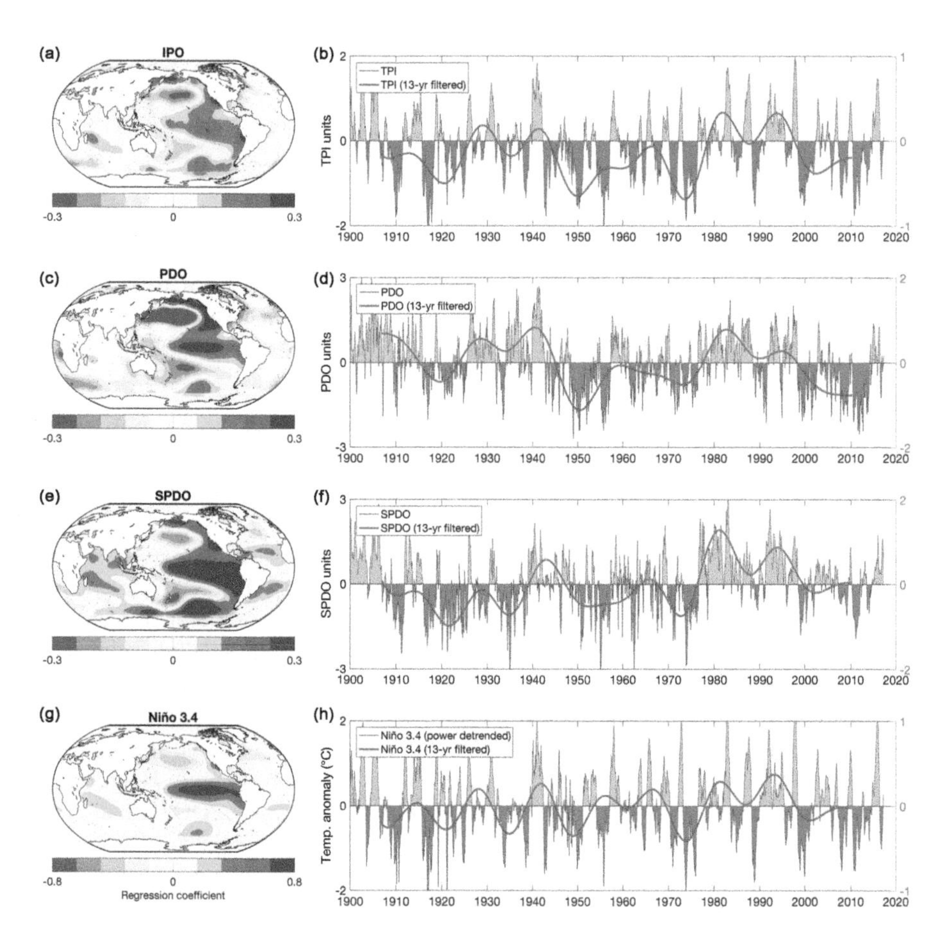

FIGURE 3.3 Spatial patterns and time series of decadal–multidecadal Pacific SST variability. (a) and (b) for the IPO, (c) and (d) for the PDO, (e) and (f) for the SPDO, and (g) and (h) for the Niño3.4. Spatial patterns are from regression of global SST anomalies onto unfiltered indices; unfiltered time series are shown as shaded anomalies, and time series are smoothed by a 13-year Chebyshev low-pass filter. (From Henley, 2017.)

Figure 3.3 were derived by regressing global SST anomalies on each filtered index. Considering the data handicaps mentioned above and consequent uncertainties in the importance of relatively small spatial and temporal differences, it can be said that the IPO, PDO, and SPDO spatial patterns (Figure 3.3a, c, and e, respectively) are similar in the first approximation. All three are characterized by variability centers of one sign spanning the tropical–subtropical Pacific and along the coasts of the Americas, and in high-latitude North and South Pacific; and variability centers of the opposite sign in mid-latitude North and South Pacific. The corresponding index time series of all three (Figure 3.3b, d, and f, respectively), especially the low-pass filtered components, are also generally similar in the post-1950 CE period. The Niño3.4 spatial pattern (Figure 3.3g), while being very generally similar to the other three patterns, is much more concentrated in the eastern and central equatorial Pacific as is well known. Unlike the IPO, PDO, and SPDO index time series, the Niño3.4 index time series (Figure 3.3h) shows more decadal, rather than multidecadal, variability. This compares well with the tropical Pacific decadal variability described in Section 3.2.2 and reinforces the hypothesis that *decadal* variability in the Pacific may be largely of tropical origin. We will pick up this line of argument again in Sections 3.2.2 and 3.4.

3.2.2 Tropical Pacific Decadal Variability

We will now see observational evidence of decadal variability in the tropical Pacific climate. Decadal ocean–atmosphere variability in the tropical Pacific region is described here with observed data, both objectively analyzed data and model-assimilated data. Hypothesized mechanisms of this variability and model simulations are described in Section 3.4.2.

3.2.2.1 Sea Surface Temperature Variability

Discoveries of the PDO and the IPO patterns added further impetus to the journey of exploration which began in the early 1990s CE. Initially, the journey consisted of applications of statistical analysis techniques to multidecades- to century-long and even longer time series of SST and SLP data. These "fly bys" revealed tantalizing hints of decadal and longer timescale patterns of variability in Pacific SST and SLP data. These patterns (see, for example, Tourre et al. (2001), White and Tourre (2003)) have stood the test of time as we will see in this sub-section. Then, guided by previous observational studies of tropical Pacific variability and its possible association with oceanic subtropical circulation cells, Lohmann and Latif (2005) began the next phase of the journey. They first found a distinct temporal pattern of decadal SST variability in the so-called Niño4 region in the central and western equatorial Pacific region (5°S to 5°N, 160°E to 150°W) from 1870 to 1999 CE using the singular spectrum analysis technique. The temporal pattern with the largest variance showed that it had a decadal period (as defined in this Chapter) (Figure 2a in Lohmann and Latif (2005)). The spatial SST anomaly pattern associated with this temporal pattern was then derived by linear regression with grid-point SST anomalies. The spatial pattern showed (Figure 2b in Lohmann and Latif (2005)) that the strongest SST variability was between 150°W and 180° longitudes in the western tropical Pacific. This spatial pattern spanned the tropical Pacific from approximately 15°S to 15°N with one sign of SST anomalies and, although it had anomalies of the opposite sign in the extratropics, it was mainly a tropical variability pattern. The temporal pattern with the second largest temporal variance of Niño4 SST anomalies showed amplitude-modulated interannual variability, and the associated spatial pattern resembled the El Niño pattern. Based on these empirical data analysis results, Lohmann and Latif (2005) proposed and tested a hypothesis about the tropical Pacific decadal variability. We will see the hypothesis and simulation results in Section 3.4.2.

Continuing with empirical analyses of tropical Pacific decadal variability, we will now see results of further analyses of SST variability and, in the next subsection, three-dimensional analyses of atmosphere and ocean observations assimilated in global general circulation models. Wang and Mehta (2008) isolated two spatial patterns of decadal SST variability in the western tropical Pacific region in ERSST version 2 data from 1952 to 2001 CE (Smith and Reynolds, 2004), using the EOF-PC analysis technique. An EOF analysis of low-pass (>7 years) filtered, monthly SST anomalies within the western Pacific region bounded by 25°N to 25°S latitudes and 60°E to 160°W longitudes revealed two EOFs which represent coherent decadal variability, and "explain" 45% and 18% of the total decadal SST variance over the region, respectively. These two EOFs were well separated from the rest of the EOFs in explained variance. Figure 3.4 shows the spatial patterns of the two EOFs in the form of regressions of filtered SST anomalies against the corresponding PC time series. EOF1 is characterized by large areas of statistically significant SST

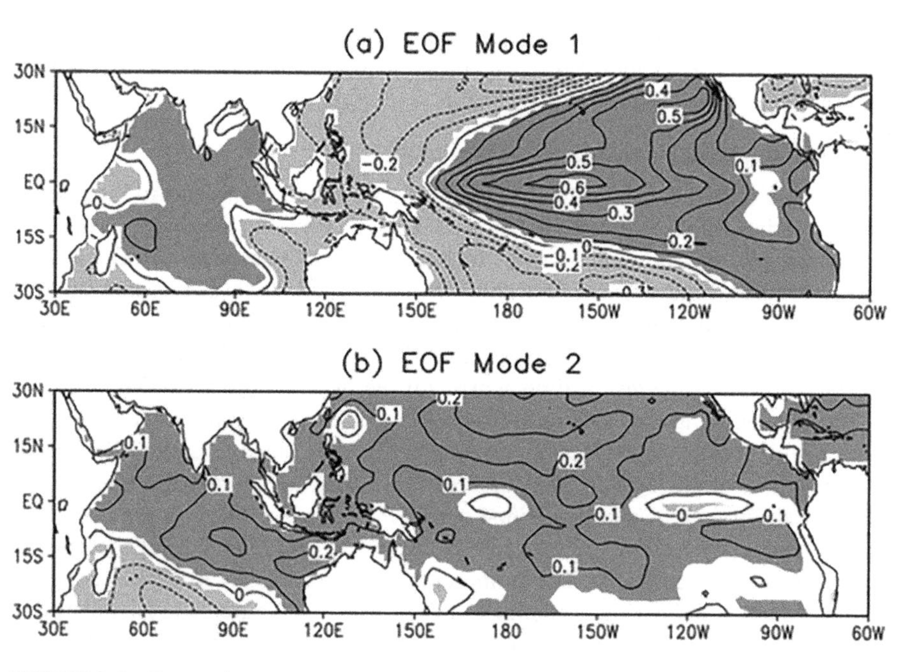

FIGURE 3.4 Regression patterns of SST anomaly (contours) associated with a two-standard deviation departure in corresponding EOF time series. Contour interval is 0.1°C, and negative contours are dashed. Dark (light) shadings indicate positive (negative) SST anomaly at the 1% significance level, estimated by Monte Carlo tests. (From Wang and Mehta, 2008).

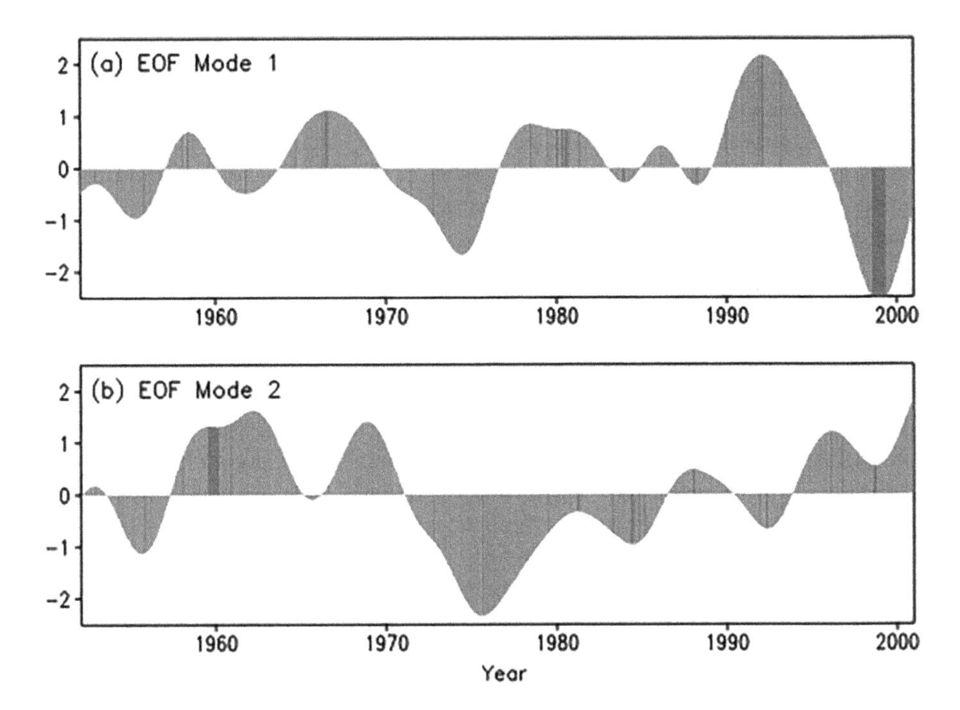

FIGURE 3.5 Normalized monthly time series of (a) PC 1 and (b) PC 2 of western tropical Pacific SST anomalies. (From Wang and Mehta, 2008.)

anomalies at the 1% level; there are warm SST anomalies in the central and eastern equatorial Pacific, cold SST anomalies in the western Pacific, and warm SST anomalies in most of the Indian Ocean (Figure 3.4a). It bears some resemblance to the El Niño SST pattern in the tropical Pacific, but with relatively weak and insignificant SST anomalies in the eastern Pacific and a much broader meridional extent. Near the Equator, the warmest SST anomalies in the central Pacific and cold SST in the western Pacific set up a strong zonal SST gradient between approximately 150°E and 180° longitudes. In the Indo-Pacific Warm Pool (IPWP[5]) region, there is an out-of-phase relationship between SST anomalies in the Indian Ocean and the western Pacific. This EOF is similar to the El Niño-like decadal SST variability pattern found by Lohmann and Latif (2005), suggesting a linkage between the IPWP and eastern equatorial Pacific SST variability on decadal timescales. The SST pattern in EOF2 shows statistically significant, coherent, warm SST anomalies in the IPWP as well as in the tropical Pacific, except for two small areas centered at 180° and 120°W where the anomalies are insignificant (Figure 3.4b). Normalized time series of the PCs corresponding to the two EOFs are shown in Figures 3.5a and b, respectively. Reconstructed total SST based on regressions of low-pass-filtered SST anomalies versus the time series of each EOF and long-term annual-average SST climatology shows that each EOF describes one aspect of the tropical Pacific decadal variability.

[5] Major attributes of the West Pacific Warm Pool and the East Indian Warm Pool are described in Chapter 5.

EOF1 mainly represents zonal and meridional spatial variations of the IPWP in the tropical central and western Pacific, whereas EOF2 represents changes in intensity of the IPWP temperature. As described briefly here, several studies found that SSTs in the tropical central and western Pacific vary at decadal timescales. The climatological SSTs in the IPWP region are nearly always at the threshold where atmospheric convection can break out, with consequent effects on local and global atmospheric circulations as we will see in the next sub-section.

3.2.2.2 Atmospheric Circulation Variability

The observed record of global atmospheric observations has been assimilated into atmospheric models by several projects, one of which is the NCEP-NCAR reanalysis system (Kalnay et al., 1996). Although there are no discontinuities related to changes in data assimilation system and model physics in the NCEP-NCAR reanalysis, problems related to discontinuities in spatial and temporal distribution of raw observational data exist. Comparisons between the reanalysis data and some independent data sources indicate (for example, Chelliah and Bell (2004)) that the NCEP-NCAR reanalysis data well represents variability of atmospheric circulations in the tropics over the 50-year analysis period. The NCEP-NCAR reanalyses have also been used to examine tropical decadal variability by Ashok et al. (2004), Chelliah and Bell (2004), and others. We will now see associations between the tropical Pacific decadal variability described in the previous sub-section and global atmospheric variability in the NCEP-NCAR reanalysis data.

Wang and Mehta (2008) examined associations between decadal variability of the tropical Pacific SSTs and variability in the thermally-direct Hadley and Walker circulations by calculating these circulations with NCEP-NCAR reanalysis data at 2.5° longitude–2.5° latitude resolution from 1952 to 2001 CE. Associations between these circulations and the two SST PC time series, based on linear regression, are shown in Figure 3.6. This figure shows the Hadley circulation and anomalous vertical velocity field associated with the positive phase of the two EOFs for DJF and JJA, respectively, which are the averages of meridional divergent flow and vertical velocity across the IPWP between 60°E and 160°W. The negative of vertical velocity in pressure co-ordinates is plotted so that it is consistent with circulations (positive upwards). In DJF (Figures 3.6a and b, upper panel), the Hadley circulation is characterized by low-level convergence, upper-level divergence, and strong updrafts in the middle troposphere between 20°S and 10°N, and downdrafts at higher latitudes. The circulation associated with the negative phase of the two EOFs also has this seasonal feature. A close inspection, however, suggests that the two EOFs modulate the Hadley circulation in different ways. The anomalous vertical velocity related to the two EOFs (Figures 3.6a and b, lower panel) indicates that in their positive phase, both EOFs contribute to the upward motion near the Equator. A much stronger and broader signal is associated with EOF1 than with EOF2. Additionally, in the northern extratropics, the descending branch is greatly enhanced by EOF1, but is weakened by EOF2. This is consistent with the spatial distribution of SST shown in Figure 3.4, in which cold SST anomalies dominate most of the northern extratropics in EOF1, while in EOF2, there are warm SST anomalies in the same region. In the Southern Hemisphere, both EOFs weaken the Hadley circulation by reducing

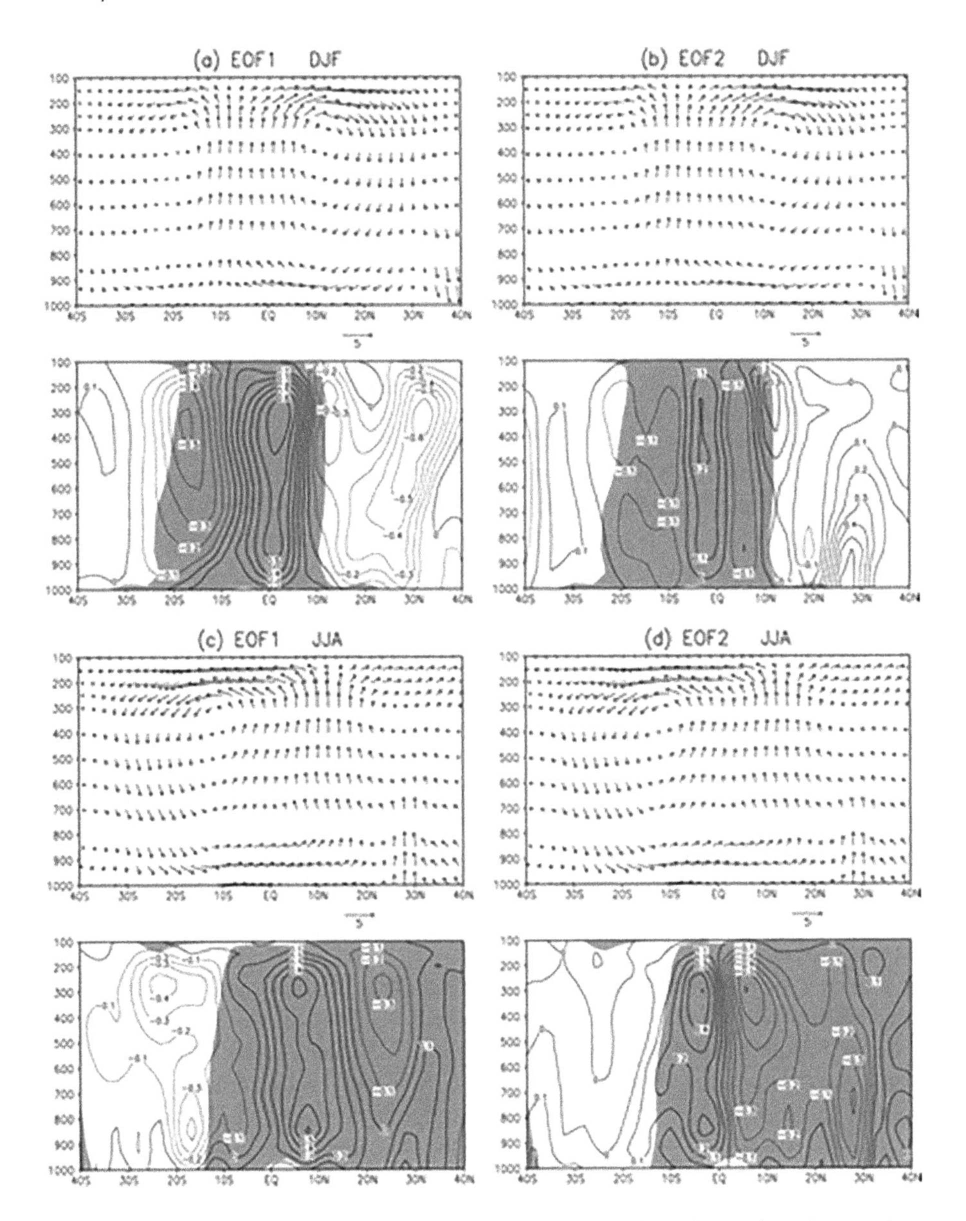

FIGURE 3.6 Total Hadley circulation (vectors; upper panel) and negative of anomalous vertical velocity (contours; lower panel) for (a)–(b) in DJF and (c)–(d) in JJA associated with the two SST EOFs, respectively. The figures show linear regressions versus each EOF time series and are averaged between 60°E and 160°W. The vertical velocity anomaly is the difference between the regressions associated with positive and negative EOF phases. Contour interval is 0.1 with a unit of -0.02 Pa s^{-1}, and negative contours are dashed. Shadings indicate updrafts in climatological seasonal mean circulation. (From Wang and Mehta, 2008.)

ascending motion in the subtropics and increasing subsidence in mid-latitudes. In general, the contrasts in the Southern Hemisphere between the two EOFs are smaller than those in the tropics and the Northern Hemisphere. In JJA (Figures 3.6c and d), low-level convergence and upper-level divergence, as well as the tropical updrafts, shift into the Northern Hemisphere, consistent with northward movement of the northern branch of the IPWP. The downdrafts in the Southern Hemisphere are much stronger in JJA than those in DJF. The anomalous vertical velocity associated with EOF1 in its positive phase enhances the Hadley circulation in the tropics in JJA. In contrast, the anomalous vertical velocity related to EOF2 in JJA only strengthens the Hadley circulation between 10°S and the Equator, but weakens the Hadley circulation between 30°N and the Equator.

Similar analyses from Wang and Mehta (2008) are shown in Figure 3.7 for the Walker circulation, that is, averages of zonal divergent flow and vertical velocity between 2.5°S and 2.5°N latitudes. In the Walker circulation, low-level cold and dry air from the tropical eastern Pacific flows westward to the IPWP where the air is heated and moistened. After moist-adiabatic ascent and rainfall over the IPWP, the upper-level outflow moves eastward, and converges and subsides over the tropical eastern Pacific region. Thus, the two decadal SST EOFs exhibit clear associations with the Walker circulation. In both DJF and JJA (Figures 3.7a and c), anomalous circulations related to the positive phase of EOF1 reinforce the updrafts in the central Pacific and Indian Ocean regions, but reduces the ascent over the western Pacific region, resulting from a thermally-direct response to the anomalous SSTs in Figure 3.4a. The anomalous vertical velocity associated with EOF2 (Figures 3.7b and d) is weaker than that associated with EOF1 in both seasons, possibly because the anomalous zonal SST gradient is much stronger in EOF1 than that in EOF2. In the western Pacific region, EOF2 strengthens the Walker circulation, which is opposite to the effect of EOF1 and consistent with warm SST anomalies in EOF2 in the same region (Figure 3.4b). Both EOFs in their positive phase weaken ascending motion over equatorial South America.

Figure 3.8 presents anomalous components of the Hadley and Walker circulations associated with the two EOFs. As Figure 3.8a–d shows, the modulation of the thermally direct Hadley circulations by the two EOFs has a strong seasonality. The statistically significant contribution of EOF1 is slightly stronger and encompasses a larger atmospheric column in DJF but is comparable to that in JJA with a northward shift, whereas the statistically significant contribution of EOF2 to the Hadley circulation in DJF is much weaker than that in JJA. The influence of EOF2 is also stronger in the Northern Hemisphere than in the Southern Hemisphere. It is interesting to note that while the circulation anomalies in the rising branch of the Hadley circulation in DJF (Figure 3.8a) are significant only in the lower troposphere near the Equator, circulation anomalies in the descending branches are significant over almost the entire troposphere in the sub-tropics of both Hemispheres.

The anomalous Walker circulation associated with EOF1 (Figure 3.8e) is large and statistically significant. This is physically consistent with the large zonal gradient of SST anomalies in EOF1 (Figure 3.4a) near the Equator. In both DJF (Figure 3.8e) and JJA (Figure 3.8g), the rising branch of the anomalous Walker circulation associated with EOF1 is near 180° because the warmest SST anomalies are

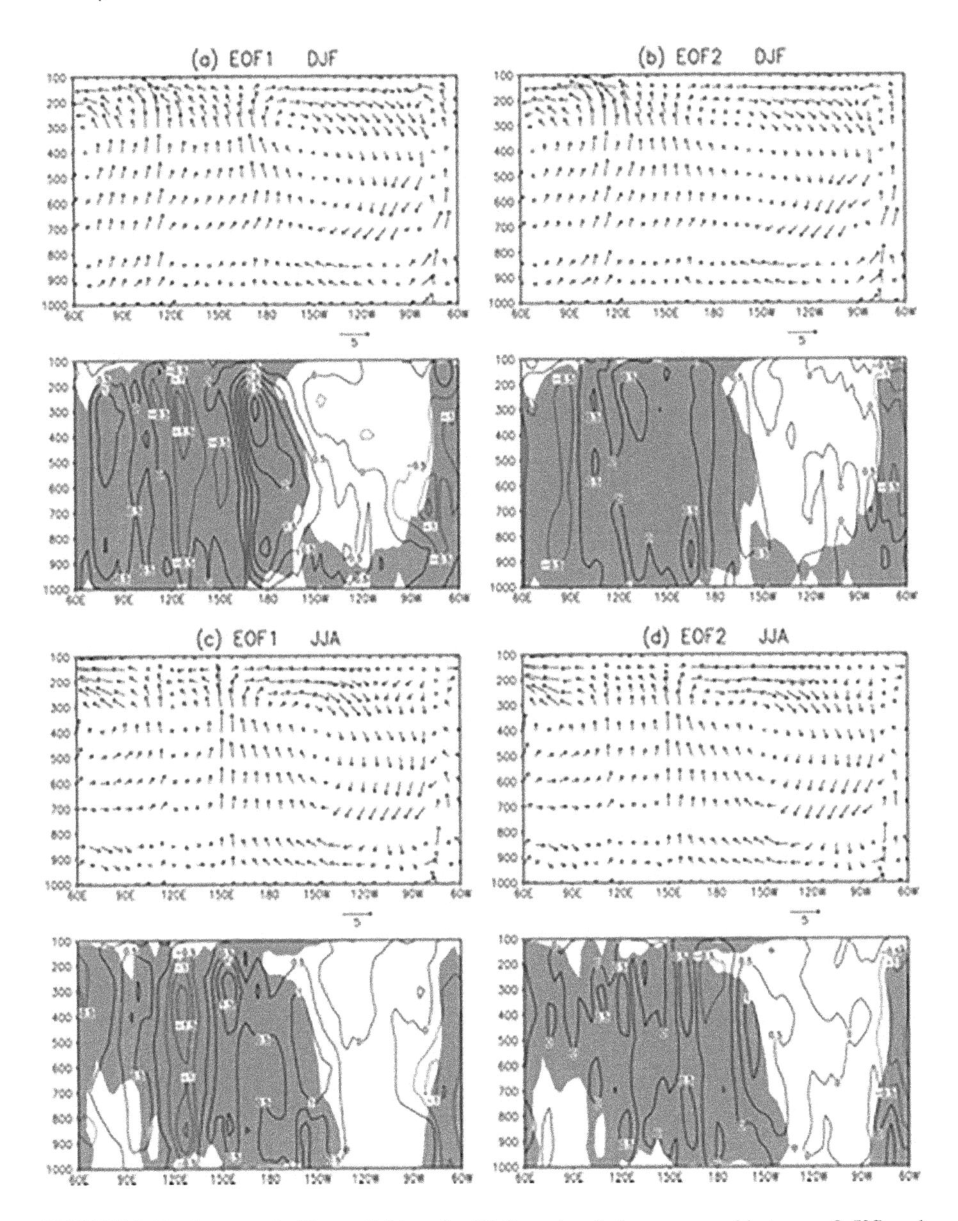

FIGURE 3.7 Same as in Figure 3.6 but for Walker circulation averaged between 2.5°S and 2.5°N. (From Wang and Mehta, 2008.)

near 180°. In contrast, the anomalous Walker circulations (Figures 3.8f and h) associated with EOF2 are weak because the zonal gradient of SST anomalies in EOF2 is weak (Figure 3.4b).

The anomalous upper-level divergent flow over the western tropical Pacific region (Figure 3.8) provides an important source for atmospheric Rossby waves to propagate from the tropics to the extratropics (Sardeshmukh and Hoskins, 1988).

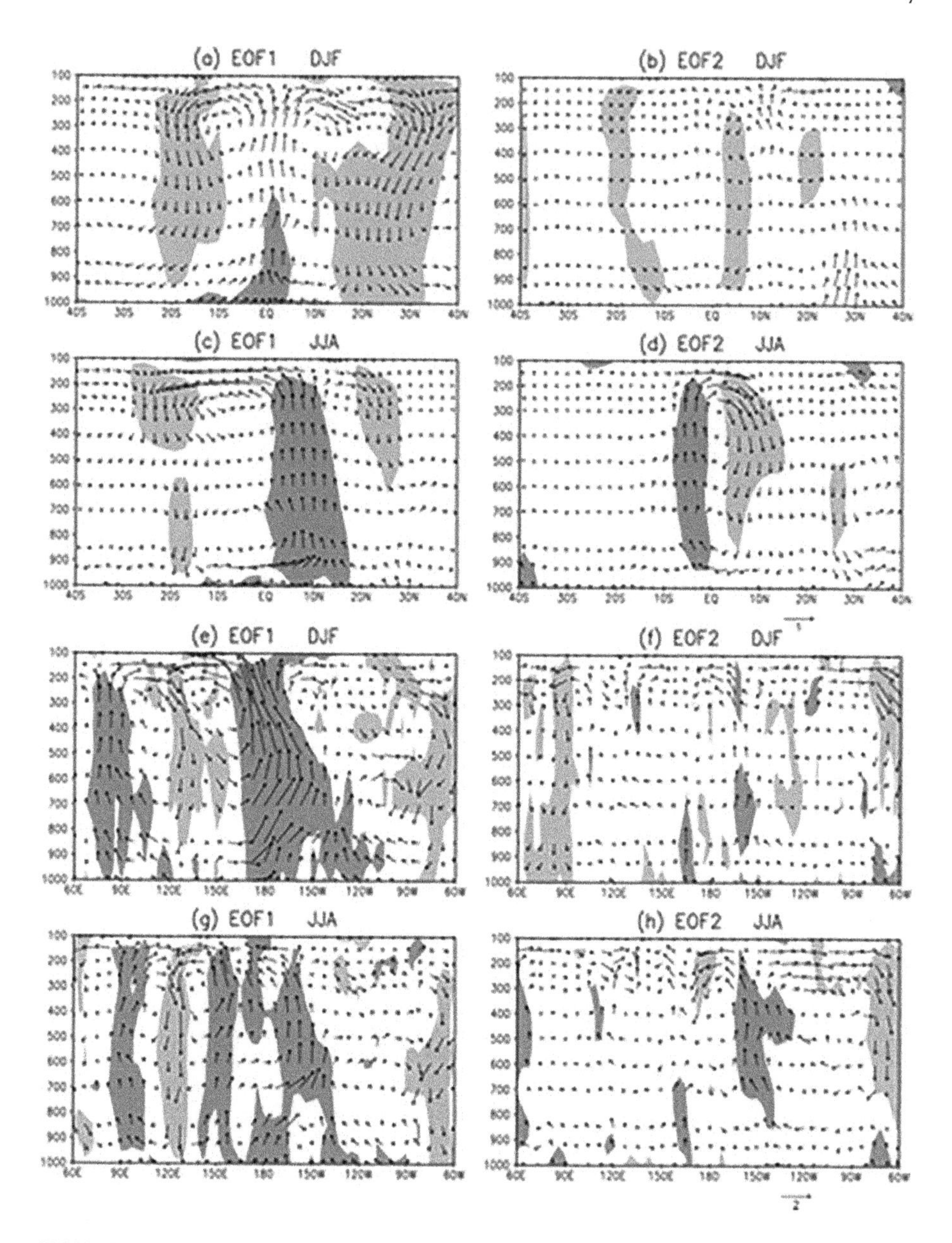

FIGURE 3.8 Same as in Figures 3.6 and 3.7 but for anomalous meridional (a–d) and zonal circulation (e–h). Dark (light) shadings indicate anomalous updraft (downdraft) at the 1% significance level. (From Wang and Mehta, 2008.)

This type of planetary waves links anomalous convection in the western tropical Pacific region to circulation anomalies in higher latitudes. Associations between the two SST PC time series and 200 hPa geopotential height anomalies (Figure 3.9) show the northern extratropical response to the heating and consequent divergent forcing in the western Pacific region. The anomalous DJF geopotential height field associated with the SST EOF1 exhibits a wave train across the Asia–Arctic Ocean–Pacific–North America regions (Figure 3.9a). In the anomaly pattern associated with EOF1 in DJF (Figure 3.9a), there are statistically significant anomaly centers over central and eastern China, northern Russia, and the Arctic, Alaska, and central North Pacific regions. To some extent, this anomalous circulation pattern resembles

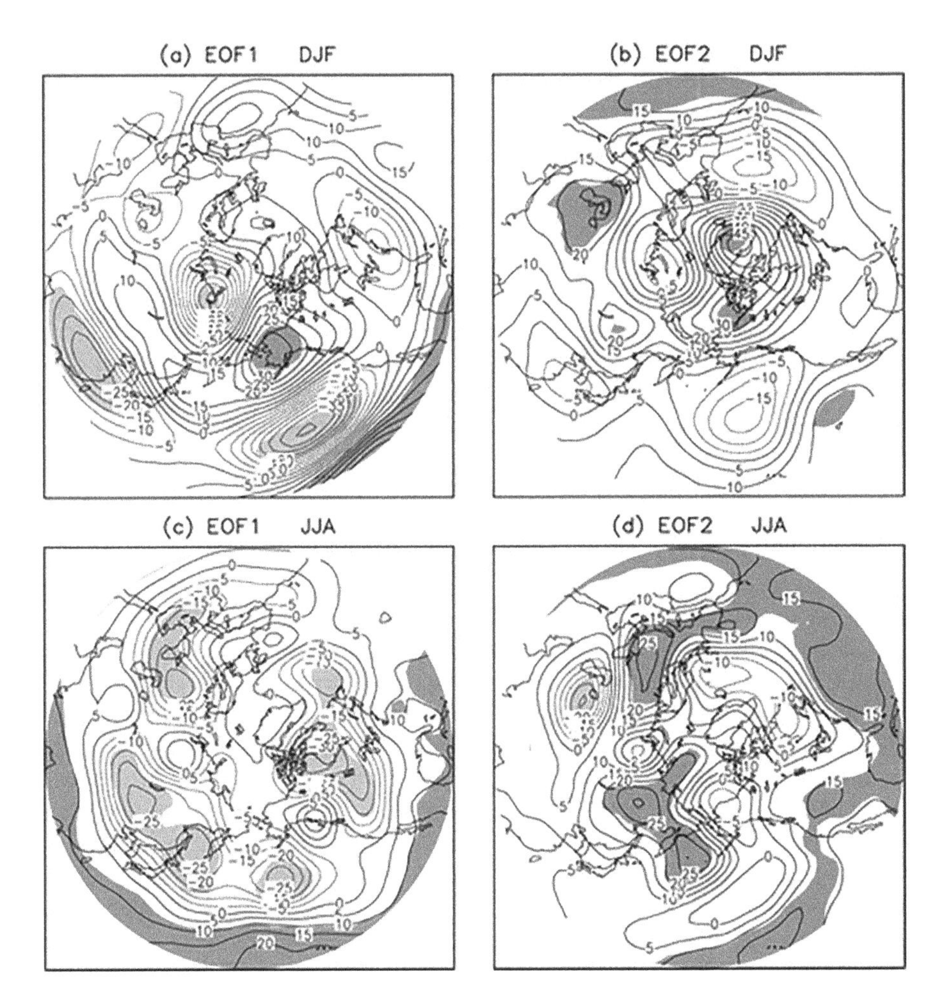

FIGURE 3.9 200-hPa geopotential height anomaly for (a), (b) DJF and (c), (d) JJA associated with two leading SST EOFs. The maps are linear regressions versus each EOF time series with a two standard deviation departure. Contour interval is 5 m, and negative contours are dashed. Dark (light) shadings indicate positive (negative) height anomaly at the 1% significance level. (From Wang and Mehta, 2008.)

the Pacific-North America pattern (Wallace and Gutzler, 1981), which is a recurrent DJF circulation pattern especially during El Niño years (Wang and Fu, 2000; Strauss and Shukla, 2002). However, there are additional centers of action over China, Mongolia, and Russia associated with the SST EOF1. In addition to its direct association with the Pacific-North America region climate variability at decadal timescale, this anomalous 200 hPa geopotential height pattern (Figure 3.9a) can also modulate seasonal to interannual ENSO impacts on the Pacific-North America region by the tropical Pacific SST anomalies on decadal timescales via anomalous extratropical pressure and circulation fields (Gershunov and Barnett, 1998).

In JJA, the 200 hPa geopotential height anomaly pattern (Figure 3.9c) associated with SST EOF1 is much more zonal and similar to the JJA North Pacific, tropical–extratropical teleconnection pattern found by Barnston and Livezey (1987). The 200 hPa geopotential height anomalies associated with SST EOF2 in DJF (Figure 3.9b) show a spatial structure similar to the Arctic Oscillation (Thompson and Wallace, 1998), with opposing geopotential height anomalies in northern middle and high latitudes. These results clearly indicate a statistically significant association between major extratropical atmospheric circulation patterns in DJF and decadal variability of the tropical Pacific SSTs. The extratropical circulation anomalies in JJA (Figures 3.9c and d) are relatively weaker compared to their DJF counterparts and also display smaller scale features, presumably due to weaker jet streams and reduced meridional absolute vorticity gradient in the northern extratropics.

The atmospheric circulation variability associated with the tropical Pacific SST EOFs also drives anomalous ocean circulations and heat transports, which can feed back to the tropical Pacific SST anomalies as we will see in the next sub-section.

3.2.2.3 Oceanic Circulation Variability

The sparsity of ocean observations was worse for sub-surface observations than that for surface observations before the 1980s CE. Despite this handicap, the sparse surface and sub-surface ocean observations have been assimilated in ocean models to produce physically consistent ocean data sets continuous in space and time. One of the longest such records was produced by the Simple Ocean Data Analysis (SODA) system (Carton et al., 2000). Although there were no discontinuities related to changes in the data assimilation system and model physics in the SODA data set, problems related to discontinuities in spatial and temporal distribution of raw observational data existed similar to those in the NCEP-NCAR atmospheric reanalysis. Comparisons between the SODA data and some independent data sources indicated (Schott et al., 2002; Xie et al., 2002) that the SODA data set well represents variability of ocean circulations in the tropics over the 50-year assimilation period. The earlier-mentioned unreliability of pre-1980s CE ocean observations, however, rendered the longer data length of the then-available SODA product problematic. Also, SODA then contained internal sources and sinks of heat generated by the assimilation procedure in addition to the usual physical terms in the heat equation. This posed a major difficulty in heat budget analysis using the then-available SODA product (Fukumori, 2006). But, Wang and Mehta (2008) used the SODA product only for an exploratory analysis of decadal variability of the tropical Pacific Ocean circulation as used by Schoenefeldt and Schott (2006) to explore decadal variability of the Indian Ocean circulation.

Figure 3.10 from Wang and Mehta (2008) shows the vertical structure of ocean temperature averaged between 2.5°S and 2.5°N, related to positive and negative phases of the two SST EOFs (Figure 3.4), respectively. The distance between two intersections of the 28°C isotherm with the ocean surface can be taken as an approximate zonal dimension of the IPWP near the Equator. Associated with EOF1, the DJF IPWP exhibited strong zonal variations between the two phases with an oscillation of thermocline (20°C isotherm) depth between the eastern and western Pacific (Figure 3.10a). This thermocline depth oscillation is associated with basin scale, out-of-phase changes in temperature anomalies between the eastern and western Pacific in the 50–150 m layer, with an approximate 3°C amplitude (Figure 3.10c, lower panel), significant at the 1% level. There are also smaller but significant temperature anomalies at 100–150 m depth between the positive and negative phases of EOF1 in the Indian Ocean. Associated with EOF2, in contrast, there is not much zonal variation (Figure 3.10b). Instead, the 28°C isotherm moves vertically between positive and negative phases, indicating significant changes in the IPWP temperature. These variations in the thermal structure are consistent with warm temperature anomalies above 100 m in the western Pacific and cold anomalies centered at 150 m in the central Pacific, and also in the Indian Ocean.

In JJA, the changes in ocean temperature between the two phases in EOF1 (Figure 3.10c) are similar to those in DJF, with an apparent zonal variation of the West Pacific Warm Pool (WPWP). The out-of-phase, mid-ocean temperature anomalies in the eastern and western Pacific between the two phases in EOF1 (Figure 3.10c, lower panel) are about a half of the magnitudes in DJF (Figure 3.10a, lower panel) but significant at the 1% level. The temperature anomalies between the two phases of EOF1 in the Indian Ocean are at 50–100 m depth in JJA. Similar to its DJF counterpart (Figure 3.10b), the IPWP shows less zonal variation associated with EOF2 in JJA (Figure 3.10d).

The vertical-meridional cross section of ocean temperatures along 150°E longitude, where the IPWP has the maximum meridional extent, is shown in Figure 3.11. Unlike its zonal variation between positive and negative phases of EOF1 (Figures 3.10a and c), the meridional extent of the IPWP changes in an opposite way between the two phases in both DJF and JJA (Figures 3.11a and c). Specifically, while the IPWP expands along the Equator, it shrinks in meridional direction and *vice versa*. This change is accompanied by a significant cooling in the tropical and subtropical western Pacific above 300 m depth, except for a narrow and shallow layer of warming above 50 m depth at the Equator (Figures 3.11a and c, lower panel). There is also significant cooling between 40°N and 60°N in the negative phase of EOF1 when the IPWP expands in the meridional direction; this cooling is coherent from the surface to 250 m depth. The IPWP also behaves differently in the meridional direction in EOF2. With little change in size along the Equator between the two phases (Figures 3.11b and d), the IPWP exhibits large variations in its meridional size (Figures 3.11b and d), especially in DJF. The changes between the two phases in EOF2, however, are opposite to those in EOF1 in the tropics and mid-latitudes, which may thereby lead to less latitudinal variations due to a mutual cancellation of the two EOFs.

Variability of near-surface currents is also associated with the two EOFs of decadal SST variability. Figure 3.12 from Wang and Mehta (2008) shows regression

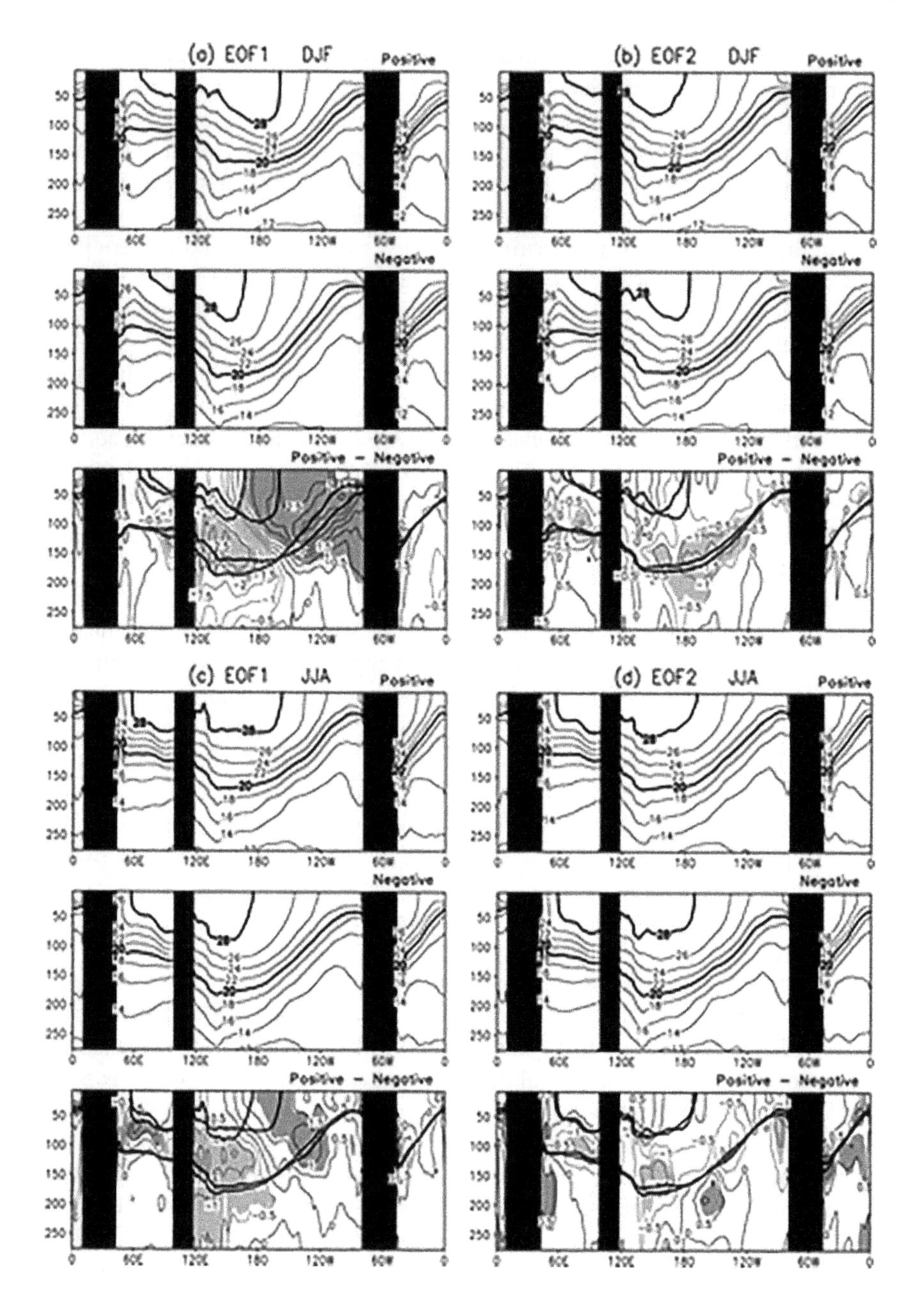

FIGURE 3.10 Ocean temperature average over 2.5°S – 2.5°N for (a), (b) DJF and (c), (d) JJA associated with positive (upper) and negative (middle) phases of SST EOFs1 and 2, and their differences (lower). The sections are linear regressions versus each EOF time series with a two- standard deviation departure. Contour interval is 2°C in upper and middle plots and 0.5°C in lower plots with negatives dashed. Thick lines are 20° and 28°C isotherms associated with the SST EOFs. Dark (light) shadings indicate positive (negative) temperature anomaly at the 1% significance level. (From Wang and Mehta, 2008.)

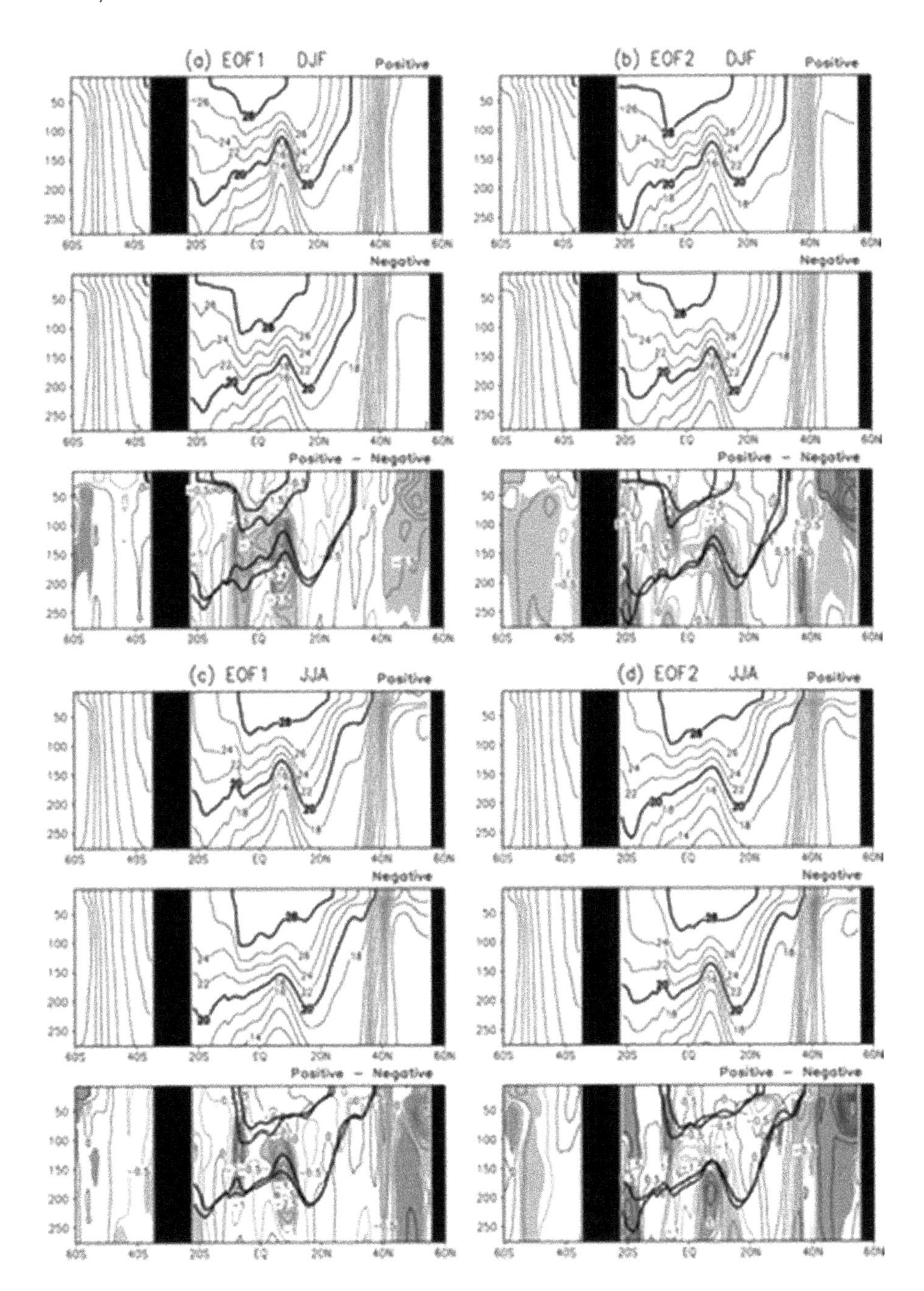

FIGURE 3.11 Same as in Figure 3.10, but for a latitude-depth cross section at 150°E. (From Wang and Mehta, 2008.)

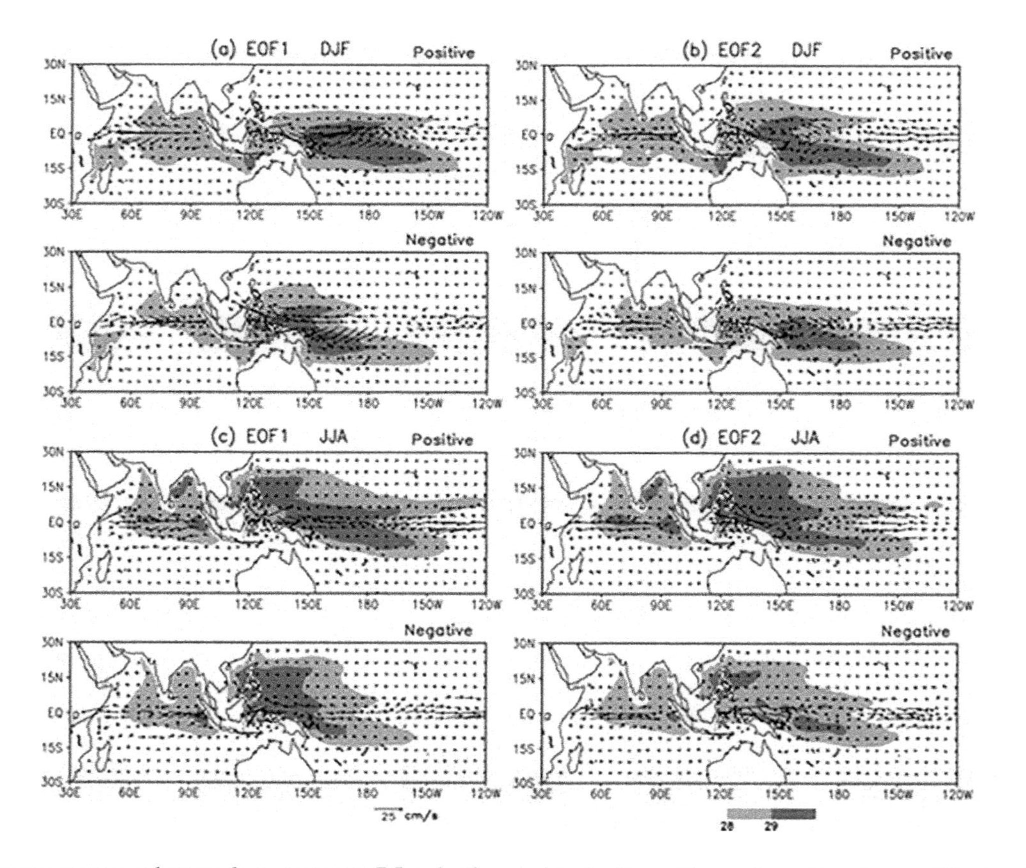

FIGURE 3.12 Ocean temperature and anomalous current at 7.5 m depth associated with positive and negative phases of the two SST EOFs in DJF and JJA. Light and dark shadings indicate temperature greater than 28°C and 29°C, respectively. (From Wang and Mehta, 2008.)

patterns of ocean temperature and anomalous ocean current at 7.5 m depth, associated with the two EOFs. The IPWP (shaded) associated with opposite phases of each EOF is similar to the annual-average patterns, but with pronounced seasonality. The JJA IPWP has a more zonally extended north branch. The changes in size and shape of the IPWP between positive and negative phases of each EOF are consistent with the results based on the vertical structure of ocean temperature in Figures 3.10 and 3.11. Associated with EOF1, the IPWP displays an out-of-phase relationship in changing zonal and meridional dimensions, indicating a strong deformation (Figures 3.12a and c). In EOF2, the variations between the two phases are dominated by changes in size and intensity of the IPWP (Figures 3.12b and d). Associated with these temperature variations, strong anomalous near-surface currents are found in the equatorial Pacific and Indian Ocean and weak anomalous currents in the subtropical Pacific. Some of the near-surface current patterns suggest that their associated thermal advection can impact the IPWP temperature. In the positive phase of Figure 3.12a, for example, strong westerly flow in the central Pacific Ocean and easterly flow in the Indian Ocean can contribute to warming and zonal expansion of the IPWP due to the advection of mean temperature gradient by anomalous currents.

Associated with the temperature and near-surface current variability is the variability of shallow tropical circulations or cells (STCs). Figure 3.13 from Wang and Mehta (2008) shows a vertical cross section of anomalous meridional velocity at 150°E associated with each EOF in its positive phase. Anomalous meridional currents are stronger in the Southern Hemisphere than in the Northern Hemisphere. Compared to seasonal-average meridional currents, significant meridional current anomalies associated with EOF1 strengthen the Southern Hemisphere STC in both DJF and JJA, and weaken the Northern Hemisphere STC in both DJF and JJA. Meridional current anomalies associated with EOF2 weaken the STCs in both Hemispheres. Stronger STCs may lead to a stronger subduction and cooling of the ocean temperature (Mehta, 1991; Mehta, 1992). Since the anomalous meridional velocities are nearly opposite between the two EOFs, the STCs may act differently in maintaining the two decadal SST EOFs. Associated with EOF1 in the positive phase, for example, a stronger Southern Hemisphere STC can cool the IPWP and may force it towards its negative phase. When the IPWP is in the negative phase, the anomalous meridional current in Figure 3.13a changes sign and thus can weaken the total Southern Hemisphere STC that, in turn, can warm the IPWP, and can force the IPWP back to the positive phase. The changes in the strength of the STCs can be induced by subtropical surface winds (McCreary and Lu, 1994; Liu, 1994). Therefore, the STCs, especially in the Southern Hemisphere, can provide a negative feedback to the IPWP in EOF1 and a positive feedback in EOF2. Since the variations of the STCs are on a meridional advection timescale of 10 to 15 or more years, which can originate from Ekman drift-related equatorial upwelling, oscillations in tropical horizontal gyres, and interactions between STC pathways and other ocean currents (Schott et al., 2002, 2004; Luebbecke et al., 2008), they provide a forcing mechanism for the tropical Pacific decadal variability.

This set of ocean and atmosphere processes responsible for tropical Pacific decadal variability is generally similar to that found by Luo and Yamagata (2001) from their analyses of ocean and atmosphere observations. They found that a positive SST

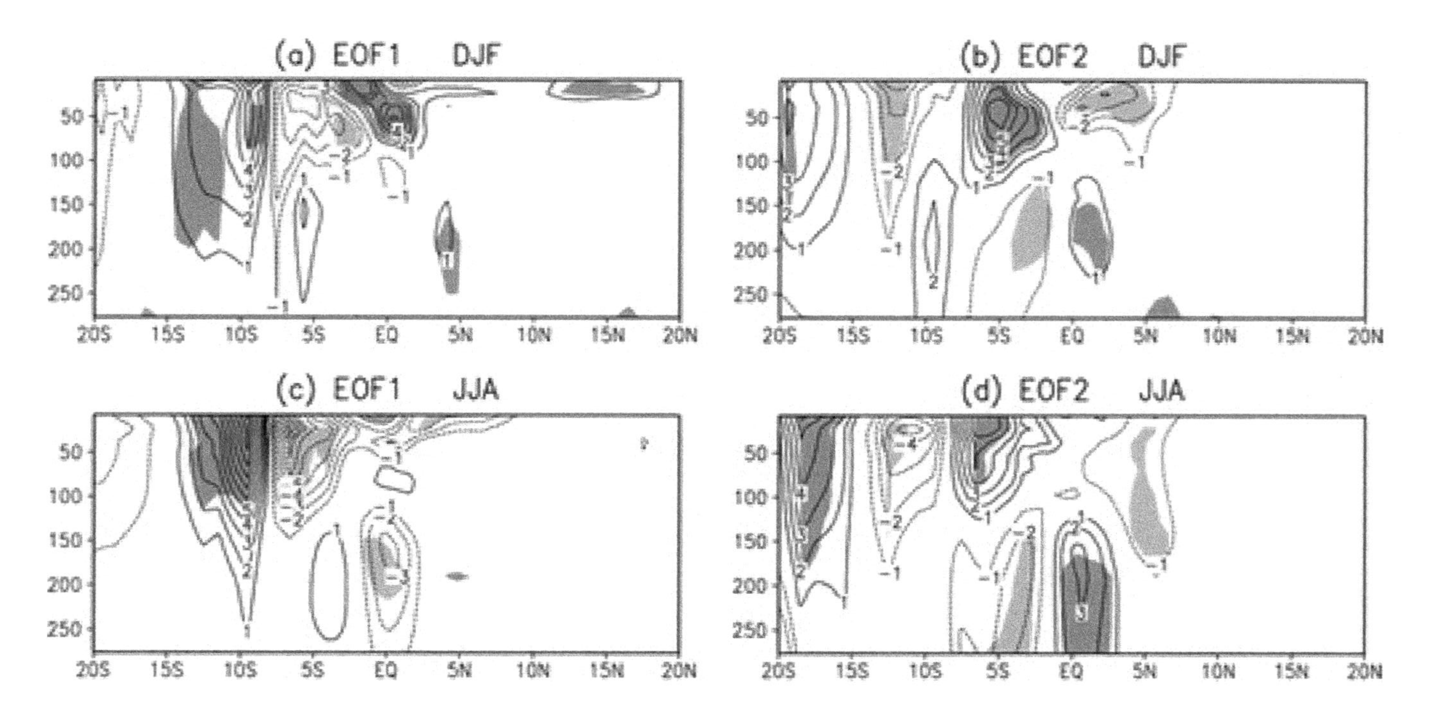

FIGURE 3.13 Same as in Figure 3.11, but for oceanic, anomalous, meridional velocity at 150°E associated with a two-standard deviation departure in the positive phase of EOF time series. Contour interval is 1 cm s^{-1} with negatives dashed and zeros omitted. Dark (light) shadings indicate positive (negative) anomaly at the 1% significance. (From Wang and Mehta, 2008.)

anomaly in the equatorial Pacific can create an anomalous, negative wind stress curl in the western tropical South Pacific, which can shoal the thermocline resulting in a negative sub-surface temperature anomaly in that region. This negative temperature anomaly moves northwestward to western and central equatorial Pacific, either as oceanic Rossby waves or as anomalous temperature advection by average currents. In the Luo and Yamagata (2001) depiction, the negative temperature anomaly then propagates eastward and grows due to the Bjerknes feedback (Bjerknes, 1966, 1969). So, from the observational evidence described here, it appears that STCs, especially in western tropical South Pacific, are playing an important role in the tropical Pacific decadal variability.

We now come to the end of descriptions of instrument-observed pan-Pacific decadal–multidecadal variability and tropical Pacific decadal variability. In the next section, we will see results of efforts to find signatures of the two classes of variability in paleoclimate proxy data.

3.3 SIGNATURES IN PALEOCLIMATE PROXY DATA

After the decadal–multidecadal patterns of Pacific climate variability were discovered in instrument-measured data as described in Section 3.2, the importance of these patterns for societal impacts motivated searches for such patterns in paleoclimate proxy data to extend the information record backwards in time and know past temporal characteristics of the Pacific decadal–multidecadal variability patterns. The problems in interpreting paleoclimate proxy data in terms of climate variables and indices were briefly described in Chapter 2 and in detail in Mehta (2017). But, despite the problems, there have been many attempts to calibrate paleoclimate proxy data, especially tree rings, in terms of the PDO/IPO index as we will see in this section.

Henley (2017) compared 12 reconstructions of the PDO/IPO/ENSO indices from North American (Biondi et al., 2001; Gedalof and Smith, 2001; D'Arrigo et al., 2001; MacDonald and Case, 2005) and Asian (D'Arrigo and Wilson, 2006) tree-ring data, documentary records (Shen et al., 2006), multiproxy composites (Verdon and Franks, 2006; Mann et al., 2009; McGregor et al. 2010; Henley et al., 2011), South Pacific Convergence Zone (SPCZ)-WPWP coral oxygen isotope[6] $\delta^{18}O$ (Linsley et al., 2008), and the Law Dome ice core from Antarctica (Vance et al., 2015). Time series of these reconstructed indices are shown in Figure 3.14. Henley (2017) has discussed data sources, quality of reconstructions, and their representativeness in detail, so only major aspects of these reconstructions are described here. In correlations among the reconstructed time series as well as in their spectral characteristics, Henley (2017) found large differences among the 12 reconstructions, especially before the advent of instrument-measured SST data which were used to calibrate/validate the reconstructions; these differences are visible in Figure 3.14. These differences are so large and irregular that Henley (2017) termed the reconstructions "paleo-spaghetti", concluding that "...paleoclimate reconstructions don't yet provide a consistent picture of the history of Pacific decadal variability in the past millennium, or even in the past 300–400 years when data availability is greatly increased". Among the possible major reasons cited by

[6] The use of this oxygen isotope as a paleoclimate indicator is explained in Section 2.3.2.

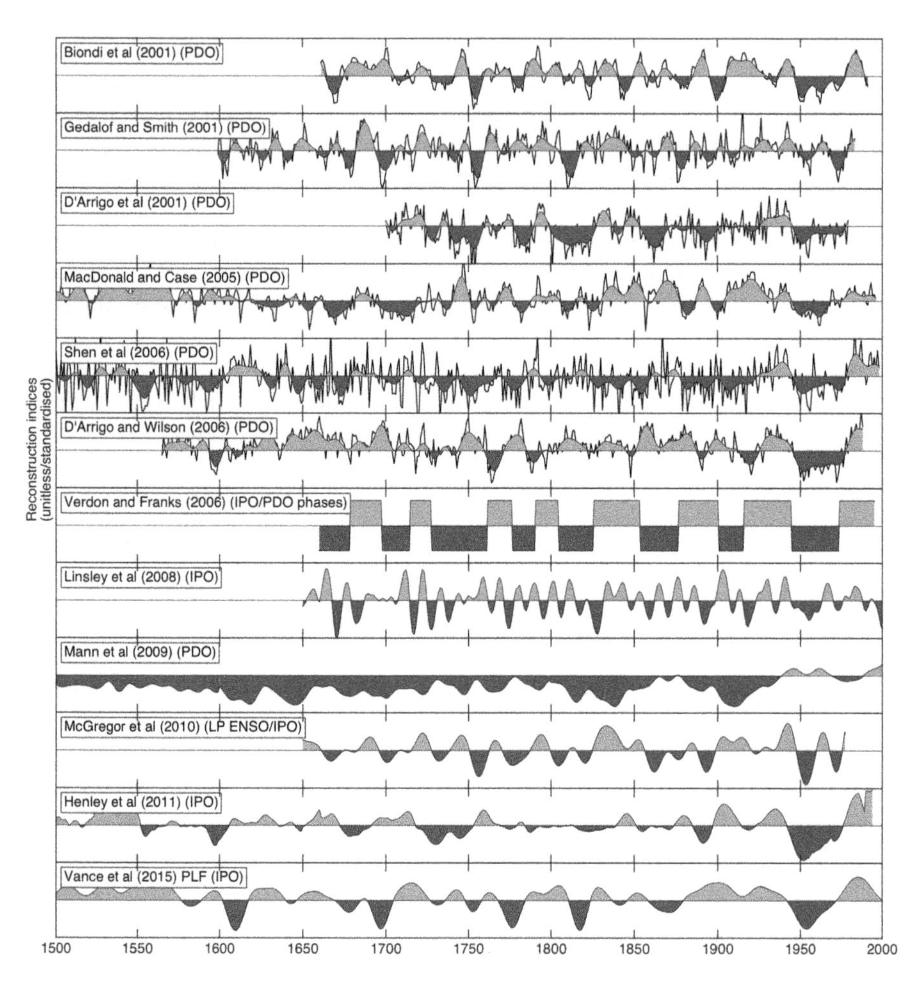

FIGURE 3.14 Paleoclimate reconstructions of indices of the PDO, the IPO, and low-frequency ENSO from 1500 to 2000 CE. Low-frequency reconstructions are shown in dark gray and light gray shaded anomalies; unsmoothed time series are shown in black where available. (From Henley, 2017.)

Henley (2017) for this unsatisfactory state are (1) non-stationarity of regional teleconnections, (2) spectral biases of paleoarchives, and (3) data processing and reconstruction techniques. Other possible major reasons may also be non-stationarity of the PDO/IPO phenomenon, and confounding factors – besides precipitation and temperature – affecting tree rings and other proxies. Newman et al. (2016)'s comparisons of seven paleo-PDO reconstructions based on tree-ring data, with six of the seven also included in the Henley (2017) study, came to generally similar conclusions.

Despite the inconsistent nature of reconstructions described above, Mehta (2017) associated two of the twelve paleoindices with indicators of climate variability such as drought indices for North America and Monsoon Asia. Biondi et al. (2001) used tree-ring information from Southern California and Baja California to reconstruct annual PDO index from 1661 CE to 1991 CE. They found that frequency spectra of the reconstructed PDO index time series had the dominant peak at approximately 17 to 23-year periods; the strength of the peak varied over the 300 years of the record. In another reconstruction of the PDO index from 1565 CE to 1988 CE, D'Arrigo and Wilson (2006) used tree-ring information from northeast Asia during the MAM season and found higher spectral density around 23-year oscillation period compared to neighboring periods. Although both sets of tree-ring information are influenced by local and/or regional weather and climate, there is a considerable degree of similarity between the two reconstructed PDO index time series in the overlap period as Mehta (2017) and Henley (2017) found. Therefore, Mehta (2017) used these two reconstructed PDO index time series to compare with dry and wet epochs as represented in a tree-ring-based reconstruction of North American Palmer Drought Severity Index (PDSI). The comparison showed that six dry epochs (1665–1675, 1752–1758, 1818–1826, 1861–1866, mid-1890s, and 1948–1958 CE) and seven wet epochs (1675–1685, 1688–1697, 1745–1750, 1793–1796, 1827–1840, 1867–1872, and 1905–1930 CE) in North America from 1665 to 1958 CE occurred when both reconstructed PDO indices were generally in a negative or a positive phase, respectively. There is an opposite-phase relationship between wet and dry epochs in North America and Monsoon Asia (Mehta, 2017), so the association between reconstructed PDO index time series and dry and wet epochs in North America applies to Monsoon Asia as well. These associations are consistent with the phases of the instrument data-based PDO index's association with dry and wet epochs (Mehta, 2017) in North America and Monsoon Asia. Thus, this extension of the PDO's association with dry and wet epochs in North America and Monsoon Asia back to the 17th century CE strengthens the credibility of at least two of the reconstructed PDO index time series for some segments of the reconstructed indices. However, it must be emphasized that both the PDO and PDSI reconstructions are based on tree-ring data with their own reconstruction problems, although the data sets are from very different locations.

One of the paleoclimate proxy-based reconstruction mentioned above is the IPO index reconstruction by Linsley et al. (2008) using annual, coral oxygen isotope $\delta^{18}O$ from 1650 to 2004 CE near Fiji and Tonga islands in the SPCZ-WPWP region. The SPCZ is also an ocean salinity front region due to heavy rainfall associated with warm SSTs and atmospheric convergence. An IPO index based on the coral record matched an instrument-measured SST-based IPO index from 1856 to 2000 CE and an SPCZ position index based on SLP from 1891 to 2000 CE (Folland et al., 2002), validating the representativeness of SST and ocean salinity variability influencing the coral oxygen isotope. A singular spectrum analysis of five isotope records then showed that an oscillation with approximately 11-year period was present going back to 1650 CE, prompting Linsley et al. (2008) to name their IPO index the Interdecadal-Decadal Pacific Oscillation (IDPO) index. A multi-taper method spectrum of this Fiji–Tonga isotope-based IPO time series in Henley (2017) confirms the prominent decadal spectral peak which can be seen clearly in Figure 3.14 (Panel 8 from top). Linsley et al. (2008) also found that coral oxygen isotope $\delta^{18}O$ records from Maiana (Urban et al., 2000) and Palmyra (Cobb et al., 2001) islands on the northern side of the SPCZ rainfall axis and very near the Equator also, showed decadal variability at a similar period as the Fiji–Tonga records, but with the opposite phase. This behavior

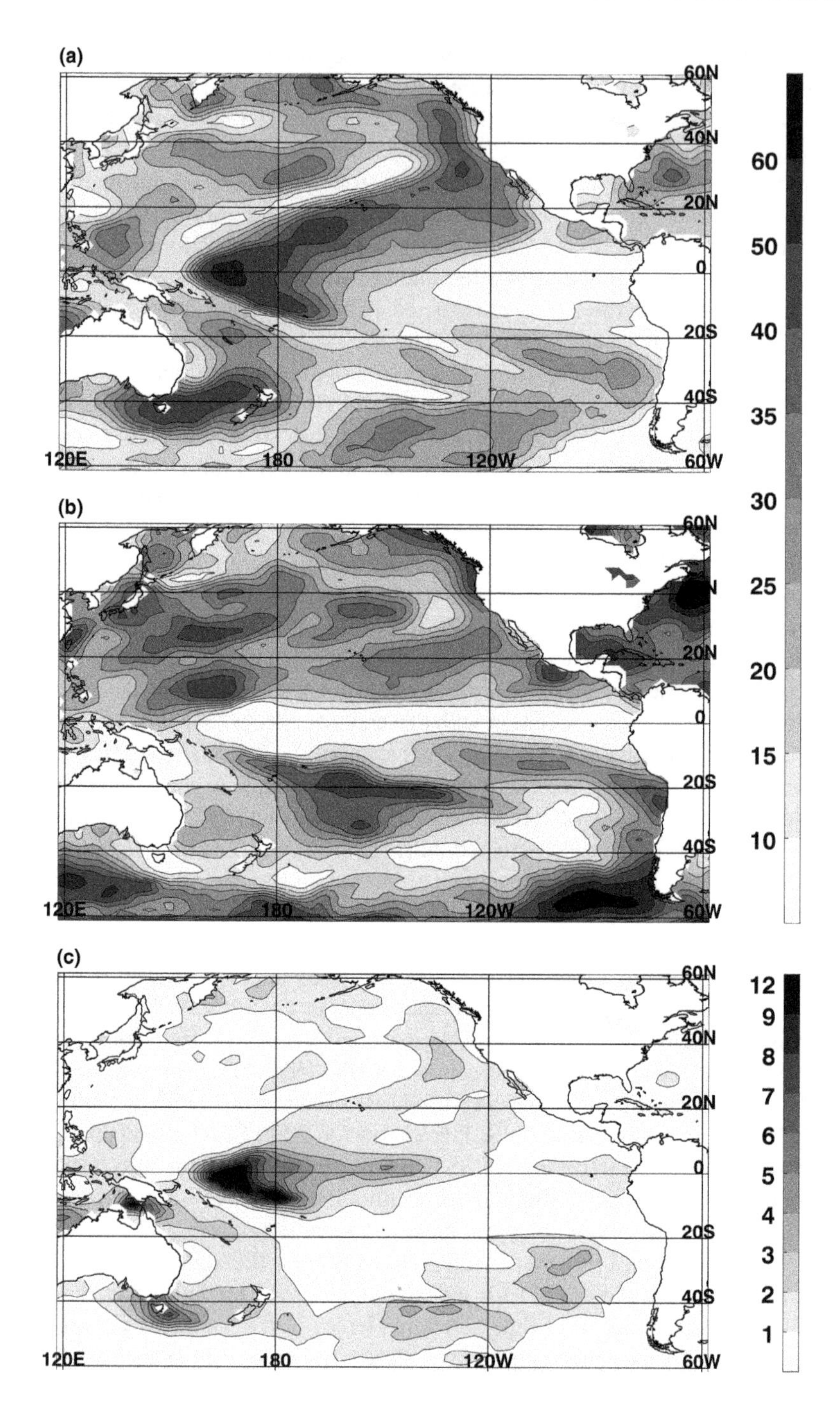

FIGURE 3.15 Fractional variances (as a percent of total variance) in (a) the 8 to 13-year period band, (b) the 14 to 68-year period band, and (c) the ratio of fractional variances in the 8 to 13-year period band and the 14 to 68-year period band.

is consistent with variability of the salinity fronts near the Fiji–Tonga islands in the SPCZ region and near the Maiana–Palmyra islands closer to the Equator as the former is in the cooler part of the IPO pattern and the latter is in the warmer part, with consequent opposite phase rainfall variability and east–west motions of salinity fronts near the two groups of islands. Thus, this IPO index reconstruction confirms the instrument measurement-based tropical Pacific decadal SST variability described in Section 3.2.2 and extends it back to 1650 CE. This paleoclimate proxy-based indication of regular decadal variability in the SPCZ region for over 350 years is consistent with the instrument data-based indications described in Section 3.2.2. All of them are important in understanding mechanisms of tropical Pacific decadal variability as we will see in Section 3.4.2.

3.4 HYPOTHESIZED MECHANISMS

Many mechanisms have been hypothesized for the PDO and the IPO. As we will see later in this section, uncoupled ocean models and global coupled ocean–atmosphere models simulate, to various extents, these two patterns and the timescales at which they vary. There is still much confusion about what these patterns represent and whether they are individual, coherent entities or composites of several entities. Unless this confusion is cleared and distinct targets for simulation and prediction are identified, progress in understanding and prediction of the Pacific decadal–multidecadal variability will not be achieved.

The origin of the confusion lies in the primary analysis technique used to isolate the space–time patterns of variability. The most popular technique is the EOF-PC analysis. This technique produces space–time patterns which contain maximum variance in the smallest number of patterns. It is a statistical technique without any dynamical or thermodynamical constraints. In the case of SST variability, while the patterns that are produced by this technique may be relevant for global atmospheric and climate variability, the lack of any dynamical or thermodynamical constraint is a handicap and produces misleading consequences if the patterns are assumed to be targets for simulation and prediction in their entirety. Sometimes, localization of time or space scales can reveal patterns which are closer to dynamical/thermodynamical modes of variability. These problems afflict understanding of decadal–multidecadal variability not only in the Pacific region, but also in the Atlantic as we will see in the next chapter. Therefore, in this section, an attempt is first made to isolate regions of high decadal and multidecadal variances, and then, mechanisms that appear to generate decadal and multidecadal variability are described.

3.4.1 PDO, IPO, What Is in the O?

As mentioned in the introduction (Section 3.1), ever since Sir Gilbert Walker's discovery of the SO, the NPO, and the NAO, atmospheric and oceanic variability patterns have been named oscillations. The PDO, the IPO, the SPDO, and some others are the more recent such names, which are given to space–time patterns revealed by EOF-PC analysis. But, what is in the O? This question is addressed here in the context of the Pacific decadal–multidecadal variability with a basic analysis of SST variances.

Figure 3.15 shows fractional SST variances in the decadal (8 to 13 years) band, in the multidecadal (14 to 68 years) band, and the ratio of the two in ERSST version 5 (Huang et al., 2017) time series from 1951 to 2018 CE. The decadal band is separated from the rest of the variability timescales because spectra of paleoclimate proxy data and instrument data show distinct peaks in the decadal band as described in Sections 3.2 and 3.3. Figure 3.15a shows that the fractional SST variance (ratio of decadal band only variance and total variance) is up to 60% in the equatorial western Pacific region. The relatively high fractional variance extends from there to the west coast of North America. There are also small areas of up to 40% fractional variances at approximately 30°N and 50°S near 180° longitude or just to the east of it. Other areas of relatively high fractional variance in the Indo-Pacific region are off the Philippine coast, off northeast Australia, and between Australia and New Zealand. It is noteworthy that there is approximately 10% decadal band variance in the eastern tropical Pacific region where the fractional variance in the multidecadal band is also approximately 10% as Figure 3.15b shows. In this band, there is approximately 40% to 50% fractional variance in a few areas in tropical and mid-latitude Pacific regions. The multidecadal fractional variance is more than 70% around the Southern Oceans, which may be due to data quantity and quality problems. The mid-latitude North Atlantic and the western Arctic have 50% to 70% multidecadal fractional variance. It is noteworthy that the Mediterranean Sea also has more than 70% fractional variance in the multidecadal band.

The ratio of the two bands (Figure 3.15c) shows that decadal band variance is several times larger than the multidecadal band variance in the tropical Indo-Pacific region, particularly in the tropical central and western Pacific region, and in the Bay of Bengal and the Arabian Sea. Decadal band variance is also several times larger than multidecadal band variance off the west coast of the USA, in the subtropical and mid-latitude South Pacific, and in equatorial Atlantic and tropical–subtropical South Atlantic. Multidecadal band variance is generally larger than decadal band variance in mid- and high-latitude regions in all oceans. Thus, decadal variability and its strength with respect to that of multidecadal variability are especially strong in the tropical central and western Pacific region. This localization of decadal signal in the tropical Pacific merits focusing on this region to understand decadal variability and the pan-Pacific region for the decadal–multidecadal variability. It is possible that the decadal timescale in the pan-Pacific pattern originates in the tropical Pacific region. The next two sub-sections describe mechanisms of tropical Pacific decadal variability and pan-Pacific decadal–multidecadal variability.

3.4.2 Tropical Pacific Decadal Variability

3.4.2.1 Role of Interactions between Meridional Ocean Circulations and the Atmosphere

We saw in Section 3.2.1 that Lohmann and Latif (2005) found tropical Pacific decadal SST variability in observed data. We also saw that Wang and Mehta (2008)'s extensive analyses of model-assimilated ocean and atmosphere data implicated interactions among SSTs, atmospheric Hadley and Walker circulations, and STCs in the tropical–subtropical Pacific Ocean in generating the tropical Pacific decadal variability. We will now see how such interactions can occur in models, and will also see comparisons of model-simulated and observed variability.

Decadal variability of the tropical Pacific climate was analyzed and modeled in several studies. Mehta (1991, 1992) hypothesized and showed with idealized, uncoupled, and coupled ocean–atmosphere models in the meridional–vertical plane that STC-like ocean circulations can generate decadal variability in tropical climate due to changes in meridional heat transport associated with them, with circulation speeds in the STC providing the decadal timescale. Changes in heat transports due to anomalous meridional velocity acting on mean meridional temperature gradient ($v'T_y$) and mean meridional velocity acting on anomalous meridional temperature gradient (VT'_y) played important roles; here, v' is the anomalous meridional velocity, T_y is the latitudinal gradient of mean temperature, V is the mean meridional velocity, and T'_y is the meridional gradient of anomalous temperature. In this idealized framework, ocean–atmosphere coupling provided non-zero growth rates of tropical decadal modes. McCreary and Lu (1994) described the dynamics and thermodynamics of STCs in a three-dimensional framework. They found that the upwelling water in the tropical region is remotely forced by wind stress in the subtropics, and that equatorial upwelling and undercurrent are forced by equatorial winds. They also found that western boundary currents play an important role in the upwelling branch of the STC. Lu and McCreary (1995) found that the InterTropical Convergence Zone (ITCZ) acts as a barrier to equatorward oceanic flow from the sub-tropics such that the water has to first flow to the western boundary poleward of the ITCZ and then towards the Equator in the boundary current. Finally, they also found that there is a convergence of equatorward flowing water in the central tropical Pacific Ocean associated with circulation cells confined to within 5° latitude from the Equator. Thus, by the mid-1990s CE, the theory stage was set for further exploration and hypothesis testing by ocean general circulation models and global coupled ocean–atmosphere models. A further impetus was provided by the increasing availability of multidecade-long time series of instrument-measured SSTs and other data, and their assimilation in ocean and atmosphere models as we saw in Section 3.2.1.

Beginning in late 1990s CE, there were several studies with idealized ocean models or ocean general circulation models to estimate contributions to tropical Pacific SST changes due to the $v'T_y$ (Kleeman et al., 1999; Klinger et al., 2002; Nonaka et al., 2002; Solomon et al., 2003) and VT'_y (Giese et al., 2002) processes. A study with a coupled ocean–atmosphere model (Merryfield and Boer, 2005) investigated the relative roles of these two processes and found that the $v'T_y$ process was much more important. Yang et al. (2004) found that both processes associated with STCs in an ocean general circulation model contributed equally. Then, Lohmann and Latif (2005) addressed mechanisms of the observed decadal SST variability in central and western tropical Pacific with an ocean general circulation model forced with winds from the NCEP-NCAR reanalysis from 1948 to 2001 CE, and with a free-running, global coupled ocean–atmosphere model. Details of both sets of results are described and discussed by Lohmann and Latif (2005), so only highlights of relationships between decadal SST variability in the central and western equatorial Pacific region and meridional ocean circulations are described here. In experiments with the ocean-only model forced by the NCEP-NCAR reanalysis data, they found that anomalous mass transported by tropical cells between 5°N and 5°S latitudes in the upper 250 m in the Pacific Ocean had a generally opposite phase compared to SST anomalies such that SST anomalies were cooler (warmer) when the tropical cells were stronger (weaker) confirming that the $v'T_y$ process played a very important role (Figure 1 in Lohmann and Latif (2005)). The tropical cell strength led the SST anomaly by a few months, implying that variability of the tropical circulations was driving the SST variability. Strengths of STCs, whose ambit was defined in this study as poleward of 10° latitude, were also found to vary inversely with the decadal SST anomalies but leading by 15 months. In general, the lag between meridional circulation changes and SST changes was found to increase with increasing latitude up to 15°, beyond which the correlation between the two became insignificant. In their free-running global coupled model, Lohmann and Latif (2005) also found that central and western equatorial Pacific SST varied with an opposite phase to that of strengths of tropical cells and STCs (Figure 9 in Lohmann and Latif (2005)).

In a more recent study, Farneti et al. (2014) simulated the tropical–subtropical decadal variability in Pacific climate with uncoupled, three-dimensional, ocean and atmosphere general circulation models. They found that tropical Pacific SST anomalies generated anomalous Hadley circulations which, in turn, changed subtropical wind anomalies which, when used to force their ocean model, generated changes in subtropical gyres and STCs such that they would provide a negative feedback to the original tropical SST anomalies due to changes in meridional heat transport. Farneti et al. (2014) also found associated subtropical Rossby waves in their ocean model experiments forced by subtropical wind anomalies. This mechanistic description of tropical–subtropical Pacific decadal variability in uncoupled atmosphere and ocean models is generally consistent with the Wang and Mehta (2008) description of decadal variability in SST, surface and sub-surface ocean currents, temperatures, and STCs; and in Hadley circulations and subtropical winds in model-assimilated observed data in Section 3.2.1. This mechanistic description is also generally consistent with Luo and Yamagata (2001)'s description of tropical Pacific decadal variability in Section 3.2.2. Thus, while details of mechanisms might differ among various types of imperfect models and their simulations might differ with imperfect observations, the hypothesized role of meridional ocean circulations and their interactions with atmospheric circulations in tropical Pacific decadal variability appears to be confirmed.

Two last notes in this story: One, ocean–atmosphere interactions in the ocean and coupled model simulations described here were only dynamical. That is, only the dynamical response of the atmosphere was included. However, as we saw in Chapter 2, the tropical–subtropical wind anomalies can also give rise to increased evaporation. This increased water vapor can be brought into the tropical precipitation zone by the return branch of the Hadley circulation and provide moisture for increased precipitation which can strengthen the Hadley circulation due to increased latent heat release. Thus, this chain of positive feedbacks can make the total (dynamical and thermodynamical) response of the atmosphere to decadal SST anomalies stronger as described in Chapter 2. The second note to this story is about the driving force(s) of the STCs. In the story narrated above, only wind stress forcing in the tropics and subtropics was included. However, we saw in Section 3.3 that variability of heavy precipitation and salinity fronts in the SPCZ-WPWP region are also associated with the tropical Pacific decadal variability. Folland et al. (2002) showed that the IPO influences the position of the SPCZ which is also shown by Linsley et al. (2008)'s coral oxygen isotope data. In this region, variability of tropical cells and STCs appears to be substantial as we saw in observational and modeling studies by Lohmann and Latif (2005), Wang and Mehta (2008), and Farneti et al. (2014). So, the involvement of buoyancy-driven changes in these meridional ocean circulations, especially in the South Pacific, as an integral part of the tropical Pacific decadal variability cannot be ruled out.

3.4.2.2 Roles of Off-Equatorial Rossby Waves and the Schwabe Cycle as a Pace Maker

Ever since off-equatorial, oceanic, baroclinic Rossby waves were implicated as a component of the so-called delayed-action oscillator theory of El Niño–La Niña events (see, for example, Schopf and Suarez (1988, 1990), Suarez and Schopf (1988), and Battisti and Hirst (1989)), the possibility that such waves may be involved in decadal–multidecadal Pacific Ocean and climate variability has been at the center of hypotheses about such variability. At the heart of this possibility is the property of Rossby waves that the farther they are excited from the Equator, the slower they are and the longer they take to reach the western boundary (the coast of Asia). We will now see how off-equatorial, oceanic, baroclinic Rossby waves can play a role in decadal, tropical Pacific variability.

McCreary (1983) was perhaps the first to find that off-equatorial, oceanic Rossby waves were one of the crucial components of self-sustained, multiyear, and longer timescale oscillations of the tropical Pacific climate system. Based on his simulations of ENSO-like variability with a simplistic ocean–atmosphere model, he identified the slow propagation of off-equatorial Rossby waves across the Pacific Ocean as a key oceanic process in ENSO-like variability. He prophesized that this process (and the atmospheric Walker circulation) would also be involved in the dynamics of future, more sophisticated models of coupled ocean–atmosphere interactions. Following the earlier-cited studies on the delayed-action oscillator mechanism, Kirtman (1997) systematically explored McCreary (1983)'s Rossby wave hypothesis with an idealized ocean–atmosphere model. While the other studies had focused on equatorial Rossby waves generated by equatorial zonal wind anomalies, Kirtman (1997) expanded the zonal wind field meridionally beyond ±7° latitude and found that oceanic Rossby waves generated by this meridionally broader wind pattern were slower and took longer to reach the western boundary (nominally, the coast of Asia), consequently increasing the period of ENSO-like oscillations in the model.

As global coupled ocean–atmosphere models started to simulate increasingly realistic climate and century-long and longer experiments with such models to study impacts of increasing greenhouse gases became more frequent in the mid-1990s, analyses of such model experiments to compare natural climate variability generated by the models with the observed variability became increasingly frequent. Knutson and Manabe (1998) was one of the first such studies about decadal variability of tropical Pacific climate. In the R30 version of the Geophysical Fluid

Dynamics Laboratory (GFDL) global coupled ocean–atmosphere model, they found that the leading SST pattern of decadal variability resembled the observed counterpart in the Global sea Ice and Sea-Surface Temperature version 2 (GISST2; Rayner et al., 1996) data set in its spatial pattern and amplitude to some extent, with a preferred oscillation period of 12 years. However, there was an important difference between the observed and simulated spatial patterns; their observed pattern had maximum amplitude in the eastern equatorial Pacific region, whereas the simulated pattern had maximum amplitude around 180° longitude. This difference may be either due to the imperfect observed data set or due to the inclusion of mid-latitude, North and South Pacific in the analysis. As described in Section 3.2.1, later analyses by Lohmann and Latif (2005) of Hadley Centre Sea Ice and SST ((HadISST1); Rayner et al., 2003), and by Wang and Mehta (2008) (ERSST version 2; Smith and Reynolds, 2004) showed that the decadal SST variability in the tropical Pacific had maximum amplitude in western and central Pacific, placing the Knutson and Manabe (1998) simulation closer to the later analyses. They also found that oceanic heat content anomalies propagated westward at approximately 12°N latitude rather than at 9°N latitude like the interannual ENSO-like variability in their model. Knutson and Manabe (1998) attributed this internally generated decadal variability to a delayed-action oscillator mechanism involving off-equatorial oceanic Rossby waves. This result was consistent also with the Kirtman (1997) study of propagation speeds of off-equatorial Rossby waves in response to meridionally broader zonal wind patterns. However, it is possible that, in view of the position of maximum SST amplitude in western and central equatorial Pacific, STCs also played an important role in the Knutson and Manabe (1998)'s simulation of tropical Pacific decadal variability.

So what is special about latitudes around 12°? Capotondi et al. (2003) investigated why Rossby wave variability is maximum at 10°S and 13°N in the Pacific Ocean in a diagnostic study with an ocean general circulation model forced with observed fluxes of momentum, heat, and freshwater from the NCEP-NCAR reanalysis from 1958 to 1997 CE. They found that anomalous, zonally coherent, Ekman forcing at these two latitudes in the Pacific, especially at the decadal timescale, is responsible for generating first baroclinic mode Rossby waves which propagate westward to the western boundary. They also found large thermocline displacements and meridional transport variations at these two latitudes. The westward-propagating heat content anomalies associated with baroclinic Rossby waves at 12°N, which Knutson and Manabe (1998) found in their global coupled model, may have been generated by the same process as in the Capotondi et al. (2003) ocean general circulation model and a Rossby wave model. In the same ocean general circulation model experiments, Capotondi et al. (2005) analyzed STC variability and Rossby wave activity in the tropical Pacific. They found that westward-propagating, baroclinic Rossby waves at the two preferred latitudes appear to be playing an important role in adjustments of interior-region equatorward transport; western boundary current equatorward transport; and equatorial SSTs. This may be a very important link between the roles of STCs described earlier (Lohmann and Latif, 2005; Wang and Mehta, 2008; and Farnetti et al., 2014) and Rossby waves described here. There are at least three possible roles of westward-propagating, baroclinic Rossby waves around 12°-13°N latitudes and, possibly, around 10°S latitude also: (1) Rossby waves may be very important in conveying information from the interior tropical Pacific to the western boundary; (2) they may be a very important component of the delayed-action oscillator mechanism proposed for naturally generated, tropical Pacific decadal variability by Knutson and Manabe (1998); and (3) they may be a very important component of the delayed-action oscillator mechanism proposed for Schwabe cycle forced, tropical Pacific decadal variability by White and Liu (2008a, b) described in Chapter 2. It is safe to conclude from the foregoing that the mechanism(s) of tropical Pacific decadal variability are within the tropical ocean–atmosphere system. It is still an open question whether this variability is excited by stochastic forcing or by the Schwabe cycle. Perhaps, in the future, as global coupled models continue to evolve and reproduce bottom-up and top-down processes described in Chapter 2 more realistically, this question will be answered unambiguously.

Two other aspects of the tropical Pacific decadal variability should be mentioned here in closing. First, low-latitude volcanic eruptions of moderate to extreme explosivity are known to influence, even change the phase of, tropical Pacific decadal variability as described in Section 3.4.4; and second, potential predictability of this variability appears to be significantly high as described in Chapter 11.

3.4.3 Pan-Pacific Decadal–Multidecadal Variability

We will now see various hypotheses proposed for pan-Pacific variability, and comparisons of model simulations and observed variability. Specifically, we will see how ocean and atmosphere processes in the pan-Pacific region can generate decadal–multidecadal variability. The focus will be on the PDO as defined for the North Pacific region and the IPO as defined for the entire Pacific region; however, as Folland et al. (2002) and Henley (2017) showed, these are essentially the same phenomena. There are several, very comprehensive reviews (Liu, 2012; Newman et al., 2016; Henley, 2017; Henley et al., 2017; and references therein) of the possible physics of the pan-Pacific variability, so only overviews and salient points are described here. The reader is referred to cited papers for mathematical details. Then, in the last part of this section, we will see how solar variability can act as a pace maker and synchronizer of pan-Pacific decadal–multidecadal variability.

3.4.3.1 Tropical Pacific Origin of Pan-Pacific Decadal Variability

As we saw in the previous sub-section, there is a reasonably convincing case for tropical origin of the tropical Pacific decadal variability. Numerous studies (many of them cited in this chapter) have shown that SST anomalies in the tropical Pacific can generate heat and momentum fluxes in mid-latitude Pacific regions via Rossby waves and anomalous Hadley circulations, and oceanic Ekman transports which can then cause SST anomalies in those regions. So, it is very plausible that the decadal component of the pan-Pacific decadal–multidecadal variability is due to the tropical Pacific decadal variability.

3.4.3.2 Multidecadal Variability in the Pan-Pacific Region

As we have seen so far, oceanic Rossby waves' pace maker role in setting the variability timescale is at the heart of (almost) all hypotheses regarding tropical Pacific and pan-Pacific decadal–multidecadal variability. If we say that oceanic Rossby waves are the heart, then integration of atmospheric stochastic forcing by the upper ocean is the lung of the variability which produces low-frequency variability. Then, there are the possible ocean–atmosphere feedbacks which regulate the variability and may generate a preferred timescale, if any, of the variability.

It is widely believed (Liu, 2012; Newman et al., 2016; and references therein) that the PDO SST pattern and associated time series are the result of a combination of different physical processes. The processes believed to be responsible for tropical Pacific decadal variability are described in Section 3.4.2. Three major groups of processes in the mid-latitude Pacific ocean–atmosphere system have been hypothesized for pan-Pacific decadal–multidecadal variability (Liu, 2012; Newman et al., 2016).

1. *Oceanic integration of atmospheric stochastic forcings*: Ocean surface heat fluxes related to atmospheric circulation around the Aleutian low-pressure center, forced by local weather noise and remote forcing from the tropics via atmospheric Rossby waves (Hasselmann, 1976; Frankignoul and Hasselmann, 1977; Frankignoul and Reynolds, 1983; Pierce, 2001; Alexander, 2010), and oceanic Ekman transport due to the same forcings (Miller et al., 1994; Alexander and Scott, 2008).
2. *Year-to-year oceanic "memory"*: Surface temperature anomalies can mix deep into the ocean in the Northern Hemisphere winter, would be insulated from the atmosphere during the following spring and summer, and re-emerge towards the surface in the following autumn (Alexander and Deser, 1995; Alexander et al., 1999, 2001; and others).
3. *Coupled ocean–atmosphere dynamics*: Changes in the Kuroshio–Oyashio ocean current system near the Asian coast, forced by winds over the North Pacific driving oceanic Rossby waves, manifesting as SST anomalies along the sub-arctic front at approximately 40°N.

We will now see how these hypothesized processes operating primarily in the tropical–subtropical Pacific region, which then affect mid-latitude Pacific Ocean via the atmosphere and feed back to the tropical–subtropical Pacific region to generate variability primarily at longer than decadal timescales. The tropical processes are the same as described in the above sub-section, except that the oceanic Rossby waves, which again act as the heart or the pace maker, are hypothesized to occur just inside or just outside the limit of the Tropic of Cancer and the Tropic of Capricorn in the North and South Pacific, respectively. The other major players in generating this rich symphony are the Bjerknes feedback among the equatorial thermocline depth-SST-trade winds in eastern Pacific; equatorial and coastal Kelvin waves, STCs, atmospheric Rossby waves, and western boundary currents in the Northern and Southern Hemispheres; and the Aleutian and South Pacific low-pressure systems and associated winds. Meehl and Hu (2006) proposed a grand hypothesis about how these players can some time come together and work cooperatively to produce the

IPO. This hypothesis, based on a 1360-year climate simulation with a global coupled ocean–atmosphere model, the Parallel Climate Model, is described here. Let us see how the IPO evolves from its positive phase (Figure 3.16a), and goes through to the negative phase (Figure 3.16b) and then again to the positive phase in this model.

The tropical Pacific SST anomalies are positive (warmer) in the positive phase of the IPO, with positive rainfall and convective heating anomalies in the tropical Pacific atmosphere. The positive convective heating anomalies excite an atmospheric Rossby wave response which intensifies the Aleutian and South Pacific low-pressure systems from where westerly surface wind stress anomalies extend to 20°N and 25°S. Due to weak tropical SST anomaly gradients, trade winds in both Hemispheres are weaker than average. As a result, the wind-driven STCs are also weaker than average. The wind forcing, supplemented by wave reflections from the equatorial eastern Pacific, produces positive upper-ocean layer thickness anomalies at 20°N and 25°S which begin to propagate westward as downwelling, oceanic Rossby waves. The Pacific Ocean–atmosphere condition, as it would be at this time, is shown in Figure 3.16a.

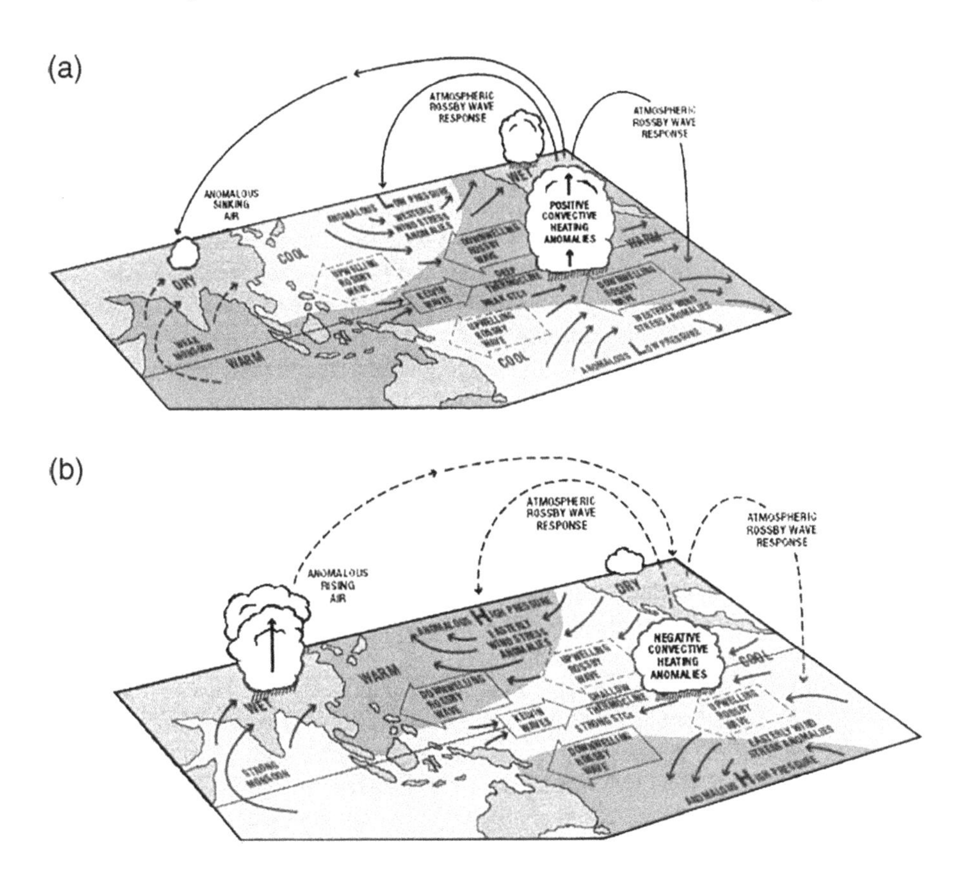

FIGURE 3.16 Schematic diagram of the coupled Pacific Ocean–atmosphere mechanisms which produce the IPO in the Parallel Climate Model. (a) The positive IPO phase and (b) the negative IPO phase. (Adapted from Meehl and Hu, 2006 © American Meteorological Society. Used with permission.)

In the meantime, negative layer thickness anomalies from the previous negative IPO phase arrive at the western boundary and propagate towards the Equator as upwelling Kelvin waves, eventually shallowing the equatorial thermocline. As a consequence of the shallower thermocline, eastern equatorial Pacific SSTs cool and trade winds in both Hemispheres strengthen, spinning up the STCs. The stronger STCs shallow the thermocline even more, and eastern equatorial Pacific SSTs cool further.

While the tropical Pacific climate system is adjusting to this interplay, the atmospheric Rossby wave response to cooler eastern equatorial Pacific SSTs weakens the Aleutian and South Pacific low-pressure systems. Consequently, easterly wind stress anomalies near the equatorward edges of these low-pressure systems reach to 20°N and 25°S. These easterly wind anomalies at the edges of the tropics weaken trade winds in both Hemispheres and excite negative layer thickness anomalies which begin to propagate westward as upwelling, oceanic Rossby waves. The Pacific Ocean–atmosphere condition, as it would be at this time, is shown in Figure 3.16b. As various instruments in an orchestra go up and down in volume and tones (or, cease to play altogether), these various components of the ocean–atmosphere system also continue to play their parts as the IPO evolves. While the upwelling, oceanic Rossby waves start their westward travel, positive layer thickness anomalies from the previous positive IPO phase arrive at the western boundary, and propagate towards the Equator and transform themselves to downwelling Kelvin waves which deepen the eastern equatorial Pacific thermocline. While this deepening happens to its deepest, eastern equatorial Pacific SSTs warm; consequently, trade winds weaken in both Hemispheres, causing STCs to spin down and warming the SSTs further. Thus, the IPO arrives at the phase where it started from and the next cycle begins.

Meehl and Hu (2006) found that such IPO cycles can have irregular, 15 to 30-year periods. Among the possible causes of this "spectral broadening" are differences in the geometries of North and South Pacific Oceans, and their respective western coastlines. Differences in first baroclinic mode Rossby wave propagation times in the two Hemispheres are due to these geometry differences, 9 to 13 years at 20°N latitude and 15 to 21 years at 25°S. Therefore, at some times in the IPO life cycles, contributions from the North and South Pacific regions create constructive interferences to produce a sizeable SST signal which can make substantial impacts on global climate as Meehl and Hu (2006) showed for South Asia and Southwest USA. At other times in the life cycles, when constructive contributions from North and South Pacific are partially synchronized or not synchronized at all, global climate impacts would be milder or even insignificant.

The ability of other global coupled models to simulate IPO-like variability and its mechanisms was investigated by Henley et al. (2017). They analyzed data preindustrial control and all-forcings (historical) experiments with 39 global coupled models in the Coupled Model Intercomparison Project version 5 (CMIP5) (Taylor et al., 2012). The models appear to simulate the overall IPO spatial pattern reasonably well; the timescales of variability (duration in positive and negative phases) are shorter in most of the models compared to the observed timescales. It must be emphasized, however, that the observed IPO's timescales are based on a relatively short record. Henley et al. (2017) also found that the extratropical signature of the simulated IPO is considerably weaker compared to the observed, which may be due to a weaker coupling of the tropical Pacific to extratropical Pacific via the atmosphere and the STCs. Again, a quantitative comparison between a relatively long-period observed and modeled phenomena is fraught with risk due to the relatively short observational record. Finally, it was shown that the IPO-like variability in a small subset of pre-industrial control simulations with the 39 models contains the ratio of decadal to total variance within 20% of the observed ratio and a generally higher simulation skill of spatial patterns, confirming that the IPO can be produced by processes within the climate system. However, considering that the IPO in instrument-measured SST data beginning in early 20th century CE (Figures 3.3a and b) seems to have occurred at decadal–multidecadal timescales with a reasonable hemispheric symmetry (the caution about relatively short observational record apply!), it is intriguing to think that perhaps there is a pan-Pacific pace maker and synchronizer which governs the intrinsically generated IPO behavior. We will see further exploration of this possibility in the next sub-section.

3.4.3.3 Solar Variability as a Pace Maker and an Inter-hemispheric Synchronizer?

In this sub-section, an intriguing hypothesis about a possible solar variability role in acting as a pace maker and inter-hemispheric synchronizer of the IPO is described. But first, we revisit hypothesized physics, as described in Chapter 2, of how the solar Schwabe cycle can influence DCV. Then, an extension to multidecadal climate variability is outlined.

In Chapter 2, we saw preliminary testing of the bottom-up and top-down hypotheses of the Schwabe cycle forcing DCV. In these hypotheses, the bottom-up mechanisms would amplify climate effects of visible solar radiation signal at the ocean surface due to positive feedbacks in the tropical–subtropical ocean–atmosphere system and the top-down mechanisms would amplify climate effects of ultraviolet solar radiation signal in the stratosphere due to positive feedbacks via ozone, meridional temperature gradients, stratospheric circulations and waves, and Hadley and Walker circulations. We also saw in Chapter 2 that the combined effects of the two sets of mechanisms in a global ocean–atmosphere model incorporating stratospheric processes can produce a significant, but weaker than observed, DCV signal. In the bottom-up hypothesis, increased solar visible radiation near maxima of the Schwabe cycle would increase evaporation in the sub-tropics and eventually increase precipitation in the tropical region by increased water vapor transport by trade winds from the sub-tropics. This increased precipitation would then increase latent heat release and drive a stronger Hadley circulation which would reinforce increased evaporation, and the response to solar forcing would be amplified. The stronger trade winds resulting from this positive feedback process would shallow the thermocline in eastern equatorial Pacific and cool SSTs due to the Bjerknes feedback around the Schwabe cycle maxima. Results of another set of model experiments were also described in Chapter 2, showing that the Schwabe cycle can act as the pace maker of DCV via its excitation of off-equatorial oceanic Rossby waves whose cross-Pacific travel would provide negative feedback and, coupled with the positive Bjerknes feedback in the equatorial Pacific, would fit the delayed-action oscillator paradigm.

Combining these results and extending to multidecadal timescales, a possible hypothesis about multidecadal solar variability acting as a pace maker and inter-hemispheric synchronizer can be formulated. In Chapter 2, Figure 2.8 shows variability of the Schwabe cycle length and the envelope of the Schwabe cycle, both varying at decadal–multidecadal timescales. Comparing variability of these two Schwabe cycle parameters with global-average SST, Weng (2005) argued that the latter appears to be influenced by the former. Mehta and Lau (1997) found that the multidecadal Niño3 SST anomaly was generally negative (positive) when reconstructed solar irradiance was positive (negative) in late 19th and 20th century CE. Thus, it is plausible that decadal–multidecadal solar variability acts on the Earth's climate via the bottom-up and top-down sets of mechanisms to force the IPO, an intrinsic mode of variability as found by Meehl and Hu (2006). Since solar radiation reaching the Earth is symmetric, in an annual-average sense, with respect to the Equator, a pace maker role of multidecadal solar variability would explain the IPO's more organized and more frequent symmetry in the North and South Pacific. Of course, given the inertia of the Pacific Ocean–atmosphere system and given this system's demonstrated (within the framework of global coupled models as described in the previous sub-section) propensity to intrinsically generate IPO-like variability, an exact match between variability of multidecadal solar forcings and the IPO should not be expected. This is an intriguing hypothesis which can be tested in a global coupled ocean–atmosphere model with stratospheric processes.

3.4.4 Effects of Low-Latitude Volcanic Eruptions

Low-latitude volcanic eruptions in Central and South America, islands in the Caribbean Sea, Philippines, and islands in the Maritime Continent inject sulfur dioxide, ash, carbon dioxide, hydrochloric acid, hydrogen fluoride, and other materials into the troposphere. Eruptions of high Volcanic Explosivity Index (VEI; Newhall and Self, 1982) can inject these materials into the upper troposphere and lower stratosphere, which then stay for months or years before they are removed by dynamical and precipitation processes. During their residence in the atmosphere, these materials reduce incoming, visible solar radiation at the Earth's surface. This reduction can influence surface energy balance and surface temperature. Materials from high VEI eruptions at low latitudes are also transported by easterly winds and can influence surface temperature thousands of kilometers downstream from an eruption location. Local and distant effects of an eruption can influence decadal variability of SSTs and sub-surface ocean temperatures, especially if the eruption is of moderate to high explosivity.

Since the 1880s CE, there have been at least seven eruptions of VEI 4 or higher. These are the Krakatoa, Indonesia, event in June–August 1883 CE (VEI 6); the Santa Maria, Guatemala, event in October 1902 CE (VEI 6); the Mount Pinatubo, Philippines, event in June 1991 CE (VEI 6); the Volcán de Colima, Mexico, event in January 1912–1913 CE (VEI 5); the Mount Agung (Bali), Indonesia, event in February–May 1963 CE (VEI 5); the El Chichón (Chiapas), Mexico, event in March–April 1982 CE (VEI 5); and the Volcán de Fuego (Guatemala) event in 1974–1975 CE (VEI 4). Observed and simulated effects of these eruptions on the PDO and the WPWP SST variability are briefly described in this section.

3.4.4.1 The Pacific Decadal Oscillation

In one of the first studies of its kind, Mehta et al. (2018) found that the time series of estimated aerosol optical depths (AODs[7]) due to these eruptions clearly showed each of the VEI 4 and stronger eruptions. A comparison of the AOD time series and the observed PDO index time series showed a substantial decrease in the index from one to a few months after each eruption. The tropical–subtropical Pacific SSTs cooled so much after the 1883 CE (Krakatoa, Indonesia), 1902 CE (Santa Maria, Guatemala), 1912–1913 CE (Volcán de Colima, Mexico), 1982 CE (El Chichón, Mexico), and 1991 CE (Mount Pinatubo, Philippines) eruptions that the observed PDO index changed sign from positive to negative for several months to a year or longer after each eruption. To verify this empirical observation and to understand the physics of how volcanic eruptions impact the PDO, Mehta et al. (2018) analyzed net surface heat flux and upper ocean temperature data from ensembles of experiments with four ESMs participating in the CMIP5 project. These four ESMs are UK Meteorological Office HadCM3, Japanese MIROC5, US NCAR CCSM4, and US GFDL CM2.1. Composites of observed and ESM variables from 12 months before to 60 months after the eruption events led to insights into how and how much these events influenced the PDO evolution. Ensemble-average variables were averaged in tropical Pacific and mid-latitude North Pacific regions where the simulated PDO patterns show largest amplitudes. Such composites were also made of eruption events which began when the PDO was in negative phase and when it was in positive phase to gain further insights into the physics of eruption effects on PDO evolution. Due to different PDO states (signs and magnitudes) at the time of each eruption of the same VEI and natural variability of the PDO, the physics of the ESMs' responses appear clearer in individual events than in composites. It was also found in these analyses (Mehta et al., 2018) that CCSM4's simulated PDO's responses to volcanic eruptions are largest in magnitude and the ensemble-average CCSM4 PDO index is the closest to the observed PDO index among the four ESMs, perhaps due to 20% to 30% larger values of AODs used in CCSM4. Therefore, the evolutions of the observed and ensemble-average simulated PDO index; and ensemble-average, anomalous net surface heat flux, SST, and sub-surface potential temperatures during two VEI 6 events (1883 CE and 1991 CE) and two VEI 5 events (1912 CE and 1982 CE) in CCSM4 experiments were analyzed in detail.

In all four eruption events (Mehta et al., 2018), there was a 5 to 10 Wm^{-2} decrease in net surface heat flux in tropical Pacific and North Pacific a few months before or at the time of peak AOD. An anomalous reduction in downward shortwave radiation was the largest contributor to net surface heat flux anomalies due to an increased albedo associated with volcanic matter injected into the atmosphere. Then, as downward shortwave radiation slowly recovered after peak AOD, downward longwave radiation continued to decrease and latent heat flux also began to decrease—perhaps due to initial SST cooling. A substantial minimum in net surface heat flux occurred approximately 12 months after the AOD peak, largely due to minima in downward longwave radiation

[7] AOD is a measure of radiation attenuation due to aerosol particles in a column of the atmosphere from the Earth's surface to the top of the atmosphere. So, the larger the AOD, the larger the particle concentration.

and latent heat. Annual cycles of SST, surface winds, and atmospheric humidity and temperatures—in conjunction with decreasing downward longwave radiation and latent heat, and the remaining AOD in the atmosphere—perhaps were responsible for the substantial decrease in net surface heat flux in all four eruption events approximately 12 months after peak eruption. The ensemble-average PDO index in all four events felt effects of these net surface heat flux anomalies at 12-month intervals. The observed index was negative when the 1883 CE eruption of Krakatoa began. At the time of peak AOD, both the observed and model indices decreased slightly in response to the net surface heat flux decrease and then both indices increased and became positive perhaps due to internal ocean–atmosphere dynamics. The initial decrease and subsequent increase were evident in SST and 100 m deep potential temperature anomalies averaged in the tropical Pacific. Ocean potential temperatures in the North Pacific felt the heat flux decrease two months after the AOD peak and perhaps responded slower due to deeper mixing of the heat anomaly. This relative warming of the tropical Pacific with respect to North Pacific was reflected in the observed and model PDO indices becoming and staying positive for over a year after the AOD maximum. The 100 and 200 m depth potential temperature anomalies were negative when the eruption began and were increasing perhaps due to internal dynamics. Then, approximately four to six months after the AOD peak, they began to decrease. It is possible that the effect of these upper ocean thermodynamics influenced the SST (0 to 5 m temperature) 12 months after the AOD peak, combined with the second decrease in net surface heat flux 12 months after the first caused the SST to begin decreasing at this time. This decrease was 0.9°C in 24 months (36 months after the AOD peak) (Mehta et al., 2018). A few months after the tropical Pacific SSTs began to decrease, the North Pacific upper ocean temperatures began to recover from the cooling they had experienced due to net surface heat flux anomalies in the first 12 months after the AOD peak. The combination of the cooling upper ocean in the tropical Pacific and warming upper ocean in the North Pacific resulted in the model PDO index to decrease substantially over almost two years. Three years after the AOD peak, the tropical Pacific upper ocean potential temperatures began to recover from the cooling and mid-latitude North Pacific upper ocean potential temperatures began to cool again; this combination caused the model PDO index to increase to positive which happened in the observed index a few months later. The VEI 6 eruption event of Mount Pinatubo in 1991 CE also had generally similar effects on net surface heat flux, tropical Pacific and North Pacific upper ocean temperatures, and the simulated PDO index as the 1883 CE Krakatoa event. The VEI 5 events in 1912 CE (Volcán de Colima, Mexico) and 1982 CE (El Chichón, Mexico) made smaller magnitude impacts on net surface heat flux, tropical Pacific and North Pacific upper ocean temperatures, and the simulated PDO index than the VEI 6 events, but the similarity of the simulated impacts on the PDO index in both VEI 5 events and their general similarity with the two VEI 6 events were unmistakable. It is, therefore, not surprising that average composites of all VEI 6 events and all VEI 4 and VEI 5 events simulated by CCSM4 depicted generally similar evolutions of heat fluxes, temperatures, and PDO indices as in the individual cases as described here. Details of the physics in the other three ESMs are described by Mehta et al. (2018). Thus, as this detailed analysis shows, low-latitude volcanic eruptions of large magnitudes can have substantial effects on the evolution of the naturally occurring PDO.

3.4.4.2 The West Pacific Warm Pool Variability

Mehta et al. (2018) also found that a decrease in the observed WPWP index could be seen immediately or a few months after each of the major eruptions mentioned earlier. Effects of these eruptions were felt by the ensemble-average WPWP index simulated by all four ESMs to various extents. After removing overall linear trends from the observed and simulated WPWP time series, the association between major volcanic eruptions and decreases in the WPWP index became even clearer. Composites of AOD and observed WPWP SST anomaly; and anomalous net surface heat flux, SST, and sub-surface ocean temperatures from each of the four ESMs during three VEI 6 and four VEI 4 and 5 events (hereafter referred to as VEI 4 + 5) led to insights into how and how much these events influenced WPWP SST index.

These composites from 12 months before eruption to 60 months after eruption showed that, in the VEI 6 eruptions, the observed WPWP SST index began to decrease as the AOD began to build up in the atmosphere approximately six to eight months before the AOD peak. Although variability due to other dynamical forcings increased the WPWP SST index for a few months, the overall decreasing trend continued for 12 months after the AOD peak when the index reached a minimum (−0.28°C) and stayed there for eight months, subsequently increasing in the next 36 to 40 months (Mehta et al., 2018). The composite, observed WPWP SST index in the VEI 4 + 5 eruptions showed a general decrease 12 to 20 months after the AOD peak, but it was less (−0.1°C to −0.2°C) than the decrease for the VEI 6 eruptions; as for the VEI 6 eruptions, effects of other dynamical forcings on the WPWP SST index were also evident. All ESMs began responding to AOD increases as eruptions began and net surface heat flux anomalies decreased to a maximum negative value approximately four months before the AOD peak. The duration and magnitude of the flux decrease varied among ESMs with the maximum magnitude (−8 to −10 Wm^{-2}) in CCSM4 and maximum duration (12 to 16 months) in CM2.1 for the VEI 6 eruptions. The largest contribution to net surface heat flux anomalies was from decreasing downward shortwave radiation due to increased albedo as mentioned earlier. For the VEI 4 + 5 eruptions, the maximum magnitude of decrease (average −3 Wm^{-2}) was found in CCSM4, and maximum duration of the decrease (12 to 14 months) was found in CM2.1. SSTs and sub-surface ocean temperatures in the ESMs responded to various extents to these net surface heat flux changes. SST decreases began as the net surface heat flux began to decrease, but the minimum SSTs were reached four to eight months after AOD peaks, except in MIROC5 the minimum SST was almost at the time of AOD peak. The SST decreases in VEI 6 eruptions ranged from −0.2°C in HadCM3 to −0.55°C in CCSM4. Sub-surface temperature anomaly composites for VEI 6 eruptions, averaged in the WPWP region, showed that the volcanic signal was mixed substantially to approximately 200 m depth and then decreased by one or more orders of magnitude at lower depths. This depth of signal penetration was consistent with 150 to 200 m mixing depths estimated from anomalous net surface heat flux and time rate of change of temperature. In VEI 4 + 5 eruptions, the depth of substantial penetration of the volcanic signal was the same as in VEI 6 eruptions, but the temperature anomalies were smaller. Thus, Mehta et al. (2018) concluded that vertical mixing of heat, forced by reduced net surface heat flux, seemed to be the primary driver of WPWP SST response during and after volcanic eruptions.

The delay and longer duration of temperature anomalies than those of net surface heat flux were due to vertical mixing of the heat anomaly in the upper ocean. It is interesting to observe that although the observed WPWP SST index showed effects of other variability, the underlying cooling of the observed index lasted as long as the cooling in the ESM WPWP SST index; the magnitude of the cooling in the ESMs, as mentioned earlier, depended on the net heat flux anomaly generated by each ESM and the response of the ESM's oceanic component to the generated flux anomaly. In the observed WPWP index and in upper ocean ESM temperatures in the WPWP region, the recovery time to average conditions was approximately four to five years. Thus, according to Mehta et al. (2018), VEI 4 and higher explosivity volcanic eruptions clearly and substantially changed the phase of the WPWP variability from positive to negative for several years. Again, this analysis shows that, as in the case of the PDO, low-latitude volcanic eruptions of large magnitudes can have substantial effects on the natural decadal variability of the WPWP.

3.5 SUMMARY AND CONCLUSIONS

Decadal–multidecadal climate variability in the Pacific region has assumed increasing importance to satisfy intellectual curiosity, to predict the variability and its worldwide impacts on vital societal sectors, and to understand and attribute the variability's role in the evolution of global warming. Description and understanding of this broad spectrum of variability are the subject of this chapter.

After a brief historical background on serendipitous discoveries of Pacific climate variability phenomena, signatures of pan-Pacific (spanning the entire Pacific Ocean region) decadal–multidecadal variability in instrument-measured data were described at length. Conclusive evidence that the PDO, the IPO, the SPDO, and decadal–multidecadal variability of the Niño3.4 SSTs are essentially the same phenomenon was then presented. It was also shown that decadal SST variability has, by far, the largest variance in the tropical central and western Pacific Ocean. Extensive analyses of instrument-measured and model-assimilated atmosphere and ocean data were presented which showed that anomalous Hadley and Walker circulations and atmospheric Rossby waves are associated with the tropical Pacific decadal variability, which can influence or even impact extratropical climate. The analyses also showed that anomalous thermocline depths and associated sub-surface temperature anomalies, and anomalous surface current and STCs, especially the Southwest Pacific STC, are also associated with the decadal SST variability in the tropical central and western Pacific. Consistently with these results, a review of a 350-year long, coral oxygen isotope $\delta^{18}O$ record showed the presence of prominent decadal variability in the SPCZ region, corroborating the likely role of the STC in this region in tropical Pacific decadal variability. The review of paleoclimate proxy-based PDO/IPO indices also showed a general inconsistency among 12 such records, with possibly two records agreeing between themselves and with dry and wet epochs in North America and Monsoon Asia indicated by a tree–ring-based PDSI record since approximately the mid-17th century CE.

Tests of hypothesized mechanisms of tropical Pacific decadal variability in uncoupled and coupled ocean–atmosphere models have identified key roles of STCs and off-equatorial oceanic Rossby waves, and their interactions with the atmosphere. The model simulations appear to be consistent with instrument-measured and model-assimilated data analysis results, both of which also identify the central-western tropical Pacific region as the source of decadal variability. Numerous global coupled ocean–atmosphere models simulate IPO-like pan-Pacific multidecadal variability. Although extratropical processes such as changes in winds around the Aleutian and South Pacific low-pressure systems reaching towards the tropical Pacific, and spectral reddening due to integration of atmospheric higher-frequency forcings by mid-latitude Pacific Ocean appear to be playing important roles, the pan-Pacific multidecadal variability appears to be primarily due to STCs, the Bjerknes feedbacks, and oceanic Rossby waves in the tropics or at the edge of the tropics. Simulations with ESMs also show that low-latitude volcanic eruptions of large magnitudes can substantially perturb surface energy balances in the tropical and mid-latitude Pacific Oceans such that not only magnitudes of the PDO and WPWP SST indices but also their phases can be changed for several years by the eruptions. These magnitude and phase changes can change multiyear predictability of these two DCV phenomena as Mehta et al. (2019) found. However, unless and until the volcanic eruptions themselves can be predicted in advance, their effects on the PDO and the WPWP variability introduce an unpredictability factor to decadal climate prediction.

The outstanding problems outlined at the end of Chapter 2 apply here also. The remaining major questions about decadal variability are the following: (1) Does stochastic forcing or coupled ocean–atmosphere interaction generate this variability? (2) Does the solar Schwabe cycle act as the overall pace maker and inter-Hemispheric synchronizer of this variability? (3) Does the tropical decadal variability force extratropical decadal variability and does the latter play any active role in the former? Question 2 is addressed to some extent in Chapter 2. Variations of these questions apply to pan-Pacific multidecadal variability as well: (1) Does stochastic forcing or coupled ocean–atmosphere interaction in extratropical Pacific generate multidecadal variability? (2) Does solar multidecadal variability act as the overall pace maker and inter-Hemispheric synchronizer of this variability? (3) Does multidecadal Atlantic climate variability associated with the Atlantic Meridional Overturning Circulation play any role in the Pacific multidecadal variability?

As we come to the end of this chapter, we are left to ponder the possibility that, despite the apparent progress described in this chapter, it should not be construed that causes of tropical Pacific decadal variability and pan-Pacific multidecadal variability are completely understood. As a matter of fact, the relatively short instrument-based observational record and the inability of paleoclimate proxy-based indicators to come to a consensus point strongly to the possibility – nay, probability – that we will not understand multidecadal variability even in the next several decades. However, we are still left with the very substantial challenge that we need to develop techniques to make probabilistic predictions of the Pacific decadal–multidecadal variability to predict their societal impacts and this broad-spectrum variability's role in the waxing and waning global warming.

4 The Tropical Atlantic Dipole
Mystery or Myth?

4.1 INTRODUCTION

Bertrand Russell, the British philosopher and mathematician (1872–1970), opined "the most savage controversies are those about matters as to which there is no good evidence either way" (Russell, 1950). He may have expressed this opinion after presciently learning about the controversy regarding the tropical Atlantic dipole! What is the tropical Atlantic dipole, and why is there a controversy? It is a spatial pattern of variability, primarily at interannual to decadal timescales, of tropical Atlantic sea surface temperature (SST) departures (called anomalies) from the average annual cycle of SSTs. This pattern emerges when the empirical orthogonal function (EOF)–principal component (PC) statistical analysis technique is applied to tropical Atlantic SST anomaly data. This pattern also emerges when a map of correlation coefficients between a time series of rainfall in northeast Brazil or West Africa and tropical Atlantic SST anomalies on a longitude–latitude grid is plotted. In this spatial pattern, SST anomalies of opposite signs appear to straddle the Equator, with strongest anomalies at approximately 15°N and 15°S latitudes. As will be described in detail in subsequent sections, this spatial pattern has been described in climate science literature as the tropical Atlantic dipole variability or the tropical Atlantic SST gradient (TAG for brevity) variability. The principle reason why this SST pattern has attracted considerable attention from the climate science and climate impacts communities is that it is associated with variability of rainfall, temperature, and other climatic variables on continents on the eastern and western boundaries of the tropical Atlantic Ocean; and also in parts of North America, South America, Europe, Asia, and Australasia. Of course, the intellectual curiosity to understand the physics of this phenomenon is also a very important reason for attracting attention from climate scientists.

A controversy arose in the 1990s Current Era (CE) in the climate science community about whether this so-called dipole or TAG pattern is a dynamical/thermodynamical mode of variability of the tropical Atlantic ocean–atmosphere system. This controversy, which rose to a level approaching "savage" in conferences and workshops on climate variability, is also partly due to the analysis techniques traditionally used in climate research and so the role of these techniques in understanding nature is also involved in this controversy. In the early 1990s CE, the major climate variability phenomenon which had seized the imagination of the climate science community was the interannual, tropical Pacific climate variability phenomenon known as El Niño–Southern Oscillation (ENSO) which was being modeled as a coupled ocean–atmosphere phenomenon. So, the newly emerged tropical Atlantic

SST dipole – especially, its decadal variability – also became a phenomenon to fit into the coupled ocean–atmosphere framework. This chapter describes the saga of manipulating newly available multidecades- to century-long time series of SST data and shorter time series of other ocean–atmosphere data to isolate tropical Atlantic climate variability patterns, the hypothesized mechanisms which supposedly cause these patterns of variability, and their simulation by a variety of models and verifications of the simulations by observations-based data.

After this brief introduction to the subject of this chapter, patterns – including the dipole – of decadal variability of the tropical Atlantic variability in instrument-measured data are described in Section 4.2. Then, hypothesized mechanisms of decadal variability of the tropical Atlantic climate, including the dipole, are described in Section 4.3. Any contemporary recounting of tropical Atlantic decadal variability would be incomplete without an exposition of the empirical relationship between this variability and the Atlantic Multidecadal Oscillation (AMO). Therefore, empirical analyses of the relationship between the two are presented in Section 4.4. Finally, the current answer to the question posed in the chapter title is described in Section 4.5, along with suggested avenues of future research for further verification and corroboration of the present evidence. The available paleoclimate records based on proxy data from the tropical Atlantic Ocean are geographically too sparse to shed any useful light on the dipole controversy. Also, there is a low degree of agreement about decadal climate variability (DCV) among available records even from the same geographic location (see, for example, Black et al. (1999, 2007)). Therefore, indications of DCV from paleoclimate records are not included in this chapter.

4.2 SIGNATURES IN INSTRUMENT-MEASURED DATA

Our quest to address the dipole mystery begins with finding signatures of tropical Atlantic climate variability in instrument-measured ocean surface and lower atmosphere data, and rainfall data on land on the African and South American boundaries of the tropical Atlantic. Then, we will see what detailed heat budget analyses using atmospheric reanalysis data and *in situ* ocean data tell us about possible roles of various physical processes in tropical Atlantic SST variability. A synthesis of results and conclusions of these analyses of observed data is presented at the end of this section.

We begin with a brief history of how the dipole mystery came to light. Long before the journey of discovery of tropical Atlantic decadal variability began, Markham and McLain (1977) analyzed sparse (in space and time) South Atlantic SST observations to gain insight into the possible physics of droughts and floods in the Ceará region of Northeast Brazil. They found two long time series of rainfall going back to 1849 and 1896 CE from Fortaleza and Quixeramobím, respectively, in the Ceará region. But, they could find monthly, tropical Atlantic SST data only from 1947 to 1967 CE, and those were mainly along shipping routes between the Cape of Good Hope and Cape Verde, and along the South American coast. Undaunted by the data sparsity, Markham and McLain (1977) pressed on with correlation analysis and found a 5° longitude–5° latitude box (approximately 5°W to 0°, 5°S to 10°S) where the correlation between the Ceará rainfall and SSTs was the highest. Encouraged by this discovery, they assembled a much longer SST record in this longitude–latitude box from shipborne measurements and devised a prediction equation for Ceará rainfall, specifically based on SST thresholds for droughts and floods. From the rainfall record, they estimated that severe droughts and floods occurred in the Ceará region approximately once per decade, thus implying DCV in this region perhaps for the first time in published literature. This was also perhaps the first instance of discovery of associations between tropical South Atlantic SSTs and Northeast Brazil rainfall.

Then, approximately 10 to 15 years after the Markham and McLain discovery, Folland et al. (1986, 1991), Semazzi et al. (1988), Hastenrath (1990), and Ward and Folland (1991), among others, found that tropical African rainfall variability was moderately to highly correlated with tropical Atlantic SST variability. Numerous studies (see, for example, Hastenrath 1991; Nobre and Shukla 1996; and references therein) also found associations among variability in the tropical Atlantic SST, winds over the tropical Atlantic and neighboring continental regions, and rainfall over Northeast Brazil at multiyear timescales. In these analyses of then-available relatively short SST time series on a regular spatial grid, the pattern of multiyear variations that attracted the most attention had a nearly constant amplitude with respect to longitude in the tropical Atlantic and opposite signs on the two sides of the Equator, with the maximum amplitudes at approximately 15°N and 15°S latitudes. This north–south, cross-equatorial pattern also resulted from an EOF-PC analysis of tropical Atlantic SST anomalies and was characterized by some researchers as a dipole mode of tropical Atlantic SST variability. Distilling the essence of these studies, Servain (1991) and Servain et al. (1999) defined a tropical Atlantic dipole index and associated it with sub-surface ocean variability in the equatorial Atlantic region and climate variability on both sides of tropical Atlantic. As mentioned earlier and shown in Section 4.2.2, this type of SST pattern also emerges when tropical Atlantic SST anomalies, and wind or rainfall anomalies over Northeast Brazil and West Africa are correlated or regressed. So, these initial analyses were the origin of the dipole mystery. These were early days in modeling interannual and longer timescale climate variability phenomena, and the climate science community's attention was focused almost entirely on modeling the ENSO phenomenon in the tropical Pacific region; hence, much of the tropical Atlantic variability research was then focused on analyses of observations.

After this initiation to the dipole mystery, we now delve deeper into observational aspects of this mystery. As mentioned earlier, records of SST measurements are the longest in the Atlantic region; therefore, analyses of these records are described first in this sub-section. Then, major results and conclusions from combined analyses of SSTs, lower atmospheric winds, and rainfall are described.

4.2.1 Sea Surface Temperature Variability

As mentioned throughout this book, inadequate spatial coverage of long (at least 100 years) observational records containing reliable measurements hampers any observational study seeking to characterize decadal and longer timescale climate variability. Since shipboard measurements were the primary source of information about SST until the advent of Earth-orbiting satellites, spatial and temporal coverages were dependent upon and limited by locations of shipping routes.

Of all the world's oceans and seas, the longest time series and reasonably dense spatial coverage combination is the richest for the Atlantic Ocean. Shipping routes between the Old and the New Worlds, and between colonizing countries in Europe and colonized countries/regions in Africa, Asia, and Australasia go back several centuries. Infrequent measurements/estimates of SST, ocean state, and surface winds were made before the 1850s CE and routinely since the 1850s CE. As a consequence, the Atlantic is perhaps the best observed ocean to study decadal variability. The available SST data after 1950 CE show that decadal (8 to 13 years) variability is prominent in the tropical Atlantic with up to 30% to 40% of annual SST variance (Figure 4.1a) and multidecadal variability (14 to 68 years) is prominent in mid- and high-latitude Atlantic with up to 70% to 90% of annual SST variance (Figure 4.1b). The ratio of the two bands is largest in the tropical Atlantic, with the decadal band variance larger than multidecadal band variance by factors of 6 to 8 (Figure 4.1c). Although the maximum variance regions in both bands are in different locations in the 1900–1950 CE period (not shown), the ratio of the two band variances is approximately the same in the tropical Atlantic. This relative prominence of decadal SST variability in the tropical Atlantic region is comparable to the relative prominence of decadal SST variability in the tropical Pacific region as we saw in Chapter 3. Therefore, it is important to understand and try to predict decadal variability in the tropical Atlantic region due to its spectral prominence as well as its societal impacts.

After this broader background into long-term SST data and importance of decadal SST variability, let us now follow the dipole mystery as it evolved in chronological order. Houghton and Tourre (1992) were the first to question the physical interpretation of the bipolar (a pattern which has positive and negative polarities) SST pattern as a dipole mode of tropical Atlantic SST variability. They used rotated EOF (REOF; Kaiser, 1958; Richman, 1986) analysis on SST data measured on merchant ships between 1964 and 1988 CE in the tropical Atlantic compiled by Servain et al. (1987). The REOF patterns showed that SSTs in the tropical Atlantic underwent independent interannual variations on the two sides of the Equator during the data record period. Soon thereafter, Kawamura (1994)'s REOF analysis of global, monthly-average SST anomalies from 1955 to 1988 CE confirmed that SST variations were independently confined largely to the North Atlantic and to the South Atlantic. Neither of these studies showed dipole-like, cross-equatorial, interannual SST variations. The analysis of SST and other data from the Comprehensive Ocean-Atmosphere Data Set (COADS) by Enfield and Mayer (1997) later confirmed the Houghton and Tourre and Kawamura results. Thus, the interannual dipole variability controversy was well and truly stoked.

Then, as longer historical SST records started becoming available in the early to mid-1990s CE, Mehta and Delworth (1995) analyzed decadal variations of the tropical Atlantic SSTs in many 100-year-long, individual SST time series from a version of the Global Ocean Surface Temperature Atlas data set (GOSTA; Bottomley et al., 1990). A variety of analyses of these time series suggested that there were two types of pronounced, quasi-oscillatory, decadal timescale SST variations in the tropical Atlantic region. One type, characterized by timescales between approximately 8 and 11 years, had high spatial coherence within each tropical Atlantic hemisphere but not between the two Hemispheres. The second type, characterized by periods between

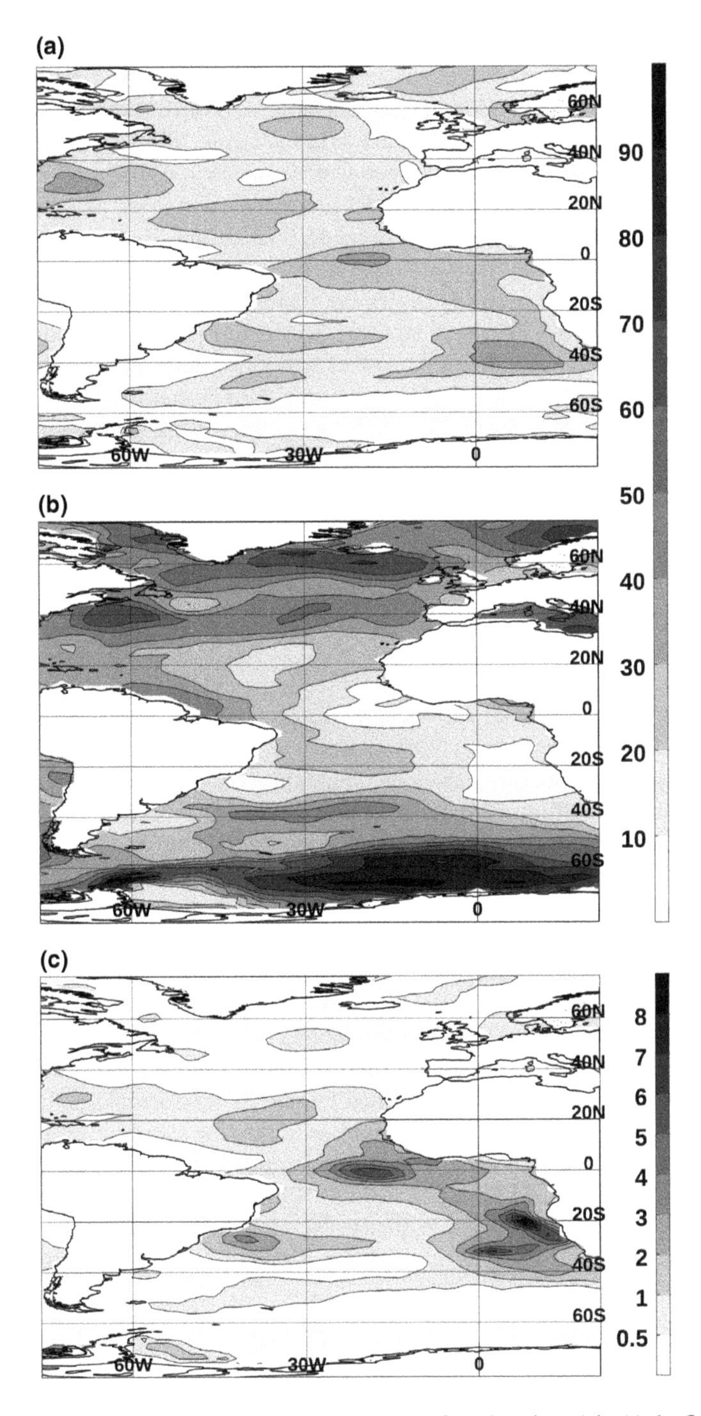

FIGURE 4.1 Fractional SST variances (as a percent of total variance) in (a) the 8 to 13-year period band, and (b) the 14 to 68-year period band. (c) Ratio of fractional variances in (a) and (b). ERSST data from 1951 to 2018 were used.

12 and 20 years, had high spatial coherence between the two Hemispheres and was considerably weaker than the first. Mehta (1998) followed up this initial study with a more detailed analysis of SST time series aggregated in 10° longitude–10° latitude boxes from 40°S to 60°N from the GOSTA data set from 1882 to 1991 CE. A variety of techniques were used to identify spatial structures and oscillation periods of the tropical Atlantic SST variations at decadal timescales, and to develop physical interpretations of statistical patterns of decadal SST variations. Real and complex EOF analyses were used to identify spatial patterns; and Fourier, Singular, and wavelet spectrum analyses were used to identify dominant oscillation periods if there were any. Tropical Atlantic SST variations were also compared with decadal variations in the extratropical Atlantic SSTs. Multiyear to multidecadal variations in the cross-equatorial dipole pattern identified as a dominant empirical pattern of the tropical Atlantic SST variations were shown to be variations in the approximately north–south gradient of SST anomalies. It was also shown that there was no dipole mode of SST variations during the analysis period. All three spectrum analysis techniques showed that there was a distinct decadal timescale (12 to 13 years) of SST variations in the tropical South Atlantic, whereas no distinct decadal timescale was found in the tropical North Atlantic SST variations. Approximately 80% of the coherent decadal variance in the cross-equatorial SST gradient was "explained" by coherent decadal oscillations in the tropical South Atlantic SSTs.

The two types of EOF analyses and time sequences of SST anomaly maps showed that there were three, possibly physical, patterns of decadal variations in the tropical Atlantic SSTs in the Mehta (1998) analysis. In the more energetic pattern of the North Atlantic decadal SST variations, anomalies traveled into the tropical North Atlantic from the extratropical North Atlantic along the eastern boundary of the basin. The anomalies strengthened and resided in the tropical North Atlantic for several years, then frequently traveled northward into the mid–high-latitude North Atlantic along the western boundary of the basin, and completed a clockwise rotation around the North Atlantic basin. In the less energetic North Atlantic decadal pattern, SST anomalies originated in the tropical–subtropical North Atlantic near the African coast, and traveled northwestward and southward. In the South Atlantic decadal SST pattern, anomalies either developed *in situ* or traveled into the tropical South Atlantic from the subtropical South Atlantic along the eastern boundary of the basin. The anomalies strengthened and resided in the tropical South Atlantic for several years, then frequently traveled southward into the subtropical South Atlantic along the western boundary of the basin, and completed a counterclockwise rotation around the South Atlantic basin. The tropical North and South Atlantic SST anomalies frequently extended across the Equator. Uncorrelated alignments of decadal SST anomalies having opposite signs on two sides of the Equator occasionally created the appearance of a dipole pattern.

Three other analyses of observed SST data independently corroborated the conclusion that decadal variabilities in the tropical North and South Atlantic were largely uncorrelated. First, Rajagopalan et al. (1998) found in their analyses of 136 years (1865–1991 CE) of Atlantic SST data (Kaplan et al., 1997; 1998), based on GOSTA, that SST anomalies north and south of the InterTropical Convergence Zone (ITCZ) were uncorrelated on all time scales. Spectral analyses of the tropical

North and South Atlantic SST anomalies showed a clear preference of variability at 10 to 20-year periods. They also found that there was a strong, broad-band coherence between the North Atlantic Oscillation (NAO) and the tropical Atlantic SST difference between northern and southern sides of the ITCZ in the 8 to 20-year band, suggesting a significant mid-latitude–tropics interaction. Moreover, tropical Atlantic SSTs on both sides of the ITCZ also exhibited significant coherence with the NAO index and sea level pressure variability over Iceland and the Azores. Then, almost at the same time, Enfield et al. (1999) analyzed tropical Atlantic SST data from 1856 to 1991 CE (Kaplan et al., 1998) and found that there were independent SST REOF patterns in the tropical North and South Atlantic. They then defined monthly SST index time series for the tropical North Atlantic (TNA; 55°W to 15°W longitudes, 5°N to 25°N latitudes) and tropical South Atlantic (TSA; 30°W to 10°E longitudes, 20°S to 0° latitudes), and analyzed their variability and co-variability with various techniques after removing linear trends and ENSO influences. The analyses showed that dipole configurations of SST anomalies in tropical North and South Atlantic were not ubiquitous and, in confirmation of the Rajagopalan et al. (1998) conclusion, that SST variabilities north and south of the ITCZ had very small correlation as shown in Figure 4.2. They also found that large-scale dipole configurations occurred 12% to 15% of the time and that chance occurrences of dipole configurations could be expected approximately 16% of the time since both tropical North Atlantic and tropical South Atlantic SST anomalies were stochastically independent. Thus, chance occurrences of cross-equatorial dipole configurations were no more frequent than those expected from stochastically independent variables. Enfield et al. (1999) also found that significant, meridional gradients between tropical North and South Atlantic SST anomalies occur almost 50% of the time. Correlation analyses in time domain, and spectral and cross-spectral analyses in frequency domain between TNA and TSA showed (Figure 4.3) prominent spectral peaks at 12 to 13-year periods, but no significant relationships between the tropical North and South Atlantic SST anomalies, except that there was a small and significant coherence between the two in the 8 to 12-year period range in boreal (Northern Hemisphere) winter and spring (December to May). Thus, results of both the Rajagopalan et al. (1998) and Enfield et al. (1999) studies agreed that there was

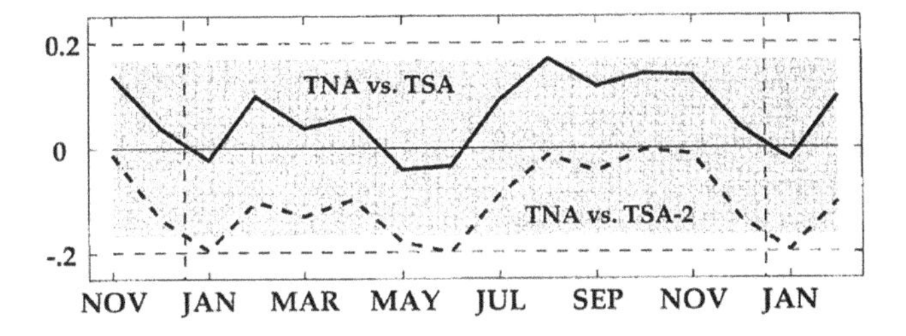

FIGURE 4.2 Correlation coefficients between sea surface temperature anomalies averaged in TNA and TSA (solid line), and between TNA and a region off the coast of Angola (TSA-2) (dashed line) for each calendar month. Gray shading and horizontal dashed lines denote values below 90% and 95% significance, respectively. (From Enfield et al., 1999.)

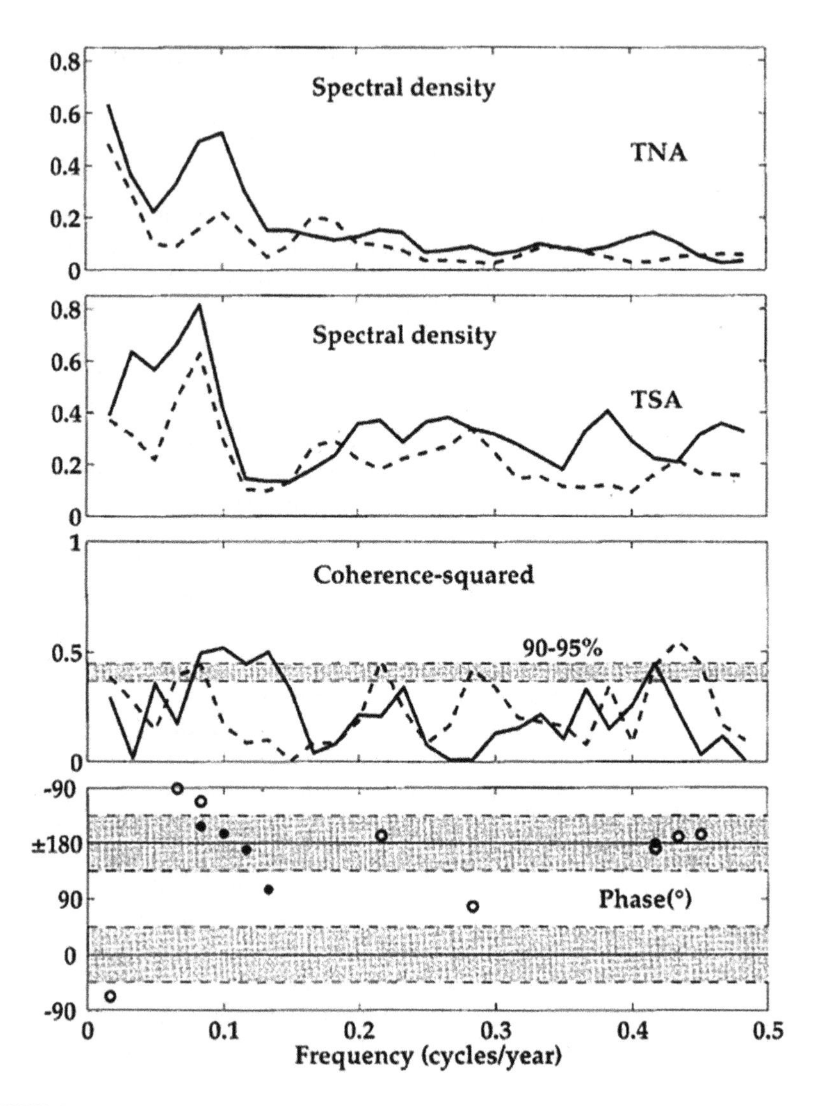

FIGURE 4.3 Fourier spectra of the tropical North Atlantic (upper panel) and tropical South Atlantic (middle panel) sea surface temperature anomaly time series from 1856 to 1991 CE. Cross-spectra (coherence-squared and phase difference) between the two time series, with 90% and 95% significance shaded in gray. Solid (open) circles denote phase difference corresponding to solid (dashed) lines in the coherence-squared curves. Phase differences between ±45° of 0° and ±180° are shaded. Solid (dashed) lines are for December to May (June to November) data. (From Enfield et al., 1999.)

no tropical Atlantic dipole mode, but SST variabilities in tropical North and South Atlantic showed a preference for 12 to 13-year oscillation periods.

Soon thereafter, Dommenget and Latif (2000) analyzed observed SST data from the Global Sea Ice and Sea Surface Temperature data set (GISST; Parker et al., 1995b) from 1903 to 1994 CE with the EOF and REOF analysis techniques, and found two SST variability patterns centered at 15°N and 15°S latitudes; the two patterns were not correlated at any timescale. From their spectral analyses of associated PC time series, they concluded that the spectra were consistent with "red noise" and that there was no preferred timescale of variability either in tropical North or South Atlantic SSTs. Dommenget and Latif (2000) also analyzed simulation experiments with several coupled ocean–atmosphere models to understand the physics of tropical Atlantic SST variability. Their results and conclusions are described in Section 4.4.

Thus, as the new millennium dawned, the evidence from analyses of century-long and longer observed SST data overwhelmingly pointed to nearly independent SST variability in tropical North and South Atlantic at decadal timescales, but whether or not there was a preferred 12 to 13-year oscillation period was data set-dependent.

To update the story of the observed tropical Atlantic SST variability to the most recent year, SST data from 1901 to 2018 CE from Extended Reconstructed Sea Surface Temperature (ERSST) version 5 (Huang et al., 2017) were analyzed. Results of EOF-PC analyses for three-month average (December–January–February, DJF; March–April–May, MAM; June–July–August, JJA; and September–October–November, SON) anomalies and for annual-average anomalies are shown in Figure 4.4. Each PC time series was normalized by its standard deviation which is given at the top of each time series panel along with the percent variance explained by each PC. At each time step, the average SST anomaly over the entire tropical Atlantic basin was subtracted from the anomaly at each grid point to remove effects of the basin-average anomaly from subsequent analyses. After this pre-processing of the SST data, the bipolar pattern is EOF 1 for all three-month average data and the annual-average data as Figures 4.4b, d, f, h, and j show. These patterns explain a minimum 24% of the SST variance in the DJF months to a maximum 39% in the JJA months; EOF 1 of annual-average anomalies explains 37% of the variance. The EOF 1 patterns show that EOF coefficients of the same sign encompass almost an entire tropical hemisphere (North or South), with opposite signs in the two hemispheres. Maximum amplitudes are near the coasts in both hemispheres as found in previous analyses described earlier. The PC time series in Figures 4.4a, c, e, g, and i are generally similar in interannual to multidecadal variability. It is noteworthy that the period from 1900 to 1960 CE shows more prominent multidecadal variability than decadal variability, and the period since 1960 CE shows less prominent multidecadal variability. The Fourier spectra of these five PC time series are shown in Figure 4.5 which show prominent and significant peaks at approximately 33 years (0.03 cycles/year) and 11–12 years (0.09 to 0.08 cycles/year) compared to the background red noise spectra (Wilks, 1995). There are differences in spectral density of these peaks among seasons and also differences between Fourier spectral density of peaks in annual and seasonal PC time series. Figure 4.5 also shows a multitude of prominent spectral peaks at interannual timescales, especially in the MAM and JJA months. To compare EOF-PCs over the entire 1901–2018 CE period with those over the pre- and post-1950 CE periods, EOF-PCs and Fourier spectra of PCs were calculated for these individual periods (not shown). These two periods were approximately before and after the Second World War. A comparison of these three periods shows that, although the EOF patterns before and after 1950 CE are generally similar, the EOF amplitudes are confined much closer to the tropical South Atlantic

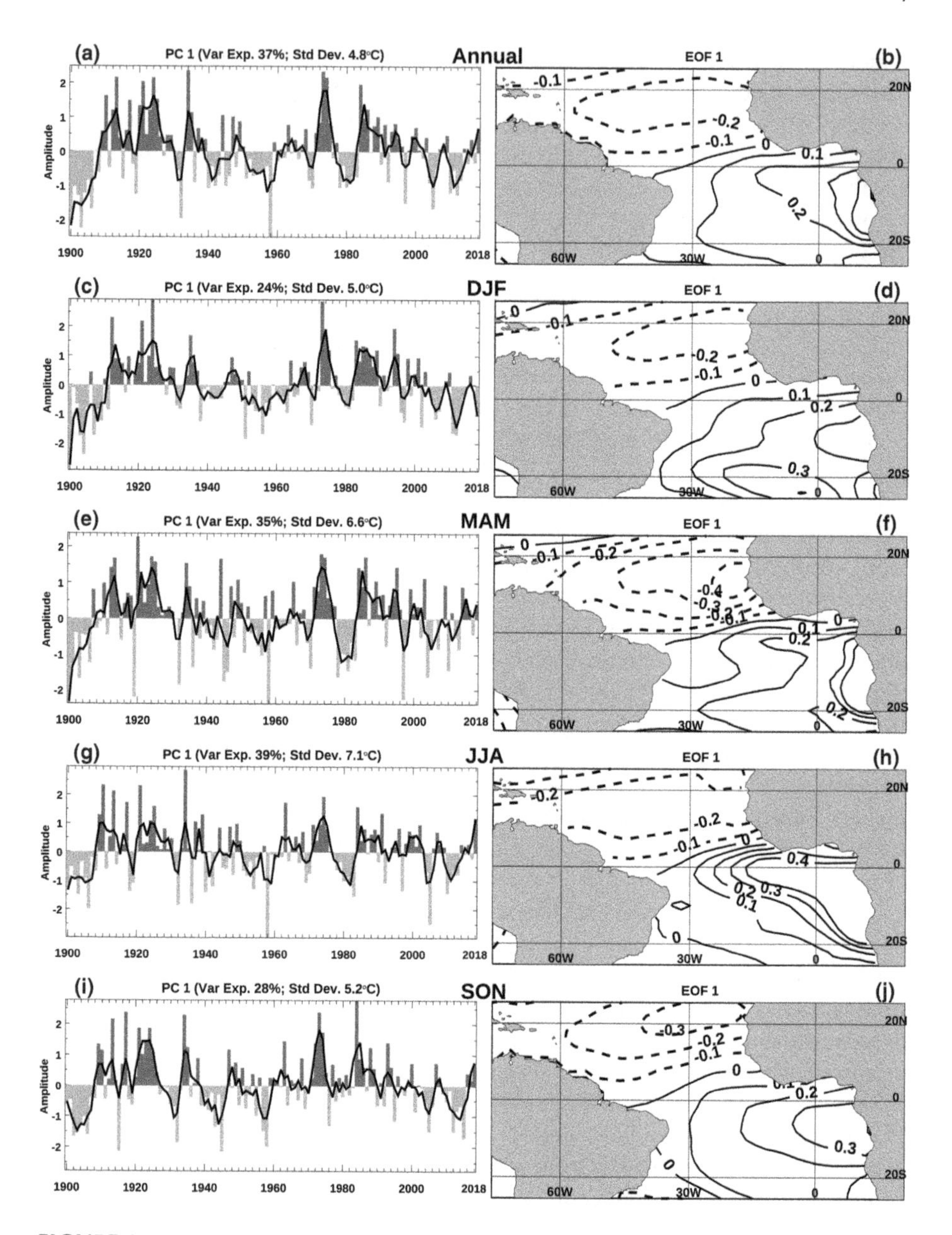

FIGURE 4.4 PC 1 (bars) and its three-point moving-window average (black line), and EOF 1 of observed sea surface temperatures from 1901 to 2018 CE. (a) and (b) annual, (c) and (d) DJF, (e) and (f) MAM, (g) and (h) JJA, and (i) and (j) SON. Each PC is normalized with its standard deviation. The variance explained by each PC and its standard deviation are shown above each PC.

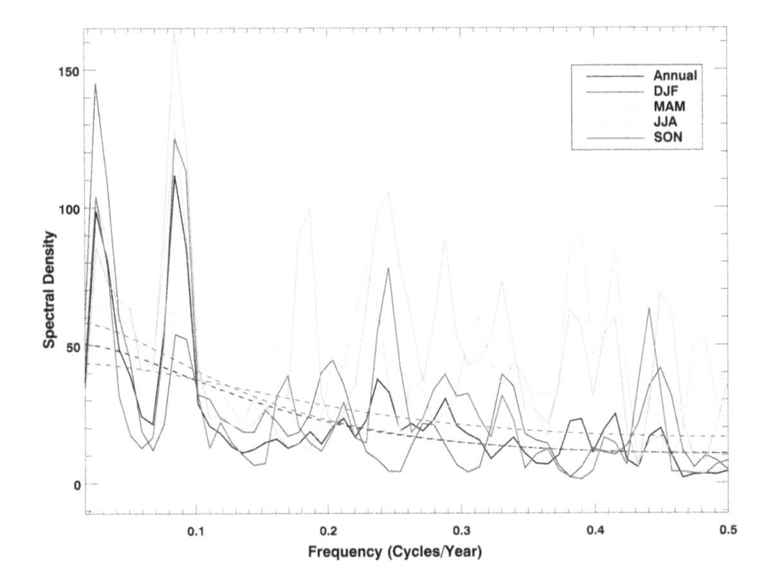

FIGURE 4.5 Fourier spectra (variance per unit frequency; solid lines) of principal components in Figure 4.4 and red noise spectrum corresponding to each Fourier spectrum (dashed lines). The color legend is shown in the box.

coast in the pre-1950 CE period compared to the post-1950 CE period. This could be because shipping routes may have been confined close to the coast in the former period and may be more spread over the South Atlantic in the latter period. Also, the availability of satellite-based SST data over the entire Atlantic since late 1970s–early 1980s CE may have contributed to a better spatial sampling of the variability signal in the tropical South Atlantic in the post-1950 CE period. As mentioned above, the prominence of multidecadal variability compared to decadal variability is present in the pre-1950 CE Fourier spectra. Thus, the addition of the most recent 25 years' data to the tropical Atlantic SST variability analyses confirms the presence of the bipolar, inter-hemispheric gradient variability as the dominant pattern and also confirms the presence of a prominent 11 to 12-year oscillation period found in most previous analyses of observed SST data. A Fourier spectral analysis of tropical North Atlantic and tropical South Atlantic SST anomaly time series, averaged from 10° to 20°N and 10° to 20°S, respectively, shows that the decadal spectral peak is much more prominent in the tropical South Atlantic compared to the tropical North Atlantic. This prominence in the SST data extended from early 1990s to 2018 CE confirms a similar prominence of the decadal spectral peak in the tropical South Atlantic found by Mehta (1998). Lastly, correlation coefficients between the annual-average and three-month average tropical North and South Atlantic SST anomaly time series from 1901 to 2018 CE are very small (approximately 0.1) and insignificant, confirming the absence of the dipole in yet another way as found by Rajagopalan et al. (1998) and Enfield et al. (1999).

After this detailed description of results of century-long and longer SST data, we will now see insights gained into the mystery of tropical Atlantic dipole variability from combined analyses of ocean and atmosphere data.

4.2.2 Upper Ocean–Lower Atmosphere Variability

Concurrently with the SST-only analyses of tropical Atlantic variability, there were several studies which combined ocean and atmosphere observations to analyze their joint variability. These non-SST data were generally over shorter periods of time. Fontaine et al. (1999) used objectively analyzed, monthly SST and pseudo-wind stress data in the tropical Atlantic region from 1964 to 1995 CE at 2° longitude–2° latitude resolution. Statistical relationships between two most energetic patterns of combined SSTs and pseudo-wind stress – identified with the singular-value decomposition technique – in the tropical Atlantic region; atmospheric variables in the tropical Atlantic region; and rainfall in West Africa and Northeast Brazil were analyzed. The most energetic SST pattern was largely in tropical North Atlantic, and the second most energetic SST pattern was largely in tropical South Atlantic, with the wind stress anomalies directed into the warmer Hemisphere from near the Equator. Their results implied that thermodynamic effects of the anomalous wind stress due to large SST anomalies in one Hemisphere generated small, opposite-sign SST anomalies in the opposite Hemisphere, but that there was no inter-hemispheric dipole SST variability at any timescale.

As observations of atmospheric winds, pressure, and temperature started becoming assimilated in general circulation models (GCMs) for longer periods of time, diagnostic studies of interannual to decadal climate variability with internally consistent data became possible. One of the first such studies on the tropical Atlantic variability was by Ruiz-Barradas et al. (2000). They conducted an REOF analysis of combined SST, oceanic heat content, zonal and meridional wind stresses, and atmospheric diabatic heating from 1958 to 1993 CE from a variety of data sources including the National Centers for Environmental Prediction (NCEP)– National Center for Atmospheric Research (NCAR) reanalysis. The third most energetic REOF (Figure 4.6) showed an SST pattern confined largely to tropical North Atlantic, with the largest amplitude at 15°N latitude off the West African coast, and a weak and localized area of opposite-sign anomalies in tropical South Atlantic; an anomalous wind stress pattern from just south of the Equator to the area of maximum SST amplitude; an anomalous oceanic heat content pattern of one sign straddling the Equator, with a small opposite-sign heat content anomaly over the opposite-sign SST anomaly area in tropical South Atlantic; and an anomalous atmospheric diabatic heating pattern straddling the Equator with opposite signs on both sides of the Equator. A spectral analysis of the corresponding PC time series showed a peak at 12-year period, although the data time series were too short to establish significance of this spectral peak. The anomalous SST and wind stress patterns found by Fontaine et al. (1999) and Ruiz-Barradas et al. (2000) were generally similar, with both SST anomaly patterns confined largely to tropical North Atlantic. However, Ruiz-Barradas et al. (2000) found that these northern ocean temperature anomalies penetrate to thermocline depth and extend southwards across the Equator, so there appeared to be a cross-equatorial, ocean dynamical expression of the tropical North Atlantic SST variability in their analyses, but without any unambiguous indication of an energetic cross-equatorial mode of variability.

The associations found by researchers in the 1990s CE between tropical Atlantic SST variability, and 850 hPa wind and rainfall variability in the tropical Atlantic

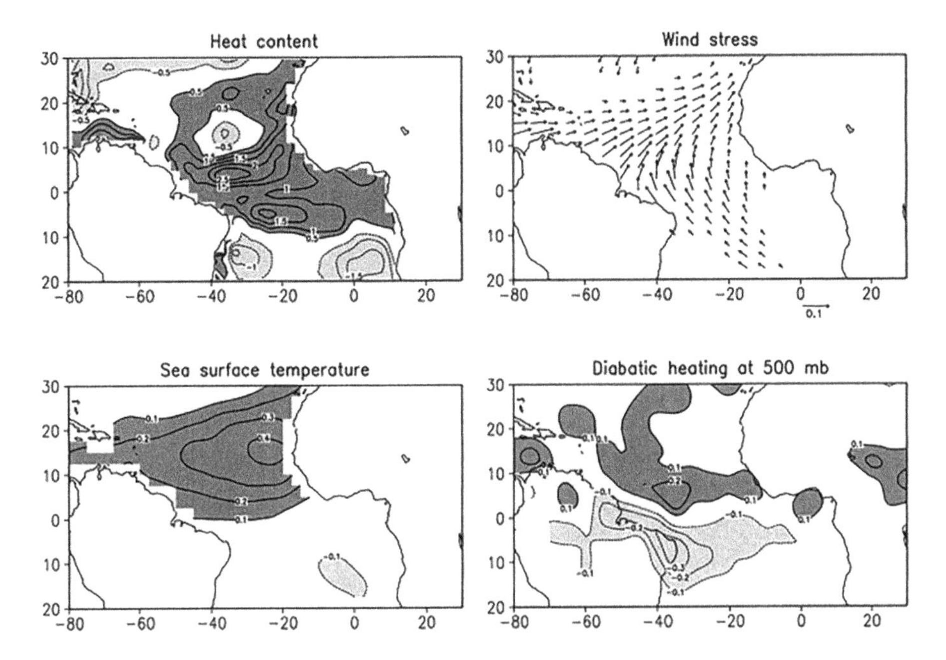

FIGURE 4.6 Third rotated empirical orthogonal function from a five-variable combined analysis: hc' (10^8 J m^{-2}) (upper left), tau' (dyn cm^{-2}) (upper right), SST' (°C) (middle left), and q' (°C day^{-1}) at 500 mb (middle right). Only wind stress anomalies larger than 0.02 dyn cm^{-2} are displayed. Explained variance is 4.9%. Dark (light) shading denotes positive (negative) anomalies. (From Ruiz-Barradas et al., 2000. © American Meteorological Society. Used with permission.)

region were mentioned in the beginning of Section 4.2.1. These associations, with data updated to 2018 CE, are shown here. Figure 4.7 shows SST and 850 hPa wind anomalies at each grid point in the positive phase of the TAG index in averaged periods DJF, MAM, JJA, and SON. SST data from ERSST version 5 (Huang et al., 2017) at 1° longitude–1° latitude resolution and 850 hPa wind data from the NCEP-NCAR reanalysis (Kalnay et al., 1996) at 2.5° longitude–2.5° latitude for the 1951–2018 CE period were used. A linear regression of SST anomalies and wind anomalies on the TAG index was fitted at each grid point, and the anomalies shown in Figure 4.7 are for one unit of the TAG index in the positive phase. In this phase of the TAG index when SST anomalies in tropical North Atlantic are warmer than those in tropical South Atlantic, 850 hPa wind anomalies are from tropical South Atlantic to tropical North Atlantic, especially in the DJF, MAM, and SON months. There are relatively weak, land-to-ocean wind anomalies over Northeast Brazil – especially in the main rainy season (MAM) – and strong ocean-to-land wind anomalies over West Africa in all months. The average climatological 850 hPa winds over Northeast Brazil are from tropical South Atlantic Ocean, bringing water vapor which condenses and rains over land. The total 850 hPa winds (anomalies added to climatological average) are thus weaker in the positive TAG phase and lead to decreased rainfall over Northeast Brazil. On the eastern side of tropical Atlantic, 850 hPa wind anomalies are from

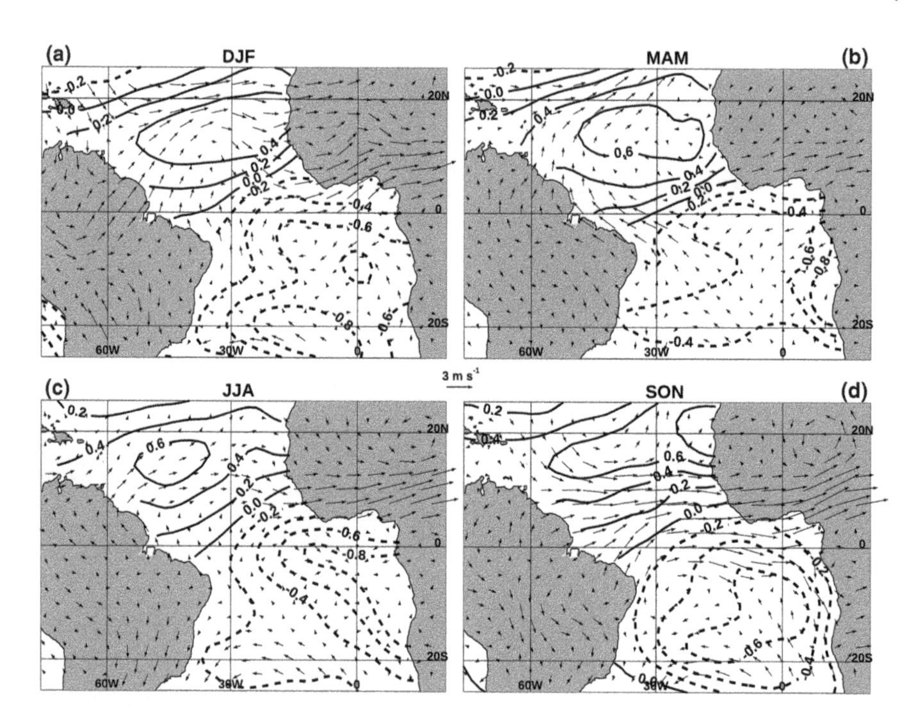

FIGURE 4.7 Tropical Atlantic SST anomalies (contours; °C) and 850 hPa wind anomalies (vectors; ms^{-1}) in positive phase of the TAG index in (a) DJF, (b) MAM, (c) JJA, and (d) SON. ERSST and NCEP-NCAR data from 1951 to 2018 were used. Scale for vector wind anomalies is shown between the two row of boxes and the contour interval for SST anomalies is 0.2 °C. Solid (dashed) contours denote positive (negative) SST anomalies.

the ocean to the land, especially in the JJA months which is the main period of the West African monsoon. Since the average climatological winds in West Africa are easterly and dry because they are from arid or semi-arid regions, these westerly wind anomalies weaken the total winds and bring water vapor from the tropical North Atlantic Ocean, which condenses and rains over land. Thus, there is an opposite-sign behavior of winds and rainfall in West Africa (above average) and Northeast Brazil (below average) in the positive TAG phase.

A comparison of Figure 4.7 with SST and 850 hPa wind anomalies for one unit of the TAG index in the negative phase shows that the SST anomalies in tropical North and South Atlantic have opposite signs in the two TAG phases, as do the wind anomalies. Wind anomalies over Northeast Brazil are from the equatorial Atlantic Ocean, especially in the MAM months, which bring a substantial amount of water vapor which condenses and rains over Northeast Brazil. Over West Africa, wind anomalies in the negative TAG phase are easterly, which strengthen the average, climatological easterly winds from semi-arid or arid land regions. Thus, wind and implied water vapor anomalies reverse signs in the negative TAG phase and, consequently, increase rainfall over Northeast Brazil and decrease rainfall over West Africa.

These implications for rainfall on the eastern and western sides of the tropical Atlantic, based on the TAG phase–wind anomalies associations, are also confirmed by direct TAG–rainfall associations. Correlation coefficients between three-month average Atlantic SST anomalies, and three-month average rainfall anomalies averaged in Northeast Brazil (Equator to 15°S, 55°W to 35°W) and West Africa (Equator to 15°N, 20°W to 10°E) from 1951 to 2018 CE are shown in Figure 4.8. Bipolar SST correlation patterns in DJF, MAM, and JJA seasons are evident for Northeast Brazil rainfall, with largest negative and positive correlations at approximately 10°–15°N and 10°–15°S latitudes, respectively (Figure 4.8a, c, e, g). Negative correlations in the tropical North Atlantic and positive correlations in the tropical South Atlantic imply that above-average rainfall in Northeast Brazil is associated with colder than average SSTs in the former region and warmer than average SSTs in the latter region. The correlations are largest in the MAM months, which constitute the main rainy season in Northeast Brazil. Figures 4.8b, d, f, and h show a generally opposite-polarity correlation pattern for West Africa rainfall, especially in DJF and MAM seasons. This implies that above-average rainfall in West Africa is associated with warmer than average SSTs in tropical North Atlantic and colder than average SSTs in tropical South Atlantic. The associations shown in Figures 4.7 and 4.8 are physically consistent.

4.2.2.1 An Oceanic Connection from Mid-Latitude North Atlantic?

To end this sub-section on an intriguing note, a combined observational analysis of a variable associated with the Atlantic Meridional Overturning Circulation (AMOC) and decadal variability of tropical Atlantic SSTs is shown. The AMOC is a major feature of the basin-scale ocean circulation in the Atlantic and plays a very important role in northward heat transport from the South Atlantic to the North Atlantic and further north. The AMOC is driven mainly by deep water formation in the Labrador Sea and in seas further north. The Labrador Sea deep water is formed because of convection in winter, which is believed to be driven by ocean–atmosphere heat fluxes associated with the NAO. The intensity of the winter convection can be estimated by Labrador Sea Water (LSW) thickness – defined as the depth within a given range of sea water densities. To verify the hypothesis that variability of the AMOC – as indicated by LSW variability – and associated northward heat transport may be involved in the tropical Atlantic decadal SST variability, Yang (1999) correlated empirical estimates of LSW thickness from 1930 to 1996 CE (Curry and McCartney, 1996) with a dipole index formed from observed tropical Atlantic SSTs from 1945 to 1993 CE (Da Silva et al., 1994a). A significant correlation with a lag of 5 years between the LSW thickness and the tropical Atlantic dipole index was found, with the LSW thickness leading. These LSW and dipole index time series are shown in Figure 4.9a. A correlation map between the LSW thickness time series and gridded SST anomaly time series in the tropical Atlantic (Figure 4.9b) shows a cross-equatorial dipole pattern with highest correlations along the western boundary of the Atlantic Ocean. But, it must be emphasized here that correlation coefficients between one reference time series (the LSW thickness or the Northeast Brazil rainfall) and spatially distributed other time series can sometimes reveal the dominant pattern of variability of the spatially distributed data rather than any physical relationship with the reference variable. Such physical relationships must be proven with other data or with model simulations. Based on these empirical associations, however, Yang (1999) further

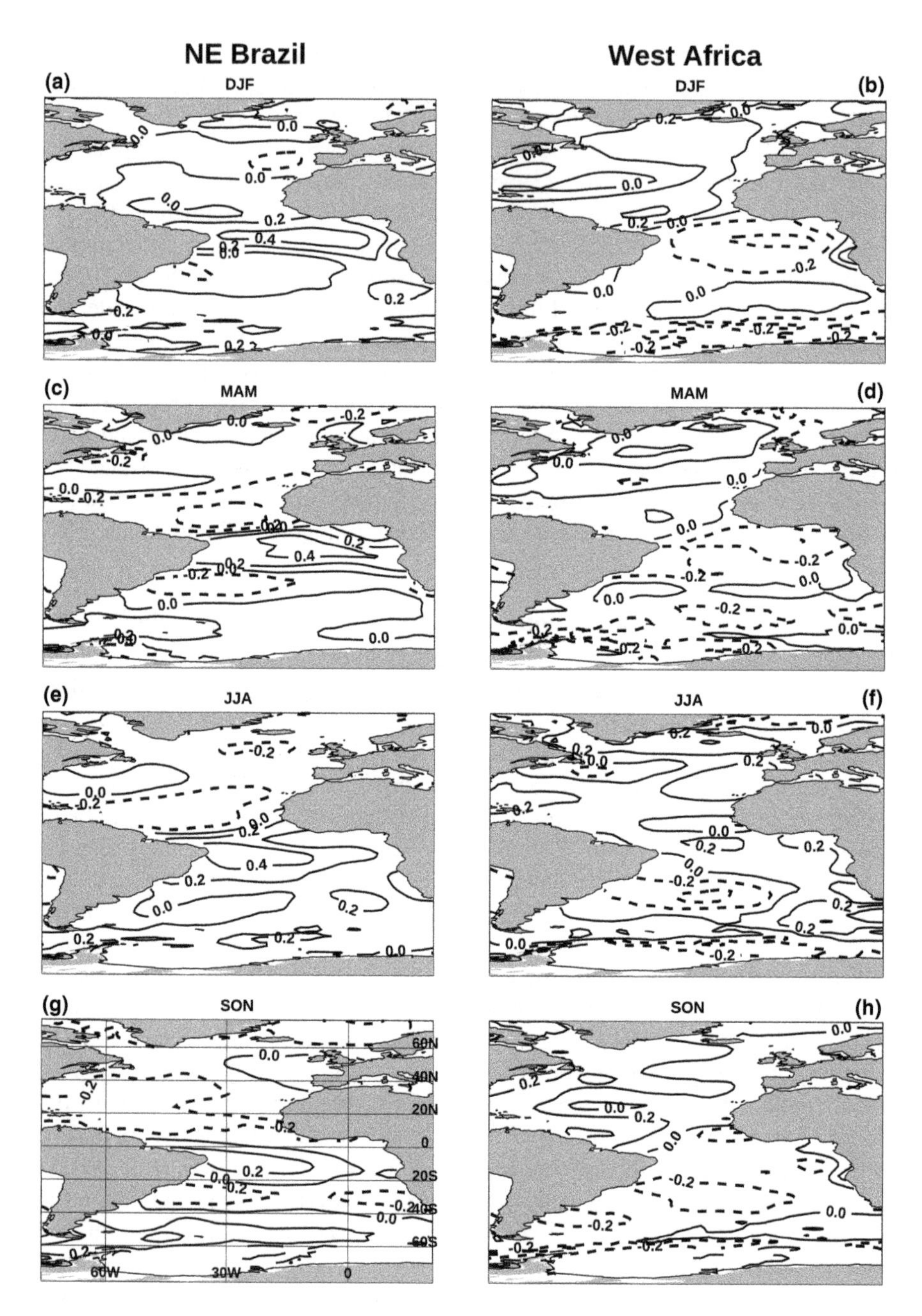

FIGURE 4.8 Correlation coefficients between rainfall anomalies averaged in Northeast Brazil and West Africa, and Atlantic sea surface temperature anomalies in (a) and (b) DJF, (c) and (d) MAM, (e) and (f) JJA, and (g) and (h) SON. Contour interval is 0.2, and solid (dashed) contours denote positive (negative) coefficients. Rainfall and ERSST data from 1951 to 2018 were used.

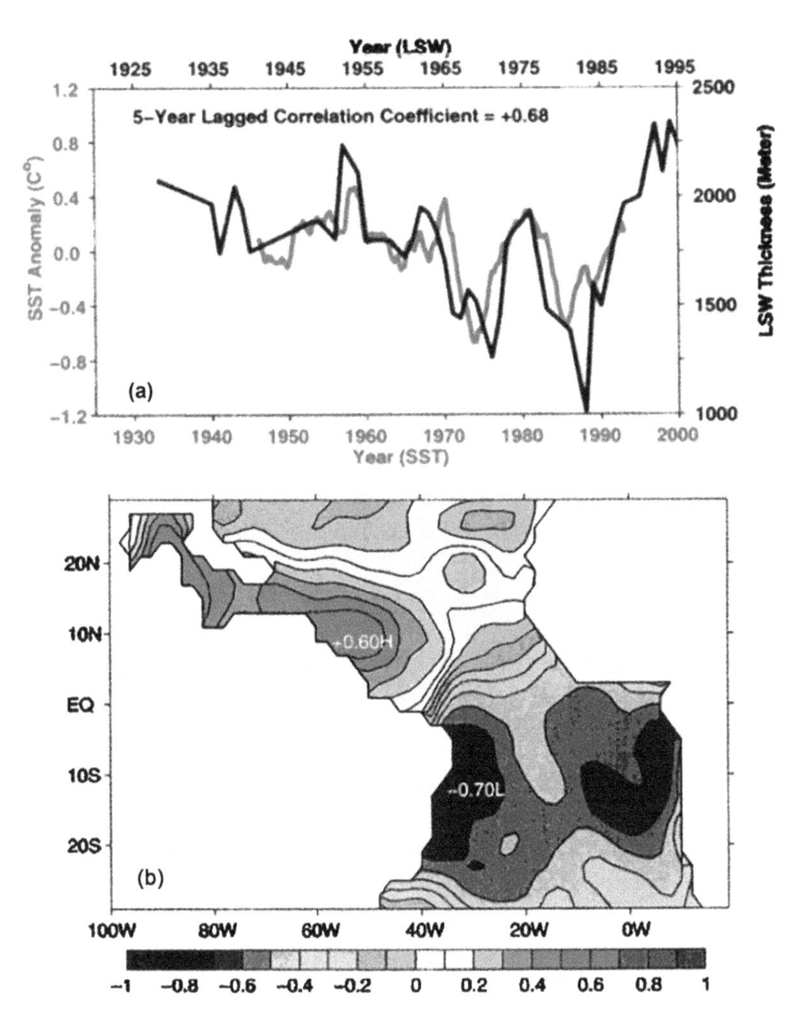

FIGURE 4.9 (a) LSW thickness (black line; right scale) and tropical Atlantic dipole index with a 5-year lag (red line; left scale). The correlation coefficient between the two is shown. (b) Correlation coefficients between LSW thickness and gridded, tropical Atlantic SST anomalies with a five-year lag. (From Yang, 1999.)

hypothesized that decadal variations in LSW thickness may be communicated to the tropical Atlantic via coastally trapped waves which could impact northward heat transport when they reach the tropical Atlantic, which can generate a dipole-like SST anomaly pattern in the tropical Atlantic. Although Yang (1999) tested this hypothesis with an ocean-only model of the Atlantic forced by idealized, cyclic forcing with a 15-year period, and found that a dipole-like SST anomaly pattern can be generated by LSW pulses, several unknowns in the entire chain of hypothesized events from LSW formation to the dipole generation were pointed out. Yang (1999) also noted that the simulated dipole pattern occurred off the western boundary of the tropical Atlantic and not off the eastern boundary as observed. So, in the absence of more convincing data and simulations of this hypothesized oceanic connection from the mid-latitude North Atlantic to the tropical Atlantic, Yang (1999)'s empirically based hypothesis must be considered suggestive rather than conclusive.

4.2.3 Synthesis

A dipole or bipolar pattern was (is) produced by EOF analyses of tropical Atlantic SST anomalies, and maps of correlation coefficients between gridded SST anomalies and rainfall anomalies in Northeast Brazil and West Africa. REOF analyses of tropical Atlantic SST anomalies produced individual patterns in tropical North and South Atlantic. A variety of spectrum and cross-spectrum analyses of century-long and longer SST anomaly time series showed that somewhat different variability spectra existed in tropical North and South Atlantic, with the presence or absence of significant spectral peaks dependent on data set and analysis techniques. However, some of these analyses showed that there was a dominant decadal timescale (as a significant spectral peak) only in the tropical South Atlantic and that a significant decadal spectral peak in the SST PC time series was largely contributed by such a peak in tropical South Atlantic. Correlation and cross-spectral analyses of time series at the tropical North and South Atlantic centers of action showed that the EOF and correlation patterns should be interpreted as variability of cross-equatorial or inter-hemispheric SST gradient and not as a dynamical/thermodynamical mode of tropical Atlantic SST variability.

These results and conclusions strongly recommend that extreme caution is required in interpreting EOF-PC patterns as modes of climate variability as demonstrated by Mehta (1998) and Dommenget and Latif (2002). Long before this tropical Atlantic dipole controversy emerged from analyses of observations, it was well known (see, for example, Horel (1981)) that despite its ability to reduce the number of degrees of freedom of a large data set into a relatively few energetic patterns of variations/variability, spatial orthogonality of the eigenvectors is a strong constraint imposed on the resulting empirical patterns. Once the pattern containing the most variance is determined by the EOF analysis technique, the subsequent patterns are often predictable because of the orthogonality constraint. This constraint is also responsible for the dependence of the empirical patterns on the geographical domain of the EOF analysis. Linearly combining a few EOF patterns into REOFs (see, for example, Horel (1981) and references therein) alleviates some of these problems, but the phase difference between coefficients within an EOF pattern can only be 0° or 180°, forcing individual EOF patterns (simple or rotated) to contain only standing oscillations. Complex EOF analyses (Rasmusson et al., 1981; Barnett, 1983; Horel, 1984; Mehta, 1998) can reveal stationary and traveling variations in the same empirical pattern, but the patterns in these three types of EOF analysis are obtained statistically and are not required to have any dynamical significance.

Major conclusions from analyses of several SST data sets employing a variety of techniques are as follows: (1) there is no dipole mode of tropical Atlantic SST variability and that the bipolar pattern should be interpreted as cross-equatorial SST gradient – defined as TAG earlier in the chapter – variability due to SST variability in tropical North or South Atlantic; and (2) the dominant decadal spectral peak in the

SST data is largely in the tropical South Atlantic. Thus, from the observational point of view, the dipole controversy is finally resolved and laid to rest. However, we will see in Section 4.4 that the controversy has instigated a variety of modeling studies which have provided valuable insights into how ocean–atmosphere systems such as the Atlantic system can generate natural climate variability.

4.3 HYPOTHESIZED MECHANISMS

Several possible mechanisms have been proposed as the cause(s) of the tropical Atlantic decadal variability. These mechanisms and their simulations by a variety of ocean and coupled ocean–atmosphere models are described in this section. First, the tropical Atlantic decadal variability as a coupled ocean–atmosphere mode and as an integrated oceanic response to atmospheric forcings is described. Then, simulations and verifications of hypothesized ocean–atmosphere feedbacks are described. Finally, after the descriptions of hypothesized mechanisms and their simulations, a possible manifestation of the centuries-old belief that solar variability forces DCV, as described and discussed in Chapters 2 and 3, is outlined in the context of tropical Atlantic decadal variability.

4.3.1 Coupled Ocean–Atmosphere Interactions in the Tropical Atlantic Region

One of the first, if not the first, mechanism proposed for decadal oscillations of the assumed tropical Atlantic SSTs was a set of unstable, coupled ocean–atmosphere interactions in which surface heat flux provided a positive feedback to SST anomalies and inter-hemispheric, meridional, ocean heat transport provided a negative feedback. This hypothesis by Chang et al. (1997) was based on a joint analysis of observed SST–surface wind stress–surface heat flux data from the COADS (Da Silva et al., 1994a) from 1950 to 1989 CE. This analysis showed a set of patterns similar to those found in SST–surface wind stress analysis by Ruiz-Barradas et al. (2000) (Figure 4.6) and also generally similar to the SST – 850 hPa wind anomaly patterns in the positive phase of the TAG index (opposite signs of anomalies to those in Figure 4.7). The high positive correlation between SST and surface heat flux anomalies, called the wind–evaporation–SST (WES) feedback in subsequent studies, prompted Chang et al. (1997) to propose that an unstable thermodynamic mode of the tropical Atlantic ocean–atmosphere system can exist due to this feedback. An intermediate coupled model consisting of an empirical wind stress pattern responding to SST anomalies and a reduced-gravity ocean model was constructed and employed to test the hypothesis. In this model, there were three free parameters – coupling parameters α for dynamical feedback and β for thermodynamical feedback, and a damping timescale γ^{-1} – which were varied to study model solutions as functions of these parameters. The model solution for $\alpha = 0$, $\beta = 1.05$, and $\gamma^{-1} = 150$ days is shown in Figure 4.10. Chang et al. (1997) found that the strength of the thermodynamical coupling parameter β controlled the oscillation period and the $\beta = 1.05$ value produced an oscillation of the bipolar SST pattern with a 13-year period without any dynamical feedback ($\alpha = 0$).

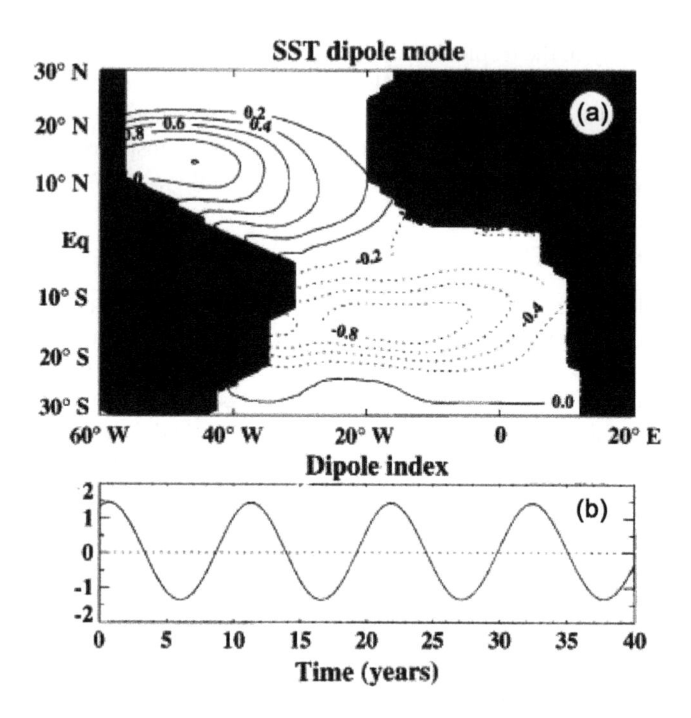

FIGURE 4.10 (a) The spatial pattern obtained by regressing simulated SST on the dipole index in (b), and (b) dipole index formed from the difference between SSTs averaged in 18° longitude bands at 15°N and 15°S latitudes. The contour interval in (a) is 0.2°C per standard deviation of the dipole index which is approximately 1°C. (Adapted from Chang et al., 1997.)

Chang et al. (2001) tested their hypothesis further by coupling an empirical atmospheric feedback model to an ocean GCM. As in their 1997 CE study, they conducted a parameter sensitivity study of this hybrid coupled model and showed that local air–sea feedbacks could support a self-sustained decadal oscillation that exhibited a strong cross-equatorial SST gradient and meridional wind variability if the ocean–atmosphere coupling is strong. As in the intermediate coupled model employed by Chang et al. (1997), the oscillation in the hybrid coupled model resulted from an imbalance between the positive surface heat flux–SST feedback and negative ocean heat transport feedback. The major imbalance between the two feedbacks in the model occurred in the tropical North Atlantic between 5°N and 15°N, suggesting that this may be one of the most important regions of tropical Atlantic Ocean–atmosphere interactions. The study suggested, however, that local ocean–atmosphere coupling in the tropical Atlantic may not be strong enough to maintain a self-sustained oscillation, and stochastic forcing may be necessary to excite and sustain the coupled variability. The Chang et al. (2001) study further showed that local ocean–atmosphere feedback and the NAO-dominated stochastic forcing were both required to simulate realistic tropical Atlantic SST variability. In the absence of the local ocean–atmosphere feedback, the stochastic forcing could produce substantial SST anomalies in the subtropical eastern Atlantic. The local ocean–atmosphere feedback appeared to be particularly important for generating the co-varying inter-hemispheric SST gradient and cross-equatorial wind in the tropical Atlantic. However, Chang et al. (2001) also found that a very strong local coupling could lead to an exaggerated tropical response and, therefore, it was more likely that the ocean–atmosphere feedback enhanced the persistence of the cross-equatorial gradient of SST and wind anomalies, while the NAO provided an important external forcing to excite the coupled ocean–atmosphere variability. As a caveat, Chang et al. (2001) cautioned that the stochastic forcing can significantly weaken inter-hemispheric SST correlation and thereby weaken the TAG pattern.

In between the two Chang et al. studies, Xie (1999) demonstrated with an idealized coupled ocean–atmosphere model of the tropical Atlantic the possibility that an unstable inter-hemispheric mode of tropical Atlantic variability could exist due to the positive WES feedback and that the coupling strength determined the oscillation period of this mode; the stronger the coupling, the longer the oscillation period due to the phase difference between zonal wind and SST anomalies. As in the Chang et al. studies, Xie (1999) also found that the amplitude growth rate of this unstable coupled mode was less than a realistic damping rate and a stochastic forcing would be required to generate the inter-hemispheric mode whose variability would then show a "red noise" spectrum rather than a preferred timescale. Thus, these three pioneering studies identified fundamental physical processes which could generate coupled, tropical Atlantic Ocean–atmosphere variability and set benchmarks for detailed observational verification as well as simulations with more complex models. However, as we saw in Section 4.2 and in this section so far, whether or not the tropical Atlantic SST variability has a preferred decadal timescale or shows a "red noise" behavior has vexed researchers and is a controversy in itself. This will be addressed further in the next sub-section.

4.3.2 Integrated Response of the Tropical Atlantic Ocean to Atmospheric Forcings

As observational evidence of tropical Atlantic SST variability at interannual to decadal timescales and its association with rainfall variability in West Africa and Northeast Brazil began to mount in the early to mid-1990s CE, tentative attempts to simulate the SST variability and understand its mechanisms also began as described in the previous sub-section. Coupled ocean–atmosphere modeling and understanding of coupled feedbacks were confined largely to the Pacific ENSO phenomenon then and modeling Atlantic variability was largely the province of stand-alone ocean and atmosphere models. Of course, the Hasselmann (1976) hypothesis about the oceans acting as an integrator of stochastic atmospheric forcings and producing a "red noise" spectrum was well known, but what was not known about Atlantic variability was whether oceanic Kelvin and Rossby waves, and zonal and meridional circulations played any role in the variability, especially to produce distinct oscillation timescales. One of the first attempts to simulate the tropical Atlantic inter-hemispheric SST variability was by Carton et al. (1996), pre-dating the idealized coupled ocean–atmosphere modeling studies described in the previous sub-section. They conducted simulation experiments with an ocean GCM forced with monthly average wind stress for the 1961–1990 CE period from COADS. The model was configured for the Atlantic Ocean from 30°S to 50°N latitudes. Carton et al. (1996) found that when forced with the observed wind stress, the model simulated observed inter-hemispheric SST variability reasonably well. They found that variability of wind-driven evaporation from the ocean was the most important mechanism driving the SST variability. In addition to the interannual and longer timescale variability of the heat flux, the integration of the forcings by the mixed layer of the ocean model generated decadal timescales. This was the beginning of the idea that the WES mechanism described in Section 4.3.1 played a major role in the tropical Atlantic SST variability.

Hot on the heels of the Carton et al. (1996) study came a generally similar study by Huang and Shukla (1997) which confirmed the conclusion that an ocean GCM configured for the tropical Atlantic Ocean between 30°S and 50°N can simulate observed inter-hemispheric SST variability when forced by monthly, objectively analyzed historical ship observations (Servain et al., 1987) of surface wind stress for the 1964–1987 CE period and parameterized surface heat flux. Additionally, upper-ocean heat content was found to play an important role in adjustments of the tropical Atlantic Ocean to basin-wide wind variations forcing the inter-hemispheric SST gradient variations. Also, this study added an important oceanic process in generating SST anomalies off the North African coast – namely, deepening and shoaling of the thermocline in response to weakening and strengthening trade winds, respectively, – and identified effects of the WES feedback mainly in the western part of the tropical North Atlantic. However, this study also showed that anti-symmetric trade wind anomalies in the tropical North and South Atlantic regions were necessary to generate the inter-hemispheric SST pattern. Consequently, the origin of such an anti-symmetric wind pattern remained a mystery until it was shown by atmospheric GCM experiments that SST anomalies even in one Hemisphere can generate such cross-equatorial anti-symmetric wind anomalies (see, for example, Chang et al. (2000)).

During the mid-1990s CE, as excitement started building about natural decadal–multidecadal climate variability generated by ocean–atmosphere interactions, several researchers started taking tentative steps in analyzing existing century-long or longer runs of global coupled ocean–atmosphere models to see if these runs showed any evidence of internally generated decadal and longer timescale variability. In keeping with hypotheses about coupled ocean–atmosphere modes of variability in the tropical Pacific and tropical Atlantic regions, the search started for such modes in global coupled models. Long control runs of such models were already available in large modeling groups in the United States of America (USA) and Europe which were engaged in simulating effects of increasing carbon dioxide (CO_2) on the Earth's climate. All external forcings such as solar radiation and atmospheric constituents were held fixed at their annual cycle levels without any interannual and longer timescale variability in the control runs, so any such variability generated by the models was entirely due to processes in the modeled ocean–atmosphere system. In analyses of last 100 years of a 200-year run of the Geophysical Fluid Dynamics Laboratory (GFDL) R15 model, Mehta and Delworth (1995) found that the model-simulated SST anomalies in the tropical Atlantic underwent quasi-oscillatory decadal variability. They also found that variability at 8 to 11-year periods were confined to each hemisphere in the tropical Atlantic and variability at 12 to 20-year periods had weak coherence between the Northern and

Southern Hemispheres. It was suggested by Mehta and Delworth (1995) that the dominant bipolar EOF pattern of simulated SSTs should be interpreted as decadal variability of inter-hemispheric SST gradient. This suggestion was later confirmed by Mehta (1998) from analyses of SST data as described in Section 4.2.1. Then, Delworth and Mehta (1998) conducted further analyses of the physics of the tropical North Atlantic decadal variability in the GFDL R15 global coupled model. They found that the northeasterly trade winds weakened over warm decadal SST anomaly pattern in the tropical North Atlantic and these weakened winds reduced evaporation and, therefore, also reduced latent heat loss from the oceans, warming the SSTs further. They also found that the weakened trade winds weakened the subtropical ocean gyre circulation, reducing (relatively) cold water advection on the eastern side of the tropical North Atlantic and also reducing coastal upwelling off the west coast of tropical North Africa. Both these oceanic processes warmed the SSTs further near the tropical North Africa coast. Finally, the Delworth and Mehta (1998) analyses revealed clockwise (anti-cyclonic) rotation of SST anomalies and even more prominent clockwise rotation of sub-surface ocean temperature anomalies in the subtropical gyre. This clockwise rotation of SST anomalies was generally similar to that seen in the Mehta (1998) analyses of observed SSTs and also to that in Hansen and Bezdek (1996) and Sutton and Allen (1997). Thus, in this relatively low-resolution, global coupled model in which average ocean heat transports were weaker than observed, the WES feedback appeared to dominate tropical North Atlantic SST variability at the decadal timescale, with contributions from ocean heat transports and coastal upwelling on the eastern side of the tropical North Atlantic. In a 500-year simulation with the same GCM of the atmosphere coupled to a 50m deep "slab" ocean which had prescribed annual cycle of heat transports, Delworth and Mehta (1998) found that the same tropical North Atlantic SST anomaly pattern as found in the fully coupled model was simulated by the slab coupled model and that the pattern was generated by ocean–atmosphere heat fluxes without any interannual variability in ocean heat transports. These experiments confirmed the main conclusion from the fully coupled model that ocean–atmosphere heat fluxes drove the observed SST pattern. Later, Seager et al. (2001) confirmed the main conclusion of the Delworth and Mehta (1998) study with a model of the atmospheric mixed layer below clouds coupled to an ocean GCM and the same atmospheric model also coupled to mixed-layer ocean models. Changes in surface heat fluxes forced by atmospheric circulation changes in the tropical North Atlantic region generated an SST variability pattern resembling the observed pattern. Seager et al. (2001) also found that ocean heat transport changes dominated by horizontal advection of anomalous temperatures by average meridional currents provided a negative feedback to SST anomalies. Without this feedback, the simulated SST anomalies were significantly overestimated, strongly implying that this feedback was an essential process involved in variability of tropical North Atlantic SSTs. From their simulation results, Seager et al. (2001) concluded that the tropical North Atlantic Ocean played a largely passive and damping role in SST variability. Almost concurrently, Dommenget and Latif (2000) analyzed tropical Atlantic variability in four coupled ocean–atmosphere GCMs (ECHAM4-HOPE2 (118 years run), ECHAM4-OPYC (240 years), ECHAM3-LSG (700 years), and GFDL-MOM (1,000 years). The SST variability was found to be strongly correlated with wind stress anomalies in the tropical Atlantic trade wind regions in these models. All four coupled GCMs simulated independent SST variability in tropical North and South Atlantic as shown by ROEF patterns in Figure 4.11 although the relative strengths of the two patterns are not the same in the four coupled GCMs. Fourier spectra of the PC time series corresponding to the two sets of REOFs show a generally "red noise" behavior without any significant decadal peak although the spectra corresponding to the tropical South Atlantic REOFs show larger spectral densities around the decadal timescale. Convinced by these tropical Atlantic SST variability patterns and spectra that these patterns can be explained by the atmosphere forcing the ocean with only a minor role played by ocean dynamics, Dommenget and Latif (2000) coupled ECHAM3 atmosphere GCM to a 50m deep ocean mixed-layer model. From their analyses of the variability simulated by this coupled model, Dommenget and Latif (2000) concluded that the variability of the upper tropical Atlantic Ocean was forced by the atmosphere, and that oceanic processes such as convection, advection, and wave propagation were not important in producing the basic spatial structure of the variability.

Xie and Tanimoto (1998) found that the tropical Atlantic decadal variability appeared to be associated with a coherent pan-Atlantic decadal oscillation characterized by zonal bands of SST and wind anomalies in the meridional direction with alternate polarities from the South Atlantic to Greenland. Experiments with a coupled ocean–atmosphere model suggested that possible interactions among wind, evaporation, and SST were proposed as important in establishing this pan-Atlantic decadal variability. Xie and Tanimoto (1998) hypothesized, based on model experiments, that extratropical wind forcing may be responsible for observed decadal variability in SST and winds in the tropical Atlantic region, in contrast to the usually invoked tropics to extratropics teleconnection.

The possible role of the WES feedback in tropical Atlantic variability in present-day climate was also the subject of another coupled atmosphere-mixed-layer ocean model study by Breugem et al. (2007). They found that in the present-day climate run, the WES feedback is confined to the north tropical Atlantic region. They also found that the temporal evolution of the meridional pattern is phase-locked with the seasonal cycle of the ITCZ. The WES feedback is positive in boreal winter and spring when the ITCZ is located close to the equator, but the feedback is negative in summer and autumn when the ITCZ is further north. Similarly, the damping of the SST anomalies in north tropical Atlantic by evaporative latent heat loss anomalies is much stronger in summer and autumn than in winter and spring, related to a change in anomalous water vapor transport.

4.3.3 Ocean–Atmosphere Feedbacks – Simulations and Verifications

We saw in sub-Sections 4.3.1 and 4.3.2 that ocean–atmosphere feedbacks are key components of the mechanisms hypothesized for the tropical Atlantic ocean–atmosphere variability. We also saw that the positive WES feedback between trade winds and tropical Atlantic SSTs, and the negative ocean heat transport feedback via meridional currents are the main proposed feedbacks. We will see in this sub-section simulations of these feedbacks by models and verifications by observed data.

Let us begin by recapitulating what the hypothesized WES feedback is. In the tropical Atlantic region, trade winds north of the ITCZ are from the northeast or the

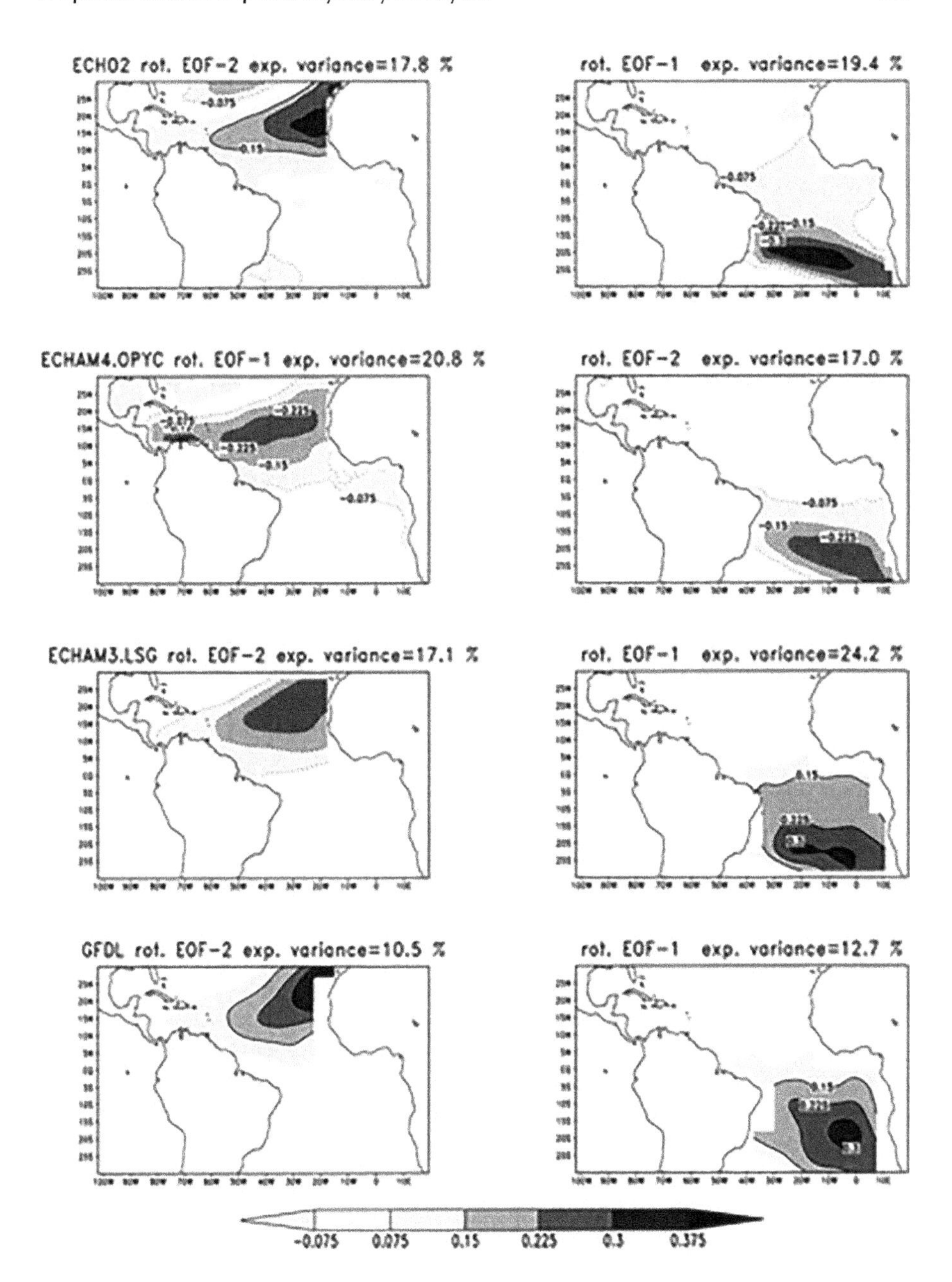

FIGURE 4.11 The two leading rotated EOFs of simulations with four coupled ocean–atmosphere GCMs. EOF patterns with maxima in the Northern Hemisphere are on the left-hand side, and the patterns with maxima in the Southern Hemisphere are on the right-hand side. (From Dommenget and Latif, 2000. © American Meteorological Society. Used with permission.)

east and from the southeast or the east south of the ITCZ. If there is a warm SST anomaly (above-average SST) in the central or eastern tropical Atlantic, there is a low-level, anomalous wind convergence over this warm SST anomaly from all sides. This converging anomalous wind pattern weakens the total trade winds which, in turn, reduces total evaporation over the warm SST anomaly. The reduced evaporation reduces latent heat loss from the upper ocean which, in turn, warms the warm SST anomaly further. Thus, there can be a positive feedback to the initial warm SST anomaly, especially in the central and eastern parts of the tropical Atlantic. This interaction among SST anomaly, winds, evaporation, and latent heat can also work in the case of a cold SST anomaly (below-average SST) in the trade wind region. This positive feedback among winds, evaporation, and SST is known as the WES feedback which was originally proposed by Xie and Philander (1994) in the context of ITCZ dynamics over the eastern Pacific and eastern Atlantic. As we saw in the previous two sub-sections, the idea that such a feedback could be operating in the tropical Atlantic began with the Carton et al. (1996) simulation with an ocean GCM, and was developed further by Chang et al. (1997, 2001) and Xie (1999) in their simulation with idealized coupled ocean–atmosphere models which produced TAG variability at decadal timescales for specific values of free parameters.

These studies set off an exciting flurry of research to estimate the WES feedback from observed data and to simulate it with other models even though it was very conclusively established by 1999–2000 CE that observed SST data did not support the assumption of a tropical Atlantic SST dipole mode at decadal timescale. In this second round of research following the initial modeling studies, Sutton et al. (2000), Czaja et al. (2002), and Frankignoul and Kestenare (2002) questioned the very existence of the WES feedback. Sutton et al. (2000) analyzed an ensemble of six simulations with the HadAM1, a GCM of the atmosphere, integrated with observed SST and sea-ice data (GISST; Parker et al., 1995b) from 1 October 1948 to 1 December 1993 CE. They found, in conformity with empirical analyses results described in Section 4.2, that simulated winds in response to the tropical Atlantic inter-hemispheric SST gradient have maximum influence on climate of the neighboring continents from March to August. They also found that simulated latent heat flux anomalies were such that they would provide a negative feedback to SST anomalies in the tropical South Atlantic, and that latent heat anomalies in the tropical North Atlantic were equatorward of the maximum SST anomalies and would not provide any feedback to the SST anomalies. Based on these results, Sutton et al. (2000) ruled out any positive WES feedback in any season. Czaja et al. (2002)'s conclusions, based on their analyses of surface heat flux data from the NCEP-NCAR reanalysis (Kalnay et al., 1996) for the 1950–1999 CE period, were similar for the tropical North Atlantic, the only region in their study. As Sutton et al. (2000) did from model simulations, Czaja et al. (2002) found from the reanalysis data that surface heat flux anomalies were too far south of the maximum SST anomalies to provide the positive WES feedback. Incidentally, they also concluded that all strong SST anomalies in the tropical North Atlantic during the study period were forced either by ENSO events or by substantial excursions of the NAO via their influences on the surface heat flux in the tropical North Atlantic. Frankignoul and Kestenare (2002, 2005)'s analyses of the 1958–2002 CE NCEP-NCAR reanalysis data also found that wind response to tropical North Atlantic SST anomalies was weak and surface heat flux feedback was negative north of approximately 10°N where the SST anomalies are largest. Frankignoul and Kestenare (2002) suggested that the apparent positive WES feedback in the Chang et al. (1997) data analysis was due to ENSO effects and statistical degeneracy. Thus, the consensus from these atmosphere-only analyses was that the WES feedback was either absent or too far away from the maximum SST anomalies to further strengthen the anomalies.

Several other researchers analyzed time series of various lengths of tropical Atlantic SSTs, gridded sea level pressure and surface wind stress data, atmospheric operational analysis and reanalysis data, and satellite-derived outgoing longwave radiation and cloudiness data to diagnose the WES feedback and other possible effects of ocean surface–lower atmosphere interactions. These time series were long enough to analyze interannual variability in detail, but not decadal variability; some of the analyses, however, did show two to three cycles of decadal variations. The inferences drawn from these analyses shed an interesting light on possible physics of interannual variability and perhaps also on longer timescale variability, especially on a positive feedback mechanism between SST anomalies, evaporation (latent heat) anomalies, and trade wind anomalies. Therefore, major results from these analyses are summarized here. Hu and Huang (2006) found, in their analyses of the NCEP-NCAR atmospheric reanalysis data from 1955 to 1996 CE, that warming or cooling of SSTs along 10°N to 15°N latitudes was associated with weaker or stronger trade winds in the boreal spring. Warmer SST anomalies increased anomalous wind convergence (westerly wind anomalies to the west and easterly wind anomalies to the east of the SST anomalies) at low levels in the atmosphere which weakened trade winds to the west of the SST anomalies because the climatological trade winds in that region are easterly. This weaker total zonal wind decreased evaporation, thereby warming the SSTs further as per the WES hypothesis. This process could also work to further cool cooler SST anomalies. Hu and Huang (2006) also found that reduced evaporation over warm SST anomalies can reduce low clouds in the trade wind region which can increase solar radiative forcing of SST anomalies, warming them further. Huang and Hu (2007) analyzed cloud estimates in the JJA season from the International Satellite Cloud Climatology Project from 1984 to 2004 CE and found that the WES feedback – low cloud – increased solar radiation relationship appeared to work in the tropical southeastern Atlantic region close to 15°S latitude near the Angola coast. These studies also found that warm SST anomalies resulting from this relationship expand westward and northwestward, possibly due to oceanic Rossby waves, and that atmospheric response to these SST anomalies can generate warm equatorial SST anomalies which can be intensified by ocean dynamical feedbacks. These equatorial SST anomalies can cause a negative feedback to the SST anomalies near the Angola coast. Thus, there appears to be a dynamical ocean–atmosphere coupling between the equatorial region and tropical southeast Atlantic region at interannual timescales.

The hypothesis that the TAG pattern was a coupled ocean–atmosphere mode of variability had another very important feedback; namely, a negative ocean heat transport feedback to drain the heat from the warm pole to the cold pole of the TAG pattern. Two groups of researchers attempted to estimate both feedbacks simultaneously from observed data to see if the two together could generate coupled ocean–atmosphere variability. Joyce et al. (2004) used sub-surface ocean temperatures measured in the upper 150 m by mechanical bathythermograph (MBT) and expendable bathythermograph (XBT) instruments. Individual temperature measurements were aggregated in 2° longitude–1° latitude grid boxes, and annual values of temperatures were calculated at 0, 50, 100, and 150 m depths (see Appendix in Joyce et al. (2004) for quality check and gridding techniques) from 1950 to 1998 CE. They also used monthly wind stress from the NCEP-NCAR reanalysis data. Output from an ocean GCM (NCOM; Doney et al., 2003; Capotondi and Alexander, 2001) of the tropical Atlantic Ocean, driven by observations-based surface heat, momentum, and freshwater fluxes, and six-hourly NCEP-NCAR reanalysis data from 1958 to 1997 CE was also used in the Joyce et al. (2004) study. They found that at the latitudes of the strongest SST anomaly signal in the tropical Atlantic, there was a weak, positive WES feedback, but the feedback may be operating only between the Equator and ±10° latitudes from December to May. This latter conclusion is the same as that by Sutton et al. (2000), Czaja et al. (2002), and the Frankignoul and Kestenare (2002, 2005) studies. The new finding in the Joyce et al. (2004) study was that the anomalous cross-equatorial wind stress is closely coupled to the cross-equatorial SST anomaly gradient, such that there is a negative wind stress curl associated with a northward-pointing SST gradient. This anomalous stress curl would drive an anomalous, cross-equatorial Sverdrup transport in the ocean. This anomalous, zonally averaged transport was found to be large (approximately 10% to 20% of the average transport) and negative which would imply heat transport from the warm pole in the north to the cold pole in the south, thus removing heat from the warm pole and providing a negative feedback. Despite this substantial negative feedback via the ocean, Joyce et al. (2004) concluded that the weak, positive WES feedback and the negative ocean heat transport feedback would not be able to generate a coupled ocean–atmosphere mode and that external forcing would be required to sustain the coupled ocean–atmosphere variability.

In another study to examine the role of horizontal oceanic heat advection in the evolution of tropical Atlantic SST anomalies, Foltz and McPhaden (2006) analyzed the mixed-layer heat balance at interannual time scales, using a combination of instrument measurement-based data, atmospheric reanalysis data, and satellite-derived data. They used surface latent and sensible heat fluxes, surface wind speed, and specific humidity from 1981 to 2002 CE; surface shortwave radiation from 1983 to 2004 CE; surface longwave radiation from 1979 to 2004 CE; SST data; and monthly mean climatological mixed-layer depth data (see Section 2 in Foltz and McPhaden (2006) for sources and details of these data). They found that mixed-layer heat storage variations at interannual timescales were driven primarily by latent heat and shortwave radiation fluxes. Latent heat loss provided a negative feedback to SST anomalies. Horizontal heat advection was also found to provide a negative feedback to SST anomalies, with the strongest effects in the 10° to 20° latitude bands. Both anomalous heat advection terms – advection of anomalous meridional temperature gradient by mean poleward Ekman currents and advection of mean meridional temperature gradient by anomalous meridional Ekman currents – were found to be similar in magnitude. Thus, both the Joyce et al. (2004) and Foltz and McPhaden (2006) studies found evidence for a substantial negative feedback due to ocean heat transports; their conclusions about the WES feedback, however, were different.

As the foregoing describes, the WES feedback was either found to be not collocated with strongest SST anomalies or too weak to further strengthen SST anomalies. The negative ocean heat transport feedback, on the other hand, was found to be substantial in transporting heat from warm SST anomalies towards cold SST anomalies after they were formed. As mentioned in Section 4.2.3, however, the modeling studies and analyses of feedbacks described in Section 4.4 have proven to be very interesting and useful in understanding the physics of idealized and actual tropical Atlantic ocean–atmosphere systems. Thus, these studies have made an invaluable contribution to the field of climate science.

4.3.4 Bottom-Up and Top-Down Response to the Schwabe Cycle as a Pace Maker?

We saw in Section 4.2 that analyses of SST data sets have shown that the tropical Atlantic variability, especially in the tropical South Atlantic, has a preferred period of 12–13 years although this conclusion is data set dependent. However, if we accept the ambiguity of this conclusion, which physical processes might give rise to this preferred period if it does exist? We saw in Section 4.3.1 that idealized coupled ocean–atmosphere models do simulate SST variability at 12 **to** 13-year periods, albeit for specific settings of free parameters for which there is apparently no physical basis. Also, more complete and more complex, global coupled models do not simulate the preferred period (see, for example, Mehta et al. (2018)).

In Chapter 2, the bottom-up and top-down hypotheses of how the Schwabe cycle with a nominally 11-year period can influence the Earth's climate and simulation studies by White and Liu (2008a) and Meehl et al. (2009) were cited in the context of the tropical Pacific DCV. In the bottom-up hypothesis, varying visible radiation would be absorbed in relatively cloud-free subtropical oceans. In the solar maximum phase, increased radiation would increase evaporation which would then increase water vapor transport by trade winds into tropical precipitation regions from subtropical regions. Increased precipitation due to the increased water vapor convergence would increase latent heat release in the tropical atmosphere, which would strengthen trade winds and equatorial ocean upwelling, resulting in cooler SSTs. The stronger trade winds would increase subsidence in subtropical latitudes, further reducing clouds and increasing absorption of visible radiation by the subtropical ocean and increasing evaporation. Thus, in the bottom-up hypothesis, positive feedbacks via ocean–atmosphere dynamics and clouds are crucial to amplify responses to original visible radiation changes. In the top-down hypothesis, ultraviolet radiation emitted from the Sun increases at the time of maximum solar activity in the Schwabe cycle. More ultraviolet radiation in the stratosphere increases ozone due to atmospheric chemistry, which, in turn, increases ultraviolet radiation absorption. This positive feedback warms the stratosphere and creates a latitudinal temperature gradient. There is further stratospheric warming due to increased wave motions which amplifies the original warming. This modified stratospheric circulation, in turn, modifies tropical tropospheric circulation causing an enhancement and poleward expansion of the Hadley and Walker Circulations and tropical precipitation maxima. According to the two combined hypotheses, solar influence can permeate from the top of stratosphere down to the surface of the Earth.

Meehl et al. (2009)'s simulations with global models showed that the two sets of physical processes can work together to amplify the Schwabe cycle signal and can generate climate variability in the tropical Pacific region with an approximately 11-year period. White and Liu (2008a) showed with an idealized coupled ocean–atmosphere model forced by an idealized Schwabe cycle that the forcing signal can excite a resonant mode of 9 to 13-year period in the tropical Pacific ocean–atmosphere system. Since their model encompasses the tropical Atlantic also, some response of the tropical Atlantic SSTs to the idealized Schwabe cycle forcing can be seen (White and Liu, 2008a; Figure 2b), although the response is more uniform (same sign, varying amplitudes) in the tropical Atlantic than the observed decadal SST variability. These model-simulated results are certainly intriguing and suggest that the Schwabe cycle operating via the bottom-up and top-down (and other, as yet unknown) mechanisms may be able to generate the observed 12 to 13-year period variability. Further research to test this intriguing and exciting possibility is highly necessary because it might provide further insights not only into the physics of the tropical Atlantic decadal variability but also into potential predictability of the variability and its societal impacts.

4.4 THE TROPICAL ATLANTIC VARIABILITY AND THE ATLANTIC MULTIDECADAL OSCILLATION

The Atlantic Multidecadal Oscillation (AMO) is prominent due to its association with the AMOC, and also due to empirical associations between the AMO and hydro-meteorological and river flow variability via the AMO's impacts on the global atmosphere. Although this chapter is about the tropical Atlantic SST dipole or TAG variability, a logical question to pose is: What is relationship between the TAG and the AMO? This short section attempts to address this question with empirical and global coupled model data analyses.

The AMO index is defined as detrended, low-pass filtered, annual-average SST anomalies averaged over the entire North Atlantic from the Equator to 60°N. (Enfield et al., 2001; Knight et al., 2005). By definition, there is a geographical overlap in the TAG and the AMO from the Equator to 20°N, so a significant correlation between the tropical North Atlantic component of the TAG and the AMO should be expected. In ERSST version 5 data from 1900 to 2018 CE, the annual and seasonal correlation coefficients between SST anomalies averaged in the tropical North Atlantic and the AMO index are moderate to high (annual 0.73, DJF 0.63, MAM 0.69, JJA 0.62, and SON 0.69) as expected. However, the annual and seasonal correlation coefficients between the AMO index and the tropical South Atlantic component of the TAG are insignificantly small (annual 0.10, DJF 0.19, MAM -0.01, JJA 0.09, and SON 0.05). These latter statistics clearly imply not only that there is no geographical overlap (by definition) between the AMO and the tropical South Atlantic, but also that the AMO does not include the cross-equatorial SST gradient contribution from the tropical South Atlantic and, hence, cannot include impacts on the tropical Atlantic atmosphere and the global atmosphere from the tropical South Atlantic.

As we have seen in this chapter, the observed TAG index's spectrum has the dominant peak at 12–13 years, although the presence or absence of this spectral peak is data set-dependent as noted in Section 4.2. In contrast, the observed AMO index time series shows, by definition, multidecadal variations (see, for example, Zhang and Delworth, 2007). Zhang and Wang (2013) show that the spectra of the AMO index, calculated with observed data and also with data from 27 global coupled GCMs participating in the Coupled Model Intercomparison Project version 5 (CMIP5) project (Taylor et al., 2012), have a barely significant peak at 9 to 10-year periods in the observed data and no peak at these periods in the average CMIP5 spectra, but have the dominant peak in all spectra at 45 to 80-year periods (Zhang and Wang, 2013). Thus, although there is a geographical overlap in the definitions of the TAG and the AMO, the TAG variability is much more representative of the *decadal* variability of the tropical Atlantic climate than the AMO.

4.5 SUMMARY AND CONCLUSIONS

Similar to the Pacific region, decadal–multidecadal climate variability in the tropical Atlantic region has assumed increasing importance to satisfy intellectual curiosity, and to predict the variability and its worldwide impacts on vital societal sectors. Specifically, the tropical Atlantic SST dipole variability at decadal timescales has been a highly controversial phenomenon for almost 30 years. Description, understanding, and resolution of this phenomenon and attendant controversy are the subject of this chapter.

A brief description of the origin of the dipole controversy began this chapter, followed by a history of the evolution of empirical knowledge of tropical Atlantic climate variability. Then, several detailed analyses of century-long and longer time series of SST data up to 2018 CE were described. These analyses showed conclusively that there is nearly independent SST variability in the tropical North and South Atlantic. These analyses also showed conclusively that there is no tropical Atlantic SST dipole mode at decadal timescales and that decadal dipole variability pattern revealed by statistical analyses should be interpreted as variability of the inter-hemispheric SST gradient or TAG. There is also strong evidence of a significant spectral peak at approximately 12 to 13-year period in the tropical Atlantic SST variability, especially in the tropical South Atlantic SSTs. However, the presence or absence of this spectral peak apparently depends on the SST data set, the spectral analysis and significance testing techniques used, and the time span of the analyses. For example, multidecadal variability appears to be stronger than decadal variability in the first half of the 20th century CE in the tropical Atlantic, and the latter appears to be stronger since then. Thus, from the observational point of view, the dipole controversy is finally laid to rest with the conclusion that there is no dipole mode of variability of the tropical Atlantic coupled ocean–atmosphere system.

Empirical analyses also showed a substantial evidence of SST anomalies traveling clockwise in the North Atlantic and counter-clockwise in the South Atlantic at decadal timescales. These traveling anomalies appear to be riding on the subtropical ocean gyres in both hemispheres. Their natural alignments on both sides of the Equator occasionally create anomalies of the same sign or opposite signs on two sides of the Equator, thus creating either a broad anomaly pattern spanning the entire tropical Atlantic or a bipolar pattern of inter-hemispheric SST gradient, the TAG. As is well known, a northward (southward) pointing TAG and the associated anomalous wind pattern draws the ITCZ northward (southward). This movement of the ITCZ at decadal timescales causes decadal variability of rainfall in West Africa and Northeast Brazil specifically, and more generally, in other parts of the world as well (Mehta, 2017). The lack of an empirical relationship between the AMO and the TAG variability at decadal timescale was briefly demonstrated.

In the early days of modeling tropical Atlantic variability, experiments with idealized coupled ocean–atmosphere models with built-in feedbacks based on empirical SST–wind stress–surface heat flux patterns showed that such models could generate dipole variability at decadal timescales due to a strong, positive thermodynamic feedback between SST and surface heat flux anomalies. However, such model solutions occurred only for specific sets of scaling parameters in the models and it was argued that damping of SST anomalies would be too strong in the real world for such unstable modes to occur. Therefore, it was argued that stochastic forcing would be needed to generate variability at decadal timescales in the real world, but that such forcing would weaken the TAG pattern and hence there would be very weak or no TAG variability. Experiments with ocean-only and coupled ocean–atmosphere GCMs showed that SST and sub-surface ocean variability can occur in the tropical Atlantic, especially in the tropical North Atlantic, due to surface wind forcing and that the WES feedback can strengthen warm/cold SST anomalies, especially on the eastern side of the basin. These coupled GCM experiments also showed that surface wind forcing can spin up/down subtropical gyres in both hemispheres and can cause SST anomalies on the eastern side of the basin due to changes in meridional heat advection and coastal upwelling. For example, westerly wind anomalies converging on warm SST anomalies in the tropical North Atlantic trade wind region would weaken the total winds and spin down the subtropical gyre. This spin down, in turn, would reduce cold water advection from the north along the West Africa coastline, and the weaker trade winds would also reduce coastal upwelling, both processes resulting in SST warming along the West Africa coast. A similar process would result in SST cooling in response to cold SST anomalies. Lastly, the model experiments also showed that a warming in one hemisphere can draw low-level wind convergence over the warm SSTs from across the Equator, thereby strengthening trade winds in the opposite hemisphere. These stronger trade winds, in turn, would produce some cooling of SSTs due to increased evaporative heat loss. This process, however, was not found to be strong enough to actually generate what was assumed to be dipole variability.

Some of the outstanding problems outlined at the end of Chapters 2 and 3 apply here as well. The remaining major questions specific to the tropical Atlantic decadal variability are: (1) Is the spectral peak at 12 to 13-year period in mid-19th to late 20th century CE instrument-measured SSTs a transient or is it a robust feature of tropical Atlantic variability? If it is a robust feature, which physical processes are responsible for it? (2) Does the solar Schwabe cycle act as the overall pace maker and inter-hemispheric synchronizer of this variability? (3) Does the tropical Atlantic decadal variability force extratropical decadal variability and does the latter play any active role in the former? (4) What roles do the AMOC and the NAO play in the tropical Atlantic decadal variability?

Effects of freshwater at mid- and high-latitudes on thermohaline ocean circulations and, consequently, on density and temperature of sea water have been demonstrated by numerous observational and modeling studies. The Yang (1999) study hypothesizing effects of LSW formation on tropical Atlantic variability was described in this chapter. Effects of interannual to decadal variations in atmospheric freshwater inputs to the tropical Atlantic Ocean as the difference between rainfall and evaporation, and of interannual to decadal variations in riverine freshwater inputs to the tropical Atlantic Ocean from the Amazon and other rivers should be studied with state-of-the-art Earth System Models (ESMs). In two exploratory studies with an ocean GCM, Huang and Mehta (2005, 2010) showed that interannual variability of atmospheric freshwater input and partial or complete blockage of outflow from the Amazon River can make substantial impacts on ocean circulations and temperatures not only in the tropical Atlantic but also in mid- to high-latitude Atlantic and in the Pacific and Indian Oceans. Mehta (2017)'s empirical analyses of the Pacific Decadal Oscillation (PDO), the TAG, and the West Pacific Warm Pool (WPWP) SST variability; and river flows in South America showed that there are substantial (10% to 20% of seasonal averages) river flow variations associated with these DCV phenomena. Therefore, possible roles of such river flow variations in the tropical Atlantic Ocean temperature variability should be studied with ESMs.

Another possible effect which should be studied with ESMs is that of low-latitude, large volcanic eruptions. We saw in Chapter 3 that the PDO and the WPWP decadal variability respond substantially to such eruptions which can even change the phase of these two phenomena (Mehta et al., 2018). Both empirical observations and ESM simulations show these responses. However, while associations between low-latitude large eruptions and changes in the TAG variability are apparent in empirical data as Mehta et al. (2018) showed, simulation experiments with four ESMs participating in the CMIP5 project do not show such associations. Analyses of net surface heat flux, ocean temperatures in tropical South Atlantic and tropical North Atlantic, and the TAG index simulated by each ESM show that quite large heat flux anomalies are generated in response to aerosol optical depth (AOD) changes in some of the eruption events, but the ocean temperature responses they evoke are very different from the observed TAG index. Even when there are net surface heat flux anomalies of opposite signs in the tropical North and South Atlantic, temperature responses do not correctly simulate the observed TAG responses (Mehta et al., 2018). So, it appears that these ESMs are deficient in their simulations of the tropical Atlantic Ocean temperature variability and its response to external forcing such as volcanic eruptions. Such a conclusion is consistent with a similar conclusion reached by Wang et al. (2013) and Xue et al. (2012) regarding tropical Atlantic Ocean simulation by ocean models. Therefore, further ESM improvements are required to accurately simulate the TAG variability and effects of external forcings on it.

In the end, we must accept that perhaps the dipole controversy will rear its head again in the future as more sub-surface ocean data and more realistic climate models become available. Also, we must accept that, despite the conclusive evidence described in this chapter, it should not be construed that causes of tropical Atlantic decadal variability are completely understood. As a matter of fact, the relatively short instrument-based, sub-surface ocean observational record and the inability of state-of-the-art climate models to hindcast this variability point strongly to the fact that we still do not understand relative roles of intrinsic variability generated by the tropical Atlantic ocean–atmosphere–land system and forcings due to extrinsic variability generated elsewhere in the Atlantic and also in other Oceans. We also do not understand how moderate-to-strong, low-latitude volcanic eruptions influence/impact the tropical Atlantic climate system. Last but not the least, we still have the unanswered question: Does variability of solar emissions play any role in the tropical Atlantic climate variability? These are the very substantial challenges we leave to future generations of climate scientists to address and possibly solve. For the time being, we can modify Bertrand Russell's opinion cited in the beginning of this chapter as "the most savage controversies can be progressively laid to rest as good evidence accumulates either way".

5 On the Waves of the *Sindhu Mahasagar*

5.1 INTRODUCTION

Inhabitants of the *Sindhu* (Indus) Valley Civilization in the 2600 to 1900 Before Current Era (BCE) period were seafarers. They traded with Mesopotamia, the Persian Gulf region, and Egypt in coastal vessels on a sea they called the *Sindhu* – a Sanskrit word for sea (Alpers, 2013). Later, when they realized the very large size of the sea, they called it the *Sindhu Mahasagar* – *Sindhu* the Great Ocean. In subsequent centuries, Arab seafarers called the northern part of the *Sindhu Mahasagar* the Arabian Sea. Since early 16th century Current Era (CE), the *Sindhu Mahasagar* is known as the Indian Ocean, named for India whose seafarers and traders united cultures on the Ocean's littoral as well as in interior regions of Asia, Africa, and Europe. Also, India's economic domination of the world from the 1st century to the early 19th century CE,[1] culminating in an almost 25% share of the world's economy and industrial output in 1700 CE (Maddison, 2003), justified the naming of the Indian Ocean.

The seafarers and maritime traders traveling in the Indian Ocean experienced seasonal reversals of winds and navigated to benefit from the winds in their voyages across and along the Ocean littoral. They called the wind reversals *mausam* – an Arabic word meaning season. Later, a degenerate form of *mausam* became monsoon as the semi-annual wind reversal is known today. Among those benefiting from the monsoon winds were seafarers from the Gujarat[2] region in western India who are believed to have brought five types of crops from Africa to Gujarat in 2000 to 1700 BCE, which then spread to the rest of the Indian sub-continent (Boivin et al., 2014). The more recent history of the Indian Ocean region is replete with stories of European navigators searching for trade routes to India, the Portuguese navigator Vasco da Gama being the first known European to reach India by the sea route in 1497 CE. Legend has it that he was shown the way to India from the east coast of Africa by a sailor from Gujarat. In Vasco da Gama's wake came spice traders and other merchants, beginning the era of colonization of East Africa, South Asia, and Southeast Asia which provided further impetus for trade and navigation along the Indian Ocean littoral, and with Europe and North America. In the era of colonization, many islands in the Indian Ocean were developed to grow sugar cane, with mass migrations from the Indian sub-continent providing the laborers who traveled on the Indian Ocean. Thus, as this very brief history shows, ancient and modern civilizations have interacted via trade and cultural exchanges on the waves of the *Sindhu*

[1] Balance of Economic Power 1 AD – 2009. http://www.pdviz.com/tag/gdp.

[2] I was born in Bhavnagar, Gujarat, a city in the region known for its ancient seafaring traditions.

Mahasagar for thousands of years. What follows in this chapter is the modern story of how physically dynamic the Indian Ocean is, especially the story of decadal variability of these waves.

Between the eastern end of the tropical Indian Ocean and the western end of the Pacific Ocean are thousands of islands known collectively as the Maritime Continent (MC; Ramage, 1968). These islands are parts of Indonesia, Malaysia, Papua New Guinea, and Philippines. Meteorological and oceanographic processes within of the MC region and the surrounding Oceans produce some of the warmest surface water in the West Pacific and the East Indian Oceans. These two regions of warmest water are known as the West Pacific Warm Pool (WPWP) and the East Indian Warm Pool (EIWP), respectively. Decadal variability of the former was described in Chapter 3. Decadal variability of the EIWP and the rest of the Indian Ocean, including the Arabian Sea and the Bay of Bengal, is described in this chapter. As explained in previous chapters, the phrase decadal variability is used here to denote variability at approximate periods from 8 to 20 years. An interannual climate phenomenon known as the Indian Ocean Dipole (IOD) came to light in the late 1990s CE (Saji et al., 1999). Attributes of the interannual IOD phenomenon are known to vary at decadal timescales. Decadal modulation of the IOD is described in Chapter 12 along with decadal modulation of other higher frequency phenomena.

After this brief introduction, decadal variability of the tropical Indian Ocean, including attributes and role of the EIWP, is described in Section 5.2. Hypothesized mechanisms of the tropical Indian Ocean decadal variability are described in Section 5.3. An exploratory analysis of statistical associations between the tropical Indian Ocean decadal variability and worldwide hydrologic variability is described in Section 5.4. The chapter is summarized, and conclusions are presented in Section 5.5.

5.2 THE TROPICAL INDIAN OCEAN AND ITS DECADAL VARIABILITY

5.2.1 Major Attributes of the Warm Pools and Important Physical Processes

Tropical Warm Pools (TWPs) contain some of the warmest upper-ocean water in the world. The TWPs are characterized by persistently warm sea surface temperature (SST) higher than 28°C, which is a threshold for atmospheric deep convection (see, for example, Fu et al. (1994)). According to this definition, the principal TWPs are located in the Indian Ocean including the Arabian Sea and the Bay of Bengal, the western Pacific Ocean, the western Atlantic Ocean, the Caribbean Sea, and the Gulf of Mexico. There are also small TWPs in the Red Sea and the eastern Pacific Ocean near the coast of Central America. Since saturation vapor pressure is an exponential function of SST (Graham and Barnett, 1987; Fu et al., 1994), the intensity of deep atmospheric convection can be very sensitive to changes in SST over the TWP regions. Even small changes in TWP SST can cause large changes in atmospheric convection, which, in turn, can dramatically alter atmospheric divergent flow locally, and further significantly modify planetary-scale wave activity and atmospheric heating globally (Sardeshmukh and Hoskins, 1988; Neale and Slingo, 2003).

The TWPs also play a very important role in ocean–atmosphere interactions via freshwater fluxes. The WPWP and EIWP together are known as the Indo-Pacific Warm Pool (IPWP) which, with the MC embedded in it, is a region of net 1 to 2 m $year^{-1}$ atmospheric freshwater flux into the ocean (Chen et al., 2004; Huang and Mehta, 2004). The resulting mass and buoyancy fluxes significantly affect the dynamics and thermodynamics of the IPWP via their effects on salinity and ocean–atmosphere heat flux (Lindstrom et al. 1987; Godfrey and Lindstrom 1989; Lukas and Lindstrom 1991; Huang and Mehta 2004, 2005). The clear correspondence of the IPWP with the location of the maximum of total atmospheric heating (Ramage, 1968; Webster and Lukas, 1992) indicates the IPWP's role in setting atmospheric heating gradients and, therefore, its importance in the general circulation of the global atmosphere. The TWPs are thus an important source of energy for driving large-scale atmospheric circulations such as the Hadley and Walker circulations, are important sinks of atmospheric fresh water, and constitute very important components of the global climate system.

The formation of the IPWP is primarily driven by ocean dynamics, especially by tropical easterly winds driving a deep thermocline on the western side of the Pacific Ocean and tropical westerly winds driving a deep thermocline on the eastern side of the Indian Ocean (Clement et al., 2005). Vertical heat advection also plays an important role in the IPWP heat balance (Schneider et al., 1996). Penetrative solar radiation, and atmospheric processes such as evaporative heat flux and cloud-radiation feedback also play important roles (Ramanathan and Collins, 1991; Waliser and Graham, 1993; Ramanathan et al., 1995; Schneider et al., 1996; Fasullo and Webster, 1999). As mentioned earlier, monsoon winds in the Indian Ocean region undergo a pronounced annual cycle of speed and direction. The EIWP feels effects of the monsoon winds and, in turn, influences the winds. Perhaps due to a strong asymmetry between southwesterly and northeasterly winds, with the former much stronger than the latter, the equatorial Indian Ocean thermocline is shallower in the west near the African coast and deeper in the east near the MC coast. This tilt is opposite to that in the equatorial Pacific and Atlantic Oceans where the prevailing trade winds are generally easterly, shallowing the thermocline in the east and deepening it in the west. The WPWP and EIWP SSTs show large variabilities on intraseasonal, seasonal to interannual, decadal, and longer time scales. These variabilities influence both average and anomalous atmospheric circulations and also interact with a number of important weather and climate phenomena, including the Madden–Julian Oscillation, the El Niño–Southern Oscillation (ENSO), the Asian monsoon, and the IOD (see, for example, Wang and Xie (1998), Fasullo and Webster (1999), Matsuura and Iizuka (2000), Tian et al. (2001), D'Arrigo et al. (2006), Ashok et al. (2004)).

5.2.2 Ancient Observers' Tales about Indian Ocean Variability

In Chapters 3 and 4 on the Pacific and Atlantic climate variability, respectively, we saw evidences of DCV in tales told by ancient observers such as tree rings, oxygen isotopes in corals, and other paleoclimate proxy data. We will now hear tales told by such observers about decadal variability of the tropical Indian Ocean in the last two centuries. The physics of how and why we can learn from these ancient observers are described in previous chapters, so they are not be repeated here.

Several $\delta^{18}O$ oxygen isotope records based on corals from locations in tropical western and eastern Indian Ocean have been analyzed to study climate variability from late 19th century CE. Although the focus of some of these studies was on describing Asian monsoon variability, or ENSO variability, or their inter-relationship, there is also interesting information in these records about decadal variability. Some of the studies have also tried to compile a record of interannual IOD events going back to the 19th century CE by combining information from coral records at locations in tropical western and eastern Indian Ocean. IOD variability as gleaned from such records is also described here because it is relevant to decadal variability of the tropical Indian Ocean. Charles et al. (1997) and Cobb et al. (2001) analyzed $\delta^{18}O$ oxygen isotope records from 1886 to 1995 CE, based on *Porites lutea* corals near Seychelles in the tropical western Indian Ocean. As mentioned previously in this book, the $\delta^{18}O$ isotope records effects of SST and isotopic composition of seawater, so precipitation anomalies also influence $\delta^{18}O$ concentrations. A multi-taper spectrum analysis (Mann and Lees, 1996) of these records showed a significant peak between 10- and 16-year periods, and a peak between 7- and 10-year periods which is below the significance level. Then, Cobb et al. (2001) compared the Seychelles records with coral-based $\delta^{18}O$ records from Palmyra Island in the tropical central Pacific Ocean. They also compared both these sets of records with rainfall time series in the Nordeste region of Brazil as an indicator of tropical Atlantic SST variability (see Chapter 4 for the relationship between the Nordeste rainfall and the tropical Atlantic SST gradient (TAG) variability). These analyses showed that the 10–16-year spectral peak was present in all three sets of records. Cross-spectrum analyses of the three sets of records showed that the 10–16-year spectral peak was coherent among them. The cross-spectrum analyses also showed that the tropical Indian Ocean warming was followed two to three years later by tropical central Pacific warming which was followed by the positive phase of the TAG variability resulting in a drought in the Nordeste. Band-pass-filtered components, in the 9 to 16-year band, of the Seychelles and Palmyra coral $\delta^{18}O$, and Nordeste rainfall from Cobb et al. (2001) are shown in Figure 5.1. The phase relationships among the three time series show that when SSTs in Seychelles and Palmyra are warmer than average, the Nordeste rainfall is below average. In another coral-based $\delta^{18}O$ record from Malindi Marine Park in Kenya in the tropical western Indian Ocean from 1801 to 1995 CE, Cole et al. (2000) found a spectral peak between 8 and 14-year periods, which was coherent with the decadal spectral peak in the Cobb et al. (2001) Seychelles $\delta^{18}O$ analysis and also with a 10 to 14-year spectral peak in an 1856 to 1992 CE Niño3.4 SST index time series based on instrument data. Thus, the Cole et al. (2000) analysis showed coherence between decadal variability in the tropical western Indian Ocean and the tropical central Pacific Ocean. Cobb et al. (2001) analyses not only confirmed this, but also that the variability in tropical SSTs in all three oceans was coherent at the decadal timescale. The presence of a preferred decadal oscillation period in the tropical western and central Pacific in proxy and instrument-measured data, and in the tropical southern Atlantic in instrument-measured data is described in Chapters 3 and 4, respectively.

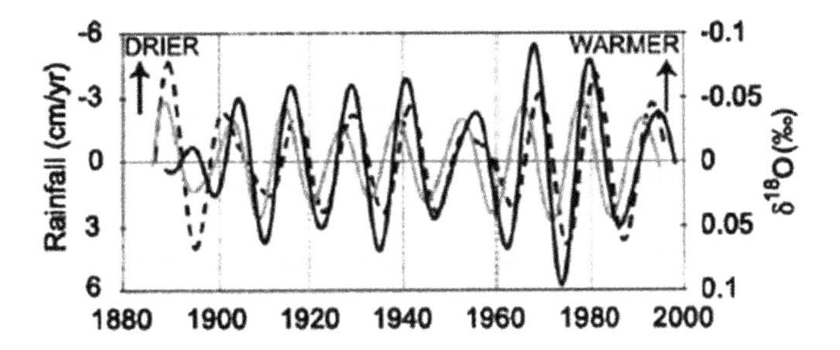

FIGURE 5.1 Band-pass-filtered (9 to 16 years, Gaussian filter) Palmyra coral (black), Seychelles coral (grey), and Nordeste rainfall (dashed). (From Cobb et al. 2001.)

Saji et al. (1999) first reported an Indian Ocean SST variability pattern which they called the IOD. In the positive IOD phase, SST anomalies were found to be warmer near the East African coast of the tropical Indian Ocean and cooler near the coast of Sumatra. In the negative phase, the converse was found. After the IOD first appeared in published literature, Abram et al. (2008) extended the instrument-based IOD time series backward to mid-19th century CE by comparing the 1846–1997 CE coral-based $\delta^{18}O$ record from Seychelles with those from Mentawai Island (1858–1997 CE) located approximately 200 km off the coast of Sumatra in the tropical eastern Indian Ocean and from Bali in south-central Indonesia (1783–1997 CE). $\delta^{18}O$ records from Seychelles and Mentawai show generally in-phase decadal and longer timescale variations. Seychelles and Mentawai are located in western and eastern regions of high IOD variance in instrument-measured SSTs. Upwelling near Bali usually precedes major, IOD-associated upwelling near Sumatra. Warming or cooling of SSTs decreases or increases, respectively, the $\delta^{18}O$ content of corals, and therefore, differences in $\delta^{18}O$ between the tropical western and eastern Indian Ocean sites can be an indicator of IOD activity in the past. After averaging from July to November west–east $\delta^{18}O$ differences to optimize correlations with instrument-measured SST data, Abram et al. (2008) found that there were 16 moderate and five strong IOD events from 1846 to 1997 CE and that 56% of the moderate-to-strong events coincided with El Niño events. Incidentally, as can be seen in Abram et al. (2008)'s results, since the IOD index is the difference between two quantities, both of which undergo generally in-phase decadal and longer timescale variability, decadal variability of the IOD index is much less than that of its two individual components. Nevertheless, the paleo-IOD index does show the presence of the decadal SST variability in the tropical eastern and western Indian Ocean.

5.2.3 Modern Observers' Tales about Indian Ocean Variability

The sparseness, in time and space, of instrument-measured ocean data has been mentioned several times in this book. The Indian Ocean is not an exception in this matter and so instrument-measured data before 1950 CE, when shipping routes and systematic ocean data measurements began to proliferate after the end of the Second World War, are not considered very reliable. Therefore, only these modern observers' tales about tropical Indian Ocean variability after 1950 CE are narrated here.

As shown in Figure 1.4 in Chapter 1, the EIWP is very large; it has covered approximately 17 million km^2 annual-average surface area[3] from 1951 to 2018 CE. As Figure 1.4 also shows, the geographical extent of the EIWP undergoes a pronounced annual cycle. Numerous studies have found that the Indian Ocean has been warming in the last 50 to 100 years (see, for example, Han et al. (2014) and references therein). Consistent with the warming, the EIWP area is also increasing as shown in Figure 5.2a, from Extended Reconstructed Sea Surface Temperature (ERSST) version 5 data (Huang et al., 2017). The EIWP area, in km^2 as well as in percent of long-term average, has almost doubled from 1951 to 2018 CE as Figure 5.2a and b, respectively, show. The annual-average EIWP area anomaly in km^2 – after removing the linear trend and subtracting the remaining long-term average – and the area anomaly as a percent of the average area are shown in Figure 5.2c and d, respectively. Figure 5.2c and d shows that there are large interannual and longer period variations in the EIWP area, with up to 30%–35% changes in area between a maximum and the subsequent minimum. It is intriguing to observe in Figure 5.2 that the area variations appear to reduce in amplitude as the EIWP area increases over the 68-year time series. A Fourier spectrum of the EIWP area anomaly time series in Figure 5.2d shows (Figure 5.2e) very prominent peaks at decadal and three- to five-year timescales. For comparison, a Fourier spectrum of the WPWP area anomaly time series is also shown in Figure 5.2e. The comparison shows that both spectra have peaks at generally the same oscillation periods. The time series of annual-average and seasonal-average SST anomalies, averaged over the tropical Indian Ocean region, including the Arabian Sea and the Bay of Bengal, from 20°S to 26°N, and their Fourier spectra are shown in Figure 5.3a to e and f, respectively. The five time series clearly show the presence of decadal variability, and the spectra confirm this presence with prominent decadal peaks in the annual- and seasonal-average SSTs, along with peaks at other timescales. The results shown here indicate that the average tropical Indian Ocean SST and the EIWP area have undergone prominent and mutually coherent decadal variability since 1950 CE. This is truly remarkable given the large size of the Indian Ocean basin. Lastly, it is interesting and important to note that the decadal timescale peaks in the tropical Indian Ocean SST spectra are at the same timescale as the peak in the WPWP SST index time series' spectrum in Mehta et al. (2018) as are the peaks in the EIWP and WPWP area spectra in Figure 5.3e. This similarity of prominent decadal timescales of variability in WPWP and EIWP is addressed further in Section 5.3.

Now, let us delve further into the decadal variability of the Indian Ocean SSTs and see its spatial and temporal patterns. After removing the basin average SST anomalies, empirical orthogonal function (EOF)–principal component (PC) analyses of annual-average and seasonal-average 1951–2018 CE data in the tropical Indian Ocean (30°E to 120°E, 20°S to 26°N) indicate (not shown) that December–January–February (DJF) and March–April–May (MAM) EOFs 1 have the same sign of SST anomalies over the entire Indian Ocean, with largest amplitudes in the tropical southern Indian Ocean. June–July–August (JJA) EOF 1 pattern begins to show one sign of anomalies along the African coast and tropical western Indian Ocean and

[3] SSTs equal to or greater than 28°C.

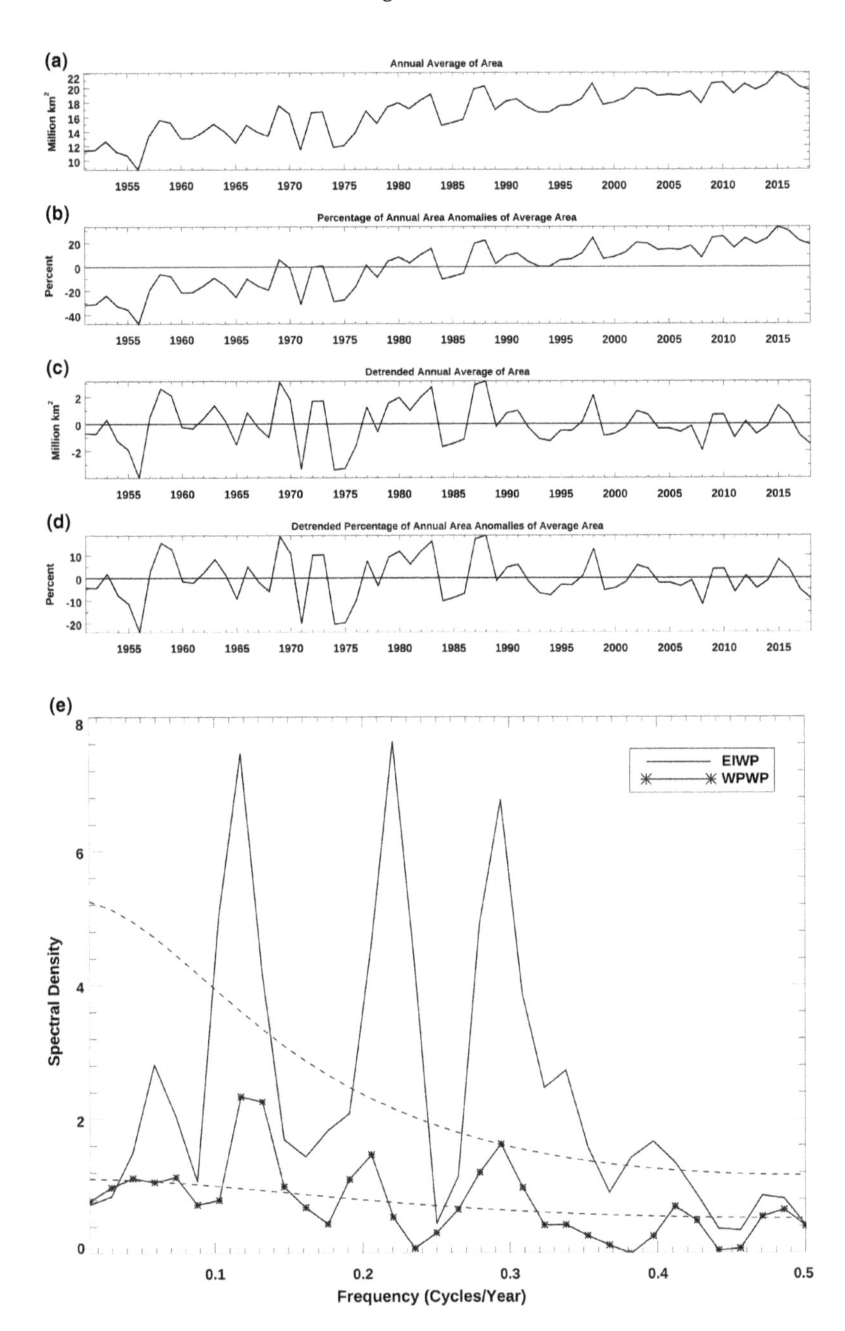

FIGURE 5.2 Annual-average EIWP area enclosed by the 28°C SST isotherm from 1951 to 2018 CE. (a) Total area (km^2), (b) area anomaly (percent of 1951–2018 CE average), (c) same as (a) after removing linear trend (km^2), (d) same as (b) after removing linear trend, (e) Fourier spectra of EIWP (line) area anomaly time series in (d) and WPWP (symbol) area anomaly time series. Dashed lines show equivalent red noise spectra. See text for details.

opposite sign in tropical central and eastern Indian Ocean. The September–October–November (SON) EOF 1 pattern shows an even stronger such pattern with the maximum amplitude in the tropical central Indian Ocean and opposite sign anomalies off the Sumatra coast and extending into the Java Sea. These seasonal EOF 1 patterns indicate that SSTs respond to the evolution of the southwest Indian monsoon winds

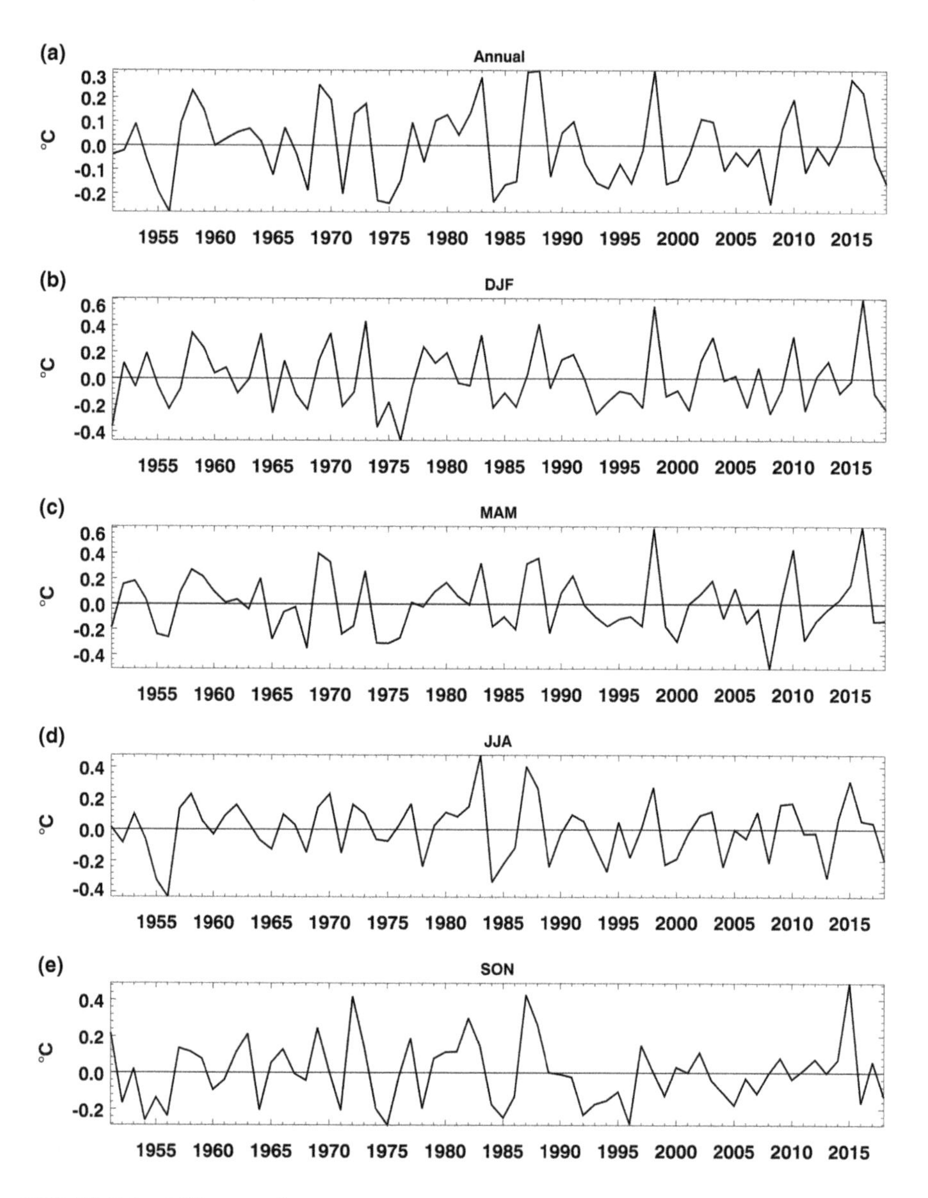

FIGURE 5.3 SST anomalies averaged over the tropical Indian Ocean in the region bounded by 30°E to 120°E longitudes and 20°S to 26°N latitudes from 1951 to 2018 CE. (a) Annual, (b) DJF, (c) MAM, (d) JJA, (e) SON.

(*Continued*)

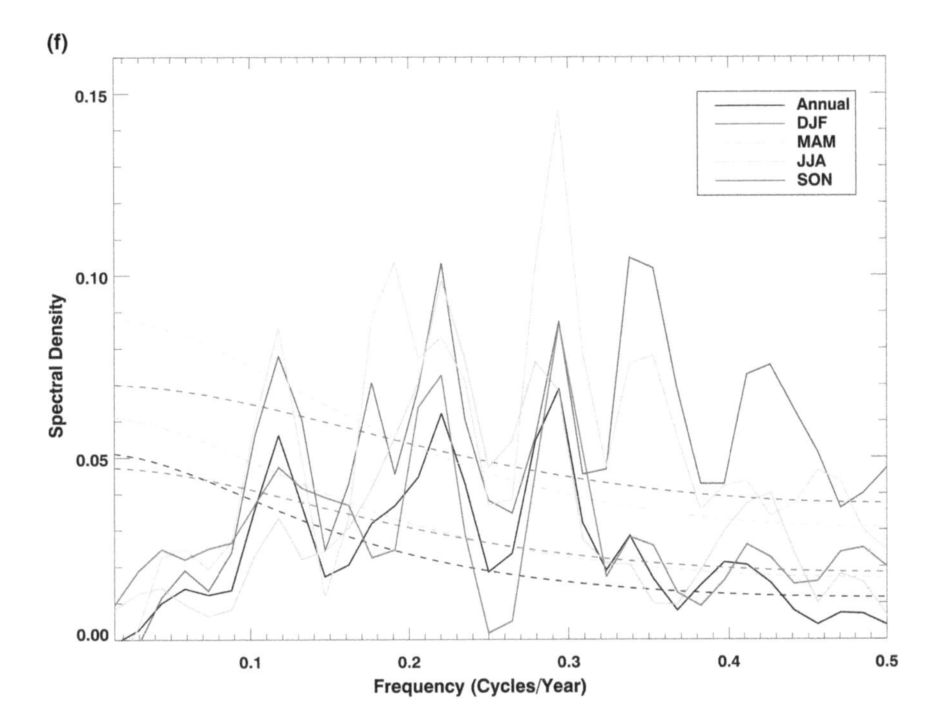

FIGURE 5.3 (CONTINUED) (f) Fourier spectra of time series in Figure 5.3a to e. Dashed lines show equivalent red noise spectra.

in late MAM season and develop as the winds strengthen during JJA and SON seasons. The JJA and SON EOF 1 patterns are reminiscent of the SST IOD patterns in Saji et al. (1999), Tozuka et al. (2007), and Han et al. (2014). EOF 1 of annual-average SST anomalies shows the same sign of SST anomalies in the entire domain except for small areas off the coasts of Madagascar and the Horn of Africa where the anomalies have opposite signs. All corresponding PC time series show interannual and longer timescale variability, and their Fourier spectra show seasonally varying peaks between 10- and 15-year periods, confirming the presence of decadal variability in the entire tropical Indian Ocean. There are also other spectral peaks in all PC 1 time series between 4 and 5.5 years, and around three-year periods.

Can instrument-measured data shed any light on possible mechanism(s) of this decadal SST variability? In Chapter 3, we saw a variety of observations of tropical Pacific decadal variability and a hypothesis, based on the observations and also on model simulations, of the mechanisms generating this variability. A possible role of the solar Schwabe cycle was also described. Among attributes of the tropical Pacific Ocean–atmosphere variability described and discussed in Chapter 3 were analyses of the National Centers for Environmental Prediction (NCEP)–National Center for Atmospheric Research (NCAR) (Kalnay et al., 1996) atmospheric data reanalysis and the Simple Ocean Data Assimilation (SODA; Carton et al., 2000) oceanic data reanalysis by Wang and Mehta (2008). The same analyses shed light on tropical Indian Ocean region decadal variability. Figure 3.8 shows that when the tropical western and central Pacific SSTs are warmer than average, the Walker circulation shifts eastward with the strongest rising motions near 180° longitude and sinking over the MC. This anomalous sinking air over the MC then generates divergence at low atmospheric levels, resulting in easterly wind anomalies over the tropical eastern Indian Ocean. Concurrently with this shifted Walker circulation over the MC, and over the tropical western and central Pacific Ocean, there is anomalous rising air motion over the tropical western and central Indian Ocean. This air diverges near the 200 hPa level, travels eastward, and sinks over the MC. Associated with the low-level easterly wind anomalies in the tropical eastern Indian Ocean near the MC are easterly upper-ocean current anomalies from SODA (Wang and Mehta, 2008; and Figure 3.12) which can transport warm water from the EIWP to tropical central and western Indian Ocean; Ekman divergence near the Equator would spread this warm water in the Northern and Southern Hemispheres. This warm water transport can increase SSTs in the tropical central and western Indian Ocean. Figure 5.4 shows that as the ocean heat transport warms tropical Indian Ocean SSTs, low-level wind anomalies associated with the shifted Walker circulation, in conjunction with the climatological annual cycle of winds (Figure 5.5), generally reduce surface heat flux from the ocean to the atmosphere as shown in Figure 5.6. Analyses of surface heat flux components from the NCEP-NCAR reanalysis show that latent and sensible heat transfers from large parts of the tropical Indian Ocean to the overlying atmosphere are reduced, causing a warming of SSTs over much of the Ocean.

Thus, as the observed atmosphere–ocean variability shown and cited here strongly indicates, the tropical Pacific Ocean–atmosphere decadal variability appears to be driving the tropical Indian Ocean–atmosphere decadal variability. Tales told by ancient observers, described in Section 5.2.1, also corroborate this indication. The role of the strong annual cycle of winds and surface heat flux in the tropical Indian Ocean variability should be further investigated because the annual cycle appears to break the east–west symmetry during the Northern Hemisphere summer and autumn seasons.

5.3 HYPOTHESIZED MECHANISMS OF INDIAN OCEAN DECADAL VARIABILITY

It is clear from the paleoclimate proxy data and instrument-measured observations described in Section 5.2 that SSTs in the tropical Indian Ocean, including the Arabian Sea and the Bay of Bengal, undergo coherent decadal variability. It is also clear that this decadal variability is associated with that in the tropical western and central Pacific Ocean described in Chapter 3. If it is assumed that the tropical Indian Ocean basin is too small in the zonal direction to support oceanic, slow baroclinic Rossby waves to propagate from the MC coast to the African coast to generate the decadal timescale intrinsically as hypothesized for the tropical Pacific decadal variability, then we must look elsewhere for the source of this variability. Given the observed variability of the Walker circulation associated with the tropical Pacific decadal SST variability, the observed variability of upper ocean currents and implied heat transports, and the observed variability of surface heat flux over the Indian Ocean, it is reasonable to hypothesize that the tropical Pacific decadal variability is driving the

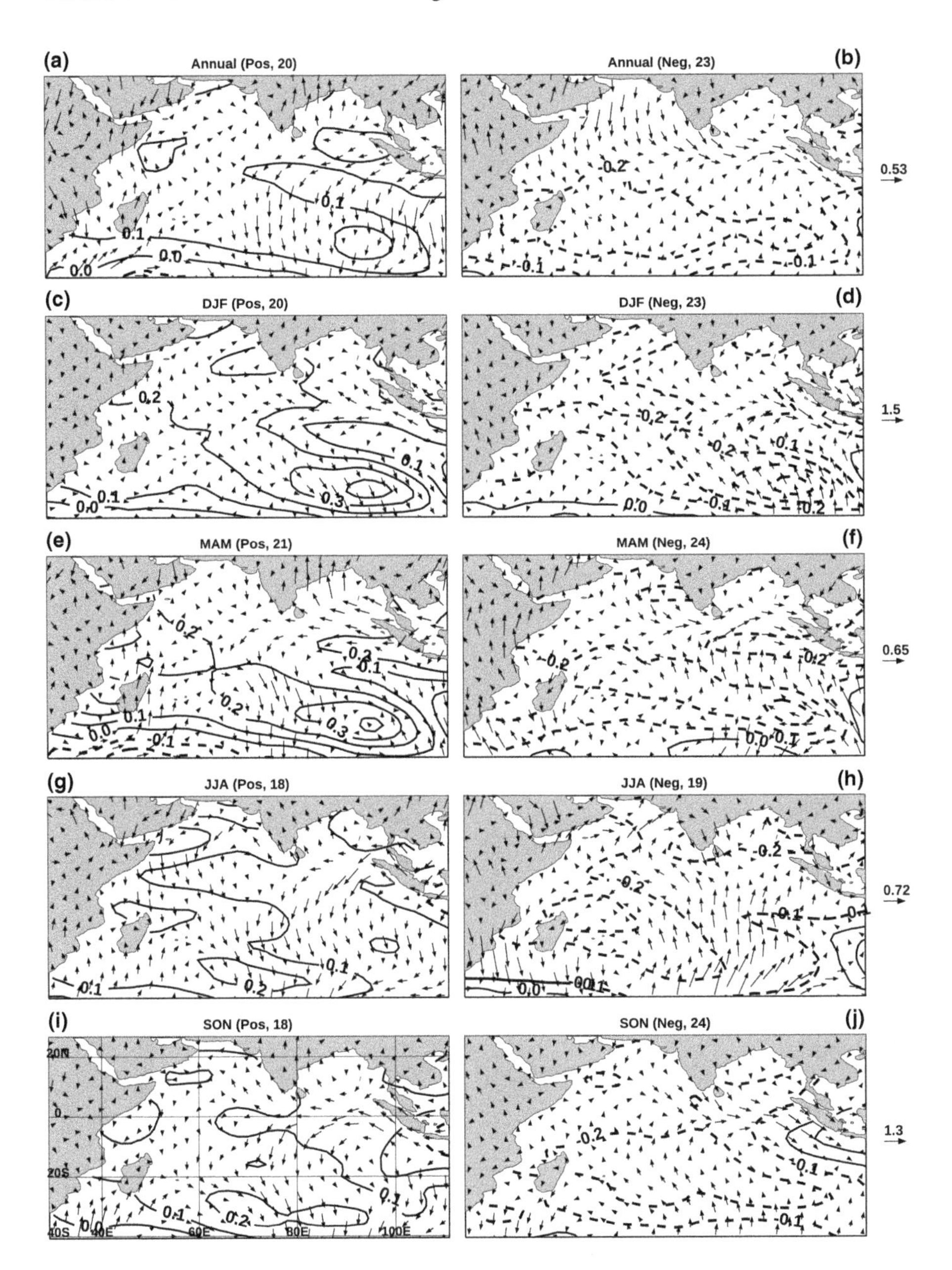

FIGURE 5.4 Composite, annual, and three-month average SST (°C) and 850 hPa wind (ms^{-1}) anomalies in positive and negative phases of the basin-average Indian Ocean decadal SST variability. (a) and (b) annual, (c) and (d) DJF, (e) and (f) MAM, (g) and (h) JJA, and (i) and (j) SON. Positive (negative) SST anomalies are shown as solid (dashed) contours. Vectors on the right of each row show wind anomaly scale (ms^{-1}) for that row. The number of samples in each composite is shown about each box.

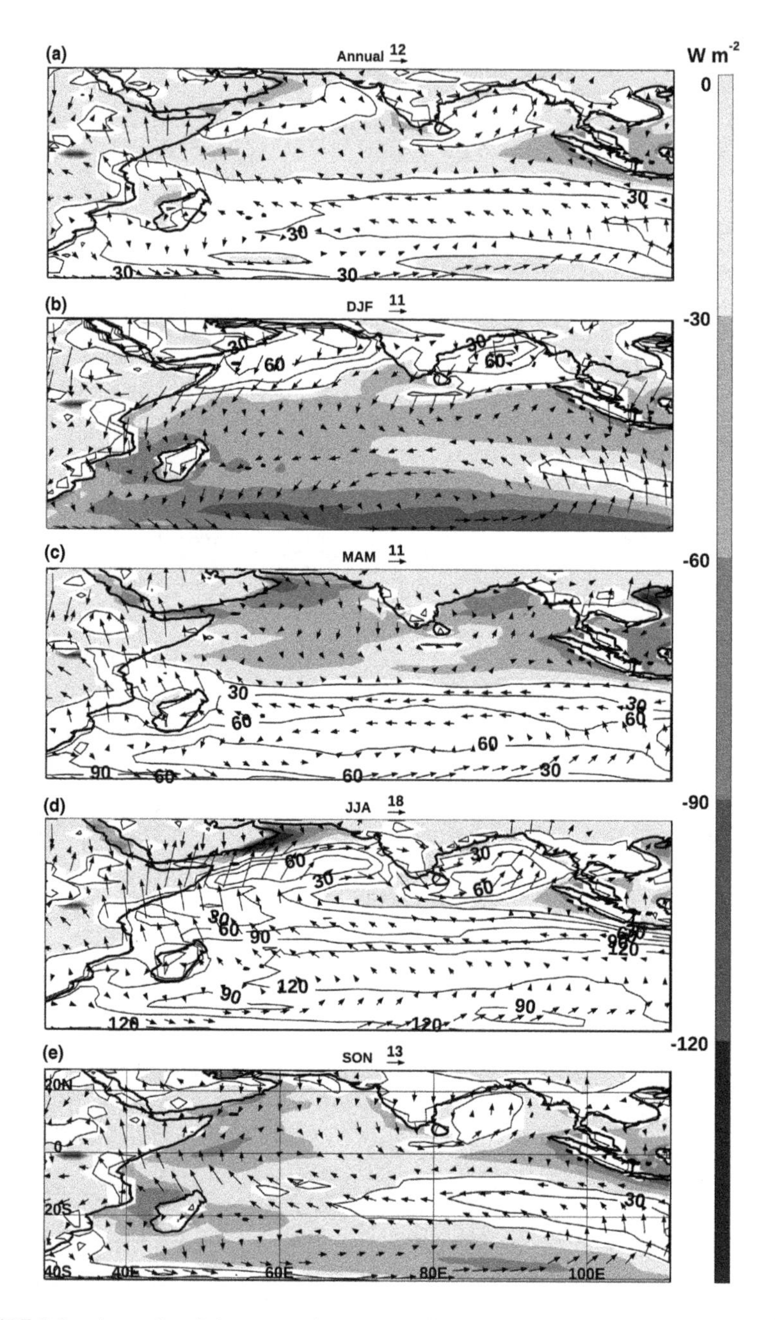

FIGURE 5.5 Annual and three-month average, climatological surface heat flux (Wm^{-2}) and 850 hPa wind (ms^{-1}) in the Indian Ocean region. Heat flux and wind data are from the NCEP-NCAR reanalysis from 1951 to 2018 CE. The gray scale on the right shows negative surface heat flux scale and the vector above each box shows wind scale. Flux going out of the ocean is denoted by positive (solid contours) values, and flux coming into the ocean is denoted by negative (shaded) values. (a) Annual, (b) DJF, (c) MAM, (d) JJA, and (e) SON.

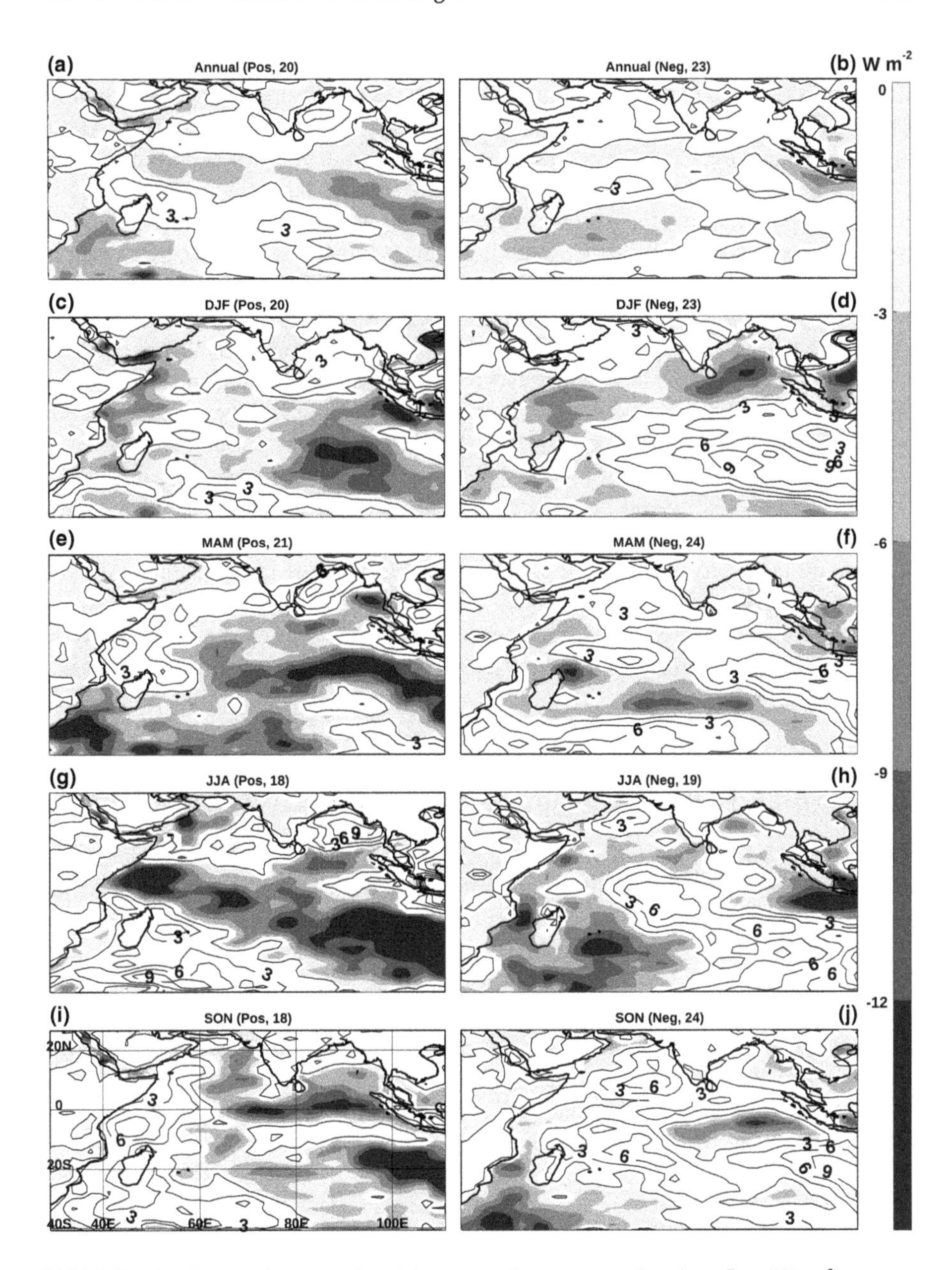

FIGURE 5.6 Composite, annual and three-month average surface heat flux (W m^{-2}) anomalies in positive and negative phases of the basin-average Indian Ocean decadal SST variability. Heat flux data are from the NCEP-NCAR reanalysis from 1951 to 2018 CE. Positive (negative) anomalies are shown in solid contours (gray scale). Gray scale on the right shows negative heat flux anomaly scale. The number of samples in each composite is shown about each box. (a) and (b) annual, (c) and (d) DJF, (e) and (f) MAM, (g) and (h) JJA, and (i) and (j) SON.

tropical Indian Ocean decadal variability as mentioned earlier. In this hypothesis, warmer SSTs in the western and central tropical Pacific shift the Walker circulation eastward which, in turn, would create sinking over the MC and easterly low-level wind anomalies over the EIWP. These easterly wind anomalies would drive easterly anomalies in upper ocean currents which would transport warmer surface water from the EIWP towards the African coast. Due to Ekman divergence, this westward heat transport would take warmer water north and south of the Equator, thus generating warm (positive) SST anomalies over the tropical Indian Ocean. Low-level wind anomalies associated with the shift in the Walker circulation drive changes in latent and sensible heat fluxes such that SSTs warm over much of the Indian Ocean. As the tropical Indian Ocean SSTs warm, they would drive upward motion and atmospheric convection, resulting in sinking motion over the MC and reinforcing the easterly wind anomalies created by the eastward shift in the Walker circulation. Thus, the Walker circulation anomalies, coupled with upper ocean heat transport and surface heat flux anomalies, would be the connection between decadal variability in the tropical Pacific and Indian Oceans. This observations-based hypothesis is plausible and should be tested further with model simulations and observed data, especially ocean heat budget analyses with model-assimilated data. The observations and the hypothesis also underscore the importance of the EIWP in decadal variability of the entire tropical Indian Ocean and its associations with worldwide hydrologic variability.

5.4 ASSOCIATIONS BETWEEN INDIAN OCEAN DECADAL VARIABILITY AND WORLDWIDE DRYNESS–WETNESS

Many studies have found statistical and dynamical associations between linear trends in Indian Ocean SSTs and climate variability in Asia, the Sahel region of Africa, Australia, the North Pacific region, and the North Atlantic region (see, for example, Han et al. (2014) and references therein). Associations between the decadal variability of the Indian Ocean SSTs and worldwide climate have not been analyzed and a detailed investigation of possible associations is beyond the scope of this book, but exploratory results are described in this section using the Standardized Precipitation Evapotranspiration Index (SPEI; Vicente-Serrano et al., 2010).

Figure 5.7a to d show composite SPEI anomalies in positive and negative phases of the area-average tropical Indian Ocean SST anomalies in DJF and JJA. The positive SST phase is defined as 0.5 standard deviation above average SST anomaly, and the negative SST phase is defined as 0.5 standard deviation below average SST anomaly. SPEI and SST data from 1951 to 2018 CE were used in this composite analysis. The number of years in each phase is also shown above each box. SPEI anomalies generally have opposite signs between positive and negative SST phases, and the differences between them are substantial and significant in DJF and JJA, especially around the Indian Ocean littoral, as Figure 5.7 shows. A warmer than average tropical Indian Ocean is associated with drier Indian and southeast Asian monsoon areas, and also a drier western Australia. This association may be due to weaker monsoons when the tropical Indian Ocean is warmer than average, weakening the meridional land–ocean temperature gradient. In the cooler tropical Indian

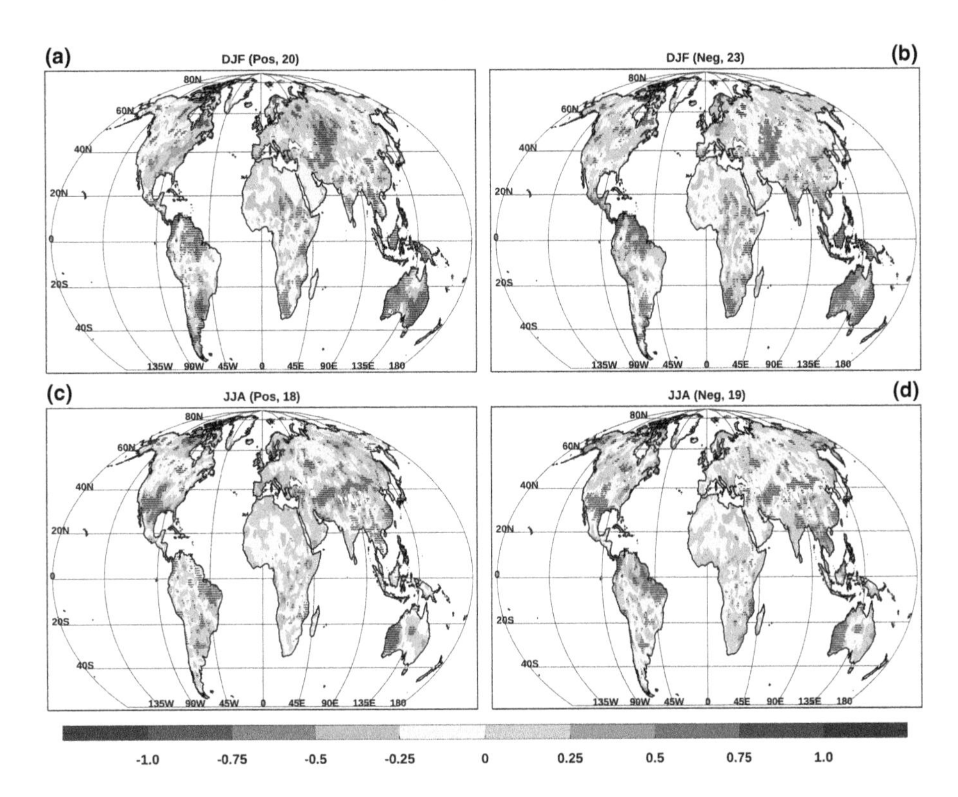

FIGURE 5.7 Composite SPEI anomalies in positive and negative phases of area-average tropical Indian Ocean SST anomalies in (a) and (b) DJF, and (c) and (d) JJA. SPEI and SST data are from 1951 to 2018 CE. Color bar below the figure shows SPEI anomaly scale, and significant differences (at or above 95%) in SPEI anomalies between positive and negative SST phases are stippled. Number of samples in each composite is shown above each box.

Ocean years, the temperature gradient and consequently the monsoons may be stronger, and the Indian and southeast monsoon areas may be wetter.

5.5 SUMMARY AND CONCLUSIONS

As we saw in Chapters 3 and 4, decadal variabilities in the Pacific and Atlantic ocean–atmosphere systems have attracted considerable attention over the last two decades, whereas decadal variability in the Indian Ocean–atmosphere system is relatively less known and much less investigated. This variability is the subject of this chapter.

In the beginning of this chapter, we saw very briefly the historical importance of the Indian Ocean. Then, decadal variability of the Ocean's SSTs, as seen in over a century-long coral-based $\delta^{18}O$ records, was described. These records indicate that the Ocean's SSTs have oscillated at a distinct decadal timescale and that this variability is coherent with decadal variability in tropical western and central Pacific SSTs. The story continued with instrument-measured ocean and atmosphere data since the 1950s CE. These data confirmed the presence of a distinct decadal timescale in basin-average Indian Ocean SSTs and also in the geographical area of the EIWP. The Walker circulation; and low-level winds, upper-ocean currents and implied heat transport, and surface heat flux in the tropical–subtropical Indian Ocean were also shown to be varying in phase with the basin-average Indian Ocean SSTs. An increase in ocean heat transport from the EIWP and a net, area-averaged reduction in ocean-to-atmosphere surface heat flux were shown to be associated with a warmer Indian Ocean in the warm or positive phase of the decadal variability. A decrease in ocean heat transport from the EIWP and a net, area-averaged increase in the surface heat flux were shown to be associated with a cooler Indian Ocean in the negative phase of the decadal variability. These observations support the conclusion that the tropical Indian Ocean decadal variability is largely controlled by the tropical Pacific Ocean decadal variability via the Walker circulation, ocean heat transport, and surface heat flux connections between the two basins. Based on this conclusion, a hypothesis about the Indian Ocean response to tropical central and western Pacific decadal variability was presented. An exploratory analysis of statistical associations between the tropical Indian Ocean SST decadal variability and worldwide dryness–wetness indicates that there are substantial and significant hydrologic anomalies associated with positive and negative phases of the SST variability. Modulations of land–ocean temperature gradients by the SST variability - resulting in modulations of the Indian and southeast Asian monsoons; and modulations of the Walker circulation by the SST variability - resulting in influences on the South American and African climates, may be responsible for the hydrologic anomalies, with consequences for water resources, agriculture, and other societal sectors.

Observed SSTs are specified as a lower boundary condition, and actual observations of winds and temperatures are assimilated in the NCEP-NCAR and other reanalysis models. The former may produce anomalous, thermally direct atmospheric circulations in response to SST anomalies in the reanalysis model even in the absence of any such signals in atmospheric observations. Therefore, multiple realizations of the atmospheric climate record should be produced by using various combinations of observed boundary conditions and observed atmospheric data. Similarly, there is an apparent physical consistency between the EIWP and WPWP SST variability and that of sub-surface temperatures and surface currents produced by the SODA and other reanalysis systems. In view of these systems' limitations, however, long records of assimilated ocean data should be produced with more accurate formulations of such systems, so that statistical characteristics of the EIWP decadal variability can be quantified as well as a more reliable insight into the physics of the variability can be derived. Also, realizations of atmospheric and oceanic climate records should be produced with "frozen" observing systems to see impacts of changing observing systems on analysis of decadal climate variability. Lastly, but most importantly, oceanic observations in the entire IPWP region before the 1950s were sparse in their spatial and temporal coverage, and the accuracy of SST and sub-surface ocean observations was unknown. In view of the suggestive but potentially very important impacts of the Indian Ocean decadal variability, continuous time series of high-quality, *in situ,* and remote-sensing-based ocean observations should be maintained in the IPWP region.

Although the results of the studies reviewed in this chapter are exciting and possibly very important for the prediction of the Indian Ocean temperatures, currents, and salinity; and their societal impacts at multiyear to decadal timescales, the results must be considered suggestive. Further observational, modeling, and reanalysis studies are necessary to understand physics of the EIWP and the rest of the Indian Ocean decadal variability and its impacts on global climate, and assess their predictability. Implications of the change and variability in the EIWP area for global atmospheric circulations and climate should be further studied with observed data and model simulations. While the importance of surface heat flux in driving the tropical Indian Ocean SST variability is indicated by the observations shown in this chapter, the importance of atmosphere–ocean freshwater flux needs to be clarified and quantified. It is possible that freshwater flux variability associated with Walker circulation and monsoon variability may also be playing an important role in generating Indian Ocean SST variability as exploratory research by Huang and Mehta (2004) indicated. So, while there is a considerable progress in understanding decadal variability of the *Sindhu Mahasagar*, it is also clear that much more research is required before the *Mahasagar* will give up all its secrets. In the meantime, modern seafarers continue to ply on the waves of the *Sindhu Mahasagar*.

6 What's in a NAM(e)?

6.1 INTRODUCTION

A carpenter-turned-missionary in Greenland, David Crantz, wrote a book (title of English translation "A History of Greenland") in 1765 Current Era (CE) in which he mentioned that when Denmark had a mild (severe) winter, Greenland had a severe (mild) winter. A missionary named Hans Egede Sabye in the 1770s CE also observed, along with other missionaries and traders, that temperatures in Greenland and Denmark appeared to vary with opposite phases. Fast forward to the early 20th century CE. Sir Gilbert Walker, a Cambridge-educated mathematician and the Director-General of the India Meteorological Department from January 1904 to December 1924 CE, was searching for mathematical methods for successful "seasonal foreshadowing" of the Indian summer monsoon rainfall which then was and still is the lifeblood of the Indian civilization[1] (J. M. Walker, 1997). Not being successful in finding such mathematical methods, Sir Gilbert turned to empirical methods, specifically to estimate lag correlations among rainfall and other meteorological variables within and outside India which could be used for seasonal foreshadowing of Indian monsoon rainfall. During this quest, Sir Gilbert serendipitously discovered three large-scale climate phenomena far away from India: the Southern Oscillation (SO), the North Pacific Oscillation (NPO), and the North Atlantic Oscillation (NAO). All three were based on Sir Gilbert's analyses of worldwide surface pressure measurements, made principally on land, and were identified as oscillations in atmospheric mass. Sir Gilbert named oscillations in surface pressure between Iceland and the Azores in the North Atlantic region as the NAO (Walker and Bliss, 1932). More broadly, the NAO is a measure of oscillations in atmospheric mass between subpolar and subtropical latitudes in the North Atlantic–Europe region. The NAO is known (see, for example, Hurrell (1995), and references therein) to play a leading role in winter weather and climate variations in eastern North America, the North Atlantic, and Europe. In the North Atlantic Ocean, NAO-associated wind and precipitation anomalies influence temperature, oceanic convection, deep-water formation, and primary production. The NAO is also known to undergo interannual, decadal, and longer timescale variability. In the last approximately 20 years, the NAO's variability at decadal timescales[2], and its interactions with the North Atlantic and Arctic Oceans, and with seas around eastern North America and western and northern Europe have attracted considerable attention from climate modelers to understand and possibly predict its variability at these timescales. Are the atmospheric phenomena first recorded by missionaries in Greenland and Denmark, the NAO discovered by Sir Gilbert Walker, and phenomenon labeled the Arctic Oscillation (AO) and the Northern Annular Mode (NAM) in the 1990s and early 2000s CE the same or dif-

[1] See Chapter 3

[2] As in the rest of this book, decadal here also refers to variability generally at 8 to 20-year periods.

ferent? What causes variability in this/these phenomenon (phenomena) at timescales of a few days to a few decades, especially at decadal timescales? Can the atmosphere by itself generate such decadal variability or do the oceans and seas underlying the Northern Hemisphere (NH) atmosphere, and sea ice in some of these oceans and seas, play any role in decadal atmospheric variability? We have already glimpsed in Chapter 2 that decadal variability of the NAO is associated with the solar Schwabe cycle with an average period of approximately 11 years. What are the physics of solar emissions playing a role in decadal variability of NAO/AO/NAM? These questions are addressed in this chapter. After this brief introduction, subsequent sections in this chapter describe instrument-measured observations of decadal variability of NAO/AO/NAM in Section 6.2. Then, decadal NAO variability seen in paleoclimate data is described in Section 6.3. Hypothesized mechanisms of the decadal variability are described in Section 6.4. Finally, a summary of this chapter and conclusions are presented in Section 6.5.

6.2 MODERN OBSERVERS' TALES

6.2.1 What's in a Name?

We saw in the introduction and in Chapter 3 how the NAO was labeled as such by Sir Gilbert Walker. But, how were the AO and the NAM identified? Let us see what underlay these two identifications. Kutzbach (1970) was perhaps the first to analyze large-scale NH sea level pressure (SLP) variability with an objective technique. Gridded, January and July SLP data from 1899 to 1969 CE were analyzed with the empirical orthogonal function (EOF)–principal component (PC) analysis technique. The first (largest variance) EOF pattern indicated that the intensity and latitudinal position of the centers of action in the North Atlantic region were associated with the intensity and latitudinal position of the Aleutian Low in the North Pacific region. Thus, the idea that there was some type of zonal symmetry in the North Atlantic and North Pacific regions was born from this SLP analysis. Then, subsequently, other researchers (see, for example, Wallace and Gutzler (1981), Trenberth and Paolino (1981), Thompson and Wallace (1998, 2000)) found further evidence of apparent zonal symmetry in NH mid- and high-latitude atmospheric circulation. Various analyses of atmospheric data over the entire NH revealed a spatial pattern extending well into the Arctic region and also apparently coherent with the same sign in the Atlantic and Pacific regions. This pattern was named the AO or the NAM because the pattern looked like an annulus if seen from above the North Pole (Thompson and Wallace, 1998, 2000). Arguments immediately broke out within the North Atlantic weather and climate communities about the relationship, if any, between the NAO and the AO/NAM. Enterprising researchers soon found a Southern Annular Mode (SAM; Chapter 7) corresponding to the NAM and serious investigations soon began to determine if these new-found patterns were truly dynamical modes of atmospheric variability or were merely artifacts of analyses.

We now come back to the beginning and trace the evolution of research in NH extratropical atmospheric variability chronologically. An earlier school of research in the 1930s CE trying to understand the general circulation of the atmosphere believed that dynamical interactions and resulting low-frequency (intraseasonal in modern parlance) variability can be understood better if the variability was divided into zonally-symmetric and zonally-asymmetric (eddies or stationary waves) components. By definition, perhaps based on their observations of weather maps which showed that winds in the mid-latitude troposphere were largely west to east (that is, zonal) around the Earth, the zonally-symmetric component had the same sign and magnitude around latitude circles, and zonally-asymmetric components varied with longitude. Rossby (1939) used these concepts to analyze weather data to describe interactions among zonally-symmetric and -asymmetric components.[3] Based on these concepts, Rossby (1939) and other researchers defined a zonal index whose variability portended the principal variability of the NH mid-latitude atmosphere. Subsequently, Lorenz (1951) further analyzed the zonally-averaged circulation at NH mid-latitudes, especially its intraseasonal ("irregular" around the seasonal cycle as Lorenz (1951) defined it) variability between high- and low-index phases. Wallace (2000) proposed that the NAO/AO/NAM paradigm and the zonal index paradigm are essentially equivalent, but that they emphasize different roles of the underlying ocean–land entities in the variability. Thus, there are at least two schools of thought about distinctness of the NAO and the AO/NAM which can be traced back to the late 1930s CE. Wallace (2000) clearly laid out both cases and opined that the two are essentially the same. Almost immediately, other studies presented contrary points of view. Ambaum et al. (2001) analyzed several types of atmospheric data (SLP; and geopotential heights, zonal wind, and stream function at various pressure levels) from the December to March period from the National Center for Environmental Prediction (NCEP)–National Center for Atmospheric Research (NCAR) reanalysis (Kalnay et al., 1996) and an extended European Centre for Medium-Range Weather Forecasts reanalysis (ERA) data set from 1979 to 1997 CE to see if NH atmospheric variability is physically consistent among these variables and among the two reanalysis systems. EOF-PC analyses of these data showed that there was a physically consistent relationship among principal patterns of the data in the Euro-Atlantic region which was essentially the NAO pattern. However, no physically consistent set of patterns emerged indicating NAM/AO variability. Ambaum et al. (2001) emphasized that although the AO pattern showed negative correlations between zonal winds at 35°N and 55°N in Atlantic and Pacific regions, climatological features in the two regions were at different latitudes, so the zonal AO pattern did not imply a simple modulation of circumpolar flow by an annular mode. For example, an increase in the NAO or AO indices was associated with strong, separated tropospheric jets in the Atlantic region, but a weakened tropospheric jet in the Pacific region, implying that variability in the NAO or the AO was not zonally uniform in an annular mode sense. Based on these results of their analyses, Ambaum et al. (2001) concluded that the NAO paradigm may be physically more relevant for NH mid-latitude atmospheric variability than the AO paradigm.

[3] The scientific interest in using weather predictions for agricultural operations as a motivation is evident in this work because it was conducted in collaboration with the US Department of Agriculture and involved Bureau of Agricultural Economics scientists.

It was time now for the other school of thought to present contrary evidence. Leading the charge from the other school, Feldstein and Franzke (2006) also analyzed several types of variables (SLP, 300- and 40-hPa stream functions, and potential temperature on the tropopause) from the NCEP-NCAR reanalysis in December–January–February for the 1958–1997 CE period to see if persistent NAO and NAM events were statistically distinguishable. In their definition, a persistent event was quantified by the pattern correlation for daily SLP remaining above a specified threshold value for five or more consecutive days. To qualify as independent events, there had to be more than 15 days' separation between two consecutive events. The analyses showed that NAO and NAM events were not distinguishable even at a relatively low confidence level. Not only that, but it was also found that wave-breaking properties of the two types of events were also the same and that positive NAO events were preceded by breaking synoptic scale waves over the North Pacific by two to three days. Thus, according to Feldstein and Franzke (2006), NAO/NAM events were not confined to the North Atlantic and they were drawn from the same statistical population, so either one of the paradigms should be regarded as redundant.

Statistical analyses results presented here lean towards treating the NAO, AO, and the NAM patterns as being the same, with the NAO being the most physical pattern of the three. Therefore, the NAO's decadal variability is the focus in the rest of this chapter, and NAO is the only name used for this phenomenon.

6.2.2 In the Beginning

It is very appropriate that the history of NAO research began near the northern center of the phenomenon. In recorded history, people from Scandinavia began colonizing Greenland towards the end of the 10th century CE (Wanner et al., 2001; Pinto and Raible, 2012). Ships plied between Scandinavian countries – especially Denmark – and Greenland frequently, exposing sailors and passengers to vagaries of weather variability. Anecdotal information, and pressure and temperature measurements began to be recorded. As we saw in the beginning of this chapter, missionaries were perhaps the first Europeans to observe and record opposite types of weather in Greenland and Denmark. But, it was not until the 18th century CE that first publications about differences between weather and hydrologic conditions in Greenland and Scandinavia began to appear (see descriptions in van Loon and Rogers (1978) and Wanner et al. (2001)). Then, in the 19th century CE, an increasing practice of measuring atmospheric pressure at the Earth's surface brought these data into use for describing weather variability over larger and larger geographic domains. From the 1830s CE to the end of the 19th century CE, many studies analyzed pressure and temperature data, and began to find geographical positions of pressure centers, among them the Azores and Russian high pressure areas (see, for example, Wanner et al. (2001) and references therein). In the 1890s CE, the first idea of potential oceanic influences on North Atlantic–Europe weather began to germinate. Pettersson (1896) and Meinardus (1898) proposed that the Gulf Stream played an important role in North Atlantic–Europe weather and that its interannual variations could be responsible for anomalous winters of opposite phases in Greenland–Western Iceland and European mainland.

We have seen many times in this book that analysis techniques have played crucial roles in the discovery of new phenomena and in developing understanding of discovered phenomena. As the 20th century CE dawned, spatio-temporal pressure and temperature data measured over and around the North Atlantic were increasing rapidly, and subjective comparisons of time series at various locations began to prove cumbersome and resulted in diminishing returns. Enter Sir Francis Galton – a polymath, statistician, meteorologist, psychologist, anthropologist, and eugenicist, along with his specialties in other fields of human endeavor.[4] He is also credited with plotting the first weather map for *The Times*, published on 1 April 1875 CE, showing the previous day's weather.[5] He also had a penchant for measuring and counting, and formulated the standard deviation as a measure of variation. Either based on the formulation of the correlation coefficient by French physicist Auguste Bravais (Bravais, 1846) or by his own data collection on length of forearm and height of humans, Sir Francis (re)formulated the correlation coefficient in 1888 CE (Galton, 1888). Interestingly, he demonstrated the usefulness of the correlation coefficient first in heredity, anthropology, and psychology (Clauser, 2007). This newly developed estimate of co-variation between two or more variables attracted the attention of several meteorologists, among them Walker (1909) and Exner (1913), the former for his efforts to predict Indian monsoon rainfall (see Chapter 3) and the latter for extratropical NH weather. Exner (1913) calculated and plotted correlation coefficients between a geographically averaged pressure measurement representing the North Pole and pressure measurements at 50 other locations in the NH from 1887 to 1906 CE and found … the Arctic Oscillation, except that he did not name it the AO. He did, however, emphasize the annular shape of the correlation coefficient contours centered at the North Pole (Wanner et al., 2001; Figure 4a). Another striking aspect of Exner's correlation pattern was that its southern center of action was located over Southern Europe and the Mediterranean Sea, perhaps because he had pressure data mainly or only from land stations.

1924 CE was a landmark year in the history of NAO research. Approximately a decade after his initial foray into mysteries of North Atlantic and European weather, Felix Exner moved his reference location from the North Pole to Stykkisholmur in Iceland and plotted a map of correlation coefficients between pressure at this reference point and pressure at approximately 70 other location for the September to March period from 1887 to 1916 CE (Exner, 1924). The resulting pattern (Wanner et al., 2001; Figure 4b) was strikingly similar to the modern-day NAO pattern. In the same year, Walker (1924) introduced the three oscillations mentioned in Chapter 3 – the NAO, the SO, and the NPO – and later proposed an NAO index in his seminal publication (Walker and Bliss, 1932). Joining these two in making 1924 CE a landmark year for NAO research was Albert Defant who formed perhaps the first index of NAO-like SLP variations from 1881 to 1905 CE. Defant (1924) subjectively related interannual variations in this index to zonally-averaged meridional SLP gradient over the Atlantic, volcanic eruptions, zonal SLP gradient between Northern Europe and the northern North Atlantic, and sea ice extent near Iceland. This study

[4] www.famouspsychologists.org/francis-galton/

[5] www.galton.org/meteorologist.html

moved the hitherto largely statistical results of analyses of extratropical NH pressure variations towards a physical hypothesis by suggesting that these variations were due to internal variability of the atmosphere at three- to five-year periods, occasionally affected by volcanic eruptions. Thus, the 1920s CE can truly be called the end of the beginning of NAO research.

This, then, is a brief history of research leading up to the modern era in extratropical NH weather and climate variability research. As this history indicates, the NAO/AO/NAM debate began in the 1920s and 1930s CE, although the abbreviations AO and NAM were not yet coined. We will continue to follow the evolution of this debate in subsequent sections. As this history also indicates, at least four physical processes were suggested as the beginning of NAO research came to an end – internal atmospheric variability; and influences of the underlying Gulf Stream variability, volcanic eruptions, and sea ice variability. We will see later in this chapter further development of these four and other possible causes of the NAO variability.

6.2.3 In the Modern Era

After the end of the beginning of the extratropical NH weather and climate variability research, this enterprise steadily accelerated as more data were collected by expanding atmosphere and ocean observing networks, and also as more complex and realistic atmosphere and ocean models were developed. This acceleration has gone into overdrive since the early 1990s CE (Wanner et al., 2001; Figure 6). Let us now see the spatio-temporal patterns of the NAO from more recent analyses. Then, observed patterns of troposphere-stratosphere interactions and patterns of oceanic variables associated with the NAO are described in the next two sub-sections.

We have seen in previous chapters that the "technique of choice" to isolate dominant spatio-temporal patterns of weather/climate variability is usually the EOF-PC technique. It is relatively easy to use, but its results are not always easy to interpret as we have seen, especially in Chapter 4. Nevertheless, the EOF-PC analysis used as the first step can be useful to reduce large spatio-temporal data sets to a few patterns carrying a sizeable fraction of the total variance. Following that path to analyze SLP data in extratropical North Atlantic–Europe region shows that the dominant EOF pattern (Figure 6.1a) of December to March data from the NCEP-NCAR reanalysis from 1958 to 1998 CE has two poles (or centers of action as mentioned in the early history), one located over and north of Iceland and the other with opposite sign located west of the Iberian Peninsula. The corresponding PC time series (Figure 6.1b) shows variability at interannual to multidecadal timescales. In Figure 6.1b, Hurrell (1995)'s NAO index from 1864 to 1998 CE is also shown. The PC time series and the NAO index time series are practically the same in the period common to both of them. It is obvious from the two time series that the NAO undergoes variability at timescales of a few years to a few decades in NH winter-average (also in other seasons or annual) time series. Of course, intraseasonal variability is dominant in daily NAO time series. Due to the geostrophic relationship between surface pressure gradients and winds, the north–south pressure gradient implied by the NAO also indicates variability of zonal winds. Numerous spectral analyses of this and other NAO index time series have shown that there are distinct peaks at decadal and

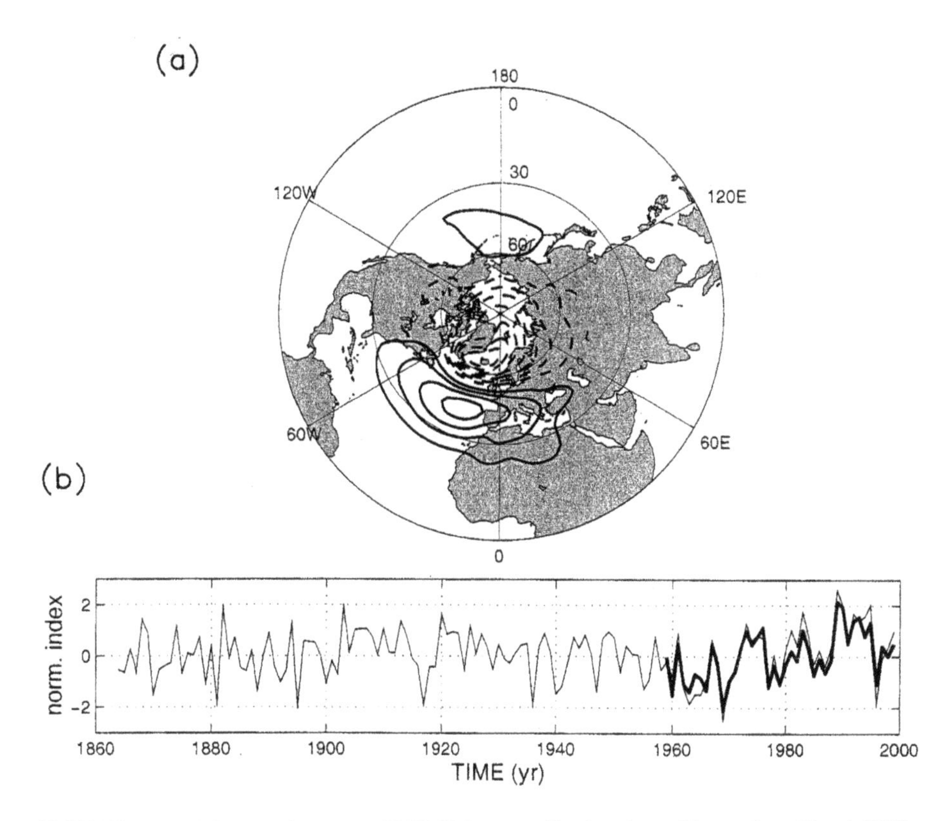

FIGURE 6.1 (a) Regression map of NH SLP anomalies in winter (December–March 1958–1998) onto the first principal component of SLP anomalies over the North Atlantic sector (20°–70°N/100°W–20°E). (b) Time series of Hurrell's NAO index (thin curve) and the first principal component of SLP (thick curve). Both time series are normalized by their respective standard deviation. The SLP data are taken from the NCEP-NCAR reanalysis. (From Marshall et al., 2001.)

multidecadal periods. For example, Marshall et al. (2001) estimated the spectrum of the winter NAO index in Figure 6.1b as well as that of PC 1 of winter SLP data from 1958 to 1998 CE with the multi-taper method. They found that the estimated spectra (Figure 6.4a) have a general "red noise" appearance, with pronounced variance at eight- to ten-year periods. Also in a recent study, Seip et al. (2019) estimated power spectral density of the 130-year NAO index time series from 1880 to 2019 CE. They found that the estimated spectrum has significant decadal–multidecadal spectral peaks centered at 13, 20, and 34 years.

The definition of the conventional NAO index, based on SLP variability at subpolar and subtropical centers in the North Atlantic region, assumes that locations of these centers are time-invariant. The same is implied by individual EOFs because they can only represent stationary patterns. Observing that the dominant North Atlantic SLP variability pattern shifts geographically at decadal timescales, Wang et al. (2012) used relative longitudinal angles between centers of SLP variability to define a decadal variability index as complementary to the conventional NAO index. This "angle index" is defined as per relative longitudinal locations of centers of SLP variability such that the index is zero if both northern and southern centers are on the same longitude, positive if the line connecting the two centers is tilted towards the northeast, and negative if the line connecting the two centers is tilted towards the northwest.

What about the AO? We saw earlier in the chapter that there is a controversy about how the NAO, AO, and NAM patterns should be interpreted, and whether they are manifestations of the same phenomenon or not. Consistently with the studies cited earlier, Deser (2000) also concluded that, in the Atlantic region, the AO pattern is nearly indistinguishable from the NAO. Monthly, gridded SLP data poleward of 20°N from 1947 to 1997 CE were analyzed with EOF and correlation coefficient techniques. Projection amplitudes and correlation coefficients of leading PC time series for NH, Atlantic-only, and Pacific-only domains on NH monthly SLP data were calculated by Deser (2000) and are shown in Figure 6.2. It is obvious from the amplitudes (upper row, Figure 6.2) that there is very little projection of the Atlantic-only PC on the Pacific region SLP anomalies and of the Pacific-only PC on the Atlantic region SLP anomalies. It is also obvious that amplitude contours over the Arctic region in all three amplitude figures form annular shapes. The conclusion from the correlation maps (lower row, Figure 6.2) is also the same. Deser (2000) then averaged monthly SLP time series from November to April for the 1947–1997 CE

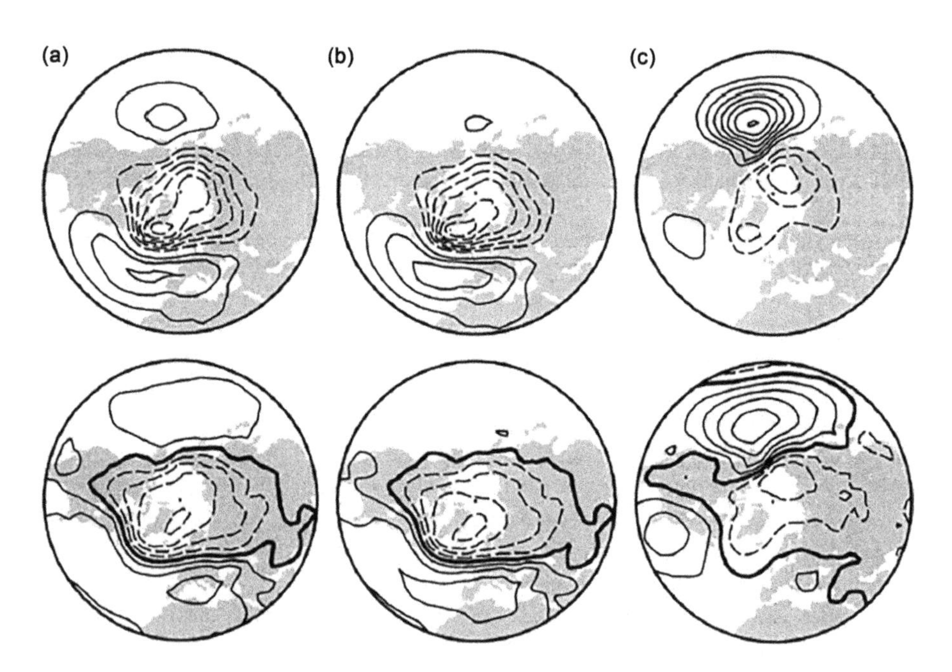

FIGURE 6.2 Leading EOF of monthly SLP anomalies poleward of 20°N based on the NH (a), Atlantic (b), and Pacific (c) domains. The patterns are displayed in amplitude (upper) and correlation (lower) form, obtained by regressing or correlating the monthly SLP anomalies over the entire Hemisphere upon the leading EOF time series from each domain. The contour interval in the lower panels is 0.2, and the zero contour is darkened. (From Deser, 2000.)

period, and calculated spatial averages within the outer-most contour of each EOF pattern. Then, the spatial SLP anomaly averages for the Arctic, the Atlantic, and the Pacific regions were cross-correlated. These correlation coefficients were −0.64 between the Arctic and the Atlantic, −0.22 between the Arctic and the Pacific, and 0.1 between the Atlantic and the Pacific, clearly indicating that there is very little correlation between SLP variability in the Atlantic and Pacific regions. Based on these results, Deser (2000) concluded that the annular shape of the SLP variability in the Arctic region dominates the AO and any co-variability between the Atlantic and the Pacific regions. In view of conclusions from Deser (2000) and other studies, it appears that the AO is mainly a statistical pattern, whereas the NAO has physical consistency across several atmospheric variables.

6.2.4 Oceans Feel the Atmosphere

We saw earlier that in the positive phase of the NAO, the Azores area high pressure is higher and the Iceland area low pressure is lower, strengthening the meridional SLP gradient. So, in physical consistency with the stronger SLP gradient, winds become more zonal, more westerly, and stronger. As a result, warmer temperatures and more water vapor from the Atlantic Ocean region are advected to northern Europe bringing wetter weather. This northward shift and weaker winds towards southern Europe make this region dryer. The strengthening of the westerly winds in the mid-latitude Atlantic region and associated strengthening of southeasterly trade winds in the tropical–subtropical Atlantic region increase ocean-to-atmosphere sensible and latent heat fluxes, cooling the sea surface temperatures (SSTs) in these two regions. A reduction in the winds and, consequently, heat fluxes at latitudes between the two stronger zonal wind regions warms the SSTs. As a result, a so-called SST anomaly tripole develops during the positive phase of the NAO such that there are negative SST anomalies in the Greenland and Irminger Seas to Newfoundland, and tropical North Atlantic near the West African coast; and positive SST anomalies in central North Atlantic at monthly to seasonal timescales. The tripole SST anomaly pattern and associated sensible and latent heat flux pattern are shown in Figure 6.3a and b, respectively. As this figure shows, there are remarkable associations between heat flux going out of the ocean and negative SST anomalies, and heat flux going into the ocean and positive SST anomalies. Surface wind stress curl anomalies in the positive NAO phase are shown in Figure 6.3c, which also shows a remarkable association with the heat flux and SST anomalies, especially in northern and central North Atlantic. In the negative phase of the NAO, the weather in northern Europe becomes dryer and colder, and that in southern Europe becomes wetter and warmer due to a weaker meridional SLP gradient and associated westerly winds. A blocking high-pressure area develops from the Ural Mountains and eastern Europe to central-western Europe, leading to a stable weather pattern over central Europe. In this (negative) NAO phase, a tripole SST anomaly pattern of opposite sign to that during the positive NAO phase develops. Marshall et al. (2001) estimated multi-taper spectra of the tripole SST anomaly pattern, a cross-Gulf Stream SST anomaly index, and a cross-equatorial SST anomaly index. These spectra are shown in Figure 6.4b. All three spectra show peak and/or generally higher variance at interannual and

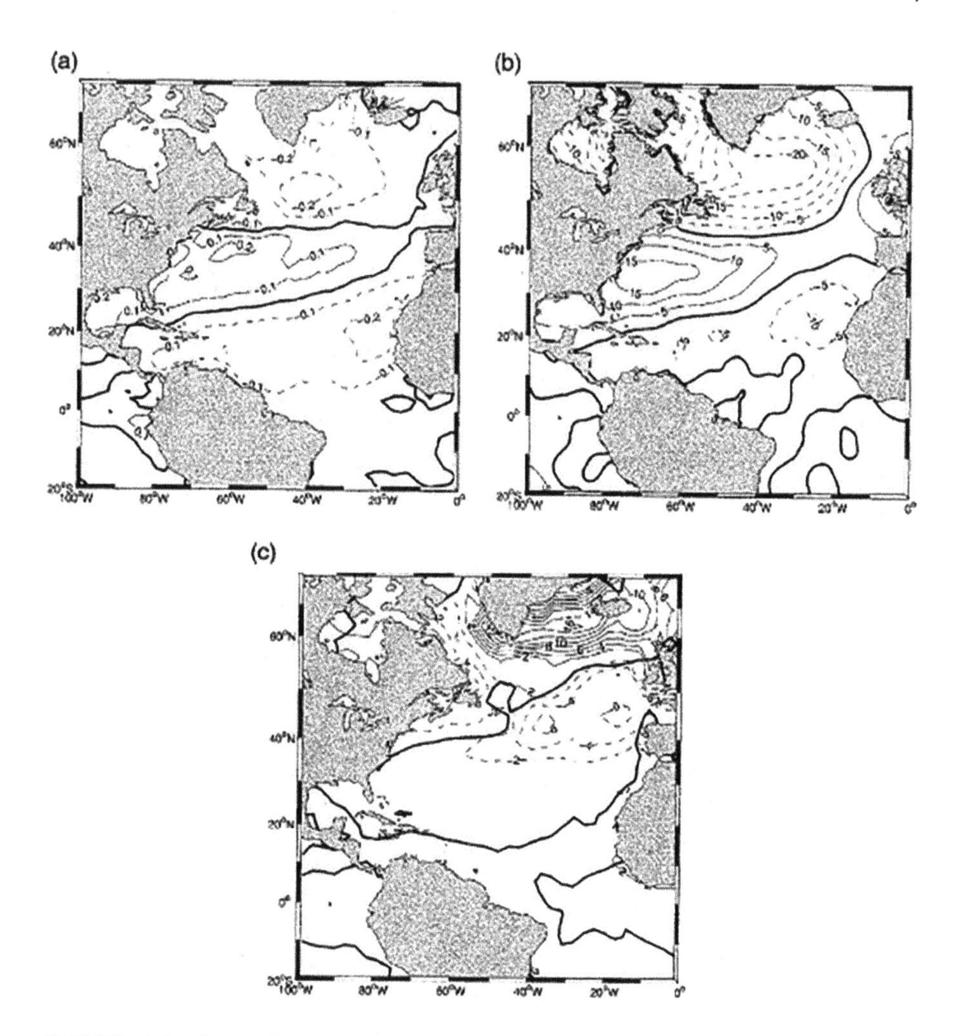

FIGURE 6.3 Regression maps of (a) SST (contour interval 0.1 K, negative dashed), (b) surface turbulent heat flux (latent+sensible, contour interval 5 Wm^{-2}, dashed out of the ocean) and (c) surface wind stress curl (contour interval 2×10^{-3} Pa m^{-1}, dashed for anti-cyclonic) anomalies onto the NAO index shown in Figure 6.1b. The thick black line is the zero wind curl line of climatological winds. SST and wind stress curl data are from the NCEP-NCAR reanalysis, and surface heat flux anomalies are from Da Silva et al., (1994a). A linear trend was removed from all datasets prior to computing the linear regressions. (From Marshall et al., 2001.)

decadal time periods. As Marshall et al. (2001) noted, however, decadal peaks in the SST anomaly spectra in Figure 6.4b extend to somewhat longer periods than those in the NAO index spectra in Figure 6.4a. In summary, the NAO forces well-defined SST anomaly patterns in the North Atlantic at interannual timescales (Wallace et al., 1990; Zorita et al., 1992; Wanner et al., 2001; Marshall et al., 2001; Pinto and Raible, 2012), and the tripole SST pattern and other SST anomaly indices also

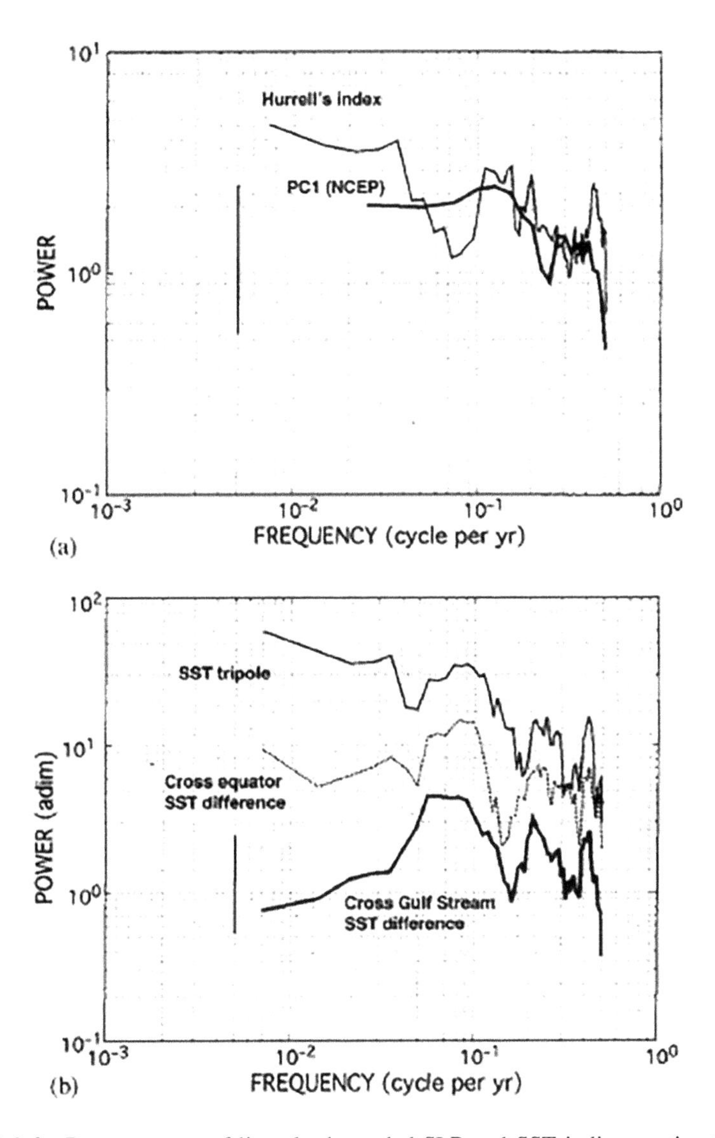

FIGURE 6.4 Power spectra of linearly detrended SLP and SST indices, estimated using the multi-taper method. (a) NAO index from Hurrell (1995; thin curve) and from a principal component analysis of winter SLP anomalies (December–March, 1958–1998) over the North Atlantic sector (20° to 70°N and 100°W to 20°E) from the NCEP-NCAR reanalysis (thick curve). (b) Indices of cross-Gulf Stream ΔT_{GS} (black continuous line), cross-equatorial ΔT_{EQ} (black dashed line) and tripole SST anomalies (gray line). The indices are ΔT_{GS} – the difference between SST averaged over [40° to 55°N and 60° to 40°W and 25° to 35°N and 80° to 60°W] (late winter-February–April), and ΔT_{EQ} [5° to 20°N and 40° to 20°W and 15° to 5°S and 15°W to 5°E] (all calendar months). The SST tripole index is defined as the first principal component of winter (December–March) SST anomalies over the North Atlantic (20° to 70°N). All SST data are from Kaplan et al. (1998). For all spectra, the 95% contour interval is shown by the vertical line, and the number of tapers has been set to K = 7. (From Marshall et al., 2001.)

undergo decadal variability as shown by Deser and Blackmon (1993) and Marshall et al. (2001).

NAO-associated SLP, wind, and temperature variability also influence sea ice extent in the North Atlantic, the Arctic, and various minor seas around the Eurasian and North American coast lines. Numerous studies of these variables in the 1990s CE alluded to decadal-multidecadal variations in sea ice (see, for example, Mysak et al. (1990) and Slonosky et al. (1997)). Then, Mysak and Venegas (1998) conducted a combined analysis of gridded, monthly sea ice and SLP data from 45°N to 90°N during the 1953–1992 CE period. Their application of the complex EOF analysis technique (see, for example, Hardy and Walton (1978), Lau (1981), Barnett (1983, 1985), Horel (1984), and Mehta (1998) for various applications of this technique) isolated sea ice anomalies propagating clockwise in the Arctic and sub-Arctic, and standing oscillation in SLP with a period ranging from eight to 14 years. The SLP oscillation was identified as phases of the NAO variability. These sea ice and SLP patterns were consistent with the decadal variability patterns of winter sea ice, SLP, 500 hPa geopotential height, and 850 hPa temperature found by Slonosky et al. (1997). Mysak and Venegas (1998) hypothesized that the decadal sea ice anomalies were generated by wind variability associated with the decadal NAO SLP variability, and interacted with ocean currents advecting them around the Arctic and sub-Arctic regions. They also hypothesized that within parts of each decadal cycle, the atmosphere can respond positively to underlying sea ice and ocean conditions, and thereby amplify the current state of the NAO. For example, negative sea ice anomalies in the Greenland Sea would increase the anomalous surface heat flux from the ocean to the atmosphere, which would deepen the Icelandic low pressure, thereby strengthening the positive phase of the NAO. Positive sea ice anomalies in the Greenland Sea can reduce the ocean to atmosphere heat flux, increasing (shallowing) the Icelandic low pressure and strengthening the negative phase of the NAO. In this hypothesis, ocean advection timescale and decadal timescale of the atmospheric component of the NAO would provide the decadal cycle time of the coupled atmosphere–ocean–sea ice oscillation. Mysak and Venegas (1998) outlined the various sea ice–atmosphere–ocean interactions based on their analysis and presented perhaps the first coherent hypothesis about decadal variability of the NAO–ocean–sea ice system as a purely internal oscillation. Major conclusions of the Mysak and Venegas (1998) study were confirmed by Deser et al. (2000) who analyzed gridded sea ice concentration data and NCEP-NCAR atmospheric reanalysis data from 1958 to 1997 CE. It was found that a multidecadal trend and decadal variability were prominent in sea ice variability. Large-scale geographical patterns of sea ice variability were found to be closely associated with dominant structures of NAO-related atmospheric circulation variability, especially in winter, which manifested itself via wind and surface air temperature variability. Deser et al. (2000) also confirmed that largest ocean-to-atmosphere heat flux anomalies were directly over reduced sea ice concentration as found by Mysak and Venegas (1998). In an intriguing extension, Deser et al. (2000)'s analyses of cyclonic storms hinted that the increased ocean-to-atmosphere heat flux over negative sea ice anomalies perhaps increased the number of such storms, providing a feedback to the original NAO forcings. Schematic diagrams, by Pinto and Raible (2012) based on Wanner et al. (2001), depicting atmospheric and oceanic features

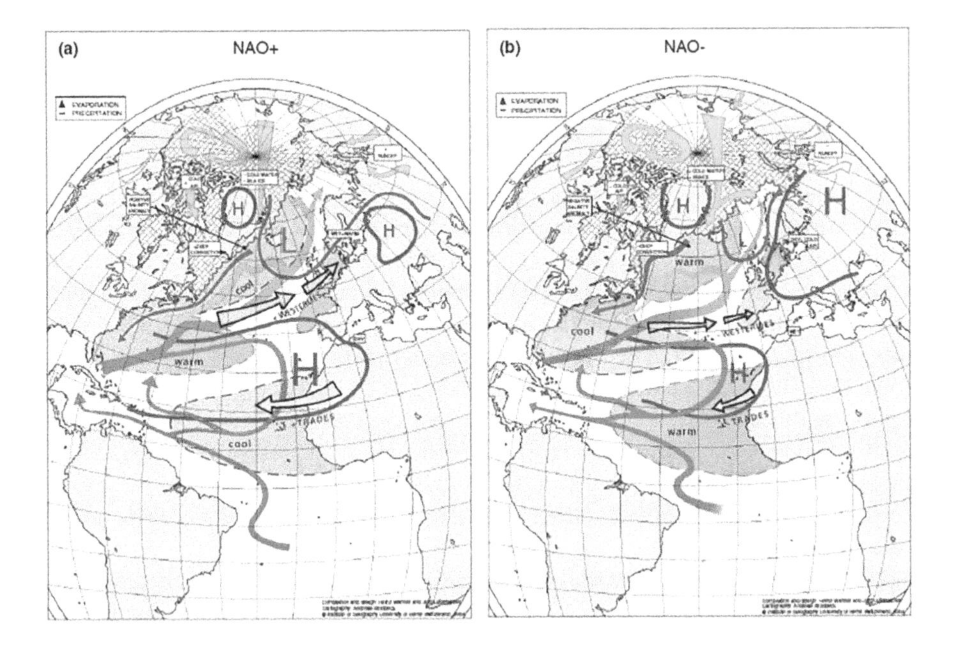

FIGURE 6.5 A schematic overview of positive and negative phases of the NAO. Shading indicates SSTs (bordered by dashed contours) and sea–ice extension; arrows show flow directions in ocean, atmosphere, and rivers; solid blue and red contours indicate sea level pressures; white rectangles describe either characteristic climate conditions or important processes: (a) the positive phase and (b) the negative phase. (From Wanner et al., 2001; Pinto and Raible, 2012.)

in positive and negative phases of the NAO are shown in Figure 6.5. As depicted in these diagrams and described here, the NAO, the North Atlantic and Arctic Oceans, the seas around western and northern Europe and northeastern North America, and sea ice undergo variability at a few days to a few decades (and perhaps longer) timescales. The decadal timescale appears to be spectrally prominent in the last several decades of the instrument-measured data. We will see in Section 6.2.4 hypothesized mechanisms of the decadal variability.

6.3 ANCIENT OBSERVERS' TALES

Earliest instrument measurements of SLP in the North Atlantic region started approximately in the 1820s CE. So, in order to estimate amplitudes and timescales of NAO variability over a longer period, many studies have reconstructed the NAO index based on ice cores, tree rings, speleothems, and documented records of weather/climate variability in and around the North Atlantic region. Such reconstructions are derived from statistical relationships – primarily, regressions of some measure of a proxy variable on instrument-measured data – between these paleoclimate proxies and instrument-measured data since the 1820s CE. A major assumption underlying these reconstructions is that the statistical relationships are stationary in time which may not be valid over multicentury periods. Another drawback of the NAO reconstructions is that regression techniques reduce variance, so regression-based reconstructions show much weaker amplitudes further back in time. There are very substantial disagreements among the numerous reconstructions of the NAO index, some of them going back to 1049 CE, as reviewed by Schmutz et al. (2000), Wanner et al. (2001), and Pinto and Raible (2012). In addition to the problems mentioned above, another possible reason for disagreements among reconstructions is that different proxies at different times of the year (for example, the use of a growing season related proxy variable) are used to reconstruct winter NAO index, and different calibration periods are used in many of the reconstructions. Uncertainties in time and amplitude of the reconstructed index can also occur due to the decreasing number and variable quality of proxy samples going back in time, locations of proxy samples, and the physical relationship between a proxy and NAO-related meteorological variability. The disagreements make it very difficult to identify any particular reconstruction as an accurate representation of interannual to multidecadal variability of the NAO. Nevertheless, some of the reconstructions are described here briefly to provide the reader an overview and references to follow up in detail.

Individual reconstructions of the NAO index are based on ice cores in Greenland (Barlow et al., 1993; White et al., 1996; Appenzeller et al., 1998); tree rings in Europe, North Africa, and North America (Cook et al., 1998, 2001; Glueck and Stockton, 2001); and corals near Bermuda (Goodkin et al., 2008). Cullen et al. (2000) used both tree rings and ice cores in a multiproxy reconstruction of the NAO index from 1750 to 1979 CE. Stalagmite growth rates in a cave in Scotland were used by Proctor et al. (2000) to reconstruct annual NAO index from the early 10th century to late 19th century CE. Luterbacher et al. (1999; 2002a, b) used descriptive documentary records to reconstruct seasonal NAO index from 1500 to 1658 CE and monthly NAO index from 1659 to 1999 CE. Rodrigo et al. (2001) also used descriptive documentary records about precipitation in Spain to reconstruct the winter NAO index from 1501 to 1997 CE. Among the more anecdote-based reconstructions was Garcia et al. (2000)'s attempt to reconstruct the NAO index from records kept on Spanish sailing ships to North America from 1551 to 1650 CE and from 1717 to 1737 CE; and Küttel et al. (2010)'s attempt to reconstruct SLP in the eastern North Atlantic, Europe, and Mediterranean Sea regions since 1750 CE using instrument measurements of surface pressure on land and marine wind information from ships' log books. Schmutz et al. (2000)'s comparison of these and other reconstructions showed that there was a low correlation between most of the reconstructed NAO indices and independent instrument-based NAO index in the early to mid-19th century CE, and among the reconstructed indices themselves. Pinto and Raible (2012)'s comparison of many reconstructions showed that there is an agreement among only normalized (by respective standard deviations of the reconstructed time series) multidecadal components of the reconstructed NAO indices and only since the late 19th century CE when instrument networks were measuring meteorological data which were used in calibration of proxy-based indices. Schmutz et al. (2000) adjudged the Luterbacher et al. (1999, 2002a, b) winter-only and annual index reconstructions closest to instrument-based NAO index, so Wanner et al. (2001) conducted a wavelet analysis of the two reconstructed indices from 1659 to 1997 CE. It was found that variability at all timescales was intermittent, with decadal variability prominent in the 18th century and in the second half of the 20th century CE. Thus, it appears from these ancient observers' tales that the only conclusive statement that can be made about proxy-based reconstructions of the NAO index is that they show substantial variability at many timescales, but do not shed any substantive light on any particular timescale!

6.4 HYPOTHESIZED MECHANISMS

After reviewing the tales told by modern and ancient observers about the NAO's decadal variability, we will now see major hypothesized mechanisms of such variability. The hypothesized mechanisms are divided into five types: (1) internally generated atmospheric variability, (2) coupled atmosphere–North Atlantic Ocean interactions, (3) coupled atmosphere–Arctic Ocean–sea ice interactions, (4) solar variability forced top-down mechanism as a pace-maker, and (5) forced by tropical west Pacific decadal variability. We saw (3) as an observations-based hypothesis in Section 6.2.2.3, and the other four possible hypotheses are described here. In addition, volcanic eruptions, especially at low latitudes, have been hypothesized to make multiyear- to decade-long impacts on the NAO. This hypothesis is described at the end of this section. As in previous sections of this chapter (and, indeed, in much of the book), major concepts and important conclusions from modeling studies of decadal NAO variability are described, and extensive references are provided for the reader to follow up further if interested.

6.4.1 Internally Generated Atmospheric Variability

In 1970s and 1980s CE, as multidecade-long time series of archived atmospheric data started to become available, researchers' attention focused on analyzing intraseasonal to interannual variability of various major features of global atmospheric circulation such as the SO, the Asian-Australian monsoons, and the NAO. Modern studies of decadal variability were about to rise above the horizon as the 1990s CE dawned.

Based on experiments with an idealized channel model of the atmosphere with fixed boundary conditions, Charney and DeVore (1979) had shown that there can be multiple flow regimes and that transitions among these regimes can cause intraseasonal atmospheric variability. In the 1980s CE, numerical models of the global atmosphere–ocean system started reaching a level of complexity and realism that allowed investigations of climate response to anthropogenic forcings. Immediately upon quantitatively analyzing decade(s)-long simulation experiments with atmosphere-only general circulation models (GCMs) with repeating, constant annual cycles of solar irradiance, several researchers found that the models simulated large-scale phenomena that exhibited variability at a variety of timescales. Among the phenomena which attracted attention was the NAO. Lau (1981) and Barnett (1985) found NAO-like variability in fixed boundary conditions simulations with the Geophysical Fluid Dynamics Laboratory (GFDL) GCM and NCAR Community Climate Model (CCM). Barnett (1985) found that the simulated NAO-like pattern exhibited two different temporal behaviors – one, continuously varying; and two, bimodal multiple equilibria. This pattern could exist in one state for some years before switching to another state in a short time. In hypotheses reminiscent of the 1930s CE, nonlinear interactions among atmospheric waves and zonally-averaged zonal winds were immediately suspected as the processes responsible for the unforced intraseasonal to interannual variability of the NAO in atmosphere-only GCMs. These simulations generally did not address decadal atmospheric variability because scientific interest was still confined to variability at timescales of a few days to, at the most, a few years. However, as we have seen in previous chapters, decadal climate variability in the context of solar variability had attracted considerable scientific interest in the previous centuries.

As the 1980s CE came to a close, two analyses of model simulations pointed to internal atmospheric dynamics as a possible cause of decadal and longer timescale atmospheric variability. James and James (1989) analyzed century-long integrations of a nonlinear, multi-layer atmospheric model, forced only by a constant representation of the annual cycle. The model had idealized representations of diabatic heating and friction, and had a uniform lower boundary without mountains, heat sources, and heat sinks. It was found that the atmosphere-only model generated maximum variability at timescales of 10 to 40 years embedded within a "red noise" spectrum, and some simulated large-scale atmospheric patterns varied with distinct 10 to 12-year oscillation periods. In an early demonstration of self-organizing chaotic variability, James and James (1989) found that there was a large-scale, decadal and longer timescale variability of the atmospheric flow and temperature entirely due to nonlinear atmospheric dynamics. Then, two years later, Mehta (1991) analyzed linear, zonally-symmetric, or annular modes of oscillation of an idealized, two-layer atmospheric model. In the presence of mean meridional circulations, this model showed oscillatory and weakly unstable modes at timescales of a week to several decades. Thus, mean meridional circulations, such as the Hadley and Ferrel Circulations, appeared to be unstable at slow oscillation times associated with their slow advection velocities. These were credible mechanisms of decadal–multidecadal variability when they were hypothesized, and they still remain possible sources of zonally symmetric and other large-scale internal atmospheric variability today.

6.4.2 Coupled Ocean–Atmosphere Variability

We saw in Section 6.2.4 that observed North Atlantic SST anomaly pattern associated with NAO variability is a tripole which has the same sign in the Greenland and Irminger Seas and the tropical North Atlantic region, and the opposite sign in central North Atlantic region. We also saw that observed surface heat flux and surface wind stress patterns are physically consistent with the tripole SST anomaly pattern. But, we saw in Section 6.2.4 that there is a distinct decadal timescale of this pattern's variability. Why is there a distinct decadal timescale? How does the ocean respond to this imprint of the NAO? Many researchers have addressed these questions with atmosphere-only and coupled ocean–atmosphere models. We will now see representative results of such studies and the insights gained from them.

Global coupled ocean–atmosphere models have been the instrument of choice to study decadal variability of the NAO and its imprint on the North Atlantic SSTs. As analyses of the archived, newly available SST and other data began to show that the NAO and its SST imprint varied at decadal timescales, described earlier in this chapter, a hypothesis emerged from global coupled model simulations to explain this decadal variability. Grötzner et al. (1998) conducted a 125-year-long experiment with the ECHO model (coupled atmosphere model ECHAM3 with ocean model HOPE; Latif et al., 1994) to see if the model simulated the observed decadal variability of the North Atlantic climate system. The atmosphere component of ECHO had a horizontal resolution of approximately 2.8° longitude – 2.8° latitude, with 19 levels in the vertical. The ocean component of ECHO had a variable meridional resolution with the highest in the deep tropics; there were 20 irregularly spaced levels in the vertical, with a realistic ocean bottom topography. The ocean component did not have any sea ice. Further details of ECHO, including ocean–atmosphere interactions, are described by Grötzner et al. (1998). ECHO simulated global climate reasonably well, except for a drift in the SSTs in the first 20 years of the simulation. It was also found that the subtropical gyre and the baroclinic Rossby wave speed in the model's North Atlantic Ocean were weaker than observed. Grötzner et al. (1998)'s analysis of the last 105 years of the simulation found an oscillation in the simulated North Atlantic ocean–atmosphere system at approximately 17-year period. The diagnosis of this oscillation revealed that a coupled ocean–atmosphere mode involving the subtropical gyre and the NAO was responsible for this oscillation. It was found that western North Atlantic SST anomalies generated an atmospheric response such that anomalous atmospheric baroclinicity, westerly winds, and associated surface heat fluxes were modified to reinforce the original meridional SST anomaly gradient via original SST anomalies' influence on statistics of transient atmospheric eddies (Palmer and Sun, 1985). This reinforcement was a positive feedback to the original SST anomalies. Then, SST anomalies of the opposite sign generated by ocean–atmosphere interactions in eastern North Atlantic were advected westward by the subtropical gyre; these SST anomalies changed the sign of the original SST anomalies in western North Atlantic, providing a negative feedback. The atmospheric response to the original SST anomalies in western North Atlantic generated a wind stress curl anomaly that modified the strength of the subtropical gyre by baroclinic Rossby waves generated in the adjustment process. This provided a delayed negative feedback. The combination of these positive and negative feedbacks formed a damped, coupled ocean–atmosphere oscillator forced by atmospheric stochastic noise, giving rise to the decadal NAO and associated SST oscillation. The somewhat longer period of the simulated decadal variability of the North Atlantic climate, compared to the observed period, may be due to deficiencies in the ECHO model, especially the weaker subtropical gyre strength and baroclinic oceanic Rossby wave speed. The reader should be reminded here that a similar set of positive and negative feedback processes has been proposed for decadal variability in the North Pacific climate as described in Chapter 3. The reader should also be reminded that decadal SST anomalies traveling in the same direction as the subtropical gyre water in the North Atlantic were observed by Hansen and Bezdek (1996) and Mehta (1998) as described in Chapter 4.

Thus, as the global coupled ocean–atmosphere model simulation-based hypothesis showed, the coupled North Atlantic climate system may be able to generate a distinct decadal timescale variability, reasonably similar to the observed variability, due to a combination of positive and negative feedbacks between the atmosphere and the North Atlantic Ocean.

6.4.3 Tropical West Pacific SST Variability

We saw in previous chapters that decadal variability in the tropical West Pacific SSTs can generate decadal variability in tropical Indian Ocean SSTs and in tropical Atlantic SSTs. Observational and modeling studies have shown that the tropical west Pacific SST variability can modulate the NAO at decadal timescales. These studies are briefly reviewed here.

Kucharski et al. (2005, 2006) and King and Kucharski (2006) found empirical evidence of an association between decadal variability in West Pacific Warm Pool (WPWP) SSTs and decadal NAO variability, and then conducted simulation experiments with an atmospheric general circulation model (AGCM) which demonstrated that a north–south SST anomaly gradient in the WPWP region can influence the NAO as observed. To begin with, Kucharski and his collaborators first analyzed surface pressure, 500 hPa geopotential heights, and surface heat flux from the NCEP-NCAR reanalysis data from 1948 to early 2000s CE; and HadISST data (Rayner et al., 2003) and the Hurrell NAO index from 1870 to early 2000s CE. All time series were filtered with an 11-year running average to focus the analyses on decadal and longer timescale variability. Regressing the NAO index on global surface pressure and 500 hPa geopotential height fields showed the NAO pattern in the North Atlantic–Europe region. More interestingly, regressing the NAO index on global SST data showed that a positive NAO index was associated with a broad Pacific Decadal Oscillation (PDO)-like pattern in the Pacific and a tripole-like pattern in the North Atlantic. Based on the western end of this PDO- like pattern in the WPWP, Kucharski et al. (2006) hypothesized that decadal SST variability in the WPWP region, especially that of a north-south SST anomaly gradient between approximately 20°S and 20°N, can potentially influence the NAO because relatively small but long-lived SST variability in this region of very warm climatological SSTs can substantially influence global atmospheric circulation and precipitation; this was also found by Wang and Mehta (2008) and described in Chapter 3. Indeed, an SST index formed to express this north–south gradient was very highly correlated (coefficient 0.82) with the NAO index, and the regression of this SST index with global SST data showed a pattern very similar to the NAO-global SST regression pattern.

To simulate this empirical association between the north–south SST anomaly gradient in the WPWP and the NAO, Kucharski et al. (2006) employed an AGCM developed at the International Centre for Theoretical Physics in Italy and named Simplified Parameterizations, primitivE-Equation DYnamics (SPEEDY). Three ensembles of simulation experiments were conducted. These three ensembles were (1) EXP1: 20-member ensemble with observed global SSTs from 1870 to 2002 CE; (2) EXP2: 18-member ensemble with observed SSTs from 1870 to 2002 CE only in the tropical West Pacific region enclosed by 140°E to 170°W longitudes and 20°S to 20°N latitudes, with climatological SSTs over the rest of the globe except for the North Atlantic basin where SST anomalies were calculated with a thermodynamic mixed-layer model; and (3) EXP3: 10-member ensemble with observed SSTs from

1870 to 2002 CE in the tropical Indian Ocean enclosed by 30°E to 120°E longitudes and 30°S to 30°N, with climatological SSTs over the rest of the globe except for the North Atlantic basin where SST anomalies were calculated with a thermodynamic mixed-layer model. Detailed analyses of the three ensembles by Kucharski et al. (2006) showed that variability of the north–south SST anomaly gradient in the WPWP region forces an NAO-like pattern in 500 hPa geopotential heights which also varies at decadal timescales. While analyses of the EXP1 ensemble corroborated the empirical analyses results, EXP2 ensemble established that the WPWP SST anomaly gradient variability can force NAO variability. The EXP1 and EXP2 ensembles showed that a wave train pattern, generally similar to the NAO-associated wave train pattern, emanates out of the WPWP region in response to the north–south SST anomaly gradient in that region. However, Kucharski et al. (2006) found that the NAO response was weaker than that observed in their empirical analysis which may be due to other regions' and other forcings' influences on the NAO, and/or due to AGCM deficiencies. It was also speculated that perhaps there is a two-way coupling between the WPWP SSTs and the NAO via the North Pacific which can generate a larger amplitude NAO response. Incidentally, Kucharski et al. (2006) also found that tropical Indian Ocean SST variability at decadal and longer timescales did not appear to influence the NAO in their AGCM experiments during the 120-year study period. King and Kucharski (2006) followed up these empirical and model studies with a more detailed empirical study to test the robustness of the earlier results. They found filtering their 120 years' empirical SST and NAO data with various moving-window widths did not change the WPWP SST gradient-NAO association. Later, Müller et al. (2008) generally confirmed the association between decadal variability of the WPWP SST anomaly gradient and the NAO by performing wavelet analysis of various SST and SLP indices for the 1900–2006 CE period. They found that there was a high coherency between NAO and Niño3 SST indices at 10 to 20-year periods. They also found via regression analysis that decadal variability of tropical Pacific SSTs was strongly associated with the North Atlantic SST anomaly tripole forced by the NAO.

It must be emphasized that the results described here may not be as conclusive as they appear. Observing that tropical SSTs warmed considerably since the 1970s CE, King and Kucharski (2006) repeated their analyses after removing the last 30 years' data. It was found that the empirical associations found using 120 years' data were not stationary and that the WPWP SST anomaly gradient's association with NAO variability at multidecadal timescales weakened and the tropical Atlantic SST anomalies' association with NAO variability strengthened. This detailed and piecewise analysis reinforced conclusions from previous studies that influences of various oceans on the global atmosphere and/or the atmosphere's responses are not stationary over multidecadal periods. Nevertheless, this set of studies demonstrated that tropical Pacific decadal variability can be a major driver of decadal NAO variability.

6.4.4 Solar Variability as a Pace Maker?

The last hypothesized cause and mechanisms of decadal NAO variability described in this chapter is the variability of solar emissions, especially the 11-year solar Schwabe cycle described in Chapter 2, and its influences on the NAO via stratosphere–troposphere–Atlantic Ocean interactions. The Schwabe cycle's possible influences on the NAO were briefly mentioned in Chapter 2 as a part of influences on the global atmosphere, and its hypothesized role in decadal variability of the tropical Pacific Ocean–atmosphere system was described in Chapter 3. As we will see, the so-called top-down influence of the solar ultraviolet (UV) radiation cycle from the upper stratosphere to the NAO in the troposphere and coupled atmosphere–Atlantic Ocean response to this top-down influence is a major candidate for generating decadal NAO variability.

As described in Chapter 2, solar UV radiation varies with an 11-year cycle. It is estimated from measurements with the Spectral Irradiance Monitor instrument on National Aeronautics and Space Administration (NASA)'s Solar Radiation and Climate Experiment (SORCE) satellite, launched in 2004 CE, that UV radiation in the 200 to 320 nm band contributes very substantially to solar heating in the middle atmosphere, largely via absorption by ozone in the stratosphere (Ineson et al., 2011). Ozone is also produced in the stratosphere by photolysis of oxygen molecules, so an increase in solar UV radiation can increase heating in the stratosphere by ozone absorption as well as by an increase in ozone concentration. Following several studies which evolved the hypothesis about how the 11-year solar cycle can influence stratospheric temperatures and circulations, Kodera and Kuroda (2002) analyzed geopotential heights and winds in the stratosphere, and the NCEP-NCAR atmospheric reanalysis data from 1979 to 1998 CE to see if there was observational evidence to support the hypothesis. They found that, in the long-term average, stratopause circulation evolves from a radiatively controlled state to a dynamically controlled state during winter in both hemispheres, accompanied by a poleward shift of the westerly jet in the stratopause region. The 11-year solar cycle changes the balance between the two types of states in winter. During the high activity phase of the cycle, the radiatively controlled state lasts longer, with stronger winds in the stratopause subtropical zonal jet. This solar influence from the stratopause and upper stratosphere regions is transmitted to the lower stratosphere via a modulation of internal variability of the polar night jet and a change in the meridional Brewer–Dobson circulation. Kodera and Kuroda (2002) found that these changes were significant in the middle and high latitudes, and prominent in the equatorial region.

Next, Kodera and Kuroda (2005) used a variety of data to delve into the possible association between solar activity and stratospheric dynamics in boreal winter (December–January–February) – geopotential height data above 100 hPa from 1979 to 1999 CE to calculate winds in the stratosphere; the NCEP-NCAR reanalysis data for the atmosphere below 100 hPa from 1958 to 1999 CE; and 10.7 cm radio flux as an indicator of solar activity. In analyzing these data, they found that, in early winter of high solar activity, the leading zonal wind pattern in the stratosphere is a meridional dipole that interacts with vertically-propagating planetary waves and extends poleward and downward to the troposphere. Absorption of solar UV radiation in the stratosphere during high solar activity would produce a meridional temperature gradient which, in turn, would produce the observed zonal wind pattern. A hemispheric scale dipole-like pattern of surface pressure anomalies between the polar region and subpolar region is produced by the extended zonal wind pattern from the stratosphere. During the low solar activity part of the 11-year cycle, the downward extension of the stratospheric zonal wind pattern is weak, producing a regional-scale surface pressure anomaly pattern. Kodera and Kuroda (2005) interpreted these results to associate decadal NAO variability with solar variability such that the NAO would be confined to the Atlantic sector during low solar activity and extend to a hemispheric pattern during high solar activity. This interpretation is consistent with Kodera (2003)'s conclusion based on analyses of SLP and Sunspot number data from 1900 to 1999 CE.

Now, it was time for modelers employing then-state-of-the-art global coupled ocean–atmosphere models to use the SORCE-measured UV data to study their models' response and compare it with observations. Ineson et al. (2011) used SORCE-measured solar UV flux estimates near a solar cycle minimum in 2004–2007 CE to drive the HadGEM3 model to study its possible response. HadGEM3 has a well-resolved middle atmosphere with the top at 85 km. SORCE-measured UV radiation in the 200 to 320 nm band during a minimum in solar activity cycle was extrapolated to the entire cycle. The difference in the 200 to 320 nm band between solar minimum and solar maximum was estimated to be 1.5 W m^{-2}. This difference was applied across the entire band, with irradiance in other bands held constant. Thus, the experiments tested HadGEM3's sensitivity only to this UV band. An 80-year reference or control simulation was conducted with the 200 to 320 nm irradiance held constant at the SORCE-measured value. Then, an ensemble of 20-year simulations with different initial conditions and the extrapolated UV band irradiance for a maximum in the solar cycle were conducted; 80 years of simulated data from this solar maximum ensemble were compared with the 80-year solar minimum control simulation. They found that, as observed in ERA reanalysis data, zonal wind anomalies descended from the stratosphere to the North Atlantic Ocean surface and caused surface pressure and temperature anomalies resembling the negative phase of the NAO. The establishment of a large enough meridional temperature gradient in the stratosphere was found to be crucial in setting off the downward and poleward propagation of zonal wind anomalies, and a large enough solar UV radiation forcing was found to be crucial for establishing the meridional temperature gradient. Thus, the SORCE-based solar UV irradiance measurements shed very important light on a possible NAO response to the 11-year solar activity cycle.

It was found in observational studies that the NAO reaches a maximum (or, a minimum) two to four years after a maximum (or minimum) in the 11-year solar cycle. Scaife et al. (2013) further analyzed the HadGEM3 experiments to study if there was a lagged response of the NAO to the solar activity cycle and found that the response increased to several hPa SLP change two to three years after the peak in the cycle. They also found that the SST anomaly tripole forced by the NAO (Section 6.2.2) also had a lagged response to the solar cycle. Scaife et al. (2013) conducted a deeper analysis to understand the lagged NAO–North Atlantic Ocean response by constructing a mechanistic, coupled NAO–SST model based on the HadGEM3 results, with adjustable parameters quantifying feedbacks between the NAO and underlying SSTs. Experiments with this two-variable mechanistic model showed that stronger ocean–atmosphere coupling was required to reproduce the observed NAO–SST tripole relationship.

We will see now how the solar cycle influence on the NAO permeates into the North Atlantic Ocean and why there can be a two to four years' delay in the atmosphere–ocean system's response to the solar cycle. As described in Sections 6.2.4.2 and 6.2.4.3, NAO-associated surface wind and surface heat fluxes drive thermodynamical and dynamical responses of the North Atlantic Ocean. In another set of experiments with the HadGEM3 coupled ocean–atmosphere model, Andrews et al. (2015) set out to study if the model reproduced the delayed response and, if it did, then how large and significant the response was in comparison with observations. The experiments were run with constant solar irradiance with the addition of spectrally-resolved, time-varying irradiance for the 1960–2009 CE period. There were 17 years each of positive and negative solar UV radiation anomalies in this period. Among other attributes of the ocean and atmosphere components of the model was the model atmosphere's highly resolved vertical levels up to 85 km height which made this so-called high-top model capable of simulating stratospheric UV heating and circulation. It was found that the model simulated zonal-average zonal wind anomalies migrating poleward and downward via interactions between the zonal-average winds and planetary waves as hypothesized in response to solar UV radiation cycles. This downward propagating zonal wind anomalies generated an NAO-like response in the troposphere and in SLP over the North Atlantic, which reached a maximum (or, a minimum) two to three years after the solar UV cycle maximum or minimum. The ocean temperature changes associated with NAO-forced surface heat fluxes went down to several tens of meters depth in winter. These fluxes generated a tripole SST anomaly pattern which extended down to greater than 60 m in the North Atlantic Ocean as shown in Figure 6.6. Responses of the upper 200 m ocean temperatures in the three poles are shown in Figure 6.7. In the northern pole of the tripole off Greenland and in the Labrador Sea, horizontal transport of temperature anomalies was substantial which disrupted vertical coherence of solar UV-forced temperature anomalies. So, in this pole, the SSTs responded passively to solar forcing without feeding back to the overlying atmosphere. The southern pole of the tripole in the subtropical Atlantic responded directly to the solar forcing variability in phase. In the middle pole of the tripole, the crucial lagged, positive feedback response occurred. The upper ocean mixed layer in the middle pole can be much deeper in winter than in summer. In the following summer, a shallow mixed layer can form on top of the previous winter's deep mixed layer and can "hide" the solar-forced temperature signal from the atmosphere. During the next winter, when the solar influence is still strong, the previous winter's temperature anomalies can "re-emerge" and begin to feedback positively to the atmosphere as proposed by Timlin et al. (2002). This summer–winter mixed layer process continued for two to three years in the middle pole in the HadGEM3 model, while the solar UV cycle forcing near a maximum or a minimum was still substantial. The maximum response of the upper ocean–atmosphere system in the model was thus lagged by three to four years after a maximum in solar UV radiation. Thus, the memory of the solar forcing resided and aggregated in the middle pole in sub-surface North Atlantic Ocean. Although, in this set of HadGEM3 experiments, the maximum lagged response of the NAO to the SORCE-observed UV radiation anomaly was 1.8 hPa whereas the observed NAO anomaly in the NCEP-NCAR reanalysis is 4.6 hPa, these experiments by Andrews et al. (2015) showed that the entire chain of hypotheses from the upper stratosphere to the sub-surface North Atlantic Ocean works in the HadGEM3

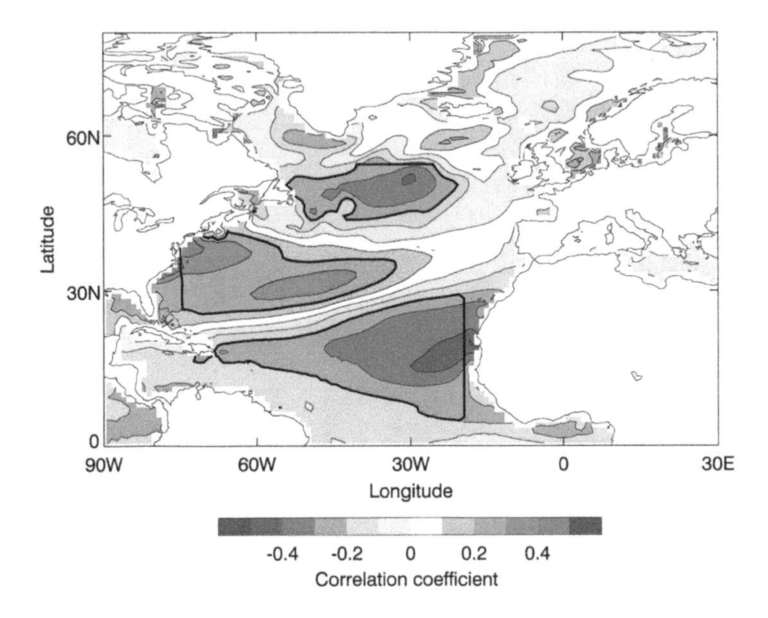

FIGURE 6.6 Correlation between the NAO index and the North Atlantic 10 m depth ocean temperature model field. The tripole regions used to generate the time-depth ocean profiles in Figure 6.7 are outlined by black contours. The regions used for analysis are within the bounds 75° to 20° W and 0° to 55°N, and have an absolute correlation greater than 0.2. (From Andrews et al., 2015; Used with permission from Creative Commons Attribution License (http://creativecommons.org/licenses/by/3.0/).)

model. These results strongly suggest that the 11-year solar cycle can be a pace-maker of decadal NAO variability.

As we saw earlier in Section 6.2.4, decadal NAO variability may be generated by internal atmospheric dynamics, integration of atmospheric noise by the North Atlantic Ocean, and coupled ocean–atmosphere dynamics. If the 11-year solar cycle can also generate decadal variability in the NAO–North Atlantic Ocean system, can there be any interaction between internally generated and externally forced decadal NAO variability? At least two studies have attempted to address this possibility with global coupled ocean–atmosphere model simulations. Incidentally, both studies have apparently used the same version of the NCAR Community Earth System Model (CESM). Thiéblemont et al. (2015) employed the CESM to see if/how the solar cycle can synchronize the CESM's internally generated decadal NAO variability. The CESM's atmosphere component had 66 levels in the vertical from the Earth's surface to 140 km and had interactive chemistry. In one set of experiments (labeled SOL), the CESM was forced by daily observed spectrally-resolved solar irradiance from 1955 to 2009 CE, and the last four solar cycles were repeated twice from 2010 to 2099 CE. In the second set, spectrally-resolved solar irradiance was held at its average value (labeled NO_SOL). Thiéblemont et al. (2015) found that the CESM generates decadal NAO variability without cyclic solar forcing. However, the amplitude of the decadal NAO variability was much smaller than observed, and the physics of

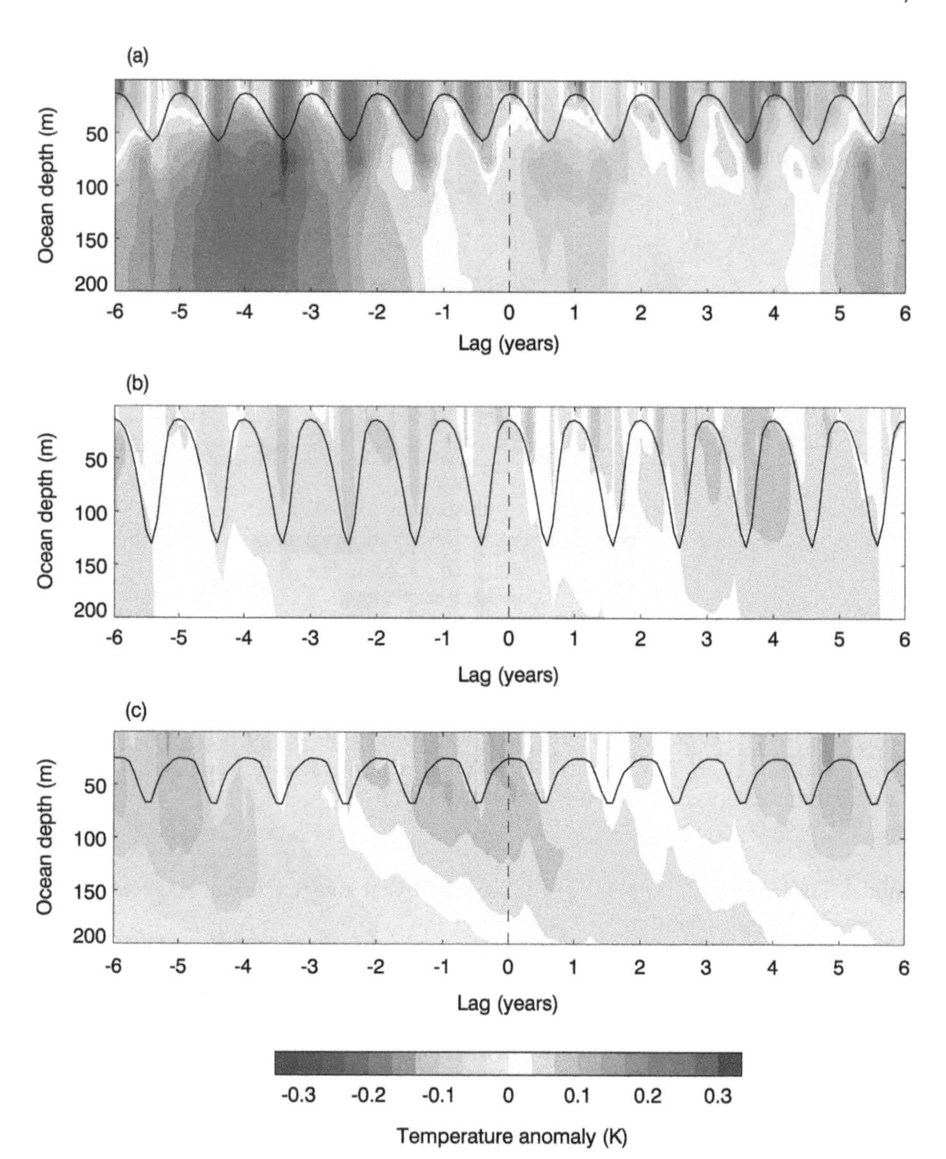

FIGURE 6.7 Upper 200 m time-depth ocean temperature for three NAO-correlated tripole regions in Figure 6.6: (a) north, (b) middle, and (c) south, for lags of −6 to +6 years with respect to solar maximum minus solar minimum. Lag 0 years correspond to the July of the solar extrema years. The seasonal cycle of the simulated monthly-mean MLD, defined as the depth where the potential density is 0.01 kg m^{-3} greater than that of the 10 m depth level, is plotted in black. Winters correspond with deep excursions of the MLD. (From Andrews et al., 2015; Used with permission from Creative Commons Attribution License (http://creative commons.org/licenses/by/3.0/).)

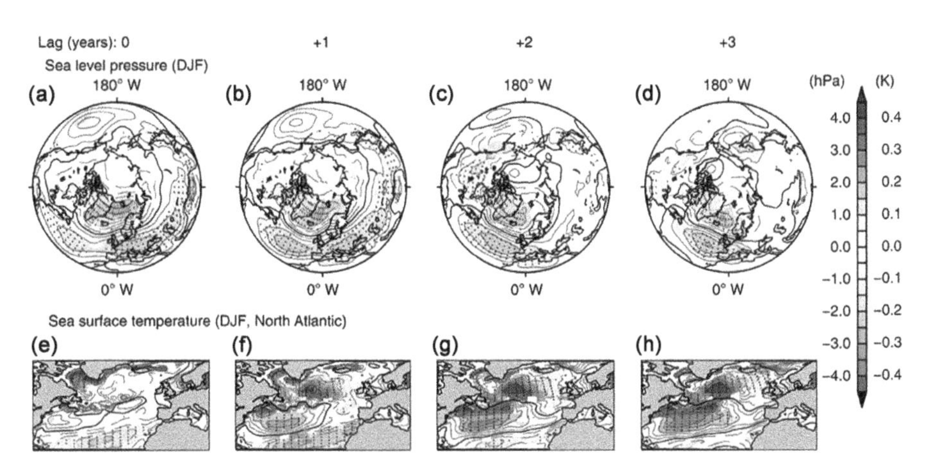

FIGURE 6.8 Consecutive lagged composite solar maximum minus minimum differences for sea level pressure (a–d) and SST (e–h) for DJF season. Sea level pressure composites are shown for all longitudes and northward of 20°N. SST composites are shown for the North Atlantic region only (80°W to 20°E and 20°N to 70°N). Significance levels are indicated by shaded contours (90%) and dots (95%). (From Thiéblemont et al., 2015; Used with permission from Creative Commons Attribution License (http://creativecommons.org/licenses/by/3.0/).)

the stratosphere–troposphere interactions were substantially different. The cyclical solar forcing experiments showed that the simulated NAO SLP and SST anomaly patterns were generally similar to observed patterns as shown in Figure 6.8. The cyclical solar forcing synchronized the internally generated NAO variability with the solar forcing with a one- to three-year lag in maximum response as Figure 6.8 shows. It was also found in the cyclical solar forcing experiments that the sequence of synchronization was consistent with the observed and hypothesized downward propagation of the solar signal from the stratosphere to the surface. The synchronization effect is shown in Figure 6.9. The Andrews et al. (2015) and Thiéblemont et al. (2015) studies substantiated hypothesized physics of how the 11-year solar cycle can generate decadal NAO variability and synchronize internally generated variability to the solar cycle or so it seemed in 2015 CE. Chiodo et al. (2019) used the same version of the CESM, but conducted their SOL and NOSOL simulations for 500 years each. In the SOL experiment, spectrally-resolved solar forcing sequence from the last four 11-year cycles (total 44 years) was repeated almost 12 times, whereas the solar forcing was held constant in the NOSOL experiment for 500 years of simulation time. It was found that while there were segments of the 500-year SOL experiment in which the simulated NAO index showed apparent synchronization with the solar cycle forcing, there were other segments in which there was no synchronization. The 500-year NOSOL experiment was also found to contain segments in which the simulated NAO index showed transient decadal variability. Chiodo et al. (2019) also showed that the relative phase between the solar forcing and 9 to 13-year band-pass filtered NAO variability was not constant in Thiéblemont et al. (2015)'s 150-year SOL experiment, implying that the apparent synchronization could be

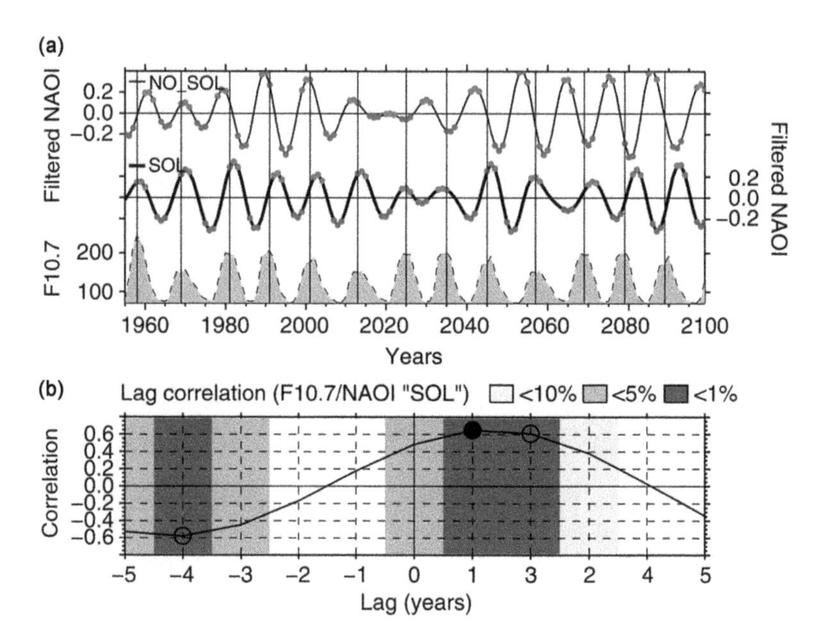

FIGURE 6.9 (a) Time series of 9- to 13-year band-pass-filtered NAO index for the NO_SOL (solid thin) and SOL (solid thick) experiments, and the F10.7 cm solar radio flux (dashed black). Red and blue dots define the indices used for NAO-based composite differences at lag 0. For each solar cycle, maximum are marked by vertical solid lines. (b) Lag correlation between F10.7 cm solar radio flux and NAO index of the SOL experiment. Significance levels are indicated by empty (90%) and filled (95%) circles. Grey shades colored stripes indicate the likelihood (<10, 5 and 1%) that the correlation coefficient between the filtered NO_SOL NAO and the F10.7 time series is higher than that between the filtered SOL NAO and the F10.7 ones. All calculations are shown for DJF. (From Thiéblemont et al., 2015; Used with permission from Creative Commons Attribution License (http://creativecommons.org/licenses/by/3.0/).)

due to chance. Chiodo et al. (2019) performed time–frequency wavelet analyses of observations-based NAO index from the HadSLP data and simulated NAO indices from the SOL and NOSOL experiments. The results are shown in Figure 6.10. It is clear that interannual, decadal, and multidecadal variabilities in the HadSLP based NAO index (Figure 6.10a) are transient as also shown by other studies. Wavelet spectra of the two simulated NAO indices also show the transience of these timescales (Figure 6.10b and c). Chiodo et al. (2019) argued that solar cycle association can appear with simulated NAO index variability when the latter is prominent, but that the decadal NAO variability can be generated by internal ocean–atmosphere processes without any synchronization with the 11-year solar cycle forcing.

Thus, the foregoing shows that insights are developing into the chemistry and physics of how the 11-year solar cycle can influence the NAO as more realistic global coupled ocean–atmosphere models and spectrally-resolved solar irradiance measurements become available and are used to simulate the influence. These insights indicate a strong possibility that the 11-year solar cycle can be a pace maker of

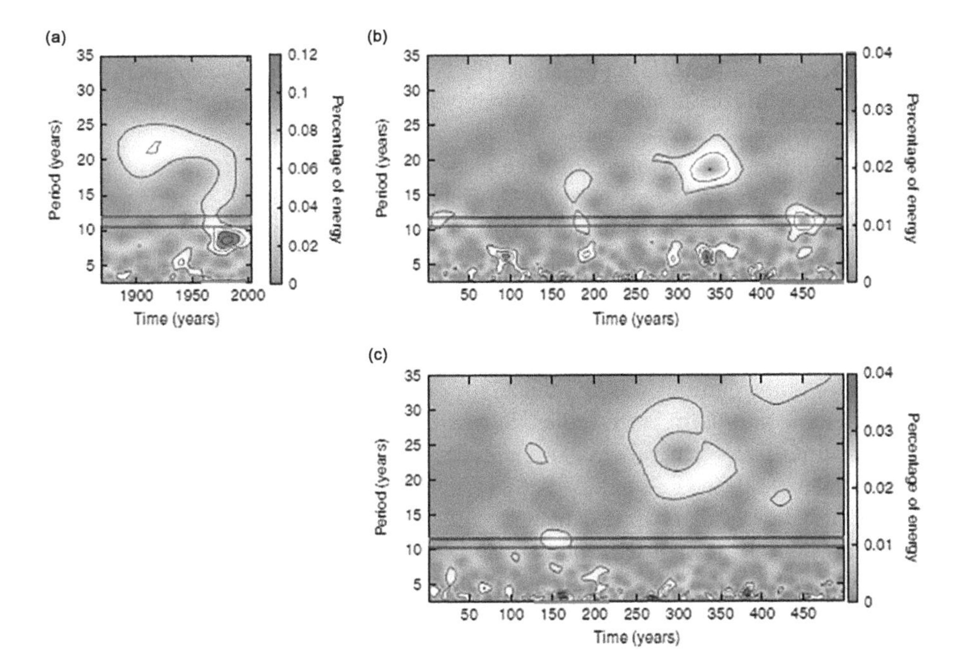

FIGURE 6.10 (a) Spectral power as a function of time and frequency for the NAO index calculated from the HadSLP reconstruction. The red box identifies the frequency range in which solar irradiance exhibits the most prominent spectral peak (10–12 years). (b) As in (a) for SOL, and (c) for NOSOL. Red lines along the x axis identify the periods over which a strong solar/NAO signal is extracted using regression analysis in each respective data set. (From Chiodo et al., 2019.)

decadal NAO variability. However, this is still not a certainty as Chiodo et al. (2019) showed and further research is required.

6.4.5 Impacts of Volcanic Eruptions

Impacts of volcanic eruptions on the Earth's climate have been studied for many decades. Long-lived temperature and rainfall anomalies in the past have been attributed to large volcanic eruptions in both low and high latitudes. Sulfur dioxide and ash from eruptions are known to influence the Earth's radiation balance and thereby the coupled ocean–atmosphere system. Sulfate aerosols formed by sulfur dioxide's reaction with water in the atmosphere scatter solar short wave radiation, thereby cooling the Earth. These aerosols, if injected into the stratosphere, can warm it by the absorption of the longwave radiation emitted by the Earth-troposphere system, with halogens among the aerosols depleting stratospheric ozone and cooling the poles. Volcanic ash, if injected high enough into the stratosphere, can absorb longwave radiation and warm the stratosphere where it resides before removal due to gravity. Thus, these ejecta from volcanic eruptions can influence zonal and meridional temperature gradients in the stratosphere, which can change stratospheric winds and can influence tropospheric winds and temperatures. The explosivity, duration, and composition of materials ejected from a volcanic eruption can determine the strength and duration of the eruption's effects on climate. Long duration effects can also influence ocean–atmosphere interactions with consequent effects on multiyear to decadal and longer timescale climate variability.

Many studies have invoked this line of possible effects of volcanic eruptions to simulate resulting evolutions of the NAO. Such studies have found either short-term (one to two years) or long-term (multidecades) impacts on the NAO, depending on whether they used an atmosphere-only model or a coupled ocean–atmosphere model. As representatives of the former (short-term) type of studies are simulations of impacts of the Mt. Pinatubo eruption in June 1991 on the NAO by Stenchikov et al. (2002), and simulations of impacts of many volcanic eruptions during late 19th and entire 20th centuries CE by Stenchikov et al. (2006). Incidentally, the NAO is referred to as the AO in both studies. Stenchikov et al. (2002) employed the GFDL SKYHI atmosphere general circulation model extending from the Earth's surface to approximately 80 km (Fels et al., 1980). Ensemble experiments with different types of forcings showed that ozone depletion in the lower stratosphere, described above, produced cooling at high latitudes, strengthened the NH polar vortex, and caused the positive phase of the NAO in late winter and early spring. Volcanic aerosols only in the troposphere produced cooling at subtropical latitudes which reduced the meridional temperature gradient in mid-latitudes and planetary wave amplitudes, and decreased wave activity flux from the troposphere into the lower stratosphere. This sequence of processes also strengthened the NH polar vortex and caused the positive phase of the NAO. Both these effects of the Mt. Pinatubo eruption lasted up to two NH winters. Stenchikov et al. (2006) expanded the scope of their investigation in several ways into effects of volcanic eruptions on the NAO. Historical runs from 1860 to 1999 CE of seven global coupled ocean–atmosphere models from the Coupled Model Intercomparison Project 3 (CMIP3) were used in their study, with estimates of forcings due to volcanic eruptions included in these simulations. The seven models were from NCAR (2), GFDL (2), NASA-Goddard Institute for Space Studies (2), and MIROC (1). The warming in the stratosphere and the cooling in the troposphere seen in observations and in the atmosphere-only study by Stenchikov et al. (2002) were also found in historical runs of these seven CMIP3 models. Composite analyses of effects of major eruptions of tropical volcanoes in the models also showed a tendency for positive NAO phase in two winters after eruptions. However, the amplitudes and zonal patterns of the models' responses differed among themselves and in comparison with observations.

In the long-term (multidecades) part of the spectrum, possible responses of NAO–North Atlantic Ocean circulation, including the Atlantic Meridional Overturning Circulation (AMOC), interactions to volcanic eruptions have been studied by many researchers with global coupled ocean–atmosphere models. In these model simulations, persistent positive phase of the NAO in response to strong tropical volcanic eruptions, as described in the Stenchikov et al. (2002, 2006) studies, forces changes in ocean–atmosphere heat transfers in the North Atlantic and Arctic regions. Due to slow response times (one to two decades) of the AMOC to the persistent NAO forcings, SST and sea ice changes in these regions occur slowly. These changes feed back to the NAO and, thus, strong tropical volcanic eruptions can generate multidecadal variations in the NAO; and in ocean circulations, temperatures, sea ice, and salinity (see, for example, Ottera et al. (2010), Zanchettin et al. (2012, 2013), and references therein). Ottera et al. (2010) have also hypothesized that tropical volcanic eruptions can act as a pace-maker of multidecadal variability in the Atlantic climate.

To conclude this sub-section on impacts of volcanic eruptions on the NAO, it should be mentioned that low-latitude, moderate-to-high explosivity eruptions have been found to influence the PDO and decadal variability of the WPWP SSTs in observed data and simulations conducted with four Earth System Models (MIROC5, NCAR-CCSM4, GFDL-CM2.1, and UKMO-HadCM3) under CMIP5, and even change the phase of these phenomena (Mehta et al., 2018). Multiyear-to-decadal predictability of these phenomena has also been found to be influenced by low-latitude eruptions in decadal hindcast experiments with the same models (Mehta et al., 2019). Since the PDO and WPWP variability have been associated with decadal NAO variability (Section 6.4.3; also, Wang and Mehta (2008)), it is plausible that tropical volcanic eruptions' effects on the PDO and the WPWP variability can, in turn, influence the NAO. This plausible connection should be investigated in future research.

6.5 SUMMARY AND CONCLUSIONS

Atmospheric variability over the North America–North Atlantic–Europe regions at intraseasonal to multidecadal timescales impacts hydro-meteorology, storms, winds, temperatures, river flows, and crop productions in these regions. There is at least two-century-long history of observations and research in the variability which has a pronounced decadal component. The primary extratropical atmospheric variability in the NH has at least three names – NAO, AO, and NAM; they are generally synonymous. It would be very useful if sources and mechanisms of NAO/NAM/AO variability can be understood, and if evolutions of these variabilities and their impacts – especially at decadal timescales – can be predicted well in advance for societal benefits. This chapter reviews what is known so far about these variabilities.

Weather observations by Scandinavian missionaries in Greenland started attempts to understand and predict variability in what was later called the NAM phenomenon. In NAO/NAM, there is an oscillation in atmospheric mass or SLP between subpolar and subtropical latitudes in NH. Wind variations are associated with the SLP variability from the Earth's surface to well over 100kms in the atmosphere. These surface wind and heat flux variations generate a tripole SST anomaly pattern from subpolar seas to subtropical North Atlantic. Frequency spectra of time series of NAO indices formed from SLP data and the tripole SST pattern show several significant peaks between eight and 20-year periods. The decadal surface wind and heat fluxes also interact with sea ice in the NH, and it is hypothesized that interactions among them can generate decadal variability. Internal atmospheric variability due to nonlinear dynamics, coupled ocean–atmosphere variability, and influences of the 11-year solar cycle are other, hypothesized causes of the decadal NAO variability. Observations and global atmosphere–ocean models extending to almost 200km show that solar UV variability cycle at approximately 11-year period can heat the stratosphere, modulate meridional temperature gradient and stratospheric circulation, and interact with planetary waves. This signal can descend to lower troposphere, influence surface winds and heat fluxes, and mixed layer ocean temperatures. Feedback processes in mid-latitude/subtropical North Atlantic can delay and aggregate the solar influence to as a pace-maker of decadal variability if the ocean–atmosphere coupling is strong enough. It is also possible that strong volcanic eruptions, especially at tropical latitudes, can directly or indirectly cause decadal NAO variations.

Both empirical and model-based approaches are needed for further progress in understanding decadal NAO variability. To describe various aspects of the variability in nature, continued measurements of solar UV and visible irradiances are required as are continued surface-based and satellite-based atmosphere–ice–ocean observations.

Analyses of Earth System Model data from the World Climate Research Program's CMIP6 project to study natural coupled ocean–atmosphere–sea ice variability and to study the role of the 11-year solar cycle and volcanic eruptions should be carried out. Decadal hindcasts of SSTs and other ocean–atmosphere variables are also available from CMIP6. They should be analyzed to estimate decadal predictability of the NAO variability, and associated variability of SST and other oceanic variables.

As we saw in this chapter, missionaries and explorers in the 18th and 19th centuries began to record their observations of weather phenomena in extratropical NH. Then, scientists working with missionary zeal and imbued by a spirit of exploration developed insights into possible causes and mechanisms of these variabilities even though they had sparse empirical data. Then, as the data availability increased in the 1990s CE, the study of decadal NAO variability expanded into a very important area of climate research. We have finally come to the time when there are actual attempts to make decadal predictions of NAO variability.

7 SAM, the Albatross

7.1 INTRODUCTION

Successes of perhaps the first international scientific research programs, the International Polar Year (IPY) in 1882–1883 Current Era (CE) and 1932–1933 CE, led to a much more ambitious and co-ordinated program, the International Geophysical Year (IGY),[1] to study 11 Earth Sciences from 1 July 1957 to 31 December 1958 CE. These 11 Earth Sciences were aurora and airglow, cosmic rays, geomagnetism, gravity, ionospheric physics, precise geo-location (longitude and latitude), meteorology, oceanography, seismology, and solar activity. The IGY period covered the maximum phase of solar cycle 19 (April 1954 to October 1964) and was thus very appropriate for an international program to study many of the 11 Earth Sciences. While 12 and 44 countries participated in the two IPYs, respectively, over 70 countries participated in the IGY. Antarctica was designated as one of the foci geographical areas. During the IGY, several research stations were established in Antarctica by, among others, the United States of America (USA), Australia, the United Kingdom (UK), France, Japan, and Belgium. Also during the IGY, the Soviet Union (Sputniks 1 and 2) and the USA (Explorer 1 and Vanguard 1) launched two Earth-orbiting satellites each, these being the first ever artificial satellites of the Earth. Research with the data collected during the second IPY in 1932–1933 CE and the data themselves were lost during the Second World War (1939–1945 CE). Therefore, preservation of data to be collected during the IGY and its free availability to scientists in all countries became a major pre-requisite of the IGY, and the first World Data Centers were established a few months before the IGY began. Each World Data Center archived the entire data set collected during the IGY and made them available to scientists from all countries.

Schwerdtfeger (1960) appears to be the first to use archived data from four stations in Antarctica established during the IGY to analyze variability of the circumpolar vortex around the South Pole. This may be the first study which ultimately led to the formal naming of the Southern Annular Mode (SAM) almost 30 years later. Unlike the history of the North Atlantic Oscillation (NAO)/Northern Annular Mode (NAM) phenomenon going back to at least the 18th century CE, the discovery of SAM goes back approximately 30 years only. However, Gong and Wang (1999) quote Walker (1928) to argue that, in his quest to find climate phenomena possibly affecting the Indian monsoon, Sir Gilbert Walker may have glimpsed a possible SH counterpart to the North Pacific Oscillation (NPO) and the NAO he discovered in the NH. The story of SAM and its decadal variability is narrated in this chapter.[2] SAM is the Southern Hemisphere (SH) counterpart to the Northern Hemisphere (NH)'s

[1] https://arquivo.pt/wayback/20160521184740/http://www.nas.edu/history/igy/.

[2] Albatrosses are among the largest birds, with their major habitat at mid- and high latitudes in the Southern Oceans region. They are known to fly in circumpolar orbits, benefitting from zonal winds associated with the SAM.

NAM/NAO – the topic of the previous chapter. The SAM is a measure of oscillations in atmospheric mass between subpolar and subtropical latitudes in the extratropical SH. The SAM is known to play a leading role in weather and climate variations in Australia, New Zealand, and southern South America. The SAM is also known to undergo interannual, decadal, and longer timescale variability as we will see in this chapter. In the last 20 years, the SAM's variability at decadal timescales[3] and its interactions with the Southern Oceans have attracted a considerable attention from researchers. It was also found in the last 20 years that the SAM variability is associated with variability in synoptic-scale storms in the SH (Rashid and Simmonds, 2004, 2005); temperatures over Antarctica (Kwok and Comiso, 2002; Marshall, 2007; Schneider et al., 2004); temperature and precipitation in the mid- and high-latitude SH (Jones and Widmann, 2003; Gillett et al. 2006; Meneghini et al., 2007; Kravchenko et al., 2011; Ho et al., 2012; Marshall et al., 2017); and sea ice, ocean circulation, and ocean temperature (Hall and Visbeck, 2002; Lefebvre et al., 2004; Sen Gupta and England, 2006; Treguier et al., 2010; Bajish et al., 2013; Yeo and Kim, 2015). Therefore, understanding and predicting future evolution of SAM at various timescales is very important.

The scientific questions posed for the NAM/NAO variability and answers as known so far were described in Chapter 6. The same questions apply to SAM variability as well. What causes SAM variability at timescales of a few days to a few decades, especially at decadal timescales? Can the atmosphere by itself generate such decadal variability, or do the Southern Oceans and sea ice in these Oceans play any role in decadal SAM variability? We have seen in some of the preceding chapters that the solar Schwabe cycle with an average period of approximately 11 years is associated with some decadal climate variability (DCV) phenomena. Is the solar cycle also associated with decadal SAM variability? If yes, what are the physics of this association? These questions are addressed in this chapter.

After this brief introduction, subsequent sections in this chapter describe instrument-measured observations of the SAM and its decadal variability in Section 7.2. Then, SAM variability as seen in paleoclimate data is described in Section 7.3, and hypothesized mechanisms of decadal SAM variability are described in Section 7.4. A summary of the chapter and conclusions are presented in Section 7.5.

7.2 MODERN OBSERVERS' TALES

7.2.1 In the Beginning

As surface and upper air data at individual stations in the extratropical SH started becoming available, researchers began to find significant features of atmospheric circulation and climate in the SH beginning in the 1930s CE. For example, a semi-annual wave in surface pressure was discovered by Reuter (1936) and Wahl (1942). Further exploration of this surface pressure wave was carried out by Vowinckel (1955), Schwerdtfeger and Prohaska (1956), Schwerdtfeger (1960, 1967), and van Loon (1967, 1971). As described in the introduction, surface and upper air measurements stations

[3] As in the rest of this book, decadal here also refers to variability generally at 8–20-year periods.

were established in Antarctica in 1957 CE in the IGY. Schwerdtfeger (1960) used geopotential height data at 700, 500, 300, 200, and 100 hPa levels from three stations around the 50°S latitude circle and also from the Amundsen–Scott station at the South Pole from April 1957 to March 1960 CE. He found that the semi-annual component of the circumpolar westerly winds' variation was dominant in the troposphere with maxima in equinoctial months when the zonally-averaged meridional temperature gradient is maximum and the annual component of the vortex variation was dominant in the stratosphere. Schwerdtfeger (1960) connected the vortex's waxing and waning with variations in the zonally-averaged meridional temperature gradient and zonally-averaged meridional circulation in the SH, and the annual cycle of solar irradiance and its absorption by stratospheric ozone. It was hypothesized that variations of the circumpolar vortex had consequences for the general circulation of the planetary atmosphere. These findings were consistent with Schwerdtfeger and Prohaska (1955, 1956)'s suggestion that the semi-annual component of surface pressure over the extratropical SH is directly related to the corresponding variation in the strength of the circumpolar westerly winds. Considering the fact that these results and conclusions were based on relatively short-term data from only a few stations, these researchers' insights into atmospheric circulations in the extratropical SH are truly remarkable.

Several researchers continued analyses of station data from Antarctica and continued to add to the slowly increasing knowledge about extratropical SH atmospheric variability. Among them was Kidson (1975) who analyzed what can be termed a worldwide data set from 1900 stations from 1951 to 1960 CE consisting of monthly-average surface pressure, temperature, and precipitation. The intention was to use empirical orthogonal function (EOF)–principal component (PC) analysis to isolate major patterns of variability. While the major EOF pattern in the NH surface pressure data reflected the Hemisphere's land–ocean contrast and the next most energetic EOF showed variability of the zonal circulation (Chapter 6), the first two EOFs of surface pressure in the SH showed vortex-like zonal circulations with oscillations between subpolar latitudes and mid-latitudes (Kidson, 1975; Figure 5). This study filled-in spatial and temporal data, and showed that there was indeed a circumpolar vortex in the SH with opposite-phase oscillations between subpolar and middle latitudes. Other data analyses in the 1970s and 1980s CE then defined various indices of these oscillations and delved further into their mysteries (see, for example, Trenberth (1976), van Loon (1967, 1971), Pittock (1973, 1980), Rogers and van Loon (1982), and Szeredi and Karoly (1987)). This, then, is a brief history of the beginning of the road that led to much more substantive research on variability in extratropical SH atmosphere in the 1990s and the 2000s CE, especially the SAM and its variability, as more data and more types of data as well as models started to become available. The interested reader can find details of this early period in the publications cited here and in references therein.

7.2.2 In the Modern Era

Time series of instrument-measured surface and upper air data lengthened in the 1990s CE, and data assimilation systems were developed to "reanalyze" these data to obtain dynamically consistent representations of the atmosphere. These new reanalysis data sets gave a major boost to researchers to build on the previous half century's attempts to describe and understand atmospheric variability in the SH using only station data. The multidecades-long time series of station data and reanalysis data also enabled detailed studies of decadal climate variability (DCV) in the extratropical SH. The story of the modern age in this DCV research is narrated here.

In the modern era, the circumpolar annular variability formally named Antarctic Oscillation (AAO; Gong and Wang, 1998, 1999) is also known as SAM (Thompson and Wallace, 2000). Gong and Wang (1999) employed the EOF-PC analysis on monthly, gridded, SH sea level pressure (SLP) data from January 1958 to December 1997 CE from the National Centers for Environmental Prediction (NCEP)–National Center for Atmospheric Research (NCAR) reanalysis (Kalnay et al., 1996). They found that a zonally-symmetric pattern dominated the EOF-PC analysis in all months; this pattern had one sign of SLP variability over Antarctica and opposite sign of SLP variability between approximately 40°S and 50°S. Thus, like the NAO/NAM phenomenon, the AAO/SAM phenomenon consists of atmospheric mass oscillations between mid- and high latitudes in the SH. During the positive phase of the AAO, SLP is generally above average in mid-latitudes and below average in high latitudes. In this AAO phase, zonal wind is stronger than average between 15°S and 30°S and between 45°S and 60°S, and weaker than average between 30°S and 45°S. In the negative AAO phase, the SLP and zonal wind anomalies have opposite signs to those in the positive phase. Guided by these dominant EOF patterns and by correlation coefficients among zonally-averaged SLP anomalies at various SH latitudes, Gong and Wang (1999) defined an index of AAO as the difference in normalized, monthly, zonally-averaged SLP anomalies at 40°S and 65°S. This index was found to explain up to 50% of monthly SLP variance over Antarctica and showed variability at months to decades periods. Later, Visbeck and Hall (2004) found that the SAM pattern explains the largest fraction of interannual variability in atmospheric circulation in extratropical SH.

Thompson and Wallace (2000) defined the dominant PC of extratropical SH SLP, geopotential height, or zonal wind anomalies directly as the SAM index rather than as the difference in zonally-averaged SLP anomalies at two SH latitudes, based on dominant EOF-PC, as Gong and Wang (1999) did. The NH counterpart was defined as the NAM index. Thus, the Thompson and Wallace (2000) definition is the same for the NH and SH annular modes. The geometric similarity of the SAM and NAM patterns is depicted in Figure 7.1 by Thompson and Wallace (2000). Figures 7.1a and b show vertical structure of north–south, opposite phase variability of zonal-average zonal winds associated with SAM and NAM variability, and Figures 7.1c and d show 850 hPa geopotential height pattern of SAM and SLP pattern of NAM. Both patterns show annular shapes, especially in polar regions. SAM is the dominant pattern of atmospheric variability in extratropical SH in all seasons. Comparing the SAM pattern in Figure 7.1c with earlier studies cited (see also Kidson (1988), Karoly (1990), Shiotani (1990), Hartmann and Lo (1998), and references therein) confirmed in hindsight that the earlier studies, using sparse data, had also found the same pattern which was later labeled AAO/SAM. These and other studies also showed that the SAM pattern has an "equivalent barotropic" (wind speed changes with height, but not the direction) structure in the vertical.

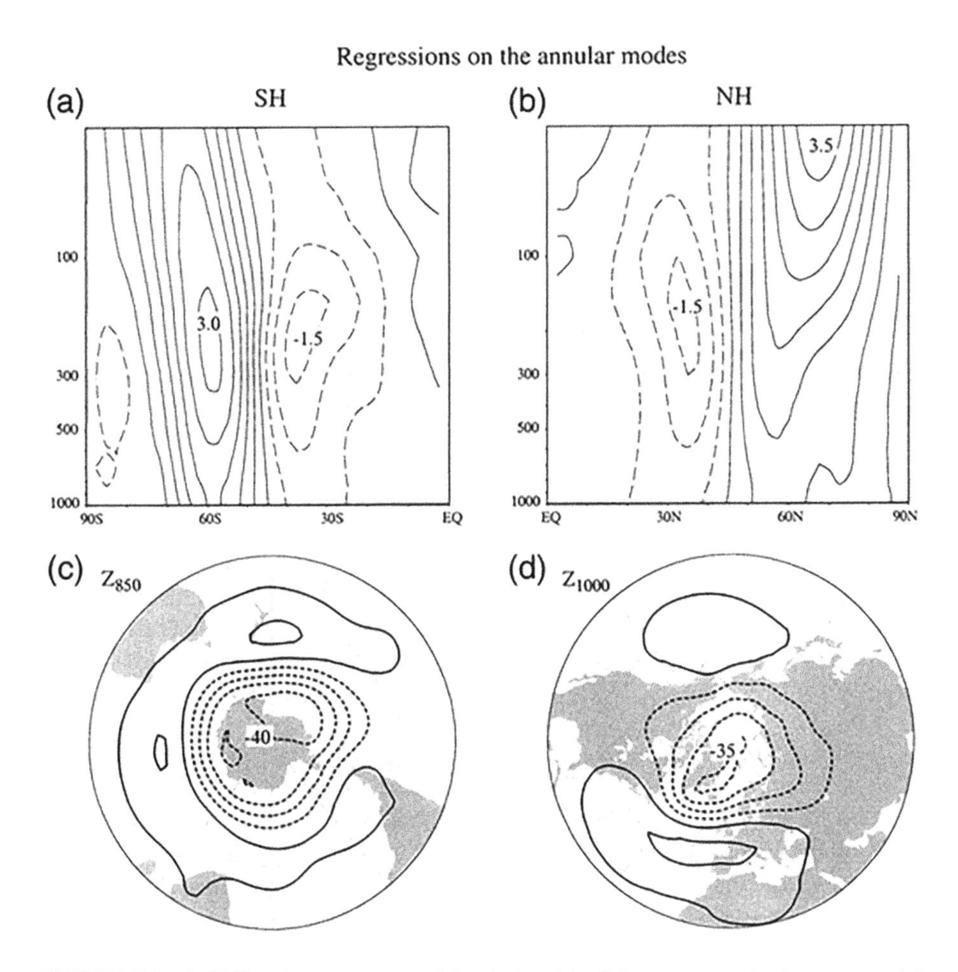

FIGURE 7.1 (a, b) Zonal-mean geostrophic wind and (c, d) lower-tropospheric geopotential height regressed on the standardized indices of the annular modes (the AO/NAM and the AAO/SAM) based upon monthly data from January 1958 to December 1997 CE. Left panels are for the SH and right panels are for the NH. Units are m s^{-1} (a, b) and m per std. dev. of the respective index time series (c, d). Contour intervals are 10 m (–15, –5, 5, ...) for geopotential height and 0.5 m s^{-1} (–0.75, –0.25, 0.25) for zonal wind. (From Thompson and Wallace (2000). © American Meteorological Society. Used with permission.)

After this brief introduction to SAM as seen by modern observers, the focus in subsequent sub-sections is on decadal variability of SAM and its associations with other atmospheric and oceanic variables as per the organizing theme of this book. In the late 20th century CE, there were several analyses of empirical station data (primarily, SLP) from SH mid-latitudes. These were focused mainly or even entirely on interannual variability (a choice forced by relatively short records) and its longer-term variability, and had too few stations in Antarctica with long enough time series to describe decadal variability of SAM. These pioneering studies, like the ones described earlier, described longer timescale variability in general terms (see, for example, Karoly et al. (1996)) and opened new avenues of research in SH climate variability. However, it was not before the first and second decades of the 21st century CE that data time series became long enough and reanalysis data started to become available to extract some quantitative information about decadal variability. A few representative studies from this post-20th century CE era are described here.

We begin with what is perhaps the most comprehensive, empirical study on decadal variability of SAM and underlying Southern Oceans. Puzzled by an empirical association between SAM and Australian rainfall at decadal timescales, and a spectral peak at 14-year period in a tree-ring-based index of the eccentricity of the SH polar vortex, but unable to attribute the decadal variability of SAM to tropical influences (Yuan and Li, 2008), Yuan and Yonekura (2011) embarked on a comprehensive analysis of station and reanalysis data sets to identify and quantify attributes of decadal variability in the extratropical SH ocean–atmosphere system. The station-based data consisted of (1) surface pressure and surface air temperature (SAT) measurements at 18 stations, with record lengths ranging from 47 to 103 years; and (2) three SAM indices from the British Antarctic Survey (BAS; Marshall, 2003), Visbeck (2009), and Fogt et al. (2009) – the last two based on weather station data going back to late 19th century CE. The Southern Oceans' associated variability was analyzed with gridded sea surface temperature (SST) data from the HadISST data set from 1950 to 2004 CE (Rayner et al., 2003). The atmospheric reanalysis data sets consisted of (1) the gridded NCEP-NCAR data from 1950 to 2004 CE (Kalnay et al., 1996) and (2) the gridded European Centre for Medium-Range Weather Forecasts reanalysis (ERA)-40 data from 1957 to 2001 CE (Uppala et al., 2005).

First, Yuan and Yonekura (2011) defined and calculated SAM indices as differences between monthly SLP anomalies between 40°S and 65°S from the NCEP-NCAR and ERA-40 data; in their assessment, the use of SLP anomalies from the climatological annual cycle reduced the influence of seasons. They also calculated SAM indices as the dominant PCs of extratropical SLP and 850 hPa geopotential height from the reanalysis data as in Thompson and Wallace (2000). A comparison of these various SAM indices showed that, although there were differences among the SAM indices due to different definitions and different data, variations with respect to time were generally in phase, especially at decadal timescales (Yuan and Yonekura, 2011; Figure 2). It was found that spatial patterns of SAM were similar in different data sets – annular around Antarctica with opposite phases in polar and mid-latitudes. The SAM indices showed two large-amplitude, decadal oscillations in the 1960s and the 1990s CE, with multidecadal modulation. Spectral analyses of the SAM index formed from the NCEP-NCAR reanalyses SLP data and zonal-average SST anomalies in 30°S to 40°S, 40°S to 50°S, and 50°S to 60°S latitude bands showed significant spectral peaks at eight to 16-year periods (Figure 7.2). It was also found that the decadal spectral peaks were not statistically stationary over the data record as also found in other phenomena described in this book. Station-based SLP and SAT showed that DCV was much more evident in mid-latitudes than over Antarctica. However, it was also found that spectral peaks in SAT and SLP time series were not exactly at the same frequency but they overlapped, perhaps due to data quality/quantity problems or due to differences in thermodynamic and dynamic atmospheric processes giving rise to SAT and SLP variability, respectively. As shown in Figure 7.3, cross-spectral analyses between SAM index, and

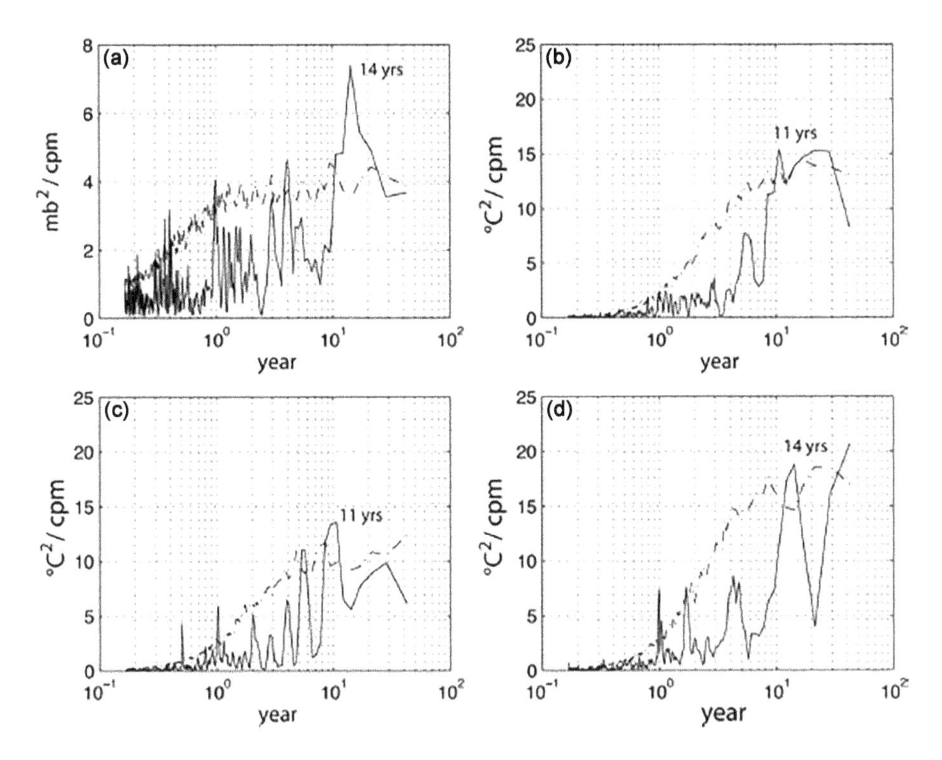

FIGURE 7.2 (a) Power spectrum density of the SAM EOF-based index constructed from 55 years of NCEP-NCAR monthly SLP anomaly. The dotted-dashed line indicates the 95% confidence level based on the red noise bootstrap significance test. The decadal peak in SLP anomaly differences-based SAM index is also significant at 95% confidence level. Power spectrum density of zonal-average SST anomalies in (b) 30°S–40°S, (c) 40°S–50°S, and (d) 50°S–60°S latitude bands. Dotted-dashed lines mark the 95% confidence level. (From Yuan and Yonekura, 2011.)

zonal-average SST anomalies in the 50°S to 60°S latitude band and zonal-average SST gradient from 30°S to 60°S showed that decadal variability of SAM was associated with SST in Antarctic subpolar seas and to meridional SST gradient at mid- to high latitudes. Composite SST anomalies during high and low SAM phases showed that SST anomaly patterns in the Southern Oceans had generally opposite signs in these two phases. Thus, Yuan and Yonekura (2011)'s comprehensive analyses of a variety of empirical data sets showed that decadal co-variability in SAM indices and mid- and high-latitude SSTs occurred at distinct eight to 16-year periods. The analyses also showed that there was a strong association between atmospheric SAM variability and mid- and high-latitude SSTs at the decadal timescales.

Time series of sea ice data are relatively short in the Southern Oceans. Satellite-based microwave sensors have provided a reasonably complete coverage of sea ice since 1979 CE. These data, along with gridded SST data and SAM index data, can be used to get an approximate indication about decadal variations in combined SAM–SST–sea ice data. Bajish et al. (2013) analyzed monthly sea ice concentration (SIC) data from

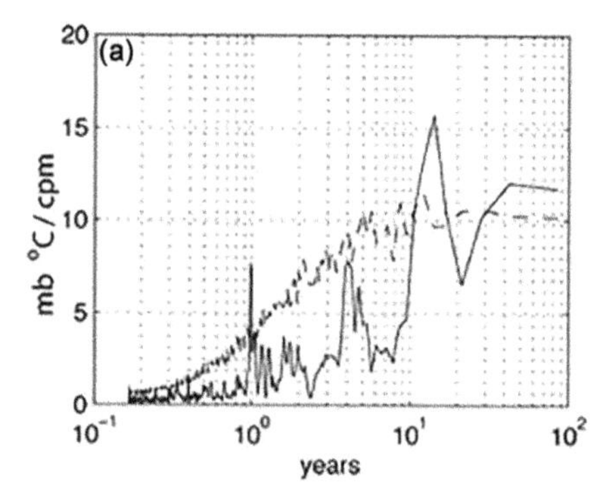

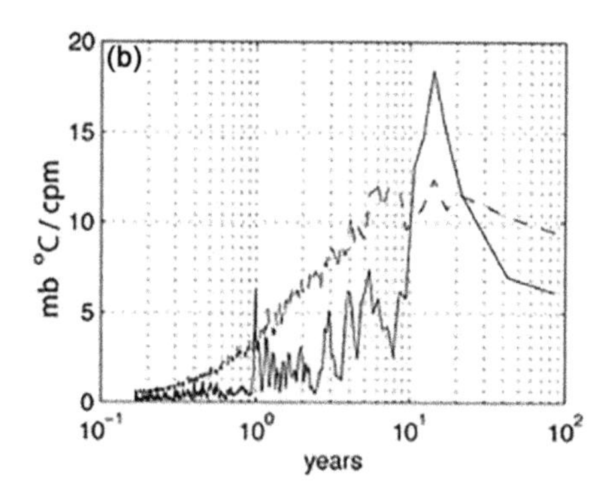

FIGURE 7.3 Cross-spectrum between the SAM index and (a) zonal-average SST anomaly at 50°S to 60°S latitude band and (b) mean meridional SST gradient from 30°S to 60°S. The power spectra were smoothed by three-point running-average filter to reduce uncertainty. Dashed line indicates the 95% confidence level. (From Yuan and Yonekura, 2011.)

passive microwave sensors on polar-orbiting satellites from 1979 to 2010 CE. These data were on a 25 km polar stereographic grid. They also used SST data from National Oceanic and Atmospheric Administration (NOAA)'s Optimal Interpolation SST V.2 data set from 1982 to 2010 CE, and the US Climate Prediction Center's SAM index from 1979 to 2010 CE. From the SIC data, Bajish et al. (2013) defined sea ice edge (SIE) as the equatorward latitudinal position of the 30% SIC isopleth. They also calculated zonally-averaged SST anomalies near the climatological SIE. Annual-average SIE position, the zonally-averaged SST anomalies, and the SAM index are shown in Figure 7.4. The circumpolar SIE position (Figure 7.4a) shows a linear trend with superimposed decadal variations shown clearly by the low-pass filtered version of the SIE time series. Zonally averaged SST anomalies (Figure 7.4b) show a slight cooling trend with superimposed decadal variations, and the SAM index (Figure 7.4c) shows a small increasing trend with superimposed decadal variations. Since all three time series are relatively short, Bajish et al. (2013) "guesstimated" oscillation periods in the three time series by the numbers of zero crossings after subtracting linear trends. They found that the SIE position has an 11 to 16-year period, the SST 10 to 13-year period, and the SAM index an 11 to 13-year period. Due to the small number of degrees of freedom in these relatively short time series, correlation coefficients among these three time series were found to be marginally significant or not significant. Despite the limitations of the data record, this study indicated the presence of physically consistent oscillatory decadal variations in SAM index, Antarctic circumpolar SSTs, and Antarctic circumpolar sea ice. Although these subjective guestimates of prominent decadal timescales from time series which spanned only three decadal "cycles" are barely indicative, these timescales are consistent with those found by Yuan and Yonekura (2011)'s analyses of relatively longer time series. Thus, the Yuan and Yonekura (2011), and Bajish et al. (2013) studies indicated that decadal sea ice and possibly ocean circulation variability, in addition to SST variability, are also associated with SAM variability, indicating that mid-latitude southern oceans may be a major component of decadal variability of SAM as we have already seen for decadal NAM/NAO variability in Chapter 6.

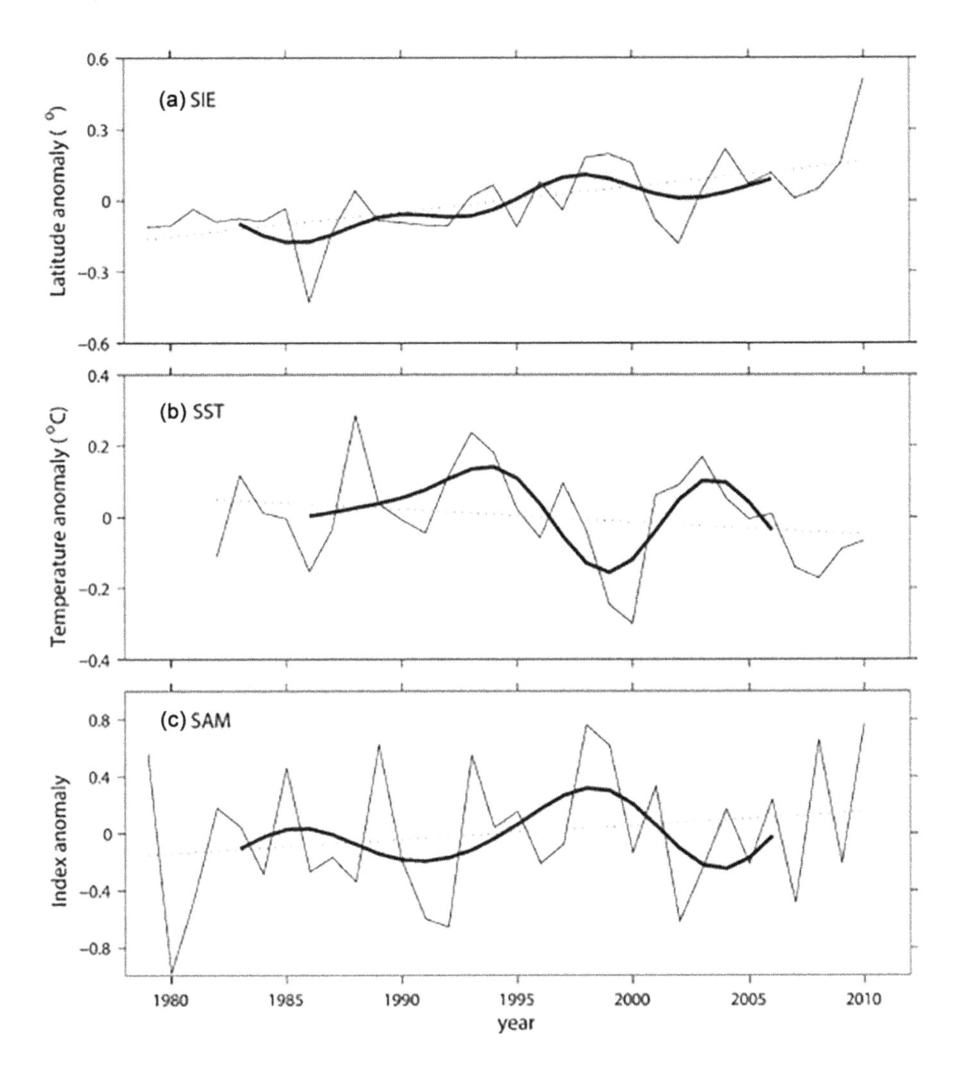

FIGURE 7.4 (a) Observed circumpolar averaged annual SIE anomaly time series (1979-2010 CE), (b) observed annual circumpolar SST (northward 5° latitude band from the SIE) anomaly time series (1982-2010 CE), and (c) annual SAM index (1979–2010 CE) from Climate Prediction Center. The bold line represents a low-pass filtered (7 years) time series. (From Bajish et al., 2013.)

We will see next if paleoclimate proxy data in the extratropical SH can shed any light on decadal variability of SAM and associated Southern Oceans variability.

7.3 ANCIENT OBSERVERS' TALES

The presence of circumpolar Southern Oceans at mid- to high-latitudes in the SH and the very cold climate of the Antarctic continent have limited proxy-climate data mainly to gases trapped in ice cores on the continent. The ice cores have been used to infer the Antarctic climate in past centuries (see, for example, Aristarain et al. (1986,1990); Mosley-Thompson (1992, 1996); Peel (1992); Thompson et al. (1994); Peel et al. (1996), and others), but the types of proxies used to infer past decadal variability at tropical latitudes such as corals and tree rings, and in extratropical NH such as tree rings are not available south of approximately 55°S. Villalba et al. (1997) made a pioneering attempt to calibrate tree-ring data from sub-Antarctic forests in Tierra del Fuego at 54° to 55°S latitudes and New Zealand between 39°S and 47°S latitudes in terms of November to February mean sea level pressure (MSLP) from 1746 to 1984 CE. Then, following Pittock (1980)'s definition of the Trans-Polar Index (TPI) as a measure of the eccentricity of the polar vortex in the SH, the normalized difference in MSLP between Tierra del Fuego and New Zealand was defined as the summer Trans-Polar Index (STPI). Villalba et al. (1997) have described details of tree-ring sites, and methods of calibration and validation of the MSLP and STPI reconstructions. Although, strictly speaking, the TPI and the STPI are not indices of the SAM, they do indicate SAM variability to some extent. A visual inspection of the tree-ring-based, November to February MSLP anomaly time series in Tierra del Fuego and New Zealand, and the STPI calculated from these two proxy MSLP data from 1745 to 1984 CE show decadal–multidecadal variability (Villalba et al., 1997; Figures 8, 9, and 11). The coherency spectrum of the two proxy MSLP anomaly time series shows that the tree-ring-based MSLP anomalies are highly coherent at 15 to 17-year oscillation periods (Villalba et al., 1997; Figure 14) and the reconstructed STPI's spectrum also shows a peak at these periods even though the peak is not highly significant. Villalba et al. (1997) also found a highly significant peak in the coherency spectrum of the STPI and Southern Oscillation Index time series from 1866 to 1984 CE, which suggested a possible association between SH polar vortex variability and tropical Pacific climate variability at decadal timescales.

Following this pioneering effort, there were several attempts to reconstruct AAO/SAM indices for longer times using several proxies of SAM and its associated variables. Among these attempts was one by Zhang et al. (2010) to use tree ring, coral, ice core, and cave sediment data to reconstruct past AAO/SAM variability. These data from 1500 to 1956 CE were used in a multivariate regression model to reconstruct the index in the December–January–February period. The Marshall AAO index (Marshall, 2003) from 1957 to 1989 CE was used for calibration and validation of the multicentury, multiproxy AAO index shown in Figure 7.5. Decadal and longer timescale variability of the reconstructed AAO index is clearly seen in Figure 7.5; however, such variability is not stationary over the 456-year record as Figure 7.5 shows. The instrument data-based AAO index since 1957 CE is also shown in Figure 7.5. The power spectrum of this time series is shown in Figure 7.6. Spectral peaks at 24- and 37 to 38-year oscillation periods are significant at the 95% level, and a 12 to 13-year oscillation peak is significant at the 90% level in Figure 7.6. Zhang et al. (2010) found, however, that their multiproxy AAO index had low-to-moderate correlations with instrument data-based AAO indices, suggesting that both sets of indices may capture decadal and longer timescale variability reasonably well, but the year-to-year variability in both may not be quantitatively reliable. But, like the Villalba et al. (1997) tree-ring-based MSLP and STPI, this multicentury, multiproxy reconstruction of the AAO/SAM index also showed the presence of substantial and

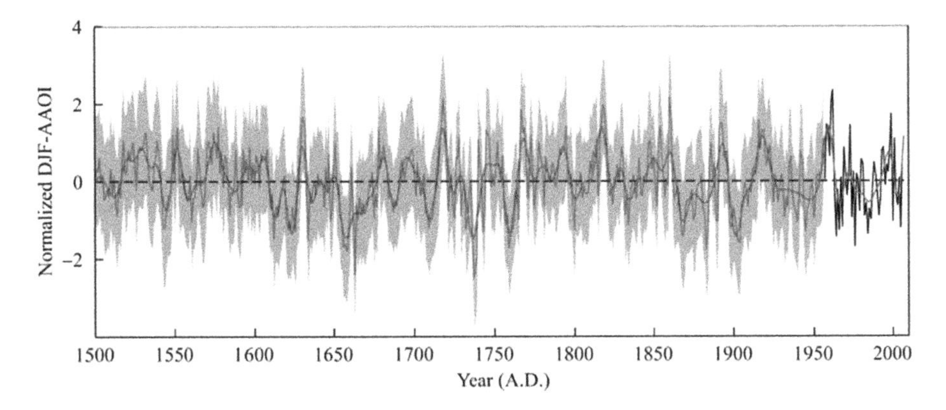

FIGURE 7.5 Time series of the DJF AAO index (blue line). The reconstruction period spanned 1500 to 1956 CE, and data since 1957 CE are observations. The red line indicates low-frequency (10 to 50 years) variations from a Butterworth filter. Shading indicates the range of ±2 x standard error derived from the unresolved variance in the calibration period. (From Zhang et al., 2010.)

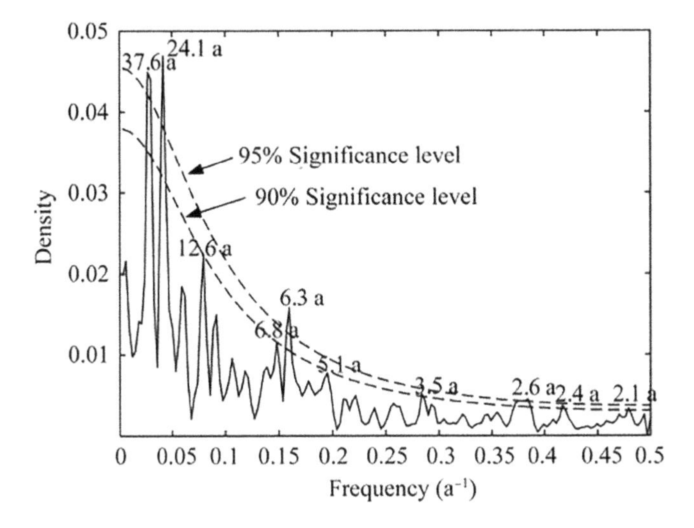

FIGURE 7.6 Power spectral analysis of the 1500 to 1956 DJF AAO time series in Figure 7.5 (dashed lines are the 90% and 95% significance levels). (From Zhang et al., 2010.)

significant decadal timescale variability. Both the reconstructed indices described and shown here have prominent decadal variability at similar oscillation periods as the shorter, instrument measurement-based data analysis results described and shown in Section 7.2. This similarity of modern and ancient observers' tales emphasizes the importance of understanding the physics of this decadal SAM/AAO variability. Steps in this direction are described in Section 7.4.

7.4 HYPOTHESIZED MECHANISMS

At least three sets of mechanisms have been proposed to understand SAM variability: (1) intrinsic atmospheric variability due to interactions among zonally-averaged zonal winds, and transient and stationary eddies; (2) intrinsic atmospheric variability driving ocean circulation, temperature, and sea ice variability, with possible positive and negative feedbacks; and (3) variability of solar emissions driving stratosphere–troposphere interactions. Effects of changing anthropogenic greenhouse emissions and stratospheric ozone have been proposed as possible mechanisms of multidecadal changes in SAM. Since intrinsic atmospheric dynamics have been explored so far only to understand intraseasonal SAM variability, (1) is reviewed briefly here. (2) and (3) are described in much more detail since they address decadal SAM variability.

We saw in Chapter 6 that nonlinear atmospheric dynamics as a possible cause of intraseasonal and longer timescale variability was being explored with idealized models of the atmosphere as the 1990s CE dawned. Robinson (1991) may have been the first to study the dynamics in a simple model and find that the results were applicable to intraseasonal variability of the zonally-symmetric circulation in the SH. He found that interactions between transient eddies and zonal-average winds generated the low-frequency variability. Other studies soon followed with increasingly more complex atmospheric models (see, for example, Yu and Hartmann (1993), Hartmann (1995), Hartmann and Lo (1998)). Taking the next step towards studying the dynamics of annular modes, Limpasuvan and Hartmann (1999) analyzed a 100-year simulation experiment with a relatively coarse resolution version of the Geophysical Fluid Dynamics Laboratory (GFDL) atmospheric general circulation model extending to approximately 10 hPa level. Realistic topography and the annual cycle of global climatological SSTs were specified in this model experiment. Using the EOF-PC technique to identify dominant patterns of extratropical 1,000 hPa geopotential height data from the experiment, Limpasuvan and Hartmann (1999) found that the dominant patterns in NH and SH were similar to the NAM and SAM patterns, respectively, identified from observed data, with zonal jets varying between mid- and high latitudes. They found these approximately zonally-symmetric patterns varying at intraseasonal timescales, with their structures extending from the Earth's surface to the tropopause level and with maximum variances during winter seasons in each hemisphere. It was also found that zonally-averaged meridional circulations varied in conjunction with the zonally-symmetric NAM- and SAM-like variability. Stationary eddies in the NH and transient eddies in the SH forced zonal jets in the respective hemispheres. A subsequent comparison of these simulated NAM and SAM, and associated eddy variabilities with corresponding patterns from the NCEP-NCAR reanalysis showed that the patterns and amplitudes of simulated NAM and SAM were similar to those derived from the reanalysis data (Limpasuvan and Hartmann, 2000). These similarities imply that NAM and SAM and their intraseasonal variabilities are generated by internal, nonlinear atmospheric dynamics. Other studies have also found that interactions among zonally-averaged zonal winds and synoptic scale eddies, stationary as well as transient, generate intraseasonal NAM and SAM variabilities (see, for example, DeWeaver and Nigam (2000a, b) for analysis of NAM variability). The foregoing, then, is a very brief summary of basic research on intrinsic atmospheric dynamics as a cause of SAM variability. As described in Chapter 6, the study by James and James (1989) is perhaps the only one so far which has argued for internal, nonlinear atmospheric dynamics as a cause of decadal and longer timescale variability, so more research is required in modeling decadal SAM variability due to intrinsic atmospheric dynamics.

7.4.1 Ocean–Atmosphere–Sea Ice Interactions

A few years after the identification of the SAM pattern and its intraseasonal variability, oceanographers began to ask: Does the SAM variability do anything to the southern oceans? Due to the Antarctic continent situated like a cap centered at the South Pole and the absence of any continent to the north for large swathes of the southern oceans, the zonally-symmetric SAM with its westerly winds at high southern latitudes can force a zonally-symmetric zonal ocean current around the Antarctic continent, the so-called circumpolar current, by applying a frictional body force to the southern oceans. What can be the ocean and climate dynamics consequences of such a current? Hall and Visbeck (2002) addressed these questions with a then-leading global ocean–atmosphere–sea ice model developed in GFDL (Manabe et al., 1991). This model had nine levels in the vertical in the atmospheric component, with an approximately 7.5° longitude – 4.5° latitude grid spacing. The oceanic component of the coupled model had 12 levels in the vertical, with a 3.8° longitude – 4.5° latitude grid spacing. The sea ice component of the coupled model estimated ice thickness from thermodynamic heat balance and advected sea ice by ocean currents, with some restrictions on advection of very thick sea ice. The coupled model was forced by the annual cycle of solar radiation. Flux adjustments were employed to prevent drift of the simulated climate, but the adjustments did not damp or amplify anomalies in model variables in a systematic way. In a crucial limitation of the coarse ocean resolution of the model, mesoscale eddies were parameterized as diffusion of potential temperature and salinity. A possible consequence of this limitation is described later. However, an advantage of the coarse resolution of such early global coupled models was the computational speed of the model, which allowed very long simulation experiments. In this case, Hall and Visbeck (2002) analyzed the middle 5,000 years of a 15,000-year-long experiment. They found that this coupled model simulated atmospheric annular modes in the NH and the SH reasonably realistically, implying that atmospheric eddies were resolved well enough for realistic interactions with the zonally-averaged zonal winds. The crucial role played by such interactions in NAO/NAM (Chapter 6) and SAM variabilities is described in this chapter. Focusing on the simulated annual cycle of atmospheric SAM, Hall and Visbeck (2002) found that it was reasonably realistic and was slightly stronger in colder months as also seen in observed data.

So far, results and conclusions about SAM from the GFDL model simulations matched those from atmosphere-only models as described in Section 7.2.2. But, what did the simulated SAM do to the Southern Oceans? After satisfying themselves that the average annual cycle of circumpolar ocean currents and other variables were reasonably well simulated by the GFDL model, Hall and Visbeck (2002) analyzed ocean variability associated with atmospheric SAM variability. They found that in the positive SAM phase, westerly winds migrated towards the Antarctic continent, which drove westerly surface currents in the southern oceans between 45°S and 70°S latitudes. Due to the zonally-symmetric geometry of the Antarctic continent, the SAM-forced surface currents were also zonally-symmetric and westerly. The Coriolis force generated by these westerly currents drove a northward Ekman drift or diverging meridional current, generating upwelling water near the Antarctic coast due to mass continuity. The SAM-associated winds are easterly between 20°S and 40°S latitudes, which drove easterly surface currents and southward Ekman drift. These two meridional currents converged approximately around 45°S latitude, generating downwelling water near this latitude. The SAM-associated surface stress on the southern oceans drove these zonal and meridional current systems down to approximately 3,000 m depth in the GFDL model, with return flows below 3,000 m. The meridional currents also transported warm and cold water such that there was an increase in northward cold water transport at southernmost latitudes and an increase in southward warm water transport at 30°S latitude. These convergence and divergence of heat transports generated SST anomalies. The meridional currents also transported sea ice northward and, consequently, ice thickness decreased near the Antarctic coast. Thus, SAM, SST, and sea ice varied coherently in the GFDL model. The spectra of simulated SAM variability and the strength of the Antarctic circumpolar current did not contain substantial decadal peaks, with the former having a generally "white noise" spectrum and the latter a "red noise" spectrum. As we saw in Section 7.2, however, both the observed SAM variability and associated SST and sea ice variabilities were found to contain substantial and distinct decadal periods. This discrepancy is addressed later. The evolution of various atmosphere, ocean, and sea ice variables in the positive SAM phase described here is shown in Figure 7.7, a schematic diagram from Hall and Visbeck (2002). Atmosphere–ocean–sea ice anomalies generally have the opposite signs in the negative SAM phase to those shown in Figure 7.7. Presciently, Hall and Visbeck (2002) predicted in their study that as coverage of all observed data increases in the future, such coherent variability would be found in the data also because the physics of SAM–ocean–sea ice variability were relatively basic and the GFDL model included most important processes to simulate the variability. Indeed, as we saw in Section 7.2.2, Yuan and Yonnekura (2011) found coherent, decadal variations in SAM and SSTs at mid- and high latitudes in the SH (Figures 7.2 and 7.3), and Bajish et al. (2013) found coherent, decadal variations in SAM, SST, and SIE in their analyses of longer and wider coverage of the observed data (Figure 7.4). In retrospect, this similarity of SAM-associated ocean–atmosphere–sea ice variability in the GFDL model simulation and later in observations speaks volumes about how well even the coarse resolution GFDL model simulated the variability.

As a part of their comprehensive analyses of decadal SAM variations (the results of empirical data analysis are described in Section 7.2.2), Yuan and Yonekura (2011) also analyzed gridded SLP and SST data from 20th-century experiments with 18 global ocean–atmosphere models participating in the Inter-governmental Panel on Climate Change (IPCC) Coupled Model Intercomparison Project 4 (CMIP4). They calculated SAM indices with SLP data from the 18 models and found that

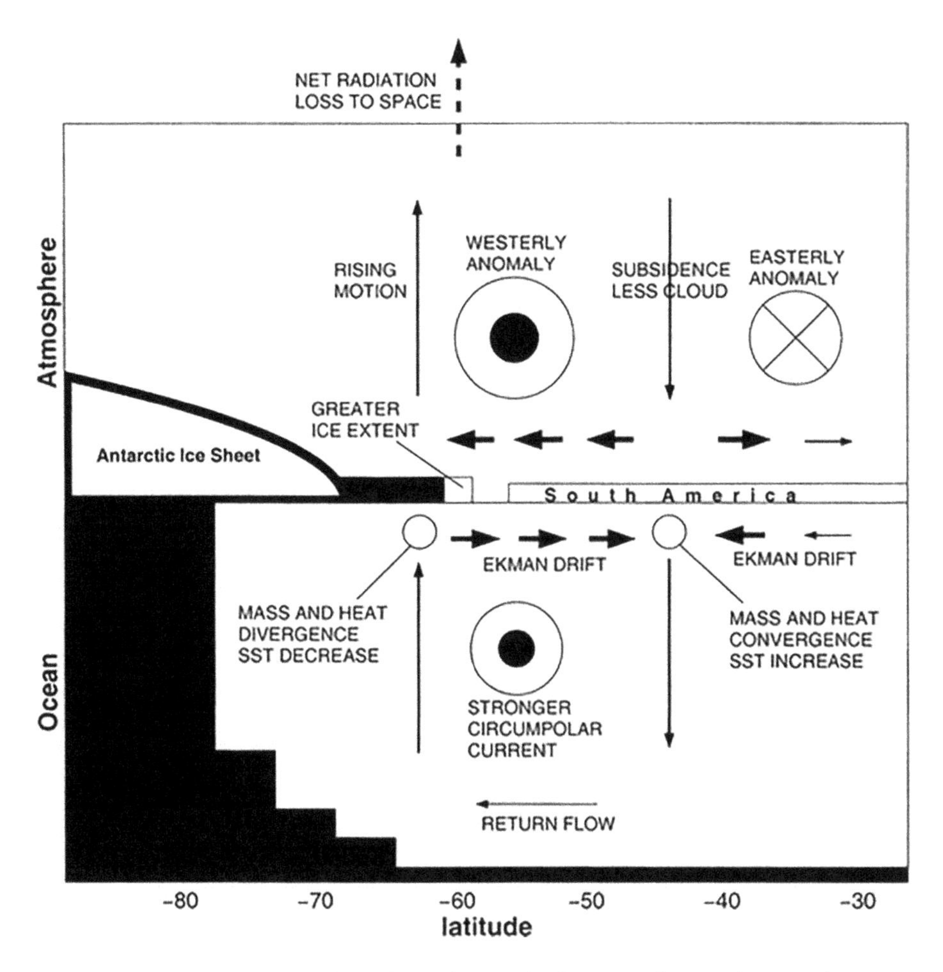

FIGURE 7.7 Schematic drawing of the changes in the atmosphere and ocean that occur when the SAM index is positive. (From Hall and Visbeck, 2002 © American Meteorological Society. Used with permission.)

16 of the indices exhibited decadal variability significant at 90% or higher level, with eight SAM indices significant at 95%. It was also found that seven models produced significant co-variability between simulated SAM index and subpolar SSTs at eight to 16-year periods, the same as the co-variability in empirical data described earlier. However, SST data from only four out of the 18 models showed decadal spectral peaks. Yuan and Yonekura (2011) speculated that perhaps oceanic eddy heat transport was important in producing distinct decadal spectral peaks and not all 18 CMIP4 models in their study resolved oceanic eddies. The reader is reminded of Hall and Visbeck (2002)'s comment that the inability of the coarse resolution GFDL coupled atmosphere–ocean–sea ice model to resolve mesoscale ocean eddies may have affected its ability to simulate realistic structure and strength of the Antarctic circumpolar current. This inability may also have affected the GFDL model's ability

to simulate ocean eddy heat transport accurately and hence distinct decadal spectral peaks as speculated by Yuan and Yonnekura (2011).

Bajish et al. (2013) also followed up their analyses of decadal variability of the SAM, SST, and sea ice with analyses of coupled ocean–atmosphere–sea ice model simulations. Their chosen coupled model was the Coupled General Circulation Model for the Earth Simulator (CFES) – consisting of the AFES3 atmospheric model (Kuwano-Yoshida et al., 2010) and the OIFES ocean–sea ice model (Komori et al., 2005, 2008). Both components of the coupled model had a relatively high resolution, approximately 1° longitude – 1° latitude in the atmosphere model and 0.5° longitude – 0.5° latitude in the ocean–sea ice model. The atmosphere model had 48 layers in the vertical, and the ocean model had 54 levels in the vertical. Without including effects of anthropogenic forcings and volcanic eruptions, a 120-year long experiment was conducted with the CFES, and the last 100 years' data were analyzed. The CFES's simulation of Antarctic SIC and the latitude of annually-averaged SIE were deemed realistic enough to analyze simulated SAM and its associated ocean–sea ice variability. Then, the 100-year long, gridded 700 hPa geopotential height, SST, and SIC data were analyzed to study co-variability of the coupled SAM–Southern Oceans–sea ice system after removing the average annual cycles of each variable and removing linear trends from the anomaly time series. The dominant EOF-PC of annual-average, 700 hPa geopotential height data in the SH revealed the SAM pattern (Figure 7.8b), similar to the observed SAM pattern described earlier and containing approximately the same amount of variance as the SAM patterns isolated from reanalysis data. The corresponding PC time series' spectrum had a prominent and broad peak at 10 to 13-year period, and secondary peaks between three- and five-year periods (Figure 7.9). The dominant EOF-PC of the annual-average SST data in the SH revealed a circumpolar pattern (Figure 7.8c), with the corresponding PC time series' spectrum containing a prominent peak at the same periods as the SAM variability spectrum (Figure 7.9). The dominant EOF-PC pattern of SIC

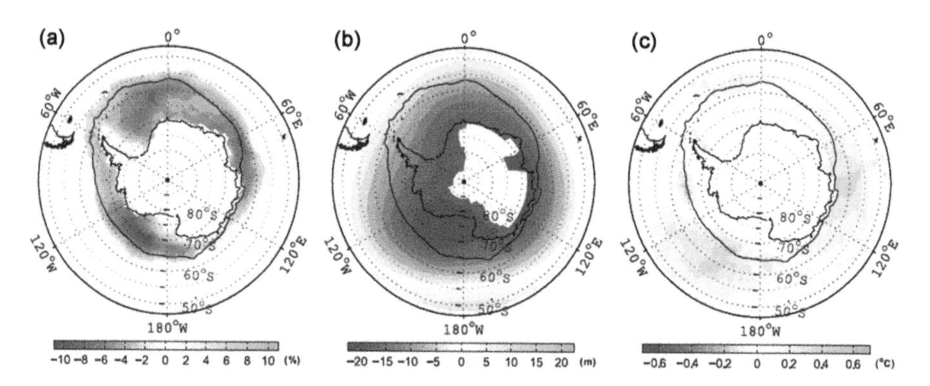

FIGURE 7.8 Spatial patterns of leading EOFs of annual-average (a) SIC anomalies, (b) geopotential height anomalies at 700 hPa, and (c) SST anomalies from CFES. The EOF patterns have been scaled by 1 standard deviation of the corresponding PCs to show the dimensional standard deviation at each grid point. The black line indicates the average SIE of the CFES. Color bars show dimensional standard deviation scales. (From Bajish et al., 2013.)

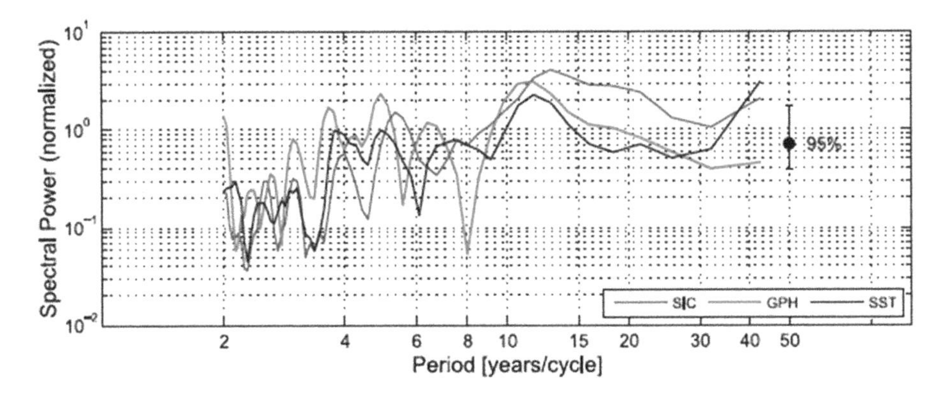

FIGURE 7.9 Power spectrum of the first PCs of SIC (blue line), 700 hPa geopotential height (GPH; red line), and SST (black line). Vertical bar indicates 95% significance threshold. (From Bajish et al., 2013.)

in the SH had a nearly circumpolar pattern (Figure 7.8a), while the corresponding PC time series' spectrum contained a prominent, broad peak at 12 to 17-year period (Figure 7.9). Thus, the simulated SAM, SST, and SIC patterns appeared to be physically consistent, and all three were found to oscillate at nearly the same decadal periods. Lagged correlation coefficients among the three PC time series confirmed the inter-relationships found in observed data (Figure 7.4).

Comparing results of this CFES simulation with the Yuan and Yonekura (2011) analyses of simulations with 18 CMIP4 models, and Hall and Visbeck (2002)'s GFDL model simulation indicates that meridional heat transport by oceanic eddies may be playing a very important role in generating a band of preferred decadal oscillation periods in the coupled SAM–SST–sea ice variability as speculated by Yuan and Yonekura (2011). This speculation was based on Meredith and Hogg (2006)'s study of eddy kinetic energy and heat transports in response to SAM variability with an eddy-resolving ocean model; it was found that the eddy transports were distributed around the Antarctic continent and responded to SAM wind stress forcing with a delay of two to three years due to the time taken to influence deep circulation of the Antarctic circumpolar current. Then, Screen et al. (2009) followed up with an ocean model study specifically designed to diagnose the role of eddies in meridional heat transport in the SH. They employed three different resolutions – 1°, 0.25°, and 0.0833° – in experiments with the Ocean Circulation and Climate Advanced Model (OCCAM; Coward and de Cuevas, 2005). Since observed mesoscale eddies at the latitudes of the Drake Passage north of Antarctica have 10 km length scale (Chelton et al., 1998), the Screen et al. (2009) study explicitly resolved the eddies at the highest model resolution. In the other two resolution versions of OCCAM, effects of eddies were parameterized. The three models were forced by 6-hourly momentum, heat, and freshwater fluxes from the NCEP-NCAR reanalysis for the 1985–2004 CE period. As in many studies described in this chapter, the SAM index was calculated as the first PC of 850 hPa geopotential height data south of 20°S. All three experiments simulated immediate cooling response of SSTs south of 45°S and warming north of this latitude to a positive SAM anomaly. As described earlier and shown in Figure 7.7 for the GFDL model, SSTs in coupled ocean–atmosphere–sea ice models also responded in the same way because the Ekman drift away from the Antarctic continent and upwelling at the continent brought up cold water which was then advected northward. Ekman drift north of 45°S was southward, which brought warmer temperatures from the north to the south. Then, something very striking occurred in the Screen et al. (2009) experiments. In the highest resolution version of OCCAM, mesoscale eddies developed two to three years after the peak positive SAM forcing, and these eddies transported kinetic energy and heat southward of 45°S. This warming was larger and longer lasting than the immediate cooling near the coast. The other two lower resolution versions of OCCAM did not simulate these eddy kinetic energy and heat transports because they could not resolve the mesoscale eddies responsible for these transports. Screen et al. (2009) showed that these eddy transports can last up to six years after the peak positive SAM forcing although the response peaked at approximately three years. Although this study demonstrated the importance of resolving transient oceanic eddies, the results can be model-dependent and ocean models with somewhat coarser resolutions may also be able to resolve eddy heat transport to some extent, depending on model formulation. The Bajish et al. (2013) CFES model's and some of the CMIP4 models' ability to produce a band of decadal oscillation periods in the spectrum of SAM–SST–sea ice variability points to this possibility.

The last question raised by these observational and modeling studies is: Which process(es) determine the decadal oscillation periods? Bajish et al. (2013) invoked positive feedback to a positive SAM phase due to cooler SSTs near the Antarctic continent and resulting stronger meridional temperature gradient which would increase atmospheric storminess and strengthen the already-positive SAM phase. This would also expand the SIE equatorward. Then, in the Bajish et al. (2013) explanation, an increased upwelling due to stronger westerly winds associated with positive SAM phase would warm the SSTs near the Antarctic continent by upwelling warmer subsurface water. The warmer water would then decrease atmospheric storminess and weaken SAM to the negative phase due to this negative feedback, bringing the SIE closer to the Antarctic continent. Thus, Bajish et al. (2013) invoked a combination of positive and negative feedbacks to explain the preferred decadal timescales in the CFES model simulation.

An alternate hypothesis can also invoke a combination of negative and positive feedbacks in a different way. The negative feedback to a positive SAM phase can work as explained by Hall and Visbeck (2002) and Bajish et al. (2013). Then, delayed eddy heat transport, as simulated by Screen et al. (2009), can bring warm water to the Antarctic continent from the north, thereby providing the negative feedback, rather than upwelling warm water at the continent's edge providing the negative feedback as in the Bajish et al. (2013) explanation. In this modified hypothesis, the multiyear delay in generating eddy heat transport may result in the decadal timescales found in observed data and coupled model simulations. Further research with coupled atmosphere–ocean–sea ice models with eddy resolving ocean–sea ice components is highly necessary to clarify details of various possible feedbacks. It is also possible that there is another mechanism of generating the preferred decadal timescales. We will see this in the next sub-section.

7.4.2 Solar Variability as a Pace Maker?

The last hypothesized cause and mechanisms of decadal SAM variability described in this chapter are variability of solar emissions, especially the 11-year solar cycle, and its influences on the SAM via stratosphere–troposphere interactions. As we will see, the top-down influence of the solar ultraviolet (UV) radiation cycle from the upper stratosphere to the SAM in the troposphere and, extrapolating from atmosphere–ocean–sea ice interactions described in Section 7.4.1, the possibility exists that the solar cycle may be acting as a pace-maker of decadal SAM variability. The reader will recall that the research to test a parallel version of this hypothesis for decadal NAO/NAM variability is described in Chapter 6.

In an admirable inter-Hemispheric harmony, Y. Kuroda and K. Kodera explored solar cycle modulations of the NAO/NAM and the SAM in parallel. Arguing from their insights into the physics of solar cycle modulation of the former and the observed extension of the SAM from the troposphere into the stratosphere, Kuroda and Kodera (2005) analyzed zonal wind, temperature, and 850 hPa geopotential height data from the ERA40 reanalysis (Simmons and Gibson, 2000) for the 1968–2001 CE period to see if there was empirical evidence of solar cycle modulation of the SAM. As a proxy of solar activity, the F10.7 cm radio flux was used for the same period. The EOF-PC definition of the SAM index was used on 850 hPa geopotential heights, and the index for 34 years was calculated. Years of high and low solar activity were identified from above- and below-average F10.7 cm flux, respectively. Fifteen SH winters (July to October) in high solar years and 18 SH winters in low solar years were identified. Significant correlation coefficients between the flux data in high solar years and low solar years with gridded 850 hPa geopotential heights from the ERA40 data set showed that a larger polar cap region was involved in the SAM variability during high solar years, with a larger extension to the Atlantic and east Pacific regions than in low solar years. The opposite-polarity regions east of Argentina and east of New Zealand were also more pronounced in high solar years than in low solar years. The correlation maps also showed that the SAM pattern persisted much longer in high solar years compared to low solar years. In a confirmation of Kuroda and Kodera (2005)'s hypothesis of solar cycle modulation of the SAM via stratosphere–troposphere interactions, they found that zonal-average zonal wind anomalies extend downwards from subtropical upper stratosphere in September (beginning of SH spring) and connects with the SAM coming up from the surface in the next few months in high solar years. No such zonal-average zonal wind anomalies extend from the stratosphere into the troposphere in low solar years. This behavior of zonal-average zonal winds in the SAM variability was similar to that in the solar cycle modulation of NAO/NAM. Kuroda and Kodera (2005) also found that implied absorption of solar UV radiation by ozone and resulting temperature anomalies also appeared to play an important role in the solar cycle modulation of the SAM during high solar years. Restricting the analysis to the satellite era from 1979 to 2001 CE generally confirmed results of the longer time series analysis. Thus, stratosphere–troposphere interactions may be the vehicle via which the 11-year solar cycle influences surface climate as we have seen earlier for the NAO/NAM variability.

Having found empirical evidence of possible solar cycle modulation of the SAM variability and formed a physical hypothesis of how this modulation can happen, Y. Kuroda and collaborators set about testing their hypothesis with a chemistry-climate model and diagnostic budget calculations. In the first exploratory study in this series, Kuroda and Shibata (2006) used a chemistry-climate model developed in the Meteorological Research Institute of Japan Meteorological Agency (Shibata et al., 1999). The combined chemistry-climate model consisted of a reasonably high-resolution atmospheric general circulation model extending to the middle atmosphere, with the upper boundary at 0.01 hPa and containing a comprehensive set of physical processes. The chemistry model (Shibata et al., 2005) contained 34 long-lived and 15 short-lived chemical species with 79 gas-phase reactions, six heterogeneous reactions on polar stratospheric clouds, and three heterogeneous reactions on sulfate aerosols – in other words, a reasonably complete set of chemical species and processes important for stratospheric chemistry. In the first study with this chemistry-climate model, Kuroda and Shibata (2006) conducted two 21-year-long experiments with climatological annual cycle of SSTs repeating every year. One experiment was forced by UV radiation corresponding to high solar activity phase and the other was forced by that in low solar activity phase. The simulated SAM in September extended to the upper stratosphere and persisted until the following March in the above-average UV flux experiment, whereas the simulated SAM was confined to the troposphere and persisted only to the following December in the below-average UV flux experiment. Differences in eddy heat and momentum fluxes were found to be the drivers of the differences in the SAM in the high and low solar activity phases. Ozone transport to lower stratosphere from middle and upper troposphere by the anomalous Brewer–Dobson circulation in the high solar activity phase and multiseason life of the transported ozone in the lower stratosphere generated the temperature, zonal wind, and meridional circulation anomalies which resulted in longer persistence of the SAM at the Earth's surface in this solar activity phase. The sequence of events occurred with opposite-sign anomalies in the low solar activity phase, resulting in a less persistent and weaker SAM. Later, Kuroda et al. (2007) confirmed with a stronger UV forcing of the same chemistry-climate model that differences in UV forcings in high and low solar activity years are responsible for stronger stratosphere–troposphere interactions in the former compared to the latter years. These differences in simulated SAM were very similar to those in empirical data analysis by Kuroda and Kodera (2005), thus verifying the hypothesized mechanisms based on empirical data analysis.

Then, almost over a decade later, Kuroda and Deushi (2016) and Kuroda (2018) extended the earlier simulations and empirical analyses to develop further insights into how the 11-year solar cycle can modulate stratosphere-troposphere interactions and the SAM. Kuroda and Deushi (2016) identified the polar-night jet stream as a very important connector between the solar activity cycle and the SAM. The polar-night jet stream consists of very strong westerly winds in autumn and winter in upper stratosphere and mesosphere near the boundary of the polar night.[4] The same chemistry-climate model used by Kuroda and Shibata (2006) was used to simulate

[4] glossary.ametsoc.org/wiki/Polar_night_jet_stream.

solar activity cycle influences at maximum, neutral, and minimum phases of the UV cycle, with the ratio of UV fluxes at maximum and minimum phases twice as large as the observed ratio to stimulate an identifiable response. The experiment was run for 42 years of model time with climatological annual cycles of global SSTs repeating every year. In this experiment, an increasing UV forcing enhanced downward-propagating solar cycle signal into the troposphere from late winter to spring. Analyses of momentum and energy budgets showed that the solar cycle UV forcing modulated interactions between zonally-averaged polar-night jet winds and planetary waves in the stratosphere, thereby influencing the SAM at the surface of the Earth.

In a subsequent study to verify results and conclusions from model simulations, Kuroda (2018) analyzed momentum and energy budgets with reanalysis data from the satellite data era because empirical data in the SH extratropics, especially in troposphere and stratosphere, were sparse and of varying quality before the advent of satellites as mentioned earlier. So, in this most recent analysis, Kuroda (2018) used 6-hourly and monthly ERA-Interim reanalysis (Dee et al., 2011) data from 1979 to 2015 CE. As in the previous analysis, the F10.7 cm flux was used as a proxy of solar UV activity and the dominant EOF-PC of 850 hPa geopotential height was used to define the SAM pattern and its variability. In these 36 years, there were 18 years each of high and low solar activity phases, respectively. After confirming that solar activity cycle was associated with decadal SAM variability in this more recent data period, diagnostic calculations of momentum and energy budgets using the ERA-Interim data with dynamical and thermodynamical equations governing zonally-averaged zonal momentum and temperatures showed with high statistical significance that the sequence of downward and poleward propagation of stratospheric zonal momentum after July was strongly connected with SAM variability during years of high solar activity, but was weakly connected with SAM variability during years of low solar activity. The rate of change of zonally-averaged zonal momentum with respect to time (acceleration of zonally-averaged zonal wind) showed that changes in zonal winds occurred during August and September in all years, but these accelerations were stronger and came down from upper stratosphere to the Earth's surface in September in high solar activity years. These zonal wind accelerations were weaker and confined to upper stratosphere in low solar activity years. The momentum budget analysis also showed that vertically propagating eddies contributed most of the zonal wind accelerations in high solar activity years. Energy budget analysis showed that the energy exchange between zonally-averaged zonal winds and vertically propagating baroclinic eddies plays a much bigger role in high solar activity years compared to diabatic heating due to the absorption of UV radiation by stratospheric ozone. Thus, this analysis of the dynamically consistent ERA-Interim data set by Kuroda (2018) and previous simulation experiments showed how the 11-year solar activity cycle can modulate SAM variability via the modulation of interactions between planetary waves and zonally-averaged zonal winds in the stratosphere; this modulated dynamical influence then propagates downward and poleward to generate decadal SAM variability.

Now, in order to address the question posed in the title of this sub-section, we combine results and conclusions from Sections 7.4.1 and 7.4.2. In the former sub-section, we saw that coupled atmosphere–ocean–sea ice interactions can generate variability at preferred decadal timescales without any interannual and longer timescale variability in any external forcings. As we saw in Chapter 6, experiments with coupled atmosphere–ocean–sea ice models forced by the 11-year solar activity cycle of UV radiation have shown that the solar cycle can synchronize intrinsic decadal variability in the NAO/NAM to the solar cycle. It is possible that the synchronization can also work in the same way for decadal SAM variability because both the intrinsic decadal variability and the SAM's response to the solar cycle via stratosphere–troposphere interactions have been demonstrated via experiments with reasonably realistic models as described in this sub-section. The insights and model results so far appear to answer the question in the affirmative. Therefore, further research on decadal SAM variability with coupled atmosphere–ocean–sea ice models forced by the 11-year solar activity cycle is necessary to understand the physics of decadal SAM variability and to clarify if the solar cycle is acting as a pace-maker of decadal SAM variability.

7.5 SUMMARY AND CONCLUSIONS

Atmospheric variability known as SAM is found to influence hydro-meteorology and storminess in southern South America, southern Africa, Australia, and New Zealand. In SAM phenomena, there is an oscillation in atmospheric mass or SLP between subpolar and subtropical latitudes in the SH. Wind variations are associated with the SLP variability from the Earth's surface to well over 100 km in the atmosphere. Decadal SST and sea ice variabilities are also associated with decadal SAM variability. It would be very useful if sources and mechanisms of decadal SAM variability can be understood, and if evolutions of this variability and its impacts can be predicted well in advance for societal benefits. This chapter reviews what is known so far about decadal SAM variability.

The time series of instrument-measured surface data in extratropical SH are relatively short, longer than 100 years only at a few stations. In these data and also in multidecades long reanalysis of surface and upper air data, decadal co-variability of SLP and SST is clearly evident at approximately 13 to 14-year period. Multiproxy-based and tree-ring-based, multicentury reconstructions of SAM/AAO indices going back to the 16th century CE corroborate the presence of decadal variability at similar periods. The solar UV cycle hypothesis proposed for decadal variability of NAM/NAO (Chapter 6) has also been proposed for decadal variability of the SAM. It has also been shown with coupled ocean–atmosphere–sea ice models that they can intrinsically generate SAM variability at preferred decadal timescales due to combinations of negative and positive feedbacks. So, it is possible that the 11-year solar activity cycle may be acting as a pace-maker of decadal SAM variability.

Both empirical and model-based approaches are needed for further progress in understanding decadal variability of the SAM. To describe various aspects of the variability in nature, continued measurements of solar UV and visible irradiances are required as are continued surface-based and satellite-based atmosphere–ice–ocean observations. Joint analyses of the extratropical NH and SH variabilities with SLP, SST, winds, temperature, and geopotential height data from reanalyses; and observed data on solar emissions might yield insights into physical relationships, if any, between the two Hemispheres and also simultaneous or nearly simultaneous influences of solar UV and other emissions on stratosphere–troposphere–ocean interactions in both Hemispheres.

Analyses of Earth System Model data from the World Climate Research Program's CMIP6 project to study natural coupled ocean–atmosphere–sea ice variability and to study the role of the 11-year solar cycle should be carried out. Decadal hindcasts of SSTs and other ocean–atmosphere variables are also available from CMIP6. They should be analyzed to estimate decadal predictability of the SAM variability, and associated variability of SST and other oceanic variables.

As we saw in this chapter, a concerted international research program during the IGY in 1957 CE began to record observations of weather phenomena at surface stations in extratropical SH. Then, scientists working with missionary zeal and imbued by a spirit of exploration developed insights into possible causes and mechanisms of SAM/AAO variability even though they had sparse empirical data. As data availability increased in the 1990s CE, the study of decadal variability of SAM expanded into a very important area of climate research. Now, we have finally come to the time when there are actual attempts to make decadal predictions of SAM. It is only a 50 to 60-year long journey in terms of time, but a remarkable leap forward in observing, understanding, and predicting a phenomenon that affects climate and many societal sectors not only around the SH, but also in other parts of the world.

8 Natural Decadal Climate Variability and Anthropogenic Climate Change

8.1 INTRODUCTION

"There is no question that climate change is happening; the only arguable point is what part humans are playing in it" – says a quotation attributed to Sir David Attenborough, the celebrated historian of nature who brought the wonderful diversity of life on the Earth into people's homes via television programs. Sir David has expressed a very well-founded uncertainty about the cause(s) of the changing climate. There are two major types of causes of the changing climate. One type consists of changes due to interactions among climate system components – the oceans, the atmosphere, the cryosphere, and the land surface–vegetation system – and due to changes in "external" forcings of the climate system – variability and changes in solar emissions, and ejections of gases and particulate matter from volcanoes. The other type consists of changes mainly due to human activities consisting of emissions of gases and aerosols from factories, power plants, and vehicles; and changes in land use and land cover such as increasing urbanization and decreasing farm land, forests, and other vegetated areas. The first type is usually referred to as natural variability and change,[1] and the second type is usually referred to as anthropogenic (of, relating to, or resulting from the influence of human beings on nature[2]) climate change (ACC). Although this book is mainly about natural climate variability at decadal timescales, it is very important to understand relationships between the two types of climate variability and change as we will see in this chapter.

Why is it important to understand relationships between the two types of climate variability and change? First, there is the human curiosity to understand the environment we live in. So, when there are intense rain or snow storms, heat or cold waves, hurricanes or cyclonic storms or tornadoes, or droughts or floods, the question arises as to what causes these weather events, which sometimes make extreme impacts. But, when such events become persistent for several seasons or several years, the question assumes even more urgency and then gives rise to inquiries as to what can be done to mitigate or adapt to such persistent events. If we know what causes such events and if we can predict them, then perhaps we can take steps to mitigate

[1] The words "variability" and "change", as used in this book are defined in Chapter 1.

[2] www.merriam-webster.com/dictionary/anthropogenic.

them or, at least, to adapt to them. Therefore, understanding impacts of each type of cause is very important for this so-called attribution of an observed phenomenon to one or more causes. Second, both types of causes work on the same Earth system and so it is very important to understand if one type influences the other type in any way – for example, if there are multidecades to century-long changes in atmospheric constituents; ocean circulations, currents, and temperatures; and precipitation patterns, atmospheric pressure, and wind patterns due to ACC, do these changes influence natural climate variability caused by interactions among Earth system components and/or that forced by variability of solar emissions or volcanic eruptions? If ACC influences natural climate variability, it is very important to understand and predict the influence. Lastly, it is conceivable that dry or wet epochs persistent for several years to a decade or longer in a region can influence the lifestyle of people living in that region and, consequently, influence amounts of emissions of carbon dioxide (CO_2), methane, and other gases and particulate matter implicated in ACC. Such a behavior change would be an influence of natural climate variability on ACC. There are exploratory efforts to analyze and understand possible interactions between decadal climate variability (DCV) and ACC, and the net effects of both on the observed changing climate. This chapter addresses these efforts. The scope of this chapter is limited to exploring DCV–ACC relationships and the importance of separating and quantifying DCV-related observed changes for ACC attribution.

After this introduction, hypothesized physics of how ACC can influence two proposed mechanisms of DCV are described in Section 8.2. Attempts to separate and quantify impacts of DCV and ACC using observed data are described in Section 8.3. Finally, the chapter is summarized and conclusions are presented in Section 8.4.

8.2 CAN ANTHROPOGENIC CLIMATE CHANGE INFLUENCE NATURAL DECADAL CLIMATE VARIABILITY?

Several hypothesized mechanisms of DCV were described in Chapters 2, 3, and 4, so hypotheses of how ACC can influence DCV will be described in this section in the context of the earlier-described mechanisms. We will first recapitulate major, hypothesized DCV mechanisms.

We saw in Chapters 2 and 3 that oceanic, baroclinic Rossby waves in the tropics and subtropics can act as "agents of change" when they arrive at the western boundary of the Pacific Ocean and that this change can then travel towards the Equator along the western boundary currents and then possibly eastward along the Equator as Kelvin waves. Due to their relatively slow propagation speeds, the Rossby waves can introduce a multiyear to multidecadal timescale in the set of oceanic and ocean–atmosphere processes which together are hypothesized as responsible for DCV in the Pacific Ocean region. The phase speed of these Rossby waves is a function of the vertical temperature stratification in the upper Pacific Ocean. Therefore, any process or phenomena that influences the stratification can potentially influence Rossby wave phase speeds and thereby their transit time across the Ocean and, consequently, the oscillation period of the Pacific Decadal Oscillation (PDO). Specifically, the baroclinic Rossby wave phase speed is proportional to static stability, but inversely proportional to the Coriolis parameter (or planetary vorticity) which increases with increasing latitude; static stability is a function of vertical density gradient (or temperature stratification). Thus, the Rossby wave phase speed decreases with decreasing stratification and with increasing latitude, and any process or phenomenon that can change stratification can change the baroclinic Rossby wave phase speed (Liu, 2012). Since ACC's manifestation as global warming can warm the upper ocean in proximity to the atmosphere, it can increase stratification (warm the already warm surface layer even more) and, consequently, increase the baroclinic Rossby wave phase speed and decrease the PDO timescale (Saenko, 2006).

Fang et al. (2014) and Zhang and Delworth (2016) investigated this hypothesis with experiments with Earth System Models (ESMs). Fang et al. (2014) studied the response of the PDO to global warming in the Fast Ocean-Atmosphere Model (FOAM; Jacob, 1997) and in 11 ESMs (Meehl et al., 2007) participating in the Coupled Model Intercomparison Project (CMIP) 3. The FOAM experiments consisted of a 400-year control run with CO_2 held constant at 355 parts per million (ppm); a 400-year 2×CO_2 (twice the concentration as in pre-industrial time) experiment; and two 400-year-long partial blocking experiments with CO_2 held constant at 355 ppm. The two partial blocking experiments (Liu et al., 2002) were conducted to investigate the role of oceanic baroclinic Rossby wave adjustment in the PDO. In these two experiments, ocean temperature and salinity were held constant at their respective climatological values from the surface to 100 m depth from 10° to 60° N latitudes and 175°W to 175°E longitudes in one experiment, and from 10° to 35°N latitudes and 175° W to 175°E longitudes in the other experiment. These constant values blocked baroclinic Rossby waves in these two regions. A comparison among these four FOAM experiments showed that the model did not simulate the PDO without the oceanic Rossby waves, demonstrating that these waves play a crucial role in the PDO in the FOAM model. The comparison between the control and 2×CO_2 experiments also showed that the PDO and associated atmospheric variability weakened in the latter experiment. While the PDO's sea surface temperature (SST) index showed a weakening and an increase in frequency, the dominant upper-ocean (400 m) heat content empirical orthogonal function (EOF) and principal component (PC) associated with the PDO showed that the heat content EOF also weakened (Figure 8.1a and d) and that the spectrum (Figure 8.1c and f), calculated with the multi-taper method, of the dominant PC's time series (Figure 8.1b and e) showed a substantial shift in the PDO's dominant timescales towards shorter periods under 2×CO_2 conditions. Comparisons of the pre-industrial control and A1B[3] scenario runs of the CMIP3 ESMs showed that an overwhelming majority of the ESMs showed a weaker and faster PDO in the A1B scenario experiments compared to the pre-industrial control experiments.

Zhang and Delworth (2016) also conducted sensitivity experiments with the Geophysical Fluid Dynamics Laboratory (GFDL) CM 2.5_FLOR ESM in which CO_2 concentrations were specified at pre-industrial, 2×pre-industrial, and 0.5×pre-industrial levels. They found that changes in vertical stratification in the Pacific

[3] In the Special Report on Emissions Scenarios, A1B is a mid-range scenario with a balanced emphasis on all energy sources. In this scenario, CO_2 concentration increases to 720 ppm by 2100 CE from an estimated 368 ppm in 2000 CE.

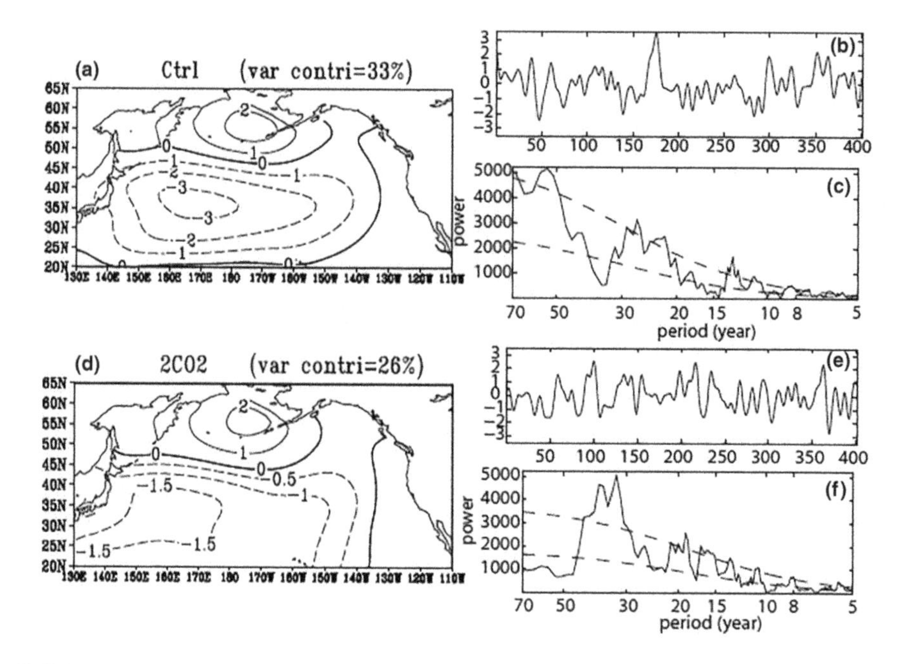

FIGURE 8.1 (a) Upper-ocean (400 m) heat content EOF 1 (10^8 J m^{-2}), (b) standardized PC 1 time series, and the (c) Multi-Taper Method raw power spectrum of PC 1 time series with 50% and 90% significance level lines from FOAM control run. (d)–(f) the same as (a)–(c), but from FOAM 2×CO_2 run. (From Fang et al., 2014.)

Ocean due to increased surface temperature associated with global warming indeed changed Rossby wave phase speeds as hypothesized by Saenko (2006) and found by Fang et al. (2014) in FOAM and other ESMs. Zhang and Delworth (2016) found that, due to this effect of global warming, the PDO period decreased to 12 years from a pre-industrial PDO period of 20 years in the GFDL CM 2.5 ESM. Global cooling associated with a decreased CO_2 concentration slowed down the Rossby waves and increased the PDO period to 34 years. These changes in dominant oscillation period of the PDO are shown in Figure 8.2, and changes in Rossby wave phases speed and Pacific basin crossing time are shown in Figure 8.3. In other effects of the CO_2 sensitivity experiments, Zhang and Delworth (2016) found that the PDO amplitude became weaker (stronger) in a warmer (less warm) world with a maximum decrease (increase) over the Kuroshio–Oyashio Extension (KOE) region. This decrease (increase) was apparently associated with a weakened (strengthened) meridional temperature gradient in the KOE region. In addition, reduced (strengthened) variability of North Pacific wind stress, partially due to reduced (strengthened) air–sea feedback, also helped to weaken (strengthen) the PDO amplitude by reducing (increasing) the meridional displacements of the subtropical and subpolar gyre boundaries. These results are generally similar to those found by Fang et al. (2014). Thus, the hypothesized effects of changes in stratification on oceanic baroclinic Rossby wave phase speeds and the consequent effect on the PDO timescales were simulated and confirmed by these two groups of researchers. One extreme

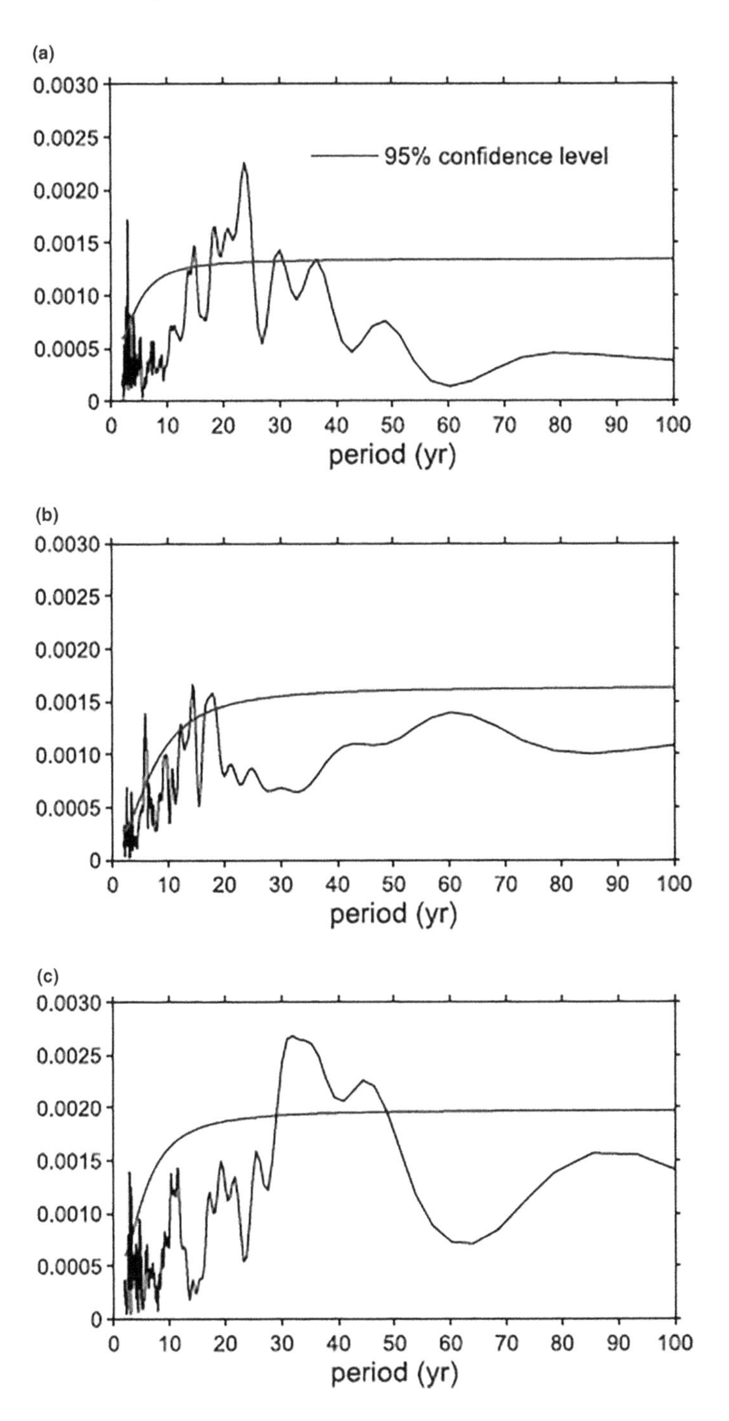

FIGURE 8.2 Power spectrum of the PDO time series in the fully coupled (a) control run, (b) 2×CO_2, and (c) 0.5×CO_2. (From Zhang and Delworth, 2016 © American Meteorological Society. Used with permission.)

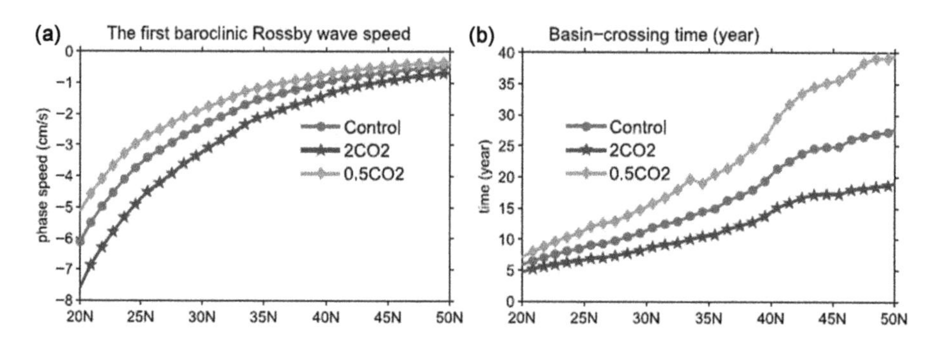

FIGURE 8.3 (a) Zonally-averaged first baroclinic Rossby wave speed (cms^{-1}), (b) Pacific basin crossing time (year) in the fully coupled control run and 2×CO_2 and 0.5×CO_2 experiments. (From Zhang and Delworth, 2016 © American Meteorological Society. Used with permission.)

scenario based on an extrapolation from this pioneering research can be that the Pacific Ocean can become very highly stratified in the case of surface warming due to a run-away greenhouse effect, and the baroclinic Rossby waves can become so fast (10,000 times faster than present speeds) that decadal variability in the Pacific Ocean climate can disappear (Fang et al., 2014)!

Another important component of DCV mechanisms we saw in Chapter 3 is the shallow tropical circulations or cells (STCs) in the Pacific and Atlantic Oceans. The basic physics of STCs and their hypothesized role in tropical Pacific decadal variability are described in Chapter 3 (Section 3.4.2). STCs are wind- and buoyancy-driven ocean circulations; therefore, any changes in these forcings due to ACC can potentially change intensities and/or speeds of STCs and influence DCV phenomena. Since the STCs are forced by tropical–subtropical winds, several studies have focused on potential changes in these winds and other atmospheric and oceanic features in scenarios of increasing CO_2. Held and Soden (2006) hypothesized, from scaling arguments about surface temperature warming, saturation water vapor pressure increase, and decreasing convective mass flux based on the Clausius–Clapeyron equation for water vapor, that tropical circulations would slow down by 2.5% to 3% corresponding to the observed warming since the mid-19th century Current Era (CE). Following up on these arguments, Vecchi et al. (2006) found that the observed, east–west sea-level pressure (SLP) gradient across the tropical Pacific has weakened in the 1861–1992 CE period; it was inferred that this weakened SLP gradient would have weakened the zonal wind stress associated with the Walker Circulation by approximately 3.5%. This weakening is approximately consistent with the Held–Soden hypothesis. Then, Vecchi et al. (2006) also found in the GFDL CM2.1 ESM that the weakening SLP spatial pattern and zonal wind stress could only be approximately reproduced if increasing CO_2 concentration was included in the ESM experiment from 1861 to 2000 CE. They also found that, as a result of this weakening zonal wind stress, the east–west thermocline tilt, surface zonal currents, and intensity and depth of equatorial upwelling have also been reduced. Subsequently, Vecchi and Soden (2007) found in 22 CMIP3 ESMs that tropical–subtropical trade winds weaken in the Pacific region in these ESMs, leading to a weakening of the Walker

Circulation, as the CO_2 concentration in the atmosphere is increased in time to 2100 CE to approximately double the pre-industrial concentration. Among other changes, these studies found that the equatorial thermocline becomes shallower and stronger, and SSTs in the eastern equatorial Pacific become warmer in increasing CO_2 simulations. Thus, there appears to be a consensus from observed data and ESM experiments that an increasing CO_2 concentration in the atmosphere from pre-industrial era to the present has weakened trade winds in the Pacific region by a few percent and that this weakening can approach 10% by the end of the 21st century CE.

Are there any potential consequences of the weakening trade winds for STCs and DCV? The weaker trade winds might lead to weaker STCs, with a possible influence on tropical DCV. Merryfield and Boer (2005), using the Canadian Centre for Climate Modelling and Analysis (CCCma) ESM, found that the pycnocline[4] transport associated with the STCs in the interior Pacific Ocean gradually weakened and decreased by over 40% as CO_2 increased to approximately double its pre-industrial concentration by the end of the 21st century. In experiments with another ESM (the ECHAM4-Ocean Isopycnal Model), Lohmann and Latif (2005) found that the Pacific STC strength decreased in the Northern Hemisphere and increased in the Southern Hemisphere as the model world warmed in response to increasing CO_2. Then, Luo et al. (2009) analyzed STC strength in experiments with 11 CMIP3 ESMs under the A1B scenario. It was found that, under this scenario, while there was no significant change in total STC transport in the Pacific, the pycnocline transport weakened and western boundary transport strengthened. Wang and Cane (2011) also analyzed STCs in experiments with seven CMIP3 ESMs under the A1B scenario. Their group of CMIP3 ESMs had some overlap with the Luo et al. (2009) group. In the Wang and Cane ESM group, off-equatorial Pacific trade winds in the Northern Hemisphere weakened and the winds in the Southern Hemisphere strengthened. This asymmetry was also reflected in STC strengths in the two hemispheres, with the Northern Hemisphere STC weakening and the Southern Hemisphere STC strengthening. The tropical Pacific Ocean circulation between ±5° latitudes showed robust weakening trends in both hemispheres, perhaps due to weakening equatorial trade winds. However, in contradiction to Luo et al. (2009), Wang and Cane (2011) found that there was insignificant change in western boundary current transport under the A1B scenario. Wang and Cane (2011) showed that this discrepancy may be because Luo et al. (2009) probably did not include shallowing of the pycnocline base which substantially reduces pycnocline transport convergence along the western boundary, resulting in a lack of compensation between interior and western boundary pycnocline transports. Thus, the consensus of various ESM experiments appears to be that while potential changes in the tropical–subtropical Pacific STCs may be too small to be measured during the first half of the 21st century CE, potential changes towards the end of the 21st century may be significant enough to be measured. It can be inferred that these potential STC changes may gradually affect tropical–subtropical Pacific DCV as the 21st century progresses. However, given the substantially variable nature of the observed DCV in the paleoclimate proxy and instrument records, it may not be possible to detect the changes associated with increasing CO_2 till late

[4] A layer in which water density increases rapidly with increasing depth.

21st century if the emissions follow the A1B scenario. If CO_2 emissions increase faster than this scenario, then their effects on DCV might become noticeable sooner.

Thus, ACC can influence DCV via at least two major mechanisms. There may be (and possibly are) many other ways in which ACC can influence DCV, but they are not yet known in these early days of understanding ACC-DCV interactions. However, it appears that this influence may not be detectable for several decades and may become noticeable only towards the end of the 21st century CE and beyond. Also, we saw in some of the previous chapters that paleoclimate proxy and instrument-measured data show that the PDO and other major DCV phenomena have varied over wide ranges of amplitudes and periodicities in the last several centuries, so attributing DCV changes observed in the future to ACC may be nearly impossible.

8.3 THE CHANGING CLIMATE – IS IT NATURAL DECADAL VARIABILITY OR ANTHROPOGENIC CLIMATE CHANGE?

Perhaps the hottest question in the global warming debate (pun intended!) is that posed in the title of this section. Are contemporary changes in precipitation, heat waves, and other weather and climate events due to ACC or to natural variability? Human memory can be fickle, and the temptation to attribute everything that occurs in nature to ACC can be irresistible for people at one end of the spectrum who believe in ACC. On the other side of the spectrum, to those who do not believe in ACC, everything is due to natural variability. The world's mitigation and adaptation efforts fall in the middle and depend critically on our ability to separate climate impacts due to natural variability and ACC. Despite handicapped by the relatively short lengths of observed data time series, some brave climate scientists have initiated this task and this section tells their unfolding stories.

8.3.1 Attribution of Changing Global Temperature

In previous chapters, we have seen that there is pronounced variability of SSTs and associated ocean dynamical and thermodynamical variables; winds, and atmospheric pressure and temperatures; and temperatures and precipitation on land at decadal and longer timescales. So, a natural question is: what role, if any, this climate and ocean variability at decadal and longer timescales plays in the changing climate we experience? How can we distinguish between internally generated DCV, climate variability forced by solar variability and volcanic eruptions; and climate variability and changes forced by anthropogenic greenhouse gases (GHGs), aerosols, changing land use patterns, and urbanization? We also have to ask what are the spatial and temporal scales at which we can detect and attribute influences of these various types of variabilities and changes. Considering that global-mean surface temperature (GMST) has been the ACC impact variable which has held the world's attention and that international climate agreements specify targets in terms of GMST, it is reasonable that attempts to attribute observed variations/changes have focused mainly on GMST. So, we will begin by looking at what such attempts have found.

We saw in Chapter 3 that the PDO (Mantua et al., 1997) or the Interdecadal Pacific Oscillation (IPO; Power et al., 1999) spans a very large horizontal area of the Pacific Ocean and extends to hundreds of meters below the ocean surface. Therefore, very large amounts of heat transfers between the Pacific Ocean and the atmosphere are involved in the IPO. It is, therefore, obvious that the IPO would contribute substantially to GMST variations/changes. The IPO's importance in this context became very prominent when it was found that GMST's rise slowed down or stopped altogether between 2001 and 2013 CE, instigating a frenetic search to understand the cause(s) of this so-called hiatus in global warming (see, for example, Easterling and Wehner (2009)). Initial studies with a variety of climate and ocean models indicated that a pronounced cooling in eastern and central tropical Pacific contributed to the hiatus in GMST. Meehl et al. (2011) attributed this cooling to a central and eastern tropical Pacific SST pattern similar to the negative phase of the IPO and to an increased uptake of heat by the sub-surface tropical Pacific Ocean. Meehl et al. (2013) later found in experiments with the National Center for Atmospheric Research (NCAR) CCSM4 ESM that accelerated warming periods are characterized by a warming GMST, increasing upper ocean heat content, and less heat content increase in the deep ocean. They also found that the positive phase of the IPO was associated, along with warming from increasing GHGs, in the accelerated warming phases. Conversely, the negative phase of the IPO much reduced or even overwhelmed the GHG-forced warming in hiatus periods. England et al. (2014) found that the trade winds were unusually strong during the hiatus period and were instrumental in cooling eastern and central tropical Pacific surface temperatures. Kosaka and Xie (2013) showed with an atmosphere-only climate model that not only the hiatus in GMST, but also associated regional changes – such as a stronger Walker Circulation, winter cooling in northwestern North America, and a persistent drought in the southern United States of America (USA) – were also generally reproduced by the model which was forced with radiative forcing and the observed history of central and eastern tropical Pacific SSTs. The latter pattern was similar to the negative phase of the IPO. Thus, these and other initial studies (Dai et al., 2015; Fyfe et al., 2016) showed that an SST pattern similar to the negative IPO phase appeared to reduce or stop warming of GMST.

Independently of this research, the Conference of Parties of the United Nations Framework Convention on Climate Change (UNFCCC) agreed in Paris, France, in December 2015 CE to pursue efforts to limit the GMST increase to 1.5°C above pre-industrial levels (UNFCCC Conference of the Parties, 2015). One of the uncertainties in implementing this GMST target is the time frame in which GMST will reach the 1.5°C warming and the role natural climate variability might play in this time frame. To address the latter uncertainty, Henley and King (2017) used data from future climate projections under the Representative Concentration Pathway 8.5 (RCP8.5[5]) scenario by 32 CMIP5 ESMs to estimate how positive and negative phases of the IPO might affect GMST in the context of stabilizing GMST at 1.5°C. Figure 8.4a shows the average GMST and an inter-quartile range[6] of GMSTs from the 32 CMIP5 ESM simulations. As per these CMIP5 ESMs, the average GMST

[5] The pathway with the highest GHG emissions.

[6] Inter-quartile range is the difference between 75th and 25th percentiles or between upper and lower quartiles.

would reach 1.5°C in 2029 CE, with the inter-quartile range from 2026 (the first year in the range) to 2032 CE (the last year in the range). Now, how to estimate the possible influence of natural climate variability, especially the IPO, in deciding the actual time frame range when GMST would reach the 1.5°C limit? It is necessary to emphasize here that the non-initialized ESM simulations, in general, cannot simulate the correct phase and amplitude of the IPO or of any other natural climate variability pattern. So, Henley and King (2017) calculated the amplitudes of positive and negative phases of the ESM-simulated IPO and added them to the range of GMSTs from the ESMs in Figure 8.4a. The resulting changes in the GMST trajectory are shown in Figure 8.4b. Under the influence of a potential positive IPO phase, GMST would reach the limit in 2027 CE (2024–2029 CE range), and under the influence of a potential negative IPO phase, GMST would reach the limit in 2031 CE (2026–2033 CE range). Thus, the evolution of the IPO pattern in the next decade can substantially influence the GMST evolution and hence the time frame to comply with the Paris Agreement. As Henley and King (2017) show, "reaching the 1.5°C limit" can be defined in various ways and the time frames are approximately the same, except for changes of two to three years, according to their various definitions. Thus, these studies clearly show that not only does the IPO influence GMST, it also has consequences for compliance with international agreements to limit ACC.

As mentioned in the opening paragraph of this section, not only the IPO and other internally generated DCV phenomena, but also other types of climate variability/changes can influence the changing GMST. Dong and McPhaden (2017) addressed the challenge of estimating GMST changes due to both internal and external causes and placing them in relative context. So, they set out to estimate contributions to GMST changes due to changes in solar emissions, volcanic aerosols, GHGs, anthropogenic aerosols, and the IPO. They used data from 129 specific forcing simulations with eight CMIP5 ESMs with four different types of experiments. They estimated relative contributions of GHGs, anthropogenic aerosols (AA), and natural (solar and volcanic) forcing to the total, externally forced GMST variations by forming a multimodel ensemble (MME) of each type of experiment. They used two observed GMST and two global SST data sets to attribute simulated GMST changes to various causes. Contributions from each cause during two accelerated warming periods (1920 to 1945 CE and 1977 to 2000 CE) and two hiatus periods (1946 to 1976 CE and 2001 to 2013 CE) of GMST are shown in Figure 8.5. A general dominance of external forcing in determining decadal variations in observed GMST was found, with the overall correlation coefficient between GMST and external forcing at 0.88 during the 1920–2013 CE analysis period. The IPO-related, internally generated GMST variations were found to be obscured at times by externally forced GMST variations. The onset of GMST cooling in 1946 CE coincided with a positive- to negative-phase transition of the IPO; then, cooling due to the Mount Agung volcanic eruption in 1963 CE contributed substantially to GMST cooling. This cooling was within a period of reduced growth in GHG concentrations during the 1940s–1960s CE period. Thus, the warming hiatus during 1946–1976 CE was due to several causes. Following this hiatus, there was an accelerated warming during 1977–2000 CE, primarily due to increased GHG concentrations and secondarily due to reduced AA emissions after

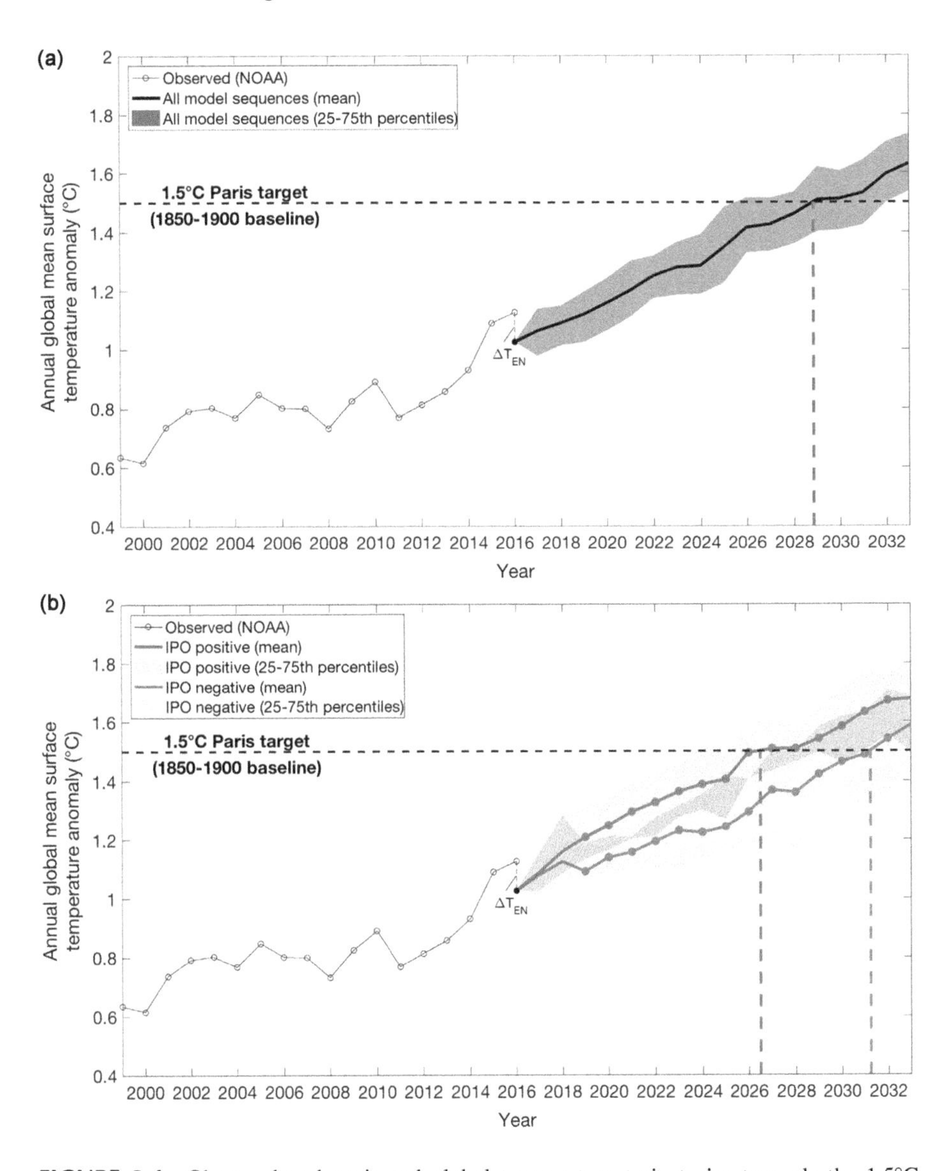

FIGURE 8.4 Observed and projected global temperature trajectories towards the 1.5°C Paris target. (a) (black) Observed global annual average surface temperature anomaly (NOAA, 2000–2016, 1850–1900 baseline), modeled plume of future temperature anomalies accounting for a temporary 0.1°C influence of the 2015–2016 El Niño, and 1.5°C Paris target and timing of the ensemble average breaching this level shown in dotted blue and red lines. (b) Same as (a), but for modeled temperature sequences of GMST in IPO positive (red) and IPO negative (blue) phases, including the preceding five years (CMIP5, RCP8.5); filled dots indicate statistically significant differences between the average in each phase (t test, 5% significance level); dotted vertical lines indicate the expected timing of breaching the 1.5°C Paris target in each IPO phase. (From Henley and King, 2017.)

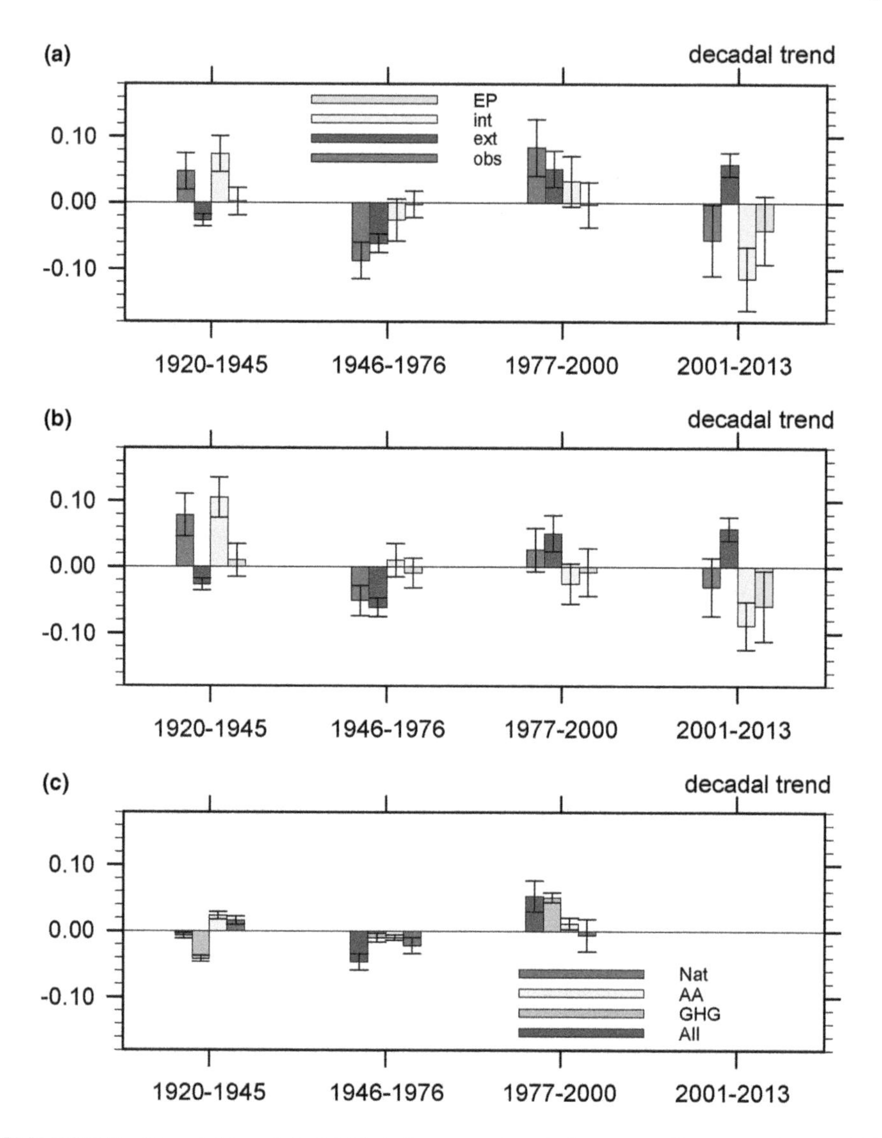

FIGURE 8.5 Decadal trends in GMST after removing the long-term linear trend over 1920 to 2013 (°C decade^{-1}) during 1920 to 1945, 1946 to 1976, 1977 to 2000 and 2001 to 2013 CE. (a) Values derived using HadCRUT, with the full observed trends in red, the externally forced trends (obtained from the MME of 18 CMIP5 models) in blue, the values of the internally generated variations (based on the difference between HadCRUT and MME) in yellow, and the values contributed by the equatorial eastern Pacific SSTs in green. (b) As in (a), but using GISTEMP. (c) The corresponding externally forced decadal trends and their components, as given by 8 CMIP5 models; full trends from external forcing in blue, GHG components in orange, AA components in light blue, naturally forced components in pink. The error bars denote 95% confidence limits for the linear trends. Note: In (c), results are given only for the first three decadal periods, since the components of the historical forcings are not available for the full 2001 to 2013 period. (From Dong and McPhaden, 2017; Figure licensed under Creative Commons Attribution (https://creativecommons.org/licenses/by/3.0/).)

the 1980s CE. The most recent GMST warming hiatus, as described earlier, was primarily due to the negative phase of the IPO. Since 2014 CE, however, the IPO phase has changed from negative to positive and GMST has reached record high values in 2014, 2015, and 2016 CE. Thus, this analysis by Dong and McPhaden (2017) showed that the IPO's contribution to GMST variations dominates in the absence of significant external forcing variations or when the amplitude of the IPO phase is very large as in the 1920 to 1945 CE accelerated warming period and the 2001 to 2013 CE hiatus period.

These studies reveal scientifically very interesting aspects of the role being played by natural DCV in determining GMST; however, this role makes it incumbent upon the climate modeling community to re-double efforts to skillfully predict the IPO and other DCV phenomena's future evolutions, not only to predict impacts of these phenomena but also to bring the maximum possible clarity to the ACC attribution problem.

8.3.2 Attribution of Changing Regional Precipitation, Temperature, and Extreme Events

As we saw in the case of the GMST, periodic or quasi-periodic natural phenomena can often confuse the attribution of observed changes to ACC or natural climate variability. Influences of DCV phenomena can be particularly confusing because record lengths of observed variables such as temperature and precipitation may be comparable to the phenomena's decadal and longer timescales. In such cases, increasing or decreasing anomalies associated with ascending or descending parts of a quasi-periodic cycle can be confused as linear trends. In addition to this confusion, attribution at regional or local scales can be confused by confounding factors such as effects of topography, land cover including urbanization, atmospheric turbulence, and others. These effects increase uncertainty at smaller spatial scales and shorter timescales, especially in inherently noisy variables such as precipitation. However, if multidecades to century-long or longer time series of observed data at weather stations are available, statistical techniques can be applied to separate multidecades to century-long trends from multiyear to decadal natural variability attributable to El Niño–Southern Oscillation (ENSO), PDO/IPO, the tropical Atlantic SST gradient (TAG) variability, the Atlantic Multidecadal Oscillation (AMO), and other phenomena. Results from two studies are described here.

Armal et al. (2018) identified linear trends attributable to ACC and quasi-periodic signals attributable to ENSO, PDO, AMO, and the North Atlantic Oscillation (NAO) in annual frequency of extreme rainfall events in the contiguous USA. They fitted Bayesian multilevel models (Renard et al. 2013) to rainfall time series from 1900 to 2014 CE at 1244 stations to combine information from across stations and to reduce parameter estimation uncertainty. After identifying stations which showed significant trends, Armal et al. (2018) tested two hypotheses for each of those stations – one hypothesis was that trends can be attributed to GMST anomalies, and the other was that trends can be attributed to a combination of GMST anomalies and the four climate variability phenomena. It was found that 742 out of 1244 stations showed statistically significant trends, out of which 409 stations showed trends attributable to

GMST anomalies. Further, 274 stations out of the 409 stations showed trends attributable to the four climate variability phenomena and to GMST anomalies; these stations are mainly in Northwest, West, and Southwest USA. Thus, Armal et al. (2018)'s analyses showed that, unless great care is applied in statistical analyses of observed data, natural decadal and longer timescales variability can easily be wrongly attributed to ACC, and that adaptation measures can be inappropriately applied.

In an example of estimating contributions from external forcings and internally generated DCV in sea level anomalies (SLA) in the Indian Ocean, Han et al. (2018) analyzed ocean model assimilated data, forced by atmospheric reanalysis, from the European Centre for Medium-range Weather Forecasts Ocean Reanalysis System version 4 (ORAS4; Balmaseda et al., 2013)) and from the Simple Ocean Data Analysis (SODA; Carton et al., 2000) version 2.2.4 from 1958 to 2015 CE. There are several approximations and differences in the types of assimilated data in these two ocean data assimilation systems as described by Han et al. (2018), but careful analysis and interpretation render results interesting and scientifically valid. In this study, external forcings include GHGs and volcanic eruptions. The Bayesian Dynamic Linear Model (Han et al., 2017), which can capture non-stationary impacts of climate variability, was fitted to the ocean data, and the resulting coefficients were analyzed to estimate contributions from various external and internal forcings to total SLA. After removing global-average sea level rise from the grid point SLAs, it was found that natural variability dominates externally forced SLA, with the former accounting for approximately 81% and the latter accounting for approximately 19% of the observed trend of falling sea level near the Seychelles where both contributions have maxima. The maximum sea level fall was found in southwestern tropical Indian Ocean–Seychelles region and weaker sea level fall in the Arabian Sea. Sea level rose during the analysis period in eastern equatorial Indian Ocean and the Bay of Bengal. Sea level rises near Sumatra, Java, and the Bay of Bengal were found to be considerably larger than the global-average sea level rise, with vertical land motions also playing an important role in the last 20 years of the analysis period. It was also found that decadal variability of the Niño3.4 index (which is nearly identical to the PDO and IPO indices as mentioned in Chapter 3), decadal variability of the Indian Ocean Dipole, and decadal variability of the Indian Monsoon explain a very large fraction of total decadal variability of SLA, with the ratio of the explained variability to total variability almost 80% and the correlation coefficient between the two almost 0.8 in major regions (southern Indian Ocean; Sumatra, Java, and the Bay of Bengal coasts; and the Indonesian Throughflow region north of Australia) of large SLA. It was also found that decadal variability of the Niño3.4 has the largest influence in southern Indian Ocean, and along the Sumatra and the Bay of Bengal coasts. Analyses of surface wind stress showed that decadal variability of the stress played a vital role in causing decadal sea level variations in most of the Indian Ocean; in the Indonesian Throughflow region, remote forcing from the Pacific Ocean also played an important role in decadal SLA variations. Thus, as for the US rainfall analysis described earlier, this study showed that unless great care is applied in statistical analyses of observed data, natural decadal and longer timescales variability can easily be wrongly attributed to changes in external forcings including ACC, and that adaptation measures can be inappropriately applied if the attribution is incorrect.

8.4 SUMMARY AND CONCLUSIONS

It is becoming increasingly important to identify and quantify contributions from DCV and ACC in the changing climate at local to global scales. It is also important to understand how ACC might influence DCV in the future to satisfy intellectual curiosity as well as make skillful near-term climate predictions. This chapter described initial research to address these very important emerging areas.

Ocean surface warming due to ACC can increase oceanic Rossby wave phase speeds and shorten DCV timescales. Weakening trade winds due to ACC can weaken STCs and perhaps influence DCV amplitude and timescales. These potential effects of ACC appear to increase as the CO_2 concentration in the atmosphere increases and may become detectable towards the end of the 21st century CE if the world follows some of the more plausible GHG emission scenarios.

Several studies of how much the IPO contributes to GMST were described and it is clear that large excursions of the IPO amplitude can make a substantial contribution to GMST and can confuse attribution by accelerating or slowing down apparent GMST warming due to ACC. It was shown with an example of extreme rainfall in the USA that the attribution problem is even harder on land at regional and local scales because noise increases as spatial and temporal scales decrease. But, Bayesian techniques can help in separating contributions from ACC and natural variability, especially from DCV. More studies are required to identify contributions from ACC and DCV in regions/areas where multidecades to century-long and longer time series of observed data are available.

The major results and conclusions described or cited in this chapter exemplify why it is very important for ESMs to hindcast/forecast major DCV phenomena and their impacts on precipitation and temperature with skill, so that contributions from DCV and ACC can be predicted with skill in the future. In hindsight looking at the importance of understanding and predicting DCV and its impacts not only for their own importance but also for attribution of ACC, it is painfully obvious now that much more emphasis should have been placed on DCV research 20 years ago to give it the necessary momentum. The confusion between DCV and ACC impacts could have been reduced, perhaps even at regional and local scales. Also, this confounding factor in ACC detection and attribution should have been brought to public notice 20 years ago to forestall the skepticism apparent now for explaining the recent hiatus in global warming. While it is possible that an accelerated pace of research in this area will result in major progress in the next few years, it is also possible that there may not be much progress in the actionable future in differentiating between DCV and ACC impacts on weather and climate – and consequently on water resources, agriculture, and other societal sectors – at regional and local scales where they are the most important. Only time will tell!

9 Modulations of Tropical Cyclones

9.1 INTRODUCTION

Tropical cyclones (TCs) are some of the most destructive weather phenomena on the Earth. They are called hurricanes in the North Atlantic region, typhoons in the western North Pacific and around China, and cyclones in the Indian Ocean–Arabian Sea–Bay of Bengal region. History is replete with stories and anecdotes of the fury of such storms and their consequences when they make landfall. Perhaps the deadliest cyclone to make landfall ever recorded was the Bhola cyclone in 1970 Current Era (CE) (Frank and Husain, 1971). It made landfall on 12 November in Bangladesh (then East Pakistan) and the Indian State of West Bengal. It is estimated that Bhola killed over 500,000 people when it made landfall, most from storm surges along and up the Ganga (Ganges) Delta.[1] Ironically, the meaning of the Hindi word *Bhola* is innocent! Among other losses attributed to this cyclone were lives of 4.7 million people affected, $63 million crop losses, hundreds of thousands of cattle and poultry lost, 400,000 houses and 3,500 schools damaged, and close to 10,000 fishing boats destroyed or damaged. These were staggering losses for the northern Bay of Bengal coastal region in 1970 CE. Partly due to a slow response of the then-Pakistan Government to this enormous humanitarian crisis, dissatisfaction with the Government resulted in a massive electoral sweep by the Awami Party in December 1970 CE in elections for the national legislature. Political consequences of this election and the ground swell of public dissatisfaction in East Pakistan with the Pakistan Government following Bhola were one of the catalysts of the subsequent civil war and India–Pakistan war in 1971 CE which resulted in the creation of an independent Bangladesh from East Pakistan. This is perhaps the only instance of a TC being a major catalyst of a civil war, an international war, and the creation of an independent country (Olson, 2005).

There were other cyclonic[2] storms also in the Bay of Bengal and the Arabian Sea in 1970 CE. Why did they occur? We saw in Chapter 5 that sea surface temperatures (SSTs) and the area of the East Indian Warm Pool in the tropical Indian Ocean–Bay of Bengal–Arabian Sea region undergo decadal variability. Is there any physical relationship between the decadal SST and East Indian Warm Pool area variability, and formation and intensification of TCs in this region? As we will see in this chapter, cyclones, typhoons, and hurricanes undergo decadal and longer timescale variability – why? Is there any physical relationship between solar cycles at 11-year and longer periods and numbers and intensities of TCs as implied by statistical associations

[1] www.cbc.ca/news/world/the-world-s-worst-natural-disasters-1.743208.

[2] Cyclonic refers to anti-clockwise or cyclonic winds around a low-pressure center in these storms and to the anti-clockwise rotation of the storms in the Northern Hemisphere.

first discovered in the 1870s CE? If causes and mechanisms of decadal variability of TCs can be understood, then perhaps numbers, intensities, and tracks of such storms can be predicted well in advance before they make landfall. Accurate multiyear to decadal predictions of these and other storm attributes such as rainfall, winds, and storm surges near sea coasts can be used to prepare affected populations and infrastructure to cope with the storms. The understanding and prediction can also be used for attributing observed variability to natural decadal and longer timescale variability and anthropogenic climate change. This chapter describes decadal and longer term variability of TCs in all oceanic regions; and the current understanding about the causes of such variability. Before we embark on the journey to see decadal variability of TCs, we will first see very briefly the basic physics of TCs and their travel.

9.1.1 Tropical Cyclone as a Heat Engine

A TC consists of a collection of thunderstorms organized in spirals around the center of the cyclone known as the "eye". The most intense thunderstorms, the heaviest rain, and the strongest winds occur in the defining boundary of the eye known as the eyewall. TCs are born over warm waters of tropical oceans where the air is humid, near-surface winds converge over warm water, and thunderstorms develop and strengthen. Atmospheric waves, the InterTropical Convergence Zone (ITCZ), and rising air motion ahead of a mid-latitude weather front traveling to tropical latitudes are prime locations for surface wind convergence and development of organized thunderstorms. The condensation of water vapor into liquid water releases latent heat which can drive further development of thunderstorms and, therefore, the cyclonic storm by warming the air which would rise and lower pressure near the center of a thunderstorm. The convergence of winds near the center would increase with increasing amounts of rising air, which also increases moisture convergence, increasing condensation and rain, and further fueling the storm with latent heat. This positive feedback can strengthen a thunderstorm as well as a cyclonic storm. Some cyclonic storms intensify into moderate-to-severe categories if they are over warm ocean water and in the absence of significant vertical shear of horizontal winds. Thus, the primary energy source which drives the development of a cyclonic storm is the underlying warm ocean water and that is why SSTs play a very important role in modulating the number and intensities of cyclonic storms. However, warming (cooling) at the top of deep convective clouds in upper troposphere can reduce (increase) the intensification potential.

Easterly winds at tropical latitudes, generally referred to as trade winds as we have seen in previous chapters, steer a TC westward and northwestward as storm nears the western side of a subtropical high pressure area with clockwise winds around it. As long as the steering winds keep a TC over warm SSTs, it can continue to intensify. In summary, warm tropical SSTs provide the water vapor (latent heat) and sensible heat which generate kinetic energy by upward motion and winds. Some of the latent heat is dissipated by outgoing longwave radiation from the top of the storm and kinetic energy of winds is dissipated by friction with the underlying surface. Thus, a TC can also be conceptualized as a heat engine (Emanuel, 1991).

After this brief introduction, decadal and longer timescale variability of TCs and their possible causes in the Atlantic, Pacific, and Indian Ocean regions are described in Sections 9.2, 9.3, and 9.4, respectively. Then, associations between decadal solar cycles and TC frequencies and intensities in the Atlantic, Pacific, and Indian Ocean regions are described in Section 9.5. Finally, a summary of the chapter and major conclusions drawn from the research reviewed here are described in Section 9.6.

9.2 HURRICANES IN THE TROPICAL ATLANTIC REGION

Bhola was perhaps the most destructive cyclone in the world's recorded history, but the tropical Atlantic has also spawned deadly hurricanes, some of which made landfall in the Caribbean, Mexico, and the United States of America (USA). The Galveston, Texas, hurricane of 9 September 1900 CE is considered to be not only the deadliest hurricane to make landfall in the USA, but also the deadliest natural disaster in the USA. Approximately 6,000 to 12,000 people were killed,[3] and physical damage worth US$35.4 million in 1900 CE dollars (over US$1 billion in 2019 CE dollars[4]) was inflicted by this hurricane. In more recent times, hurricane Katrina in 2005 CE and hurricanes Maria and Harvey in 2017 CE also wreaked havoc where they made landfall. Katrina made landfalls in Bahamas and the Gulf Coast of the USA, especially New Orleans and neighboring areas in Louisiana, on 23 August 2005 CE as a Category 5 hurricane (sustained winds at 280 km hour^{-1}). Deaths of over 1800 people and physical damage worth US$125 billion were directly or indirectly attributed to Katrina. Hurricane Maria made landfalls in Lesser Antilles and Puerto Rico as a Category 5 hurricane on 16 September 2017 CE and continued to wreak havoc through 2 October 2017 CE. Deaths of over 3000 people and physical damage worth US$96 billion were directly or indirectly attributed to Maria.[5] Other hurricanes in recorded history going back to the early 17th century CE have caused untold amounts of death, physical damage, and human misery as well as destroying flora and fauna in affected regions. But, hurricanes do not occur in the same numbers and strengths year after year. There is interannual, decadal, and multidecadal variability of hurricane attributes as we will see in this section.

Observations of hurricanes in the tropical Atlantic region[6] were sparse and inaccurate before the 1950s CE. Observations of landfalling hurricanes, especially in the USA, go back to the beginning of the 20th century CE. Since many coastal regions in the USA have weather stations since then, accurate estimates of wind speed and direction, rainfall, and surface pressure associated with landfalling hurricanes are also available. Another source of information about landfalling hurricanes in the USA is estimates of financial costs incurred due to damage to public and private property. Since 1970 CE, satellite data are available as a continuous record. Analyses of these various data have yielded very interesting and important information about Atlantic hurricanes.

---

[3] www.npr.org/2017/11/30/566950355/the-tempest-at-galveston-we-knew-there-was-a-storm-coming-but-we-had-no-idea.

[4] www.in2013dollars.com/us/inflation/1900?amount=1.

[5] en.wikipedia.org/wiki/List_of_deadliest_Atlantic_hurricanes.

[6] Tropical Atlantic region here includes the Gulf of Mexico and the Caribbean Sea.

Interannual and longer timescale variability of hurricanes is quantified by various measures based on a hurricane's physical attributes (surface pressure, winds) and their evolution during the life cycle of the hurricane. The number of hurricane days and the category of hurricane strength on the Saffir–Simpson scale (Simpson, 1974) are some of the other measures of a hurricane. Analyses of these various data have yielded very interesting information on decadal and longer timescale variability of tropical Atlantic hurricanes. Results of these analyses and conclusions drawn from them are described in this section. Associations of decadal and longer timescale variability of hurricane attributes with the tropical Atlantic SST gradient (TAG) variability and the Atlantic Multidecadal Oscillation (AMO) (see Chapter 4), and the possible physics of these associations are also described. In the last 10 to 15 years, actual and potential changes in hurricane attributes associated with global warming have attracted considerable attention from researchers, policymakers, and the general public. Analyses of hurricane observations to identify and quantify changes, as compared to decadal and longer timescale variability, are also briefly mentioned in this section as appropriate.

9.2.1 Empirical Analyses

There are documentations of decadal and longer term variability of hurricanes in reports by the National Oceanic and Atmospheric Administration and a book titled *Early American Hurricanes 1492–1870* (Ludlum, 1989), but Gray (1990) was perhaps the first to show that there were multidecadal variations in the frequency of intense (categories 3, 4, and 5; Simpson (1974)) hurricanes and also in the number of intense hurricane days. Specifically, Gray (1990) found that there were twice as many intense hurricanes in the epoch from 1947 to 1969 CE compared to the 1970–1987 CE epoch. Also, there were four times as many intense hurricane days in the first epoch compared to the second epoch. A comparison of the intense hurricane frequency with rainfall in West Sahel in West Africa showed that the rainfall also varied between the two epochs such that it was substantially above average when the intense hurricane frequency was much higher and substantially below average when the intense hurricane frequency was much lower. The intense hurricane frequency and West Sahel rainfall time series from 1940 to 1988 CE are shown in Figure 9.1. Both time series show five-year running averages, so decadal variations are smoothed and multidecadal variations are enhanced, but the decadal variations are still evident. Intense hurricane tracks during the wet and dry West Sahel years also show (Figure 9.2) a very substantial difference between the two epochs. These epochal differences were also seen in US landfalling intense hurricanes on the East Coast, Florida, and the Caribbean Sea as Figure 9.2 shows. Gray (1990) also showed that the number of intense hurricane days in the ten wettest years in West Sahel was approximately ten times larger than the number of intense hurricane days in the ten driest years, with four times as many intense hurricanes in the wet epoch compared to the dry epoch. In addition to identifying a global SST pattern associated with the epochal variations in intense hurricane frequency and West Sahel rainfall, Gray (1990) also found that vertical wind shear in tropical North Atlantic was weaker when the intense hurricane frequency and the West Sahel rainfall were much above average.

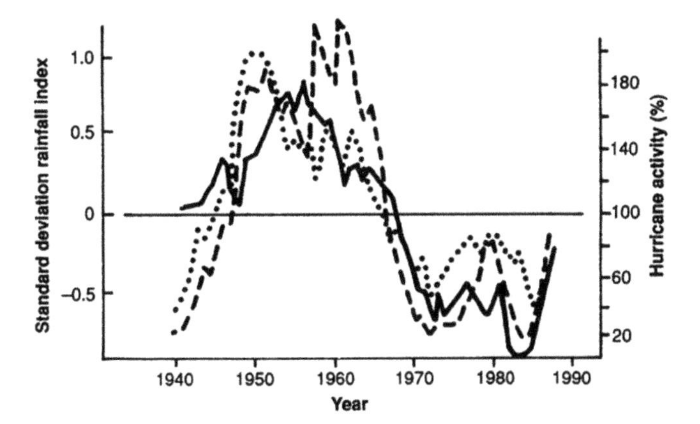

FIGURE 9.1 Comparison of five-year running averages of standard deviation of rainfall index from June to September in Western Sahel (solid line), percent variability of category 3, 4, and 5 hurricanes (dashed line); and percent variability of category 3, 4, and 5 hurricane days (dotted line). (From Gray, 1990.)

This pioneering research on variability of hurricane attributes, its association with rainfall variability in West Sahel, and hypotheses regarding causes and mechanisms of both phenomena was followed by other studies in the 1990s CE and later. Gray et al. (1992, 1994) and Landsea (1993) combined six measures of seasonal tropical Atlantic cyclone activity into an index of net tropical cyclone (NTC) activity. The six measures are the seasonal total numbers of named storms, hurricanes, intense hurricanes, named storm days, hurricane days, and intense hurricane days. The NTC index is an arithmetic average of the percentage departure of each of the measures from their respective long-term averages. As Landsea (1993) pointed out, the NTC index before the mid-1940s is not as reliable as the post-1940s CE period because of changes in observation platforms/techniques. Analyses of the NTC index time series by Mehta (1998) showed that decadal variations in the NTC index from 1886 to 1991 CE were physically consistent, but only moderately coherent with decadal variations in tropical North Atlantic SSTs. However, both Fourier and Singular spectra of the NTC time series showed a decadal oscillation period which was coherent with tropical North Atlantic SST variability, suggesting an association between the NTC and decadal SST oscillations. Mehta (1998) also found physically consistent multidecadal (~40–50 years) variations in both the NTC index and the tropical North Atlantic SSTs as in the Gray (1990) analysis. However, in view of the shortness of both NTC and SST records, and the unreliability of the NTC index before the mid-1940s CE, the association at the multidecadal timescale was mentioned as only suggestive. Mehta (1998) hypothesized that decadal variations in tropical North Atlantic SSTs can slowly modulate the amount of tropospheric moisture in the tropical North Atlantic atmosphere and also slowly destabilize the lower troposphere above warmer SSTs, thereby modulating development of hurricanes.

Continuing the development of the hypothesis about causes and mechanisms of decadal–multidecadal variability of Atlantic hurricanes, Landsea et al. (1999) also

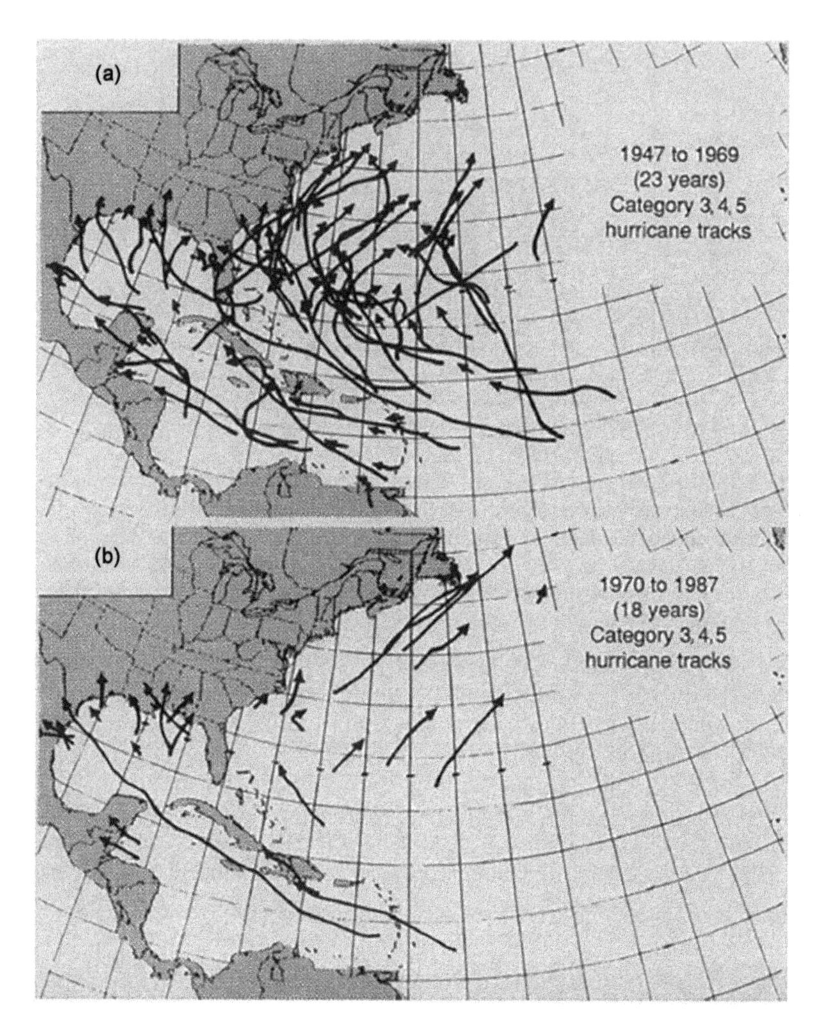

FIGURE 9.2 Comparison of categories 3, 4, and 5 hurricane tracks in (a) wet (1947 to 1969 CE) and (b) dry (1970 to 1987 CE) epochs in West Africa. (From Gray, 1990.)

suggested that tropical North Atlantic SST variations can influence moist static stability in the atmosphere and that vertical wind shear can also be influenced by the SSTs via an inverse relationship with surface pressure in the Main Development Region (MDR) of tropical Atlantic hurricanes between 10°N and 20°N. Surface pressure would decrease over warmer SSTs, which would weaken lower tropospheric trade winds, which would weaken vertical wind shear, providing an environment conducive to hurricane genesis. As Gray (1990) did, Landsea et al. (1999) also found that the number of intense (sustained wind speed $50 \mathrm{m} \cdot \mathrm{s}^{-1}$ and greater) hurricanes and the number of intense hurricane days underwent decadal–multidecadal variability and that it was not clear from the data analyses that SST variations influenced frequencies of all categories of hurricane. Landsea et al. (1999) supported Gray

(1990)'s conclusions regarding an association between variations in intense hurricane frequency and rainfall in West Africa. It was hypothesized that in epochs of above-average rainfall, vertical wind shear over the MDR was reduced, trade winds were weaker, and upper tropospheric westerly winds were also weaker. Such epochs were also associated with warmer than average tropical North Atlantic SSTs which, as we saw earlier, were associated with increases in intense hurricane frequency. We saw in Chapter 4 that decadal variability of rainfall in West Africa and northeast Brazil is statistically and physically related to the TAG which influences the north–south migrations of the ITCZ, low-level trade winds, implied water vapor transport, and rainfall. Landsea et al. (1999) also showed with US landfalling hurricane data going back to the late 1890s CE that 1899 to 1925 and 1971 to 1994 CE were less active periods when tropical North Atlantic SSTs were cooler than average, and that 1926 to 1970 CE was a more active period when the SSTs were warmer than average. These less and more active periods of hurricane activity indicated by US landfalling hurricane data generally overlap with those found by Gray (1990) in the post-1940 CE tropical Atlantic hurricane data. Incidentally, Landsea et al. (1999) also pointed out that there were only weak linear trends in Atlantic hurricane attributes and that decadal–multidecadal variability was much more prominent.

Thus, results and conclusions of observed data analyses by Gray (1990), Mehta (1998), and Landsea et al. (1999) converged to showing that the tropical Atlantic SST variability modulates climate on land on the east and west sides of the tropical Atlantic, and also modulates the development of intense hurricanes in the MDR. This was a remarkable convergence of insights based entirely on analyses of multidecades to century-long time series of observed data into very powerful and consequential phenomena. As the Sun set on the 1990s CE, a hypothesis based on data analyses was developing about the role of tropical North Atlantic SST variations in decadal–multidecadal variations of tropical Atlantic hurricanes via dynamical and thermodynamical influences on the troposphere in the MDR. As has been mentioned in several other chapters, the 1990s CE was the decade in which increasing availability of archived ocean–atmosphere–land data facilitated the research on which the field of modern decadal climate variability (DCV) research is founded. The work built on these beginnings in the 1990s CE has yielded substantial insights into a multitude of DCV phenomena and is now resulting in operational multiyear to decadal climate outlooks and forecasts as the 2010s CE come to a close as described in Chapter 10.

But, we are getting ahead of the story of decadal–multidecadal Atlantic hurricane variability. As the new millennium dawned, debates about anthropogenic climate change and its influences on natural weather and climate phenomena began to heat up. Atmospheric data measured by satellite- and aircraft-borne instruments began to show what appeared to be substantial changes not only in Atlantic hurricane activity but also in TC activity in other ocean basins. Goldenberg et al. (2001) pointed out that Atlantic hurricane activity was higher in the 1995–2000 CE period compared to previous 25 years. They found that there were two and a half times more intense hurricanes and five times more hurricanes affecting the Caribbean region. They also found that tropical North Atlantic SSTs were warmer and vertical wind shear was weaker in the 1995–2000 CE period. It was reiterated that most tropical storms in the Atlantic region form from easterly atmospheric waves propagating westward from

Africa to the MDR and these waves are seeds of 85% of intense hurricanes. However, Goldenberg et al. (2001) found that the number of easterly waves from Africa was almost constant from year to year and so the modulation of the ocean–atmosphere environment by SST variations in the MDR modulated the formation of intense hurricanes. When the tropical North Atlantic SSTs are warmer than average at decadal and longer timescales (or, TAG is positive, pointing northward), lower atmospheric stability is reduced, so storms can develop coherently in the vertical direction and are more resistant to vertical wind shear which is also reduced by warmer SSTs. Although Goldenberg et al. (2001)'s focus was on multidecadal hurricane and SST variability, they also mentioned that there was a substantial amount of variability at the decadal timescale and that El Niño–Southern Oscillation (ENSO) events also influenced the formation of hurricanes by modulating low-level winds and vertical wind shear over the tropical Atlantic. More importantly, Goldenberg et al. (2001) moved the physical hypothesis, which had evolved in the 1990s CE, forward by suggesting that the multidecadal variability in the Atlantic SSTs due to the thermohaline circulation (THC; see, for example, Rahmstorf (2003)) – a globe-girdling ocean circulation transporting heat and salt from the South Atlantic to the North Atlantic – known as the AMO (see Chapter 4) was associated with multidecadal variability of intense hurricanes. THC, and associated SST, variability have been shown to influence not only Atlantic hurricane activity but also rainfall in the African Sahel region and in other parts of the world (Zhang and Delworth, 2006). Using a much longer record of US landfalling hurricanes, they showed (Figure 9.3) that very few intense hurricanes made landfall in the USA in negative (cool) phases of the AMO from 1903 to 1994 CE, but a substantially large number of intense hurricanes made US landfall in positive (warm) phases of the AMO from 1926 to 2000 CE. The degrees of separation of intense hurricane statistics between opposite phases of two reference time series, West Sahel rainfall in Figure 9.2 and the AMO SST index in Figure 9.3, are truly remarkable and indicate the fundamental role of SST variability and the ocean–atmosphere interactions giving rise to the SST variability in modulating the West Sahel rainfall and intense hurricanes. As we have seen in this section and in Chapter 4, similar degrees of separation exist at decadal timescale as well where the underlying driver is SST variability generated by ocean–atmosphere interactions. While this comprehensive study advanced the understanding of decadal–multidecadal variability of Atlantic hurricanes, Goldenberg et al. (2001) could not shed any light on their initial observation that the hurricane activity was higher in the 1995–2000 CE period compared to the previous 25 years. The analysis of that possibility was carried forward by another study.

As mentioned several times in this chapter, data on Atlantic hurricanes at sea are sparse before the mid-1940s CE. The satellite observation era began approximately in 1970 CE, and there is now a continuous record, augmented by aircraft-borne measurements. We have seen the dilemma several times in previous chapters that multidecades to century-long time series of a climatically important variable on land, such as precipitation or temperature, may be available, but observed data on many types of ocean–atmosphere–land variables which may be involved in precipitation or temperature variability are usually not available for longer than a few decades. In the case of Atlantic hurricanes, this dilemma also exists, and researchers must work

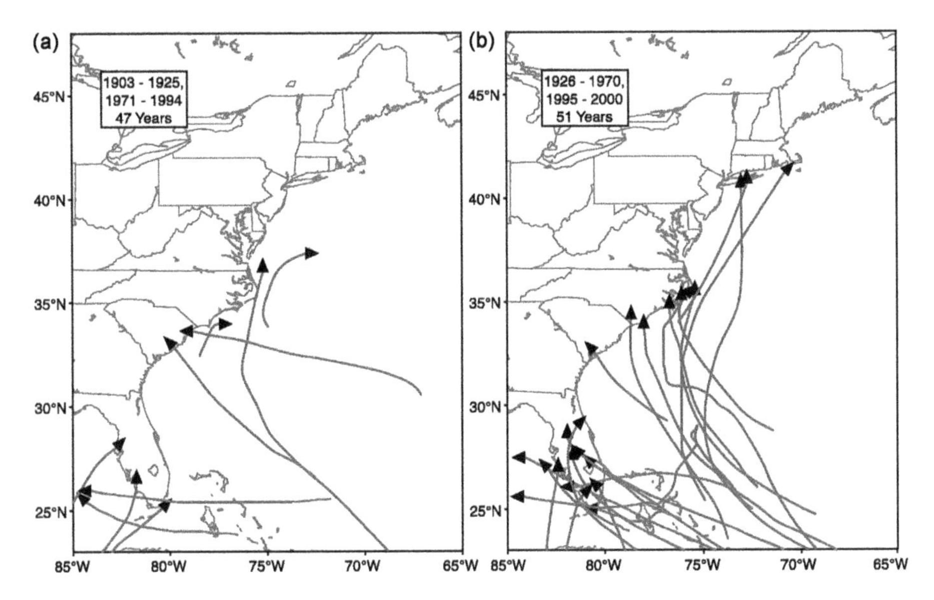

FIGURE 9.3 Difference in US East Coast intense hurricane landfalls between (a) colder and (b) warmer SSTs associated with the AMO. The solid red lines indicate where the storms were at major hurricane intensity. Ranges of colder and warmer years are mentioned. (Adapted from Goldenberg et al., 2001.)

to combine the best information available from all types of data or use a much more complete data set even if shorter. Webster et al. (2005) followed the latter path in their analysis not only of Atlantic hurricanes, but also of TCs in other ocean basins. They used satellite data to analyze numbers of TCs and cyclone days, and cyclone intensity from 1970 to 2004 CE in the Atlantic, Pacific, and Indian Oceans. Webster et al. (2005) found that there was no global, increasing or decreasing, trend in the numbers of TCs even though the background SSTs have been warming. They did find, however, that the numbers of category 4 and 5 cyclones increased from the late 1970s to 2004 CE in almost all ocean basins. The largest increases were in the North Pacific Ocean, the Indian Ocean, and the Southwest Pacific Ocean. The smallest percentage increase was in the North Atlantic Ocean. They also found that the numbers of cyclones and cyclone days decreased in all basins except in the North Atlantic during the 1995–2004 CE period. The conclusion that numbers of category 4 and 5 cyclones had increased from the late 1970s to 2004 CE in almost all ocean basins stimulated many researchers (see, for example, Wu et al. (2006), Klotzbach (2006), Landsea et al. (2006), Kuleshov et al. (2010), and Kossin et al. (2013)) for the next ten years to verify this conclusion because it implied that a possible impact of anthropogenic climate change on TCs was detected. Finally, Klotzbach and Landsea (2015) in a landmark study not only analyzed ten more years of cyclone data than the Webster et al. (2005) study, but also investigated the provenance of the data to see if data collection and/or processing techniques inadvertently introduced changes. In their analyses of global TC data till 2014 CE, Klotzbach and Landsea (2015) found that global frequency of category 4 and 5 cyclones showed a small but insignificant downward trend and the percentage of category 4 and 5 cyclones showed a small but insignificant upward trend between 1990 and 2014 CE. In addition, it was also found that accumulated cyclone energy (a measure of accumulated kinetic energy) showed a large and significant downward trend during this period. Why were these conclusions different from the Webster et al. (2005) conclusions? Klotzbach and Landsea (2015) answered this question by analyzing how TC warning centers around the world collected and processed data. They found that the apparent increase in category 4 and 5 cyclones in the Webster et al. (2005) study was due to improvements in observations at the warning centers in the 1970s and the 1980s CE. When these changes were accounted for, no significant upward trends in category 4 and 5 cyclones were found. This "decadal" group of studies from 2005 to 2015 CE clearly showed the oft-repeated folly of interpreting decadal–multidecadal variations in a climate or weather phenomenon as climate change. This group of studies, as countless ones before them, also highlighted the folly of accepting observed data at face value without investigating their provenance; the history of climate science is replete with repetitions of this folly as this quotation attributed to Aldous Huxley captures the essence of such follies – "That men do not learn very much from the lessons of history is the most important of all the lessons of history". However, as the foregoing amply demonstrates, observed data on landfalling hurricanes, satellite remote sensing data, a composite cyclone index of tropical North Atlantic hurricanes, and shipborne measurements of tropical Atlantic SSTs clearly show the unmistakable presence of decadal–multidecadal variability associated with SST variability at these timescales. These data also show that SST variability and associated rainfall variability in West Sahel modulate vertical shear of zonal winds and vertical profile of atmospheric water vapor and heat, and lower tropospheric stability in the tropical North Atlantic which, in turn, modulate the frequency of intense hurricanes.

9.2.2 Simulations with Models

The observed decadal–multidecadal variability in Atlantic hurricanes, and its connections to large-scale atmospheric circulation and SSTs in the tropical Atlantic region have been simulated in several studies with atmosphere-only and coupled ocean–atmosphere models. These global scale models have relatively coarse horizontal grid spacing and so cannot resolve TCs and their life cycle. However, such models shed light on how SST variability influences large-scale atmospheric variability and can simulate cyclone-like vortices which might be a reliable indicator of actual cyclone behavior. Vitart and Anderson (2001) used a global atmospheric general circulation model, forced by observed SSTs, to study interannual and decadal–multidecadal variability of tropical Atlantic cyclonic storms. The model had 18 levels in the vertical and a relatively coarse horizontal grid, with a reasonably complete set of parameterizations of physical processes. To minimize effects of initialization, sets of ten simulations were conducted with the same SSTs but with different initial conditions. Observed effects of interannual El Niño and La Niña events on atmospheric circulation in the tropical Atlantic region were simulated by the model, and it was found that tropical Pacific SST anomalies influenced vertical wind shear in the tropical North Atlantic and, thereby, influenced TC formation and frequency.

Ensembles of model experiments with SSTs in the 1950s CE (warmer) and 1970s CE (less warm) simulated almost the same average TC frequencies as observed in each period and indicated that warmer tropical Atlantic SSTs in the 1950s CE generated more TCs in the Atlantic than the less warm SSTs in the 1970s CE because vertical wind shear was less strong in the former period compared to the latter period. Vitart and Anderson (2001) also found that convective available potential energy, a measure of buoyancy force pushing air parcels upwards, was increased when the underlying SSTs were warmer (1950s CE) compared to when they were less warm (1970s CE). Thus, these atmosphere-only model experiments supported the observed association among tropical Atlantic SST, vertical wind shear, and cyclone frequency. These observed associations and their simulations with an atmosphere-only model were confirmed by Zhang and Delworth (2005) with a global, hybrid, coupled ocean–atmosphere model in which the dynamic Atlantic Ocean from 34°S to 66°N was replaced by a slab ocean model and ocean–atmosphere heat flux corresponding to multidecadal SST variability in the Atlantic was specified to simulate effects of this multidecadal variability on global climate and tropical Atlantic cyclones. It was found that this hybrid coupled model simulated effects of multidecadal Atlantic SST variability not only on TC frequency in the Atlantic but also on rainfall in the Sahel and the Indian monsoon region. As in the Vitart and Anderson (2001) atmosphere-only experiments, Zhang and Delworth (2005) also found that less warm tropical Atlantic SSTs in the 1970s CE caused stronger vertical wind shear and reduced cyclone frequency whereas warmer tropical Atlantic SSTs in the 1950s CE caused the shear to weaken, increasing the cyclone frequency. Thus, both observed data and model simulations firmly point to decadal–multidecadal SST variability in the tropical Atlantic as the main modulator of TC activity in this region.

9.3 TYPHOONS IN THE TROPICAL WESTERN AND CENTRAL PACIFIC REGIONS

Cyclone Bhola, landfalling in Bangladesh (then East Pakistan) – India and the Galveston hurricane, landfalling in the coast of Texas in the USA, were very destructive TCs as we saw in previous sections. The Western Pacific–South China Sea region also has a typhoon story to match. Typhoon Haiphong[7] (named for its landfalling location like the Galveston hurricane) originated in the Philippine Sea in early October 1881 CE, crossed Philippines, traveled westward in the South China Sea gaining strength, curved to the northwest, and made landfall in Haiphong in modern-day Viet Nam. In its wake, this typhoon left 10,000 to 20,000 people dead in Philippines, and over 300,000 people dead in and around Haiphong. This very destructive typhoon destroyed towns and flooded agricultural fields. Although such a destructive typhoon has not occurred since Haiphong, typhoon activity has undergone interannual to multidecadal variability. What follows is a description of typhoon activity since the beginning of the 20th century CE, its association with decadal–multidecadal SST variability, and the likely physics of this association. Systematic typhoon observations in the oceanic parts of these sub-regions date back to early

[7] www.britannica.com/event/Haiphong-cyclone.

1950s CE. As in other oceans, the record of observations of landfalling typhoons is much longer and goes back to the beginning of the 20th century CE in some countries. Before the advent of surface-based and airborne typhoon observing systems, the number of cyclones over oceanic regions and the estimated intensity of each cyclone were recorded in descriptive terms. Other attributes, such as wind speed and direction, and rainfall rates and amounts were recorded quantitatively only for landfalling typhoons. In this section, the tropical West Pacific region refers to western North Pacific, the Philippine Sea, the East China Sea, and the South China Sea. Observations of typhoon variability and their possible causes are described for each sub-region separately.

9.3.1 Western and Central North Pacific

Yumoto and Matsuura (2001) analyzed decade-to-decade changes in frequencies of typhoons in the western North Pacific from 1951 to 1999 CE, and associated the changes with large-scale oceanic and atmospheric conditions in the region via analyses of observations. Yumoto and Matsuura (2001) found that there were two epochs of higher frequency period (HFP) from 1961 to 1972 CE (HFP1) and from 1986 to 1994 CE (HFP2). The average numbers of cyclones in these two HFPs were 30.2 per year in HFP1 and 29.8 per year in HFP2. They also classified two other epochs from 1951 to 1960 CE and 1973 to 1985 CE as low-frequency periods (LFPs). In the two LFPs, the average numbers of cyclones were 24.6 in LFP1 (1951 to 1960) and 25.3 in LFP2 (1973 to 1985). Thus, there were approximately five fewer storms per year in the LFPs compared to the HFPs. This difference is clearly evident in Figure 9.4 for both typhoon season (July to October) and annual cyclone numbers. Analyses of SST and atmospheric data showed that SSTs were substantially warmer east of 150°E in the tropical–subtropical western North Pacific during the two HFP epochs compared to the two LFP epochs. Positive (cyclonic) relative vorticity anomalies at 850 hPa and positive divergence anomalies at 200 hPa were also found between 10°N and 20°N in the western North Pacific during the two HPF epochs, indicating that atmospheric conditions were more conducive for cyclogenesis.

Chan (2008) essentially confirmed the Yumoto and Matsuura (2001) conclusions, based on analyses of category 4 and 5 typhoon frequency and large-scale ocean–atmosphere conditions from 1960 to 2005 CE. It was found that typhoon frequency was substantially larger than average in the 1960–1970 CE and 1987–1997 CE epochs, and smaller than average in the 1971–1986 CE epoch. When the category 4 and 5 typhoon frequency was larger, SSTs in the southeastern part of the western North Pacific were warmer, atmospheric moist static energy was larger and its vertical gradient was more negative, and vertical wind shear was small. These conditions were conducive to increased convection and cyclogenesis in this region as indicated by increased precipitation rate. The large-scale atmospheric flow was such that the developing cyclones moved northwestward, and the steering winds were such that the cyclones would stay over warmer water for a longer time, resulting in strengthening to category 4 and higher typhoons. In the lower-than-average typhoon frequency epoch, ocean–atmosphere conditions were less favorable for cyclone formation and strengthening. Thus, Yumoto and Matsuura (2001) and Chan (2008) established a

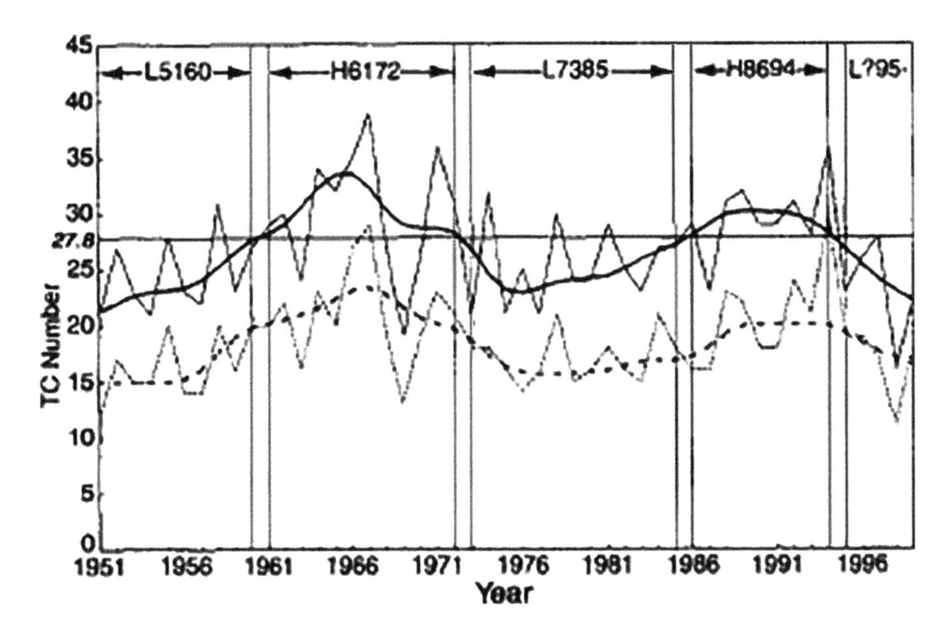

FIGURE 9.4 Time series of numbers of TCs from 1951 to 1999 CE in Western North Pacific. Solid and dashed lines show annual and typhoon season (July to October) storms, respectively. Bold solid and dashed lines show smoothed numbers of annual and typhoon season storms, respectively. L5160 refers to low-frequency period from 1951 to 1960, H6172 to high-frequency period from 1961 to 1972, L7385 refers to low-frequency period from 1973 to 1985, H8694 refers to high-frequency period from 1986 to 1994, and L95 refers to low-frequency period after 1994. (From Yumoto and Matsuura, 2001.)

physical relationship between decadal and longer timescale ocean–atmosphere variability and typhoon frequency in the western North Pacific region.

Kubota and Chan (2009) then extended the perspective on decadal and longer timescale variability of cyclones in the western North Pacific region by analyzing data on landfalling cyclones in Philippines. The Philippine Weather Bureau started publishing monthly bulletins in the late 19th century CE, and cyclone data were published in these bulletins after weather stations were established in Philippines around 1900 CE as mentioned by Kubota and Chan (2009) and references cited by them. Data from these bulletins for the 1901–1940 CE period were used in the Kubota and Chan (2009) analysis. After the Second World War, data from 1945 CE onwards are available from the US Air Force–US Navy Joint Typhoon Warning Center; these cyclone data were independently validated by data from the Japan Meteorological Agency. Kubota and Chan (2009) have described the procedure for identifying landfalling cyclones in Philippines. The time series of the annual number of landfalling cyclones, with a gap between 1940 and 1945 CE during the Second World War, is shown in Figure 9.5. Decadal and longer timescale variability of cyclones is clearly seen in this time series. Kubota and Chan (2009)'s wavelet analysis of this time series indicated that the dominant timescale varied from 32 years before 1920 to 10 to 22 years around 1960s and after 1990s CE. Timescales shorter than ten years

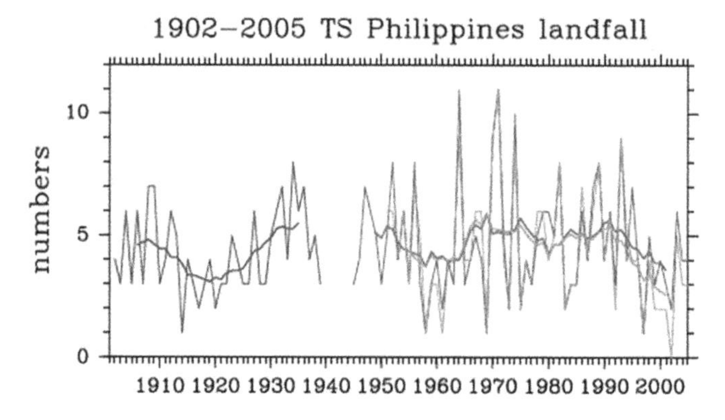

FIGURE 9.5 Annual number of landfalling TCs in Philippines using data from the Philippine Weather Bureau (from 1902 to 1939; red lines), Joint Typhoon Warning Center (from 1945 to 2005; blue lines), and Japan Meteorological Agency (from 1951 to 2005; green lines). Thick lines show 10-year running average numbers. (From Kubota and Chan, 2009.)

dominated between mid-1960s and mid-1980s CE. This analysis established the presence of decadal and longer timescale variability of the western North Pacific TCs back to the beginning of the 20th century CE.

Further east and northeast in central North Pacific between 180° and 140°W longitudes and Equator to the coasts of North America and Asia in latitude, Chu (2002) found decadal variability of cyclones and associated it with variability of atmospheric conditions. TC data, based on maximum sustained winds, for the 1966–2000 CE period from the National Hurricane Center (Brown and Leftwich, 1982; Chu and Wang, 1997; Chu and Clark, 1999); sea-level pressure data and wind data at various levels in the atmosphere, relative vorticity at 1,000 hPa, and total precipitable water (vertically-integrated, atmospheric water vapor) from the National Centers for Atmospheric Prediction (NCEP)–the National Center for Atmospheric Research (NCAR) reanalysis (Kalnay et al., 1996); and SSTs from Smith et al. (1996) were used in this analysis. Chu (2002) documented two types of cyclones in central North Pacific in the July to September period – in one type, 30 cyclones originated in central North Pacific and 68 cyclones traveled westward from eastern North Pacific. The number of cyclones in central North Pacific is shown in Figure 9.6. A change-point analysis (Elsner et al., 2000) showed that there were two changes in the annual cyclone number in 1982 and 1995 CE. There were fewer cyclones from 1966 to 1981 and from 1995 to 2000 CE epochs, and more cyclones from 1982 to 1994 CE epoch. Thirty cyclones occurred in the first inactive epoch (16 years) and 56 cyclones occurred in the first active epoch (13 years). So, almost twice as many cyclones occurred in the first active epoch compared to the first inactive epoch, with approximately the same number of years in both epochs; only 12 cyclones formed in the second inactive epoch in six years from 1995 to 2000 CE.

Further analyses by Chu (2002) of the cyclone frequency data shown in Figure 9.6, in conjunction with atmospheric and SST data, showed that there were warmer SSTs, lower sea-level pressure, stronger low-level anomalous cyclonic vorticity, reduced

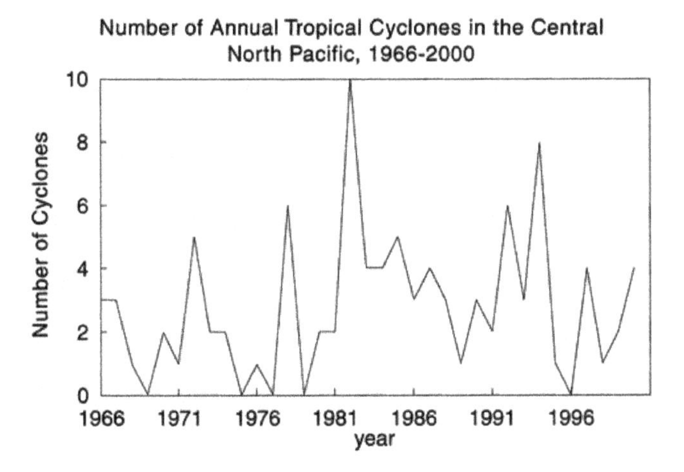

FIGURE 9.6 Time series of TCs in the central North Pacific from 1966 to 2000 CE. (From Chu, 2002 © American Meteorological Society. Used with permission.)

vertical wind shear, and increased precipitable water during the 1982–1994 CE active epoch in the major activity area between 10° and 20°N latitudes. The observed atmospheric and SST changes were established prior to the peak cyclone season, and were over a much larger spatial scale than the cyclones, implying that they contributed to rather than resulted from the cyclones. The steering-level (500 hPa) atmospheric winds were also conducive to cyclone formation and growth in the active epoch, such that there were easterly winds in eastern North Pacific steering cyclones westward and southwesterly winds in western Hawaiian Islands steering cyclones to the north. Thus, as Yumoto and Matsuura (2001) and Chan (2008) did for the western North Pacific typhoons, Chu (2002) established a physical relationship between large-scale, decadal ocean–atmosphere variability and cyclone frequency in the central North Pacific.

9.3.2 South China Sea

Typhoon attributes in the South China Sea have also undergone decadal–multidecadal variability. Specifically, there were two epochs of frequency of typhoon formation in and typhoon tracks passing through the South China Sea between 1951 and 2001 CE as Ho et al. (2004) found. In the 1951–1979 CE period, more typhoons formed in the East China Sea and the Philippine Sea than in the South China Sea. After 1980 CE, more typhoons formed west of 135°E longitude and fewer formed east of this longitude, and fewer typhoons passed from the Philippine Sea and the East China Sea to the South China Sea. These changes in typhoon attributes in the South China Sea are physically consistent with warmer SSTs after 1980 CE in the South China Sea. Ho et al. (2004) attributed the shift in typhoon tracks to a westward expansion of the subtropical northwestern Pacific high-pressure area in the late 1970s CE. Contrary to these indications of changes in the South China Sea typhoon attributes, Wang et al. (2013) found that warming SSTs in the Indian Ocean may have actually reduced TC formation in the South China Sea from the mid-1970s to late 1990s CE. In this hypothesis, based on analyses of the European ReAnalysis (ERA) 40 (Uppala et al., 2005) reanalysis of atmospheric data, lower level atmospheric convergence over warming Indian Ocean SSTs would cause upper level divergence which, in turn, would cause sinking over the South China Sea, forming a lower level anti-cyclone and suppressing TC formation.

The consensus from the observational studies emerged that decadal–multidecadal SST variability in the Western and Central tropical Pacific appears to modulate typhoon frequency by modulating large-scale atmospheric circulation, especially vertical shear of zonal winds, lower level relative vorticity, lower level convergence and upper level divergence, and vertical water vapor and temperature profiles. A few modeling studies have attempted to simulate mechanisms which create atmospheric conditions modulating typhoon frequency at decadal–multidecadal timescales. However, possible influences not only of the Pacific Ocean but also of the Indian Ocean in modulating typhoon frequency in the Western and Central North Pacific, and in the South and East China Seas complicates simulating such influences. The Pacific Decadal Oscillation (PDO) and decadal–multidecadal variability of tropical Indian Ocean SSTs were hypothesized as the drivers in observational studies as we saw earlier. The North Pacific Gyre Oscillation and SST gradient between the Western tropical Pacific and the tropical Indian Ocean have also been hypothesized as the drivers of decadal–multidecadal typhoon frequency variability. Modeling results to test the hypotheses are mixed so far as reviewed by Li and Zhao (2018). So, the quest to understand and predict decadal–multidecadal variability of typhoons must go on because landfalling typhoons can be very destructive and their skillful prediction at multiyear to decadal lead times can be used to prepare human populations, and built infrastructure to cope with them.

9.4 CYCLONES IN THE TROPICAL INDIAN OCEAN, THE ARABIAN SEA, AND THE BAY OF BENGAL REGIONS

At the beginning of this chapter, we saw how one particular cyclone, Bhola, in the Bay of Bengal in November 1970 CE devastated large areas in what was then East Pakistan and in India, and killed approximately 500,000 people. In this section, we will now see attributes of decadal variability of cyclones not only in the Bay of Bengal, but also in the Arabian Sea and the tropical North Indian Ocean regions.

Cyclones in the Bay of Bengal and the Arabian Sea affect India (pre- and post-Independence in 1947 CE), so Indian meteorologists have been at the forefront in cyclone research in these regions. While early studies focused on understanding the physics of cyclone formation and on prediction of cyclone intensities and tracks, Singh et al. (2000) and Patwardhan and Bhalme (2001) were perhaps the first studies to analyze linear trends and variability of frequencies of cyclone at interannual to multidecadal timescales in century-long time series. Singh et al. (2000) analyzed cyclone frequencies separately in the Bay of Bengal, the Arabian Sea, and the North Indian Ocean from 1877 to 1998 CE. They found a significant, increasing trend in Bay of Bengal cyclones in May and November – the main cyclone months in this region – and a significant, decreasing trend in June and September – months of the beginning and end of the Indian summer monsoon. Singh et al. (2000) also found an increasing trend in the frequency of severe cyclones in the North Indian Ocean in May, October, and November. No significant long-term trends were found in the Arabian Sea cyclone frequency. Spectrum analyses of the cyclone frequency time series showed significant oscillation periods of 29 years in the May cyclone frequency and 44 years in the November cyclone frequency in the Bay of Bengal. In the Arabian Sea cyclone frequency time series, significant oscillation periods of 13 years in May and June frequency, and ten years in the November frequency were found. Singh et al. (2000) also found that the El Niño–Southern Oscillation (ENSO) phenomenon was associated with cyclone frequencies in the Bay of Bengal and the North Indian Ocean, such that there were significant spectral peaks at approximately two- to five-year periods, and there was a reduction of cyclone frequency in the Bay of Bengal during El Niño events. Patwardhan and Bhalme (2001), on the other hand, analyzed total TC frequencies from 1891 to 1970 CE over a very large region (5°N to 30°N latitudes, 50°E to 100°E longitudes) encompassing the Arabian Sea, the Bay of Bengal, and the North Indian Ocean. Also, they analyzed only seasonal and annual data. As the Singh et al. (2000) study showed, however, linear trends and variability in cyclone frequencies were different in these three regions and among months, so analyzing seasonal and annual combined data from all three regions resulted in different conclusions in the Patwardhan and Bhalme (2001) study. The latter study found that there was only a decreasing trend in cyclone frequency during the Indian summer monsoon and that there were no significant oscillation periods in the spectra of the seasonal and annual cyclone frequency time series.

As interest in analyzing and predicting long-term trends and their attribution to anthropogenic climate change, and interannual to multidecadal variability of cyclones in the general Indian Ocean region grew, the available cyclone data in the Bay of Bengal, the Arabian Sea, and the North Indian Ocean regions were further analyzed by many researchers. Attempts also began to diagnose causes of the trend, if any, and variability of cyclone frequency. Pattanaik (2005) found high and low cyclone frequency periods in combined data over the Bay of Bengal and the Arabian Sea from 1891 to 2003 CE. Applying a ten-year moving average to the cyclone frequency time series during the June to September monsoon season, it was found that 1969 to 1978 CE was a HFP with 79 cyclones and 1993 to 2002 CE was a LFP with 26 cyclones. So, this was a substantial and significant difference in cyclone frequencies in two periods with equal numbers of years. It was also found that the Bay of Bengal and the Arabian Sea SSTs were approximately 0.2°C–0.4°C warmer during the LFP. Analyses of 850 hPa winds, vertical shear of zonal wind between 850 and 200 hPa, specific humidity at 500 hPa, and horizontal wind shear from the NCEP-NCAR reanalysis (Kalnay et al., 1996) showed that the anomalous, stronger low-level winds generated anomalous cyclonic vorticity over the Arabian Sea and the Bay of Bengal during the HFP; there was anomalous anti-cyclonic vorticity over these regions during the LFP due to weaker winds. Further, Pattanaik (2005) found that anomalous, mid-tropospheric specific humidity was above average over the entire Indian region during the HFP and below average during the LFP. Anomalous vertical shear of zonal wind was found to be smaller during the HFP compared to the LFP. These differences in the cyclone environment indicated that differences in large-scale atmospheric circulation, rather than regional SST differences, were the main cause of the observed decadal variations in cyclone frequencies. But, what caused these decadal-scale circulation differences? It is well known that the PDO is associated with global atmospheric circulation variability (see, for example, Mehta (2017)); the Indian June to September monsoon winds are weaker in the positive phase of the PDO and stronger in the negative phase of the PDO. A comparison of the HFP and LFP years with Figure 3.1b (Chapter 3) shows that the PDO was in the negative phase during the HFP epoch and in the positive phase (on the average) during the LFP epoch. So, it is possible that the negative PDO phase caused the stronger, low-level monsoon winds and anomalous cyclonic vorticity which, in turn, facilitated many more cyclones. The positive PDO phase during 1993–2002 CE may have caused weaker, low-level monsoon winds and anomalous anti-cyclonic vorticity which, in turn, facilitated substantially fewer cyclones. Thus, in this observations-based hypothesis, the PDO may be the "pace maker" of cyclone frequency in the Bay of Bengal and the Arabian Sea regions. Simulation experiments with global climate models are needed to test this hypothesis. If this hypothesis is found to be largely true, then multiyear to decadal lead time PDO predictions can be used to predict cyclone frequencies in the Indian region at these long lead times.

A generally similar conclusion about causes of variability in cyclone activity was reached by Ng and Chan (2012) via their analyses of cyclone attributes, SSTs, and NCEP-NCAR atmospheric reanalysis data, even though this study was about interannual variability and not decadal–multidecadal variability. Their cyclone attribute data consisted of number of cyclones in the North Indian Ocean region, number of intense cyclones, and accumulated cyclone energy from 1983 to 2008 CE. It was found that during the post-monsoon months (October–November–December), over the Bay of Bengal, atmospheric flow patterns and moist static energy, not local SSTs, affect cyclone activity. The interannual ENSO phenomenon was found to modulate 850 hPa relative vorticity, vertical shear of zonal wind, 200 hPa divergence, and moist static energy such that fewer intense cyclones occurred in the Bay of Bengal during El Niño events. During La Niña events, the opposite atmospheric conditions facilitated a larger than average number of intense cyclones. In addition to the two- to seven-year ENSO timescale, Ng and Chan (2012) also found that the time series of accumulated cyclone energy in the Bay of Bengal showed variations at eight to 16-year timescales. About the Arabian Sea cyclone activity, it was concluded that a relatively short time series of data (only 25 years) and very few cyclones annually made any conclusion from the statistical analysis unreliable. The conclusion of this study about the Bay of Bengal TC activity clearly pointed to tropical Pacific variability as the source of modulation of the atmospheric circulation and water vapor at interannual timescale and is, thus, physically consistent with the Pattanaik (2005) study described earlier. More recently, Rajeevan et al. (2013)'s analyses of cyclone attributes, TC heat potential (a measure of upper ocean heat content) from Goni and Trinanes (2003), and NCEP-NCAR atmospheric reanalysis data from 1955 to 2011 CE also found that there was a larger number of intense cyclones in July and August (the main summer monsoon months) when the vertical shear of zonal wind was weaker. It was also found that the TC heat potential was marginally larger when the number of intense TCs was larger.

Unlike decadal–multidecadal variability of TCs in the Atlantic and, to some extent in the Western North Pacific and Seas bordering China and Southeast Asia, such variability in the tropical North Indian Ocean–Bay of Bengal–Arabian Sea region has not attracted any significant attention from modelers to understand causes and mechanisms the variability.

9.5 SOLAR VARIABILITY AS A PACE MAKER?

As we have seen in several previous chapters, variability of solar emissions at decadal timescales has been invoked for centuries as a cause of climate variability. Variability of TC numbers and intensities is no exception to this general invocation. If there were statistical inferences or speculations about decadal solar variability as a cause of TC variability before the 19th century CE, they are lost in the mists of time. Apparently, the first studies published in journals were in the 1870s CE by Meldrum (1872, 1885) and Poey (1873). The former two studies were about variability in the frequency of TCs in the southern Indian Ocean, and the latter study was about TCs in the North Atlantic–Caribbean and the southern Indian Ocean regions. Sunspot numbers have been the solar variable of choice in these early and later modern statistical studies. Mr. Charles Meldrum was a professor of mathematics in the Royal College of Mauritius in Port Louise, Mauritius, from 1848 and founded the Mauritius Meteorological Society in 1851 CE. In 1862 CE, Mr. Meldrum was appointed the observer in charge of the meteorological observatory in Port Louise. Among his duties were to study meteorological data from ships docking in Port Louise and to identify characteristics of cyclones in the Indian Ocean. About his data collection system, Meldrum (1885) wrote "Two clerks were constantly occupied in tabulating the meteorological observations contained in the log-books of vessels that arrived in the harbour of Port Louis from different places". Daily photographs of the solar disk were also recorded in the Port Louise observatory from 1880 CE. Such photographs were also recorded in observatories in Greenwich, United Kingdom (UK), and Dehra Dun, India. These photographs probably showed a cyclical behavior of Sunspots, which Mr. Meldrum associated with his observations of cyclical behavior of southern Indian Ocean cyclones. Following up on his speculation (Meldrum, 1872) that variations in number, extent (duration), and intensity of TCs south of the Equator in the Indian Ocean were associated with the 11-year solar cycle, Meldrum (1885) scrutinized annual maps of cyclone tracks based on ship data from 1856 to 1884 CE between the Equator and 34°S and saw cyclic variations of the tracks, attributing them to the 11-year cycle of Sunspot numbers. There is also anecdotal evidence that there was an 11-year cycle in the number of ship wrecks in the Indian Ocean (Hoyt and Schatten, 1998). The Mauritius meteorological observatory's data show that attributes of TCs in the southern Indian Ocean had an 11-year cycle. Regardless of whether this cycle was physically related to the 11-year solar cycle, the 11-year cycle in the Indian Ocean ship wrecks sounds plausible. It is truly remarkable how little information was available then, but how meticulously it was recorded and transcribed to develop an ocean basin wide picture of TC variations spanning decades. It is also remarkable that Mr. Charles Meldrum and other weather observers applied whatever little knowledge and insight they had about the atmosphere to build further understanding of weather phenomena. It is very surprising that this pioneering research in a data-sparse era has apparently not been followed by other researchers in the modern, data-rich era.

It is also very surprising that possible associations between decadal solar cycles and tropical Pacific cyclones have not been explored but for one solitary published study. In this study, Kim et al. (2017) analyzed best-track data of West North Pacific TCs from 1977 to 2016 CE, with intensities defined as a maximum sustained wind speed of at least 17 m s^{-1}. For comparison, associations between El Niño and La Niña events, and TCs with the same definition were also analyzed. It was found that in solar maximum years without El Niño and La Niña events, tropical Pacific cyclones form at low latitudes and undergo their entire life cycle at or near the latitude where they form. In these years, the cyclones form in more eastern parts of the Pacific basin and travel westward where they finish their longer life compared to those forming in solar minimum years. These attributes were found to be similar to those of TCs during El Niño events. Kim et al. (2017) also found that the formation and evolution of TCs differed in the ascending phase of solar cycles compared to the descending phase. Among the differences was a higher maximum sustained surface wind speed in the latter phase. As this empirical study clearly demonstrated, El Niño and La Niña events and solar cycle phases appear to confound each other's associations and possibly physical effects on tropical Pacific cyclones. These possible effects should be studied with appropriate coupled ocean–atmosphere models.

As mentioned in Chapter 4 and earlier in this chapter, the Atlantic basin and its littoral region are relatively rich in atmosphere, ocean, and landfalling TC data going back to at least the 19th century CE. Many empirical studies have exploited these long time series to extract decadal variability information and associate it with Sunspot numbers. After Poey (1873), Cohen and Sweetser (1975) may have been the first to associate the annual number of TCs in the North Atlantic and the length of the cyclone season with Sunspot numbers. They calculated the maximum entropy spectrum of the three time series – the first two from 1871 to 1973 CE and the third from 1750 to 1963 CE – and found several sharp peaks at approximately the same periods in all three spectra from eight to 22 years. Cohen and Sweetser (1975) associated spectral peaks at approximately 11- and 22-year periods in the TC data with Sunspot variability at these two periods without offering any physical explanation for the association. Since then, there have been sporadic attempts to push this frontier of solar–North Atlantic TCs forward, resulting in a hypothesized role of solar ultraviolet (UV) radiation in influencing TCs. Notable among these attempts are statistical analyses of US landfalling hurricanes with respect to Sunspot cycles, and of Caribbean and Gulf of Mexico hurricanes with respect to solar UV radiation cycles. Elsner and Jagger (2008) constructed regression models of US landfalling hurricane numbers from 1866 to 2006 CE as a function of SSTs, the North Atlantic Oscillation index, the Southern Oscillation index, and the Sunspot number. It was found that the probability of US landfalling hurricanes decreases with increasing Sunspot numbers, a conclusion later supported by analyses of hurricane and Sunspot number records from 1749 to 1850 CE (Hodges and Elsner, 2011). Based on this conclusion, Elsner and Jagger (2008) hypothesized that, in the western Atlantic including the Caribbean Sea and the Gulf of Mexico where the upper ocean heat content is sufficiently large, hurricanes may be prevented from reaching their maximum potential intensity by outflow temperature near the tropopause as mentioned in Section 9.1.1. As described also in several chapters in this book, solar UV radiation may warm the lower stratosphere and upper troposphere near a maximum phase of solar activity, indicated by Sunspot numbers in the Elsner and Jagger study, thereby preventing hurricanes in these regions from reaching their maximum potential intensity as enabled by upper ocean heat content, and favorable phases of the North Atlantic Oscillation and the Southern Oscillation. This hypothesis was tested, to some extent, by correlating Sunspot numbers with upper troposphere and lower stratosphere temperatures from the NCEP-NCAR reanalysis data set during the August to October hurricane season. The correlations were found to be positive, implying that the near-tropopause temperatures were indeed warmer during high Sunspot epochs. Repeating the correlation analysis with radiosonde data from around the world confirmed this conclusion. Analyses of daily data on US landfalling and Caribbean storms showed that storms were more intense when Sunspots were fewer than average. Elsner and Jagger (2008) cautioned that the lengths of the temperature time series from the NCEP-NCAR reanalysis and radiosonde data were only approximately six solar cycles long; the data near the tropopause level may not be very reliable; and tall TCs may influence temperatures near the tropopause. Despite these cautions, however, the emergence of this testable, physical hypothesis was a refreshing contribution to the field as was the later verification of this hypothesis by Elsner et al. (2010) using data on solar Magnesium II core-to-wing ratio, an indicator of solar UV radiation at wavelengths less than 240 nm which is important for ozone production and heating in the upper stratosphere (see Chapter 2).

Other solar influences besides UV radiation undergo cyclic variability as described in Chapter 2. Among them are solar wind and cosmic ray fluxes, and their influences on geomagnetic activity. Basing their study on possible effects of solar cycle influenced geomagnetic activity on the freezing rate of supercooled water droplets by ionization processes in clouds at high levels in the atmosphere hypothesized by Tinsley and Dean (1991), Elsner and Kavlakov (2001) explored possible association between geomagnetic activity and intensity of Atlantic hurricanes. As applied to extratropical cyclonic storms in winter, the freezing would increase ice crystals, increasing latent heating of the storms and consequent strengthening according to this hypothesis. To see if this possible effect may be at work in intensity changes in North Atlantic hurricanes, Elsner and Kavlakov (2001) analyzed hurricane intensity as indicated by maximum sustained 1-minute wind speed at 10 m height and the Kp index of geomagnetic activity (Chapter 2) from 1950 to 1999 CE. These analyses were carried out for three categories of Atlantic hurricanes as (1) tropical-only cyclones which form from tropical disturbances, such as an easterly wave whose role was mentioned in Section 9.2.1; (2) baroclinically enhanced, which originate from tropical disturbances, but reach hurricane intensity resulting from favorable mid-latitude baroclinic influences; and (3) baroclinic or baroclinically initiated, which originate from atmospheric processes that derive energy from thermal gradients on constant pressure surfaces (Elsner et al., 1996). It was found that, out of these three categories, the maximum intensities of hurricanes of the third category show a significant association with the Kp index averaged over the 10-day preceding and the day of the hurricane intensity (total 11 days). While acknowledging that the geomagnetic processes which possibly influence baroclinically initiated hurricanes are poorly understood, Elsner and Kavlakov (2001) speculated that the geomagnetic field-related ionization of upper levels of a cyclonic storm can lead to additional latent release, subsequent warming of the cyclone core, leading to a reduction in the central pressure of the storm and its intensification. This heating of the storm core may also lead to its maintaining itself against vertical shear of horizontal wind which, as we have seen earlier in this chapter, can prevent the strengthening of a cyclone.

Finally, we come to the question posed in the title of this section – Solar Variability as a Pace Maker? The empirical evidence presented here is credible, and the specific physical hypotheses proposed for the role of solar variability as a pace maker of decadal variability of TCs are testable with appropriate models based on first principles of physics and chemistry. So, the answer to the question is – possibly.

9.6 SUMMARY AND CONCLUSIONS

TCs are essentially weather systems which bring destructive winds and storm surges ashore when they make landfall after their formation and strengthening over the oceans and seas. These weather systems also bring copious amounts of rain not only to the coasts but also to interior regions, causing flooding which accounts for a substantial fraction of the rainfall in these regions. Thus, the cyclones, especially intense cyclones, have both destructive and beneficial effects on human society, including the built environment, and natural and managed ecosystems. That is why skillful predictions of cyclone tracks and intensities even a few days in advance are very important. Understanding and skillful prediction of interannual to multidecadal variability of cyclone attributes are also very important to satisfy intellectual curiosity and to plan adaptation measures to minimize damage and maximize benefit of landfall cyclones as well as attribute observed cyclone attributes to natural variability or to anthropogenic climate change. This chapter reviews what is known so far about decadal and longer timescale variability of cyclone attributes in the Atlantic, Pacific, and Indian Ocean regions.

As in research on decadal variability of other weather and climate variables and systems, empirical associations of observed TCs, especially seasonal/annual numbers, with the 11-year Sunspot cycle in the Indian Ocean and the Caribbean Sea regions began in the 19th century CE. In the 20th and 21st centuries CE, however, such associations with hurricanes in the North Atlantic Ocean, the Caribbean Sea, and the Gulf of Mexico regions, and landfalling hurricanes in the US have led to the hypothesized role of 11-year solar UV radiation cycles influencing temperatures in the lower stratosphere and upper troposphere and thereby influencing hurricane development in these regions. Further research on the solar cycle–TC relationship in all ocean basins with empirical data and global coupled ocean–atmosphere models reaching up into the upper stratosphere and including ozone chemistry and solar wind and geomagnetic field processes which influence clouds are required to pursue this very interesting line of research to push the knowledge frontier forward as well as to improve multiyear to decadal predictability of TCs. Possible confounding effects of El Niño and La Niña events should also be studied with appropriate models.

Data on cyclone attributes in formation and strengthening stages are available since the 1950s CE for all oceans and seas. Data on landfalling TCs, however, go back to mid- or late 19th century CE. For the Atlantic region, these data show that the frequency and number of intense hurricane days vary at decadal–multidecadal timescales. Data on landfalling hurricanes along the US coast and along the Caribbean Sea coast also show such variability over a longer period of time. Easterly atmospheric waves coming off West Africa, when they arrive over tropical North Atlantic Ocean, develop into TCs and strengthen if SSTs and atmospheric conditions such as vertical shear of zonal winds in the so-called MDR are favorable. Tropical Atlantic SSTs undergo decadal and multidecadal variability (Chapter 4), and these SST variability phenomena modulate conditions in the MDR, thereby modulating the frequency of hurricanes, especially those of intense (category 4 and 5) hurricanes.

Analyses of TC frequencies in the West North Pacific Ocean, and the East and South China Seas also show that the frequencies vary at decadal–multidecadal timescales as does the frequency of landfalling TCs in Philippines. As in the case of Atlantic hurricanes, the decadal–multidecadal variability of the West North Pacific and the China Seas typhoons also shows physically consistent association with the PDO at decadal and longer timescales. The vertical shear of zonal wind, moist static energy, and lower atmospheric stability are modulated by the SST variability, and typhoon frequency is then affected by this modulation.

Decadal variations in the Bay of Bengal TC frequency are associated with large-scale atmospheric conditions, possibly modulated by the PDO via vertical wind shear, moist static energy, and low-level relative vorticity. TC frequency variability in the Arabian Sea has a significant spectral peak at 10–13 years which is the same timescale as that found in tropical Indian Ocean SST variability (Chapter 5). Further research is required to investigate possible physical connections between them.

Thus, it can be strongly concluded that there is decadal–multidecadal variability of TC attributes, especially those of intense TCs, in all oceans and major seas. Observational studies also show that large-scale SST variability in the tropical Atlantic and tropical Pacific Oceans modulates atmospheric conditions in these regions such that there are more TCs or more intense TCs in the positive phase of the SST variability and fewer TCs or fewer intense TCs in the negative phase of the SST variability. In the Bay of Bengal, the major influence on TC frequency is atmospheric conditions which are apparently modulated by the PDO. Local SST variability appears to play a minor role in variability of the Bay of Bengal TCs.

Observational studies, supported to some extent by modeling studies, show that SST variability modulates vertical shear of zonal winds, lower level cyclonic vorticity, lower level convergence and upper level divergence, and vertical profiles of atmospheric water vapor and temperature. This modulation of some or all of these attributes of the atmosphere in the TC development regions modulates cyclone frequency, especially that of intense hurricanes in the tropical Atlantic region where the MDR is relatively well defined compared to other oceans and seas. Also well defined in the tropical Atlantic case are the easterly atmospheric waves coming from Africa, some of which develop into tropical storms of varying intensity. In other oceans and seas, however, cyclone development regions are not as well defined, perhaps due to the sizes of the other oceans and seas, and atmospheric attributes in these regions are under the influences of not only local SSTs but also remote SSTs. One observational study cited in this chapter showed that warming SSTs in the tropical Indian Ocean can cause large-scale atmospheric subsidence over the South China Sea and thereby inhibit cyclone formation there. Low-level winds in the Western tropical North Pacific to the South China Sea region can also be modulated by zonal SST gradients in the Indo-Pacific Warm Pool region. These and other influences can act at various times to influence cyclone formation and evolution. Therefore, much observational and modeling research is needed to unravel these influences to understand and predict TC variability in these regions. There is some progress in quantifying TC variability in the tropical North Indian Ocean–Bay of Bengal–Arabian Sea region and in identifying possible roles of environmental variables such as local and remote SSTs and vertical shear of zonal winds. However, simulation studies with atmosphere-only and coupled ocean–atmosphere models are required to begin to unravel these and other possible influences. As mentioned several times in this chapter, the sheer destructive power of landfalling TCs and the need to develop adaptation options to cope with decadal–multidecadal variations in such destructive cyclones make it imperative to place modeling and prediction of TCs in all oceans and seas on an urgent footing. The research done so far is only the end of the beginning of the endeavor to understand and predict decadal–multidecadal variability of TCs.

10 Looking Through a Cloudy Crystal Ball

10.1 INTRODUCTION

"Prediction is very difficult, especially if it is about the future" – says a quotation attributed to Professor Niels Bohr, the Danish physicist who made fundamental contributions to atomic structure and quantum theory, and was awarded the Nobel Prize for Physics in 1922 Current Era (CE). If a physicist of Professor Bohr's understanding of nature expressed this profound truth, why has the humankind obsessed about predicting what might happen in the future? If repeated attempts to predict the future in many areas of natural and human-induced phenomena were always doomed to failure, why have scientists, economists, priests, and other specialists (and non-specialists as well) persisted in their efforts? The motivations are many – predictions' practical uses in testing theories in all disciplines of science, adapting to weather and climate phenomena for societal well-being, making efficient and sustainable uses of natural resources, investing money in stock markets and other financial sectors, and conducting religious rituals, to name a few. Predictions also play a very important role in understanding how natural and human-made systems work, and also how they interact. These and other motivations have driven scientists and non-scientists to attempt to predict natural phenomena and their purported effects on human societies for many centuries as mentioned in Chapter 1. Attempts to predict purported effects of solar and lunar cycles on the Earth's climate started many centuries ago as described in Chapter 2. We will pick up this story in the 20th century CE and follow it to the present time in this chapter.

Predictability and prediction are two words used mostly interchangeably. But, they are not synonyms, and it is very important to understand their true meanings before proceeding further. Predictability means capable of being predicted, and prediction means either the act of predicting or something that is predicted.[1] A phenomenon or event which is not capable of being predicted has no predictability, and therefore, its predictions would always be wrong or correct completely by chance, whereas some attribute or property of a phenomenon or event may be predictable, but its prediction would be correct with some accuracy only at certain times or only in certain circumstances. Predictability may depend on how well the forces underlying a phenomenon or event are understood and quantified in a mathematical or statistical model. Predictability may also depend on the complexity of the forces driving a dynamical system and its responses. According to the Chaos Theory propounded by Henri Poincaré in 1890 CE (Wolfram, 2002), nonlinear, deterministic dynamical systems may respond very differently to small changes in forces driving the system

[1] www.merriam-webster.com/dictionary.

or by small changes in initial conditions. Henri Poincaré proposed that meteorological phenomena may be sensitive to initial conditions and may be chaotic (Steves and Maciejewski, 2001). Meteorologist Edward Lorenz demonstrated in 1963 CE with a relatively simple model (Lorenz, 1963) that a nonlinear dynamical system representing the atmosphere can have non-periodic and chaotic solutions which can be sensitive to initial conditions of the dynamical system. The evolution of such a system would have very limited predictability. As we will see in the next section, Lorenz's research defined an upper bound for predictability of atmospheric motions; however, it is not obvious that actual predictions have even reached this upper bound on predictability; that is, predictions lose their skill even before the lead time of the prediction reaches the predictability limit.

On the other hand, there are dynamical systems whose evolution can be predicted far in advance. For example, the laws of physics governing orbital and rotational motions of the Sun, planets, and their major satellites are well understood, and their dynamics are quantified in equations of their motions. Even though the stability of these orbits is in doubt beyond a few tens of millions of years due to interactions among the orbits and responses to small changes in gravitational effects, these equations can be and are used to predict solar and lunar eclipses, and motions and locations of these solar system bodies in the interim period. These equations are also used to calculate exact positions of these bodies at any given time so that spacecrafts launched from the Earth can be steered to orbit or land at a precise location on a body in the solar system. Thus, the motions of these solar system bodies and spacecrafts can be predicted with very high accuracies. So, this is an example of very high predictability and very high accuracies of predictions.

In weather and climate prediction, the accuracy of a prediction is referred to as skill. There are many measures of prediction skill, two of the most popular being the correlation coefficient and the root-mean-square error (RMSE) between the observation and the prediction (see, for example, Murphy (1988), Wilks (2011)). According to these measures, the higher the prediction skill, the larger the correlation coefficient, and the smaller the RMSE. The appropriateness of a skill measure depends on the type of application of the prediction as we will see later in this chapter. Assuming that appropriate prediction skill measures are found/devised for various applications, there are many societal sectors where a decadal climate prediction can be very useful in decision-making processes. There are substantial impacts of decadal dry and wet epochs on transportation infrastructure, recreation, electricity generation, navigation on inland waterways, state and national legislation for water management and efficiency, urban water systems, crop production, hay and grass production, livestock, ecology, and the economy (Mehta, 2017). Intrastate and interstate conflicts can also arise about the use of water in transboundary rivers and reservoirs. Long-term decision-making, especially in infrastructure investments, the agricultural sector, and water and land management, would benefit greatly by 5 to 10-year climate outlooks. Prediction of correct phase (positive or negative) of average anomaly in precipitation and temperature, river flow, drought index, and other quantities over the next 2–10 years can be very useful for management decisions in water and agriculture sectors if the information is provided at the spatial resolution required for each sector (Mehta et al., 2013a; Mehta, 2017). The correct prediction of important decadal climate variability (DCV) phenomena one year in advance can be worth approximately $80 million per year to the agricultural economy of the Missouri River Basin (MRB) – a major breadbasket of the United States of America (USA) and the world; even the correct prediction of just the phase of the DCV phenomena one year in advance can realize a sizeable fraction of this monetary value (Fernandez et al., 2016). Therefore, accurate predictions of DCV phase transitions sustained for several months to a year or longer can be useful in the adaptation of worldwide agriculture and water resources to DCV-related hydro-meteorological conditions, with important consequences for water and food securities. Understanding and prediction of DCV phases is also important for attribution of DCV phase transitions to internal ocean–atmosphere processes or changes in external forcings as we will see later.

After this introduction, a brief history of weather and climate prediction and sources of potential decadal climate predictability are described in Section 10.2 as a prelude to reviews of empirical predictability and predictions in Section 10.3 and dynamical predictability and predictions in Section 10.4. A summary of this chapter and conclusions are presented in Section 10.5.

10.2 A CONTEMPORARY HISTORY OF WEATHER AND CLIMATE PREDICTION

Contemporary history is a record of events described by a witness (Mitford, 1808). It is subjective and not all-encompassing. No observer can record everything that happens in his/her life, but records – somewhat subjectively – what appears to be important and relevant to the narrative. In that spirit, a brief history of numerical weather and interannual climate prediction, along with important basic concepts, is described first in this section because generally similar models and initialization techniques are used so far in the fledgling decadal prediction efforts. Then, a brief introduction to potential sources of decadal predictability is given in this section before statistical and dynamical decadal predictability and prediction research is described in the next two sections.

10.2.1 Are Weather and Climate Predictable?

Portuguese sailors had recognized the importance of favorable winds in their travels from Europe to southern Africa, Asia, and the New World in the 15th century CE. More importantly, they knew that their sailing ships would reach home only if they found favorable winds on their return journeys. They named these winds *volta do mar* – meaning turn of the sea as well as return from the sea. Perhaps because the northeast to southwest winds in eastern North Atlantic charted a path or track of the ships going from western Europe to Africa and North and South America, the winds were named trade winds, with the word "trade" meaning path or track.[2] Later, as these trade winds were found to be important to all seafaring countries bordering the Atlantic and as the sailing ships plying the Atlantic were increasingly carrying raw materials or finished products, the phrase "trade winds" began to imply winds

[2] Random House Webster's College Dictionary (Second Edition).

favorable for commerce.[3] This sea-borne commerce provided an impetus to scientists to understand and predict trade winds. In that quest, Halley (1686) proposed that solar heating is the cause of atmospheric motions, specifically trade winds and monsoons, based on atmospheric observations during his expedition to and in St. Helena in the South Atlantic. Then, Hadley (1735) included another important force, namely, Earth's rotation, in his explanation of trade winds and what was later named in his honor as the Hadley Circulation. Although these two studies initiated the beginning of understanding the origin of trade winds and also of the general circulation of the atmosphere, the picture was still not complete until Ferrel (1858) propounded that the conservation of angular momentum via the Coriolis effect is a very important constraint on atmospheric motions. In retrospect, these three studies can be said to have placed the understanding and prediction of weather and climate on a solid, scientific base.

Then, a unique event occurred in the history of weather and climate sciences as the 20th century CE dawned – the establishment of the science of meteorology was proposed by Vilhelm Bjerknes in 1904 CE (Bjerknes, 1904). His pioneering paper proclaimed that the aim of this new science would be to predict the future states of the atmosphere from the present state. Bjerknes further stated that "if it is true, as every scientist believes, that subsequent atmospheric states develop from the preceding ones according to physical law, then it is apparent that the necessary and sufficient conditions for the rational solution of forecasting problems are the following:

1. A sufficiently accurate knowledge of the state of the atmosphere at the initial time.
2. A sufficiently accurate knowledge of the laws according to which one state of the atmosphere develops from another." (Lynch, 2004)

This manifesto can be said to be the first "roadmap" of numerical weather and climate modeling and forecasting, which has been generally followed to this day.

After V. Bjerknes's pioneering paper was published in 1904 CE, the British mathematician, physicist, and meteorologist L.F. Richardson was perhaps the first to propose that weather can be modeled with differential equations expressing the laws of physics governing motion, pressure, and temperature of air in the atmosphere (Richardson, 1922). Consistently with the Bjerknes conditions, Richardson also proposed methods to solve these equations at points on a grid spanning a finite area subject to specified boundary conditions. He envisaged an arrangement of human computers – literally, persons who calculate – sitting in a tiered auditorium, each computer representing one grid point (Figure 10.1). In Richardson's scheme, primary variables (velocity, pressure, temperature, humidity, etc.) in the discretized equations would be "initialized" – that is, given beginning values – with observed data at time t and then values of these (and other) variables would be calculated a short time increment later at $t+\Delta t$. Even though Richardson was aware of errors due to approximating a differential equation by an equivalent finite-difference equation, he was apparently unaware of a fictitious (computational) solution of the finite-difference

[3] Oxford English Dictionary (Second Edition).

FIGURE 10.1 L.F. Richardson's conceptual weather forecast laboratory. (Reproduced from Lynch, 2004; original from Bengtsson, 1984.)

equation which is not a physical solution. So, when he applied his method to make a 6-hour weather forecast on 20 May 1910, initializing variables measured at 7 AM, the forecast after 6 hours was for a dramatically large pressure increase due to the fictitious solution, whereas the actual pressure change was quite small. As Lynch (2004) showed, Richardson's 6-hour forecast predicted a very large pressure change because of the difficulty of calculating pressure tendency (change of pressure with respect to time) from small differences between large terms using unbalanced observed data. Lynch (2004) also showed that Richardson's time increment Δt was too large, resulting in the fictitious solution overwhelming the physical solution. If not for this fictitious solution, Richardson's forecast might have been close to being correct as Lynch showed. Although Richardson was not successful in making an accurate weather forecast, his assumptions of a set of differential equations governing the atmosphere behavior, boundary conditions, numerical techniques for obtaining solutions on a discretized grid, and the use of computers to obtain solutions set the template to be followed by subsequent generations of weather and climate forecasting efforts. Balanced or initialized data replaced raw observations to begin the forecast process, filtered equations to remove very large oscillations due to gravity waves were developed, and human computers were replaced by electronic computers in subsequent decades. Thus, Richardson's dream was finally realized (Lynch, 2006).

But, we have run ahead of the history, so let us go back to the 1910s CE. Weather forecasting then consisted entirely of making maps of available wind, pressure, and temperature data from a relatively small number of weather stations, and then looking for analogues in past maps to project or extrapolate the weather forward by a few hours or a day or two to make a forecast. Due to a relatively small observations network and the divergence in weather trajectories from almost the same patterns, this technique was not very successful. As the number of weather observing stations increased during the First World War (1914-1918 CE), there was a marginal improvement in forecasting, especially because observations and bombing by fledgling combat flying services, poison gas attack and defense, and trench warfare in inclement weather demanded more accurate weather forecasts over battlefields.[4] Then came L.F. Richardson's revolutionary but unsuccessful experiment to make a 6-hour forecast numerically, and it was obvious to meteorologists that better balanced initial data, governing equations which did not have rapidly growing solutions, and much faster computing technology were required before a successful attempt at numerical weather prediction could be made. So, in the 1920s and 1930s CE, meteorologists continued with their forecasting-by-analogue techniques. A noteworthy development during this period was the mathematical derivation of the condition which decided if a solution of a discretized system of time-dependent partial differential equations with specified grid resolution and time step would be stable or not (Courant et al., 1928).

To augment the weather-observing networks during the Second World War (1939-1945 CE), all sides deployed ship-based and land-based observing stations. Since the weather in North Atlantic and Europe usually traveled from west to east, Germany, the United Kingdom (UK), and the USA deployed surface ships and submarines in the western and northwestern Atlantic Ocean to make weather observations and transmit the data to their respective military weather organizations. Perhaps the most famous weather forecast in history was made by Allied meteorologists for the D-Day landings in France on 6 June 1944 CE, using the analogue technique and weather observations in North Atlantic and elsewhere in the Western Hemisphere. The Allied meteorological team forecast a break in stormy weather for two days beginning 6 June. On the other side, German military meteorologists forecast strong storms in the English Channel till mid-June. The Supreme Allied Commander, US General Dwight D. Eisenhower, decided to launch the D-Day invasion, and the rest is history! On the other side, the German military was lulled by the forecast made by their meteorologists and was not on high alert when the invasion began. There is an apocryphal story about this difference in forecasts. While riding to the Capitol on 21 January 1961 CE for his inauguration, President-elect John F. Kennedy is supposed to have asked President Eisenhower why the Normandy invasion had succeeded. President Eisenhower is supposed to have replied, "Because we had better meteorologists than the Germans"! Whatever the authenticity of this story, four developments occurred during the Second World War which gave a major boost to numerical

[4] By analogy to battle fronts between armies shown on military maps during the First World War, the weather front (generally, between cold and warm air masses) became a staple of weather maps subsequently to this day.

weather forecasting efforts stagnant after L.F. Richardson's unsuccessful attempt. These developments were (1) installation and operation of a rapidly growing, worldwide weather observing network of regional and national networks; (2) advances in communication technologies; (3) development of electronic computers; and (4) employment of a very large number of meteorologists, mathematicians, physicists, and other scientists in operational weather forecasting for war-time requirements. The last gave deep insights to the forecasters into weather evolutions over the entire world and into how physical solutions of dynamical equations used by Richardson and others could relate to weather phenomena.

In late 1940s CE, the Meteorology Research Group at the Institute for Advanced Studies in Princeton, New Jersey, USA, began to plan a systematic approach to the problem of numerical weather forecasting. The Group was founded in 1948 CE by John von Neumann, who was a great polymath and whose contributions to many fields of mathematics, sciences, computers, nuclear weapons, and geopolitical strategy are too numerous to mention here[5] (McRae, 1999). Jules Charney, a physicist who was the first to explain the role of long baroclinic waves in atmospheric motions, was appointed to head the Group. Charney had simplified differential equations of motion with the quasi-geostrophic approximation applicable to the relationship between pressure and winds in the extratropics. He had also assumed that the extratropical atmosphere was mainly barotropic on the large scale, justifying the use of a barotropic, quasi-geostrophic system of equations to launch a new, numerical weather forecasting effort (Charney et al., 1950). So, the team consisting of Charney, Ragnar Fjörtoft, and von Neumann simplified the barotropic, quasi-geostrophic system of equations to one vorticity equation and discretized it as a finite-difference equation on a grid spanning approximately 10°N to 70°N latitudes and 120°W to 55°W longitudes. The grid spacing and time step were decided based on the Courant et al. (1928) stability condition. The team was then instructed by Mrs. Klara Dan von Neumann, who was a computer scientist and one of the first computer programmers, to program this discretized system for the Electronic Numerical Integrator and Computer (ENIAC) (Charney et al., 1950). This first, post-Richardson numerical weather prediction system was used to produce four, 24-hour, vorticity forecasts initialized with observed data at 500 hPa level and were compared with observations. Charney et al. (1950) described the results and discussed possible causes of differences between the forecasts and observations. Thus, the road to modern weather and climate forecasting was opened!

One of the mathematicians who worked in the US military for weather forecasting during the Second World War was Edward Lorenz. After the War, Lorenz earned Master's and Ph.D. degrees in meteorology at the Massachusetts Institute of Technology (MIT). While working at MIT on numerical weather prediction with relatively simple but nonlinear mathematical models, Lorenz discovered that very small changes in initial data appeared to make very large differences in trajectories of weather evolution predicted by these models. He explored this phenomenon, first proposed by Henri Poincaré in 1890 CE as mentioned in Section 10.1, giving birth to the science of Chaos Theory – also known as "the butterfly effect" (Palmer, 2008).

---

[5] www.ias.edu/von-neumann

In a landmark paper, Lorenz (1963) illustrated his discovery more formally with a relatively simple, three-component system of nonlinear, ordinary differential equations modeling a convective process which may be applicable to the atmosphere. This system of equations was found to have non-periodic solutions which were unstable with respect to small perturbations such that small differences in initial states could evolve to very different states. With the aid of phase space diagrams depicting evolutions of the three components, Lorenz illustrated the non-periodic and unstable nature of the solutions. Then, extrapolating to the actual atmosphere and its predictability, Lorenz (1963) suggested the possibility that very long range prediction of atmospheric flow may be impossible unless the initial (that is, present) state of the atmosphere is known exactly which may be impossible in view of observational inaccuracies and incompleteness. This and other follow-on research established a limit to weather predictability, with the actual predictability limit dependent on the scale of a weather phenomenon – for example, a few hours for the cumulus scale, a few days for the synoptic scale, and up to a few weeks for the longest scale (Lorenz, 1969; Shukla, 1981).

This initial exploration of dynamical weather predictability and the computational aspects of making weather forecasts with numerical-dynamical models generated insights and hope that it may be possible to make semi-operational or operational forecasts in real time which can be used for various applications. In the next step up from the Charney et al. (1950) formulation in model complexity came a two-layer model of the atmosphere by Phillips (1951, 1956) which simulated/predicted major features of the atmospheric general circulation reasonably realistically. Following these model and computational developments, operational numerical weather prediction centers were founded in Europe and the USA as described by Harper et al. (2007) on the 50th anniversary of operational numerical weather prediction.

As outlined in the beginning of Section 10.2.1, the foregoing history of numerical weather prediction led to the next milestone in the journey towards climate prediction. This part of the journey was initiated by a meteorologist from India – Jagadish Shukla – who was very passionate about predicting the Indian monsoon rainfall for societal benefits, following in the footsteps of Sir Gilbert Walker – an early Director-General of the India Meteorology Department (see Chapter 3). After graduate studies in India, Shukla arrived in MIT in the early 1970s CE to work with Charney, Lorenz, and other stalwarts of atmospheric dynamics and numerical weather prediction research. Undaunted by the apparent limit to weather predictability as found by Lorenz and others, Shukla (1981) argued that this limit applied only to synoptic-scale weather disturbances on timescales of a few days, but perhaps not to predictability of monthly averages which depended on predictability of low-frequency planetary scale waves. He showed with numerical prediction experiments using a global general circulation model with fixed boundary conditions (such as sea surface temperatures, soil moisture, snow, and sea ice) that the evolution of long planetary waves remained sufficiently predictable at least up to 30 days and perhaps up to 45 days so that monthly averages can also be predicted for these durations. This was a major development and had the effect of removing an insurmountable block on the road towards climate predictability. Then, at approximately the same time, Charney and Shukla (1981) also realized that the tropical atmospheric dynamics were linear (or at least much less nonlinear than extratropical atmospheric dynamics) and so the tropical predictability should not suffer from the same limit as imposed on extratropical predictability by nonlinear dynamics. Charney and Shukla (1981) hypothesized that due to this simpler linear dynamics and due to slowly varying sea surface temperatures (SSTs) of the tropical oceans, the tropical monsoons and other tropical weather phenomena should have much longer-term predictability. Shukla (1996) later demonstrated with a global general circulation model that even for very large differences in initial atmospheric conditions, the final wind and rainfall solutions in the tropics did not diverge substantially if the same SST field was prescribed. These results implied that there did not appear to be any sensitivity of monthly to seasonal tropical climate prediction to initial conditions. This was in marked contrast to extratropical weather as described earlier. These results also showed that the extratropical weather may also benefit from an enhanced predictability associated with large SST anomalies in the tropical Pacific Ocean.

While the ideas about weather and climate predictability based on internal atmospheric dynamics were evolving from the mid-1940s to the mid-1990s CE, a stream of thought was also gathering momentum from another direction. As described in Chapter 3, the El Niño and Southern Oscillation phenomena in the tropical Pacific region were motivating many oceanographers and meteorologists to understand possible causes of these phenomena. The possibility that these two phenomena were the oceanic and atmospheric components, respectively, of the same coupled ocean–atmosphere phenomenon El Niño–Southern Oscillation (ENSO) was turning into high probability. Simulation experiments with non-coupled, idealized ocean and atmosphere models strongly suggested that this was indeed the case and the first fledgling coupled ocean–atmosphere models emerged in the mid-1980s CE. Among pioneers of this line of thought were a student–teacher pair in the MIT – Stephan Zebiak and Mark Cane. They developed an idealized coupled model of the tropical Pacific Ocean–atmosphere system, and used the model to study El Niño-like events and understand their physics (Zebiak and Cane, 1987). This was one of the most consequential Ph.D. dissertations in the history of climate science (Zebiak, 1984). The roles of equatorial Kelvin and Rossby waves in the Pacific Ocean, previously hypothesized and studied in non-coupled models by other researchers, were clarified in the framework of the coupled system. This clarity also identified the source of a quantum leap in climate predictability – namely, these two types of waves which acted as the repository of the memory of earlier ocean–atmosphere interactions in the equatorial Pacific region. This model was then used to make experimental predictions of El Niño events from 1970 to early 1980s CE, and such events were found to be predictable one to two years in advance (Cane et al., 1986). It was nearly 40 years since the first numerical forecast with the barotropic vorticity equation to this moment when truly multiseason, even multiyear, predictability appeared possible. Along the way, the identification of atmospheric planetary waves as the repository of memory over 30 to 45 days and the linear dynamics of the tropical atmosphere and their response to slowly varying underlying SSTs were also very important milestones in this journey. The demonstration of multiseason ENSO prediction skill by the Cane et al. (1986) and other follow-on studies to clarify the role of the annual cycle of tropical climate in this skill generated substantial excitement and led to the founding of many modeling groups around the world for seasonal to interannual prediction.

Last, but not only not the least but perhaps the most important, aspect of all timescales from weather prediction to climate prediction is how the predictable signal is extracted. We have seen that the chaotic nature of the atmosphere in particular and the climate system in general generates unpredictable noise beyond a certain point in time even though there may be a predictable signal associated with synoptic or planetary wave scale weather, slowly varying tropical SSTs, or oceanic Rossby and Kelvin waves. If trajectories from initial states diverge after some time due to chaotic dynamics, how can the predictable signal – however small it may be – be extracted? This question has vexed weather and climate forecasters for over 50 years since Lorenz's pioneering work on chaotic atmospheric dynamics described in Section 10.2.1. Enter a method known as ensembles – which can be defined as a group producing a single effect or emphasizing the roles of all performers as a whole rather than a star performer.[6] Both definitions are apt when applied to the concept of ensembles applied to numerical weather or climate prediction. The origin of this application appears to have been lost in the mists of time, but Shukla (1981) had employed a multi-member ensemble to extract a monthly average atmospheric signal. A major effort to apply this method to short- to medium-range and seasonal weather prediction was begun by Timothy Palmer and associates in the 1980s CE at the UK Meteorological Office and later at the European Centre for Medium-Range Weather Forecasts (ECMWF) (see, for example, Palmer (2018) and references therein for a history of ensemble prediction). Although there are several ensemble-based methods to extract a predictable signal, the application of the basic method is simply to average multiple realizations started from significantly different initial conditions to extract the predictable signal that may be common to some or all of the so-called ensemble members. Molteni et al. (1996), and Toth and Kalnay (1997) have described applications of ensemble methods to medium-range weather prediction at ECMWF and the US National Centers for Environmental Prediction (NCEP), respectively.

While ensemble methods became routine and widespread in the 1990s CE in numerical weather prediction, the development of more detailed, increasing resolution models of the global atmosphere, and increasing computing capabilities spurred the search for understanding and predictability of seasonal and longer-term climate, first with "perfect" oceans and then with dynamical, coupled ocean–atmosphere models. In "perfect" ocean experiments, histories of observed SSTs were specified, and reproducibility and predictability of atmospheric phenomena was studied. Several studies of the role of global SSTs, especially Atlantic SSTs, in variability and potential predictability of the North Atlantic Oscillation (NAO) began to find that intrinsic, chaotic variability of the mid-latitude atmosphere obscured the SST-forced NAO signal and that applications of the ensemble approach were needed to extract the potentially predictable signal (see, for example, Rodwell et al. (1999), Mehta et al. (2000a), and Bretherton and Battisti (2000)). These and other such studies initiated research in how to improve potentially predictable climate signal to unpredictable

[6] www.merriam-webster.com/dictionary/ensemble.

noise ratios and how large ensemble sizes should be; the approaches developed for seasonal climate prediction have now carried over to decadal climate prediction with Earth System Models (ESMs), especially at the UK Meteorological Office – Hadley Centre for Climate Science and Services, the international leader in decadal climate prediction (see, for example, Scaife et al. (2014), Eade et al. (2014), Smith et al. (2019)). The importance of a multi-member ensemble for decadal climate prediction has been recognized and the Coupled Model Intercomparison Project (CMIP) 5 and CMIP6 have required participating modeling groups to carry out ensembles of decadal hindcasts and forecasts as a condition of participation. As we will see in Section 10.4.2, ensemble sizes for decadal prediction are now becoming seriously large. However, it is also obvious that future progress in extracting whatever potential decadal predictability there might be will depend substantially on improving signal-to-noise ratios by increasing ensemble sizes along with improving ESMs and initialization systems.

10.2.2 Potential Sources of Decadal Climate Predictability

ENSO and global warming due to increasing greenhouse gases (GHGs)-dominated climate science as the 1990s CE dawned. Then, as described in earlier chapters, gradual releases of ocean data archives and hypotheses of decadal and longer timescale climate variability extrapolated from coupled ocean–atmosphere interactions in ENSO physics resulted in the re-birth of natural DCV research in the early 1990s CE. It was believed that understanding the physics of natural DCV was important not only for possibly predicting it and its societal impacts, but also for attribution and prediction/projection of anthropogenic climate change – how this belief has come true! This newly emerging field was "jump-started" and pushed along to its present state by a series of workshops from 1992 to 2014 CE when the first multi-model, but still experimental, decadal predictions began to emerge from CMIP5. Reports and journal articles describing proceedings of these workshops make fascinating reading and give a very vivid account of the progress of this field (CRC, 1995; Mehta and Coughlan 1998; Mehta et al., 2000b, 2006, 2011; Meehl et al., 2009, 2014; Rosenberg et al., 2007). Physics of DCV and potential sources of decadal predictability were at the heart of the discussions in these and other workshops and conferences from the beginning.

We saw in Section 10.2.1 that there must be sources of weather and climate predictability which can be harnessed to some extent for actual prediction. For daily weather, the sources are synoptic scale waves; for monthly average weather, the sources are planetary scale waves; and for seasonal to interannual climate, the sources are oceanic equatorial Rossby and Kelvin waves, their interactions with the atmosphere, and the coupling between the equatorial thermocline and trade winds in the eastern equatorial oceanic regions. But, what does decadal predictability mean? As we saw in Chapter 2, ever since the Schwabe cycle began to be used as a climate predictor in the early 19th century, the assumption (and hope!) was that Sunspots were somehow causing climate variability and that the 11-year cycle can be used for year-to-year, or at the most multiyear, climate prediction. As an aside, average life expectancy at birth was 30 to 40 years and so one Schwabe cycle or a decade was approximately 30% of a human being's entire life; or, conversely, an average human life was approximately three Schwabe cycles long. In comparison, depending on the country, average life expectancy today ranges from approximately 60 years to 80+ years which span five to seven Schwabe cycles. Therefore, long term had a much shorter-term meaning in the early 19th century than it has now. In contemporary parlance, decadal predictability continues to imply multiyear predictability, but it also implies other, more nuanced interpretations. These are (1) seasonal or annual predictability due to decadal and longer timescale phenomena such as the Pacific Decadal Oscillation (PDO); (2) predictability at seasonal or annual (even monthly) resolutions over one decade, including over one cycle of a decadal oscillation; (3) average predictability over the better part of a decade or an entire decade; and (4) average predictability from one decade to the next decade. Predictability and prediction studies have addressed all of these interpretations of decadal predictability as we will see in the next two sections.

Why do we expect any predictability at multiyear to decadal and longer timescales? Are there possible sources of predictability at these timescales? We saw in Chapter 2 that the Schwabe and Hale cycles of solar variability were believed to be a source of climate predictability, and were used to make predictions for over two centuries. Then, there is the oceanic integration of atmospheric noise, which can generate a red noise spectrum and provide a source of predictability due to persistence as we saw in Chapters 3 and 4. We also saw in Chapter 3 that off-equatorial and tropical–subtropical oceanic Rossby waves, perturbations in shallow tropical ocean circulations, perturbations in subtropical gyre circulations, and interactions of all of these phenomena with the atmosphere have been hypothesized to be important components of tropical Pacific and Atlantic decadal variability. Due to the slow speeds of these waves and perturbations, and their interactions with the atmosphere, they can be potential sources of decadal predictability. As we move to timescales longer than a few years, changes in external forcings also become increasingly important. Principal external forcings are atmospheric constituents, and solar emissions and galactic cosmic rays as we saw in Chapter 2. Atmospheric constituents can change due to anthropogenic emissions such as carbon dioxide (CO_2), methane, and other so-called GHGs; and due to gaseous and particulate matter injected into the atmosphere–ocean system by volcanic eruptions and anthropogenic activities. These external forcings can also be potential sources of predictability. Thus, as we go from weather prediction at timescales of hours to days, to climate prediction at multidecadal to century timescales, the prediction problem evolves from initial-value problem to forced boundary value problem as schematically shown in Figure 10.2. The decadal prediction problem, for all interpretations of decadal predictability, is at the overlap between the two as the mixed initial-value and forced boundary value problem. Exactly how long the time span of this overlap is and at what rate the initial-value problem tends toward the forced boundary-value problem is an evolving topic of research and is described later in this chapter. Suffice it to say that there are numerous potential sources of multiyear to decadal and longer timescale predictability as mentioned here. The questions then are: (1) What are the limits on skills and lead times of decadal predictability? (2) Is it possible to realize any of the potential predictability? (3) If yes, how much can be realized in terms of skill and lead time?

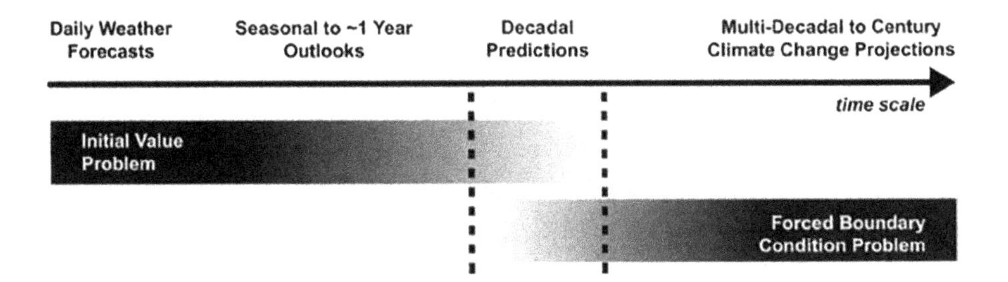

FIGURE 10.2 Foretelling the future: From initial-value prediction problem at hourly to weekly timescales to forced boundary value projection problem at multidecadal to century timescales. Note that the two types of problems overlap at the decadal timescale. (Adapted from Meehl et al., 2009 © American Meteorological Society. Used with permission.)

(4) What are the best tools to realize whatever prediction skill there may be? These questions are addressed in the next two sections.

10.3 EMPIRICAL PREDICTABILITY AND PREDICTION

After familiarizing ourselves with a brief history of numerical weather and interannual climate prediction, we now turn our attention to encouraging efforts to estimate decadal climate predictability and to realize whatever predictability there may be with empirical and dynamical modeling methods in this and the following sections. Both classes of methods have various approaches to estimate predictability, and both sets of methods have positive and negative aspects to them. Empirical methods use observed climate data, mainly SST, surface air (or, 2 m) temperature, and sea level pressure (SLP), so the predictability estimated with these methods and data have some relevance to the actual climate system. However, relatively short time series of these observed data, compared to timescales of decadal variability and predictability, suffer from shortcomings of inadequate sampling and representativeness. On the other hand, there are very long (hundreds or even thousands of years) ensemble runs of state-of-the-art ESMs,[7] so the time series of model SST, SLP, and other variables and integrated quantities (such as total heat content in the upper several hundred meters of the ocean) are very long compared to the timescales of interest. However, these models have substantial shortcomings in simulating the actual climate variability, especially spatiotemporal patterns of major DCV phenomena, so the representativeness and usefulness of predictability estimates made with these models may be of limited applicability to the actual climate system. Nevertheless, both sets of methods applied to respective data sets are revealing very interesting characteristics of decadal climate predictability and so both are very useful as idealizations of whatever the actual decadal climate predictability may be. There are also efforts to combine empirical and ESM data to exploit positive aspects of both types. In the next two sections, results from both sets of methods and also some examples of actual predictions are described and discussed.

[7] Here, ESM refers to any global, coupled ocean–atmosphere–land model that is used to simulate/predict/project climate.

10.3.1 Estimation of Empirical Predictability

Estimates of predictability using empirical data can be called empirical predictability. There are various methods for estimating empirical predictability of any data time series. These methods are used widely, not only in weather and climate prediction but also in estimating predictability of stock market prices, and of water and electricity consumptions, to name a few applications. Three types of empirical climate predictability methods and their results are described here: (1) linear multivariate methods; (2) nonlinear local Lyapunov exponent method; and (3) persistence-based, phase transition probabilities of climate indices. In these descriptions, the emphasis is on predictability of SST and SLP fields, and indices of major patterns of DCV. The three types of methods are described very generally and briefly here since the goal is to see what applications of these methods say about decadal predictability. The reader is referred to cited literature for details of methods and data.

10.3.1.1 Linear Multivariate Methods

There are many examples of empirical weather and climate predictability studies and their applications, too numerous and not very relevant to mention here, for actual predictions. Generally in such studies, some type of a statistical model is fitted to past observed data of a variable, and then, the model is used for future prediction. There are methods for application to a univariate time series (such as a climate index time series) and for application to multivariate time series (such as time series on a spatial grid). Also, it is believed that a significant peak in the spectrum of a time series implies predictability of the process giving rise to the time series over the entire preferred oscillation period (for example, skill over years 1 to 10 if there is a significant spectral peak at 10-year oscillation period). But, the climate system is multivariate and nonlinear, and prediction errors build up faster than the oscillation period (also damping time is typically shorter), so that a spectral peak does not guarantee predictability over the entire period. Two linear methods developed for estimating climate predictability and applied to decadal climate predictability are briefly described, and their results are compared.

In Chapters 3 and 4, we saw the hypothesis that the ocean can integrate high-frequency atmospheric noise and generate univariate red noise due to its large thermal inertia (Hasselmann, 1976). The actual climate system can generate many interacting patterns of variability at various space and timescales. An extension of the univariate red noise hypothesis to the actual climate system can be called multivariate red noise (Newman, 2007). In this hypothesis, multivariate red noise can have both stationary and traveling patterns with transient growth, which can interact and produce spectral peaks in area-averaged or empirically derived climate indices. Newman (2007, 2013) applied the Linear Inverse Model (LIM; Penland and Sardeshmukh, 1995) method to estimate interannual to decadal climate predictability, assuming that climate variability is generated by one or more multivariate red noise processes. The LIM method attempts to empirically extract the linear dynamical system from simultaneous and time-lag covariance statistics of the system variables (Newman, 2007, 2013; and references therein). Therefore, Newman (2013) called the nonlinear climate system, approximated by multivariate red noise, as predictably linear.

The LIM system can be formulated as a multivariate, nonlinear time-tendency equation for a state vector **x** equaling a linear stable and predictable linear dynamics term and an unpredictable white noise term. The linear dynamics term consists of a linear matrix **L** operating on the state vector **x**. The noise term consists of outcomes of all rapidly de-correlating nonlinear processes as well as atmospheric variability. Among interesting properties of **L** is that its eigenvalues can be complex, implying propagating eigenvectors with defined oscillation periods. Also, even though the real parts of L's eigenvalues are generally negative, implying damped oscillations, **L** can be non-normal and so its eigenvectors can be non-orthogonal and interacting. This last property of **L** implies transient growth of perturbations. **L** and its associated operators can be determined from observed data by lagged-covariance matrices as described by Newman (2007). Then, the LIM model can be used for predictions. Prediction skill (expressed as average anomaly correlation) and RMSE can be calculated using the LIM predictions and corresponding verification data.

In applications of the LIM method to estimate decadal predictability of surface temperatures, Newman (2013) used gridded SSTs and 2 m surface air temperatures on land from 1901 to 2009 CE to calculate the required LIM operators. Then, these LIM operators were used to hindcast SST and 2 m temperature anomalies from 1960 to 2000 CE and prediction skills for various lead times were calculated. The LIM prediction skills were compared with prediction skills based on persistence and with decadal hindcast skills of three CMIP5 ESMs for the same period. The three ESMs are UK Meteorological Office – Hadley Centre HadCM3 Decadal Prediction System (identified as DePreSys), Max Planck Institute ESM – Low Resolution (identified as MPI), and Geophysical Fluid Dynamics Laboratory (GFDL) CM2.1. Comparisons of anomaly correlations are shown in Figure 10.3 for skills in two- to five-year lead time and six- to nine-year lead time. The comparisons indicate that there were pronounced similarities in geographical variations of skill between LIM and CMIP5 ESM hindcasts, implying perhaps that both sets of skills have similar sources. It was also found that LIM skills at two- to five-year and six- to nine-year lead times were comparable to or better than the three CMIP5 ESMs' hindcast skills, and much larger than damped persistence skills. It is noteworthy that the highest skills in the Indo-Pacific Oceans are in the East Indian Warm Pool (EIWP) and West Pacific Warm Pool (WPWP) regions where the SSTs are some of the warmest in the world and whose interannual to decadal and longer timescale variability is associated with climate variability in many parts of the world. Relatively high skills in the North Atlantic are also encouraging because multidecadal variability of SSTs in this region is also associated with worldwide climate variability. An index of the North Atlantic SSTs' multidecadal variability, known as the Atlantic Multidecadal Oscillation (AMO), also has high prediction skill in this comparison as described later.

Prediction skills of the PDO are generally low as shown in Figure 10.4a. This is not surprising given the low prediction skills in almost the entire Pacific region except the West Pacific as mentioned above. LIM's PDO hindcast skills were higher than the three CMIP5 ESMs for two-year lead time, but damped persistence had higher skills than all of them. An intriguing feature of the LIM's PDO hindcast skill was an increasing skill after two-year lead times, reaching almost back to its one-year lead time skill at nine years. Figure 10.4d shows that RMSEs are generally

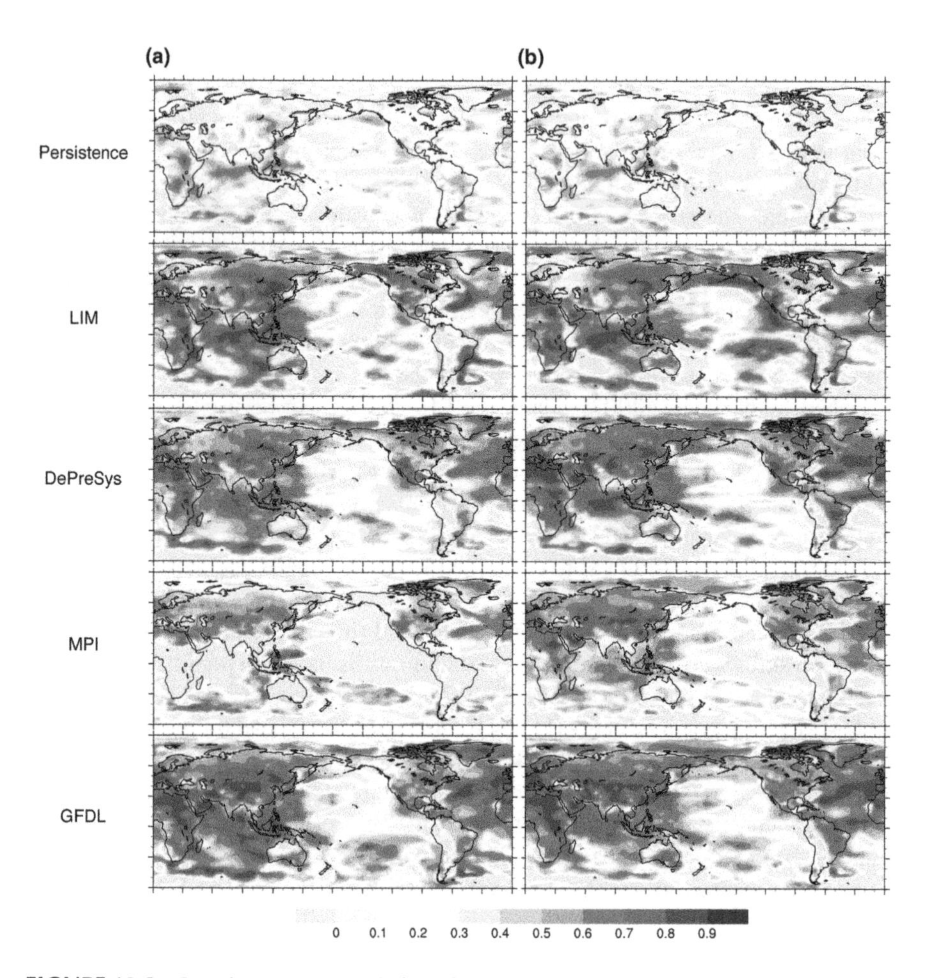

FIGURE 10.3 Local anomaly correlation of hindcasts averaged over lead times of (a) two to five years and (b) six to nine years, for damped persistence, the LIM, and the CMIP5 models; hindcasts were initialized annually from 1960 to 2000 CE. Contour interval is 0.1 with negative values indicated by blue shading. Shading of positive values starts at 0.1; warmer shading denotes larger values of correlation. (From Newman, 2013 © American Meteorological Society. Used with permission.)

constant at all lead times except for the MPI and GFDL ESMs. LIM's AMO correlation skill (Figure 10.4b) is the highest among all models and generally increases with lead time as do the RMSEs (Figure 10.4e). A comparison of hindcasts of global-average surface temperatures (Figure 10.4c) shows that prediction skills are high and nearly constant for all except the GFDL ESM, while the RMSEs increase with lead time (Figure 10.4f). Thus, Newman (2013) showed that the empirical, LIM-based decadal predictability skill is at least as high as that of three of the ESMs participating in the decadal hindcast component of CMIP5. The PDO's relatively low empirical predictability beyond two-year lead time is consistent with other studies as we will see later in this and next sections.

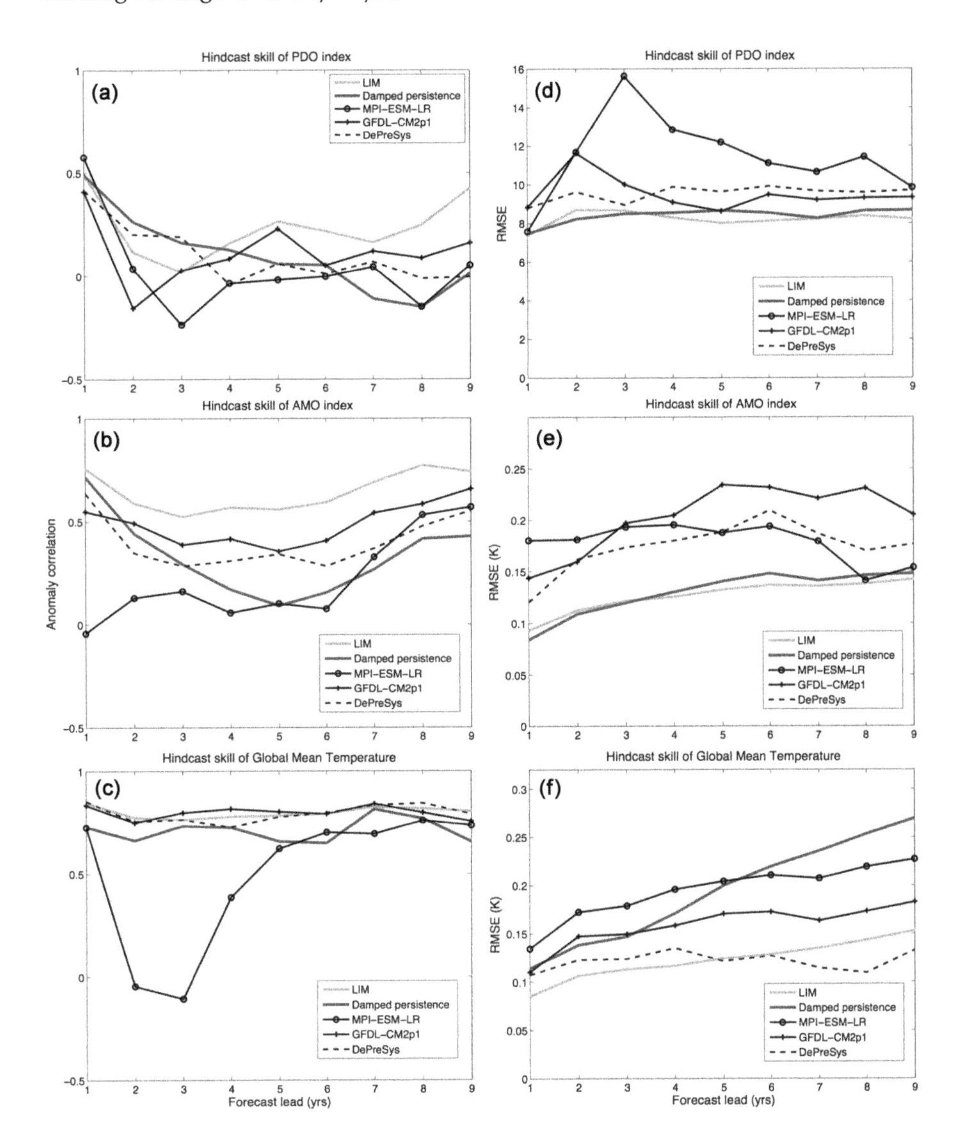

FIGURE 10.4 (left) Anomaly correlation skill comparison for (a) PDO and (b) AMO indices, and (c) global-average temperature, from hindcasts initialized in the 1960–2000 CE. Using the lag-1 autocorrelation to approximately estimate degrees of freedom, a value of correlation coefficient; (0.4, 0.55, and 0.55) is 95% significant (different from 0) for (a)–(c), respectively. (right) As in (left), but using RMSE as the skill measure for (d) PDO, (e) AMO, and (f) global-average temperature. Damped persistence is determined from the lag-1 autocorrelation of the observed index time series, not from local gridded values (as in Figure 10.3), which would yield lower index skill. (From Newman, 2013 © American Meteorological Society. Used with permission.)

The LIM-based predictability studies showed the lead time dependence of traditional measures of predictability such as anomaly correlation and RMSE while exploiting the multivariate red noise concept and statistically explaining how multiple spatiotemporal variability patterns can sometimes generate the appearance of stationary and traveling anomalies and can give rise to spectral peaks. As mentioned earlier, the LIM-based method also suffers from a small number of samples of decadal variability and very few samples of multidecadal variability due to the relatively short observed data time series. Therefore, can a measure of decadal predictability applicable to multivariate systems and independent of lead time be developed? What if the most predictable space–time structures can be identified in multicentury ESM simulations, modeled with a statistical method, and then used in conjunction with observed data to make predictions and estimate predictability? Would such an approach increase overall predictability? Also, what is the relationship among oscillation period, damping time, and prediction skill? Timothy DelSole and associates (DelSole and Tippett, 2008, 2009a, b; DelSole et al., 2013) addressed these questions with a strategy to define a new measure of predictability, called the average predictability time (APT), which would be applicable to a multivariate system and which would clarify the dependence of predictability on damping time and preferred oscillation period. DelSole and Tippett (2009a) showed that predictability is limited by the damping time rather than by the oscillation period. They then applied their APT formulation to estimate decadal predictability of gridded temperatures and temperature indices of decadal and longer timescale climate phenomena.

We saw in Section 10.2.2 that predictability at decadal timescale involves an overlap between unforced initial-value and forced boundary-value predictability problems. DelSole et al. (2013) separated the two problems and estimated each contribution to total predictability. They formulated the unforced initial-value problem as a linear regression system. Since the observed SST record is relatively short (as mentioned many times in this book), regression coefficients for the prediction system were estimated from pre-industrial control and 20th century climate simulations with eight CMIP5 models. Data from last 300 years of each control run and 35 ensemble members of 20th century runs were used to estimate the regression coefficients after re-gridding all data to the same 5° longitude – 5° latitude grid. The prediction system was applied to gridded data as well as to PDO and AMO indices. Following DelSole and Tippett (2009a), prediction skill was defined as 1 minus the ratio of mean-square prediction error and climatological variance; the skill equals 1 for a perfect forecast and is negative if the mean-square error is larger than climatological variance. Skills dependent on lead time for area-average SSTs during 1910–2004 CE in the North Atlantic and North Pacific (averaged from 37.5°N to 72.5°N latitudes in both) are shown in Figure 10.5. Global-average skill is also shown in Figure 10.5. As the figures show, the prediction skill of the unforced variability is positive in the first five years. However, the skill is only slightly larger than persistence forecast skill in the North Atlantic at most lead times; there is larger than persistence skill in the North Pacific, but only at shorter lead times as Figure 10.5 shows. It is noteworthy that the skill of persistence forecast decreases much more rapidly with increasing lead time in the North Pacific than in the North Atlantic. This difference in skills is also reflected in the skills of the AMO and the PDO indices;

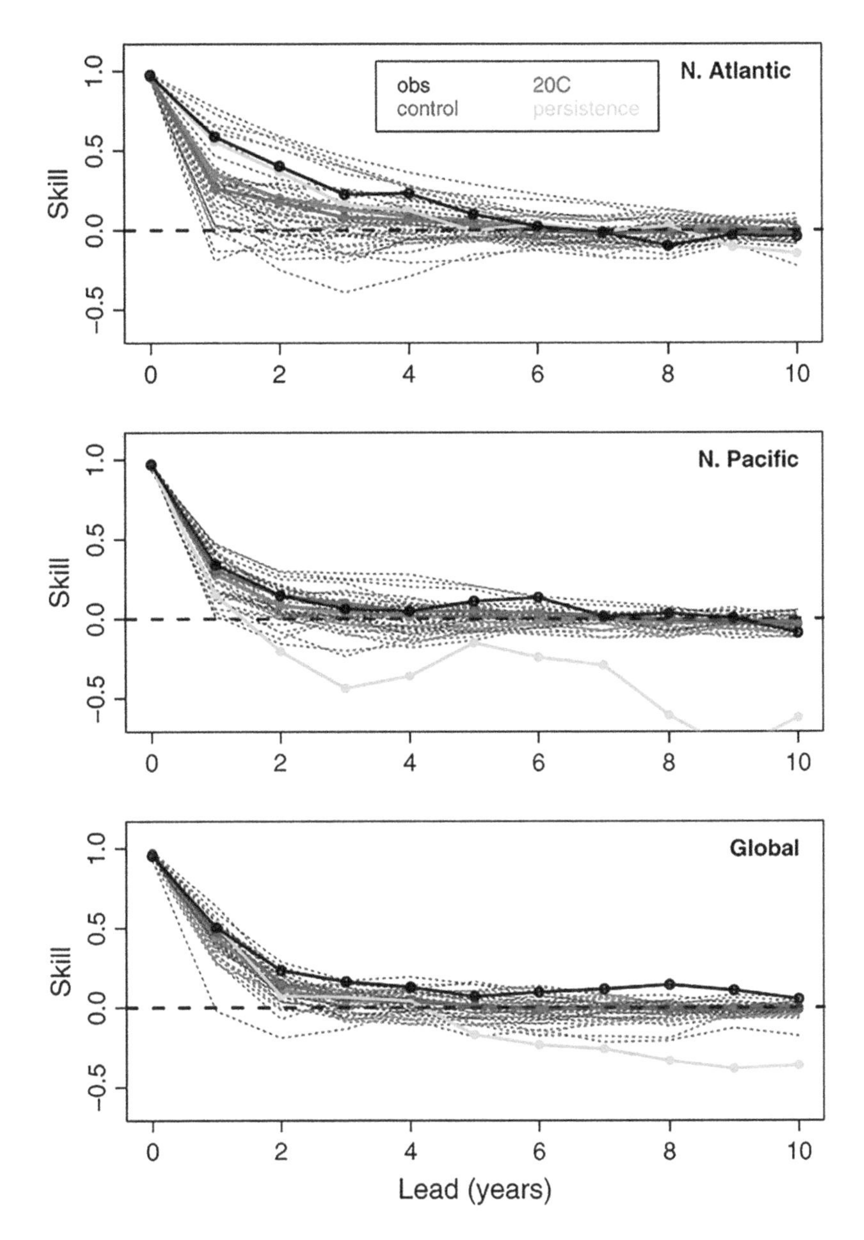

FIGURE 10.5 The skill of a multivariate regression model in predicting the unforced component of observed, annual-average SST during 1910–2004 CE (black) in the North Atlantic, North Pacific, and global domains. Also shown is the skill of the same regression model in predicting temperature anomalies in single realizations of the control simulations (blue) and the unforced component of temperature anomalies in single realizations of the 20th century runs (red). The corresponding skill of a persistence model, in which a prediction of the index equals the initial value of the index, is shown as the thick green curve with dots. Thick curves with dots show the skill over all realizations of the control (blue) and 20th century (red) runs. (From DelSole et al., 2013.)

the PDO index's skill decreases rapidly in the first two years and is positive but very small in subsequent years. It is possible that the AMO's longer dominant timescale, mentioned in Chapter 4, and the presence of higher frequency variability in the PDO index are responsible for this difference. DelSole et al. (2013) estimated prediction skill of the forced boundary-value component from the multi-model ensemble average. Since the forced component skill is from a model simulation, it is constant with respect to lead time as shown in Figure 10.6 which also shows the sum of the two skill components. It is very clear that the forced component of the total skill in the

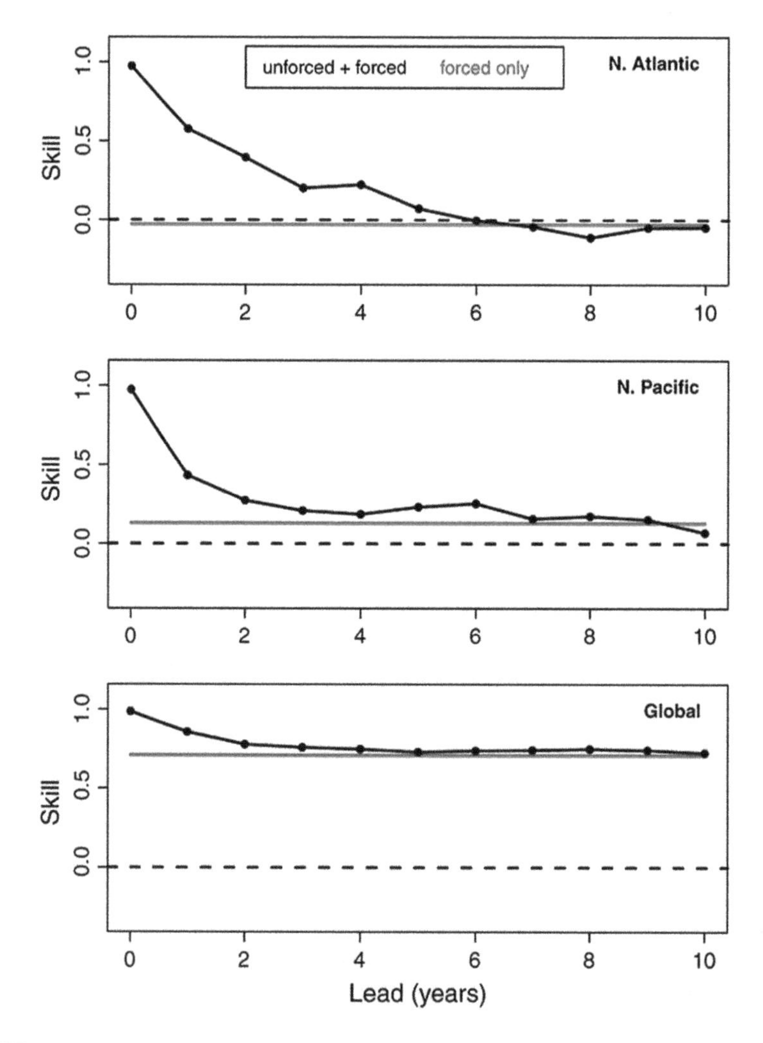

FIGURE 10.6 The skill of predicting observed annual mean SSTs during 1910–2004 CE using an estimate of the forced component derived from 35 20th-century simulations from eight CMIP5 models (red) and using this estimate of the forced component and a multivariate regression model to predict unforced variability (black). The black dashed line denotes zero skill. (From DelSole et al., 2013.)

North Atlantic has no skill, leaving the initial-value skill as the only contributor to total skill. The forced component is small but positive in the North Pacific and is almost as large as the initial-value component skill of global-average SST forecast.

10.3.1.2 Nonlinear Local Lyapunov Exponent

We saw in Section 10.2.1 that two infinitesimally close trajectories of a nonlinear dynamical system can diverge and this can lead to loss of predictability. A quantity that characterizes the rate of separation of such trajectories is known as the Lyapunov exponent (LE) (Boeing, 2016). Trajectories converge at an exponential rate if this exponent is negative and there is a bifurcation if the exponent is zero. A positive exponent is an indicator of a chaotic system whose trajectories are very sensitive to initial conditions and diverge at the rate of the largest Lyapunov exponent. The future behavior of such systems may have very limited or no predictability. Therefore, LEs are estimated as an indicator of predictability limit (Ding and Li, 2007).

The largest LE is defined as the long-term average growth rate of an infinitesimally small initial error. This is also referred to as the global LE (Ding and Li, 2007). Local LEs estimate short-term growth rates of infinitesimally small initial errors. According to Ding and Li (2007), the local LEs characterize non-uniform spatial organization and indicate predictability variations on chaotic attractors. Estimates of the local LEs are based on linear error dynamics because they assume infinitesimally small initial errors. If the errors are large enough to invalidate the application of local LE formulations, they cannot be used to estimate predictability and new formulations of local LEs should be defined for behaviors of nonlinear dynamical systems. Ruiquaing Ding and associates formulated a nonlinear local LE (NLLE) and applied it to estimate predictability of weather and climate variability at various timescales, including the Madden–Julian Oscillation and decadal climate (Ding et al., 2008, 2010, 2011, 2016). The NLLE formulation can estimate predictability limits over various timescales by quantifying the evolution of the distance between initially local dynamical analogs (LDAs) from observed data.

The NLLE formulation and its computational aspects are described by Ding et al. (2016), and their application of the NLLE method to worldwide SST and SLP data, and to indices of the PDO, the AMO, and the Northern and Southern Annular modes (NAM and SAM, respectively) to estimate decadal predictability limits is described here. In these estimates, 95% of error saturation level was defined as the predictability limit. Extended Reconstructed Sea Surface Temperature (ERSST) version 3 data from 1854 to 2011 CE, extended Kaplan SST data from 1856 to 2011 CE, HadSLP data from 1850 to 2011 CE, and reconstructed SLP over eastern North Atlantic and Europe from 1659 to 1999 CE were used to estimate potential predictability. In addition to these gridded data, PDO and AMO indices from the ERSST data, and NAM and SAM indices from the HadSLP data were calculated and used. Linear trends and the average annual cycles were calculated from all data time series and were removed to calculate SST, SLP, and index anomalies. To remove high-frequency noise from the anomaly time series, a 9-year Gaussian filter was applied, and the NLLE method was applied to the low-pass filtered time series.

The predictability limit at grid points is shown in Figure 10.7. In the filtered, worldwide SST data, depending on location, the predictability limit ranges from

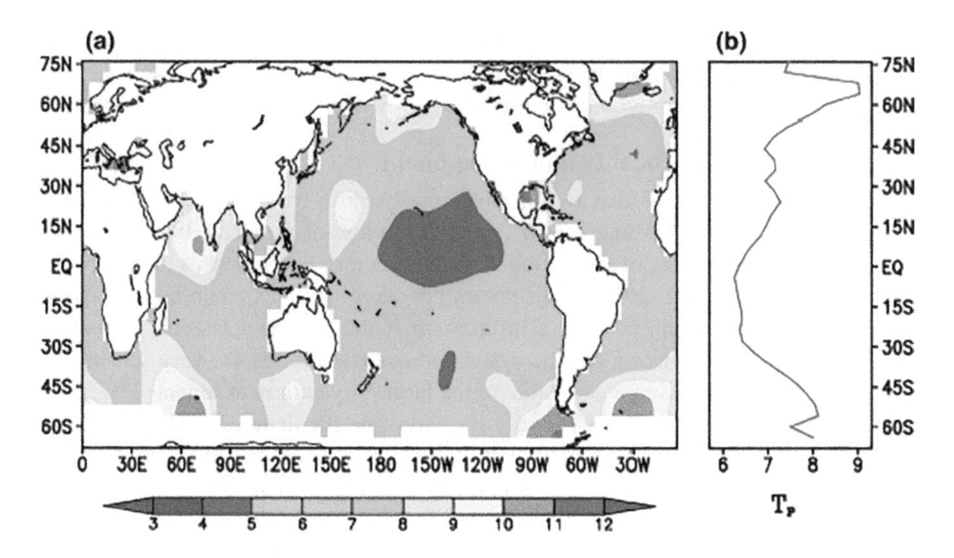

FIGURE 10.7 (a) Spatial distribution of the predictability limit (T_P, in years) of the 9-year low-pass-filtered SST and (b) its zonal-average profile based on the ERSST data set. (From: Ding et al., 2016.)

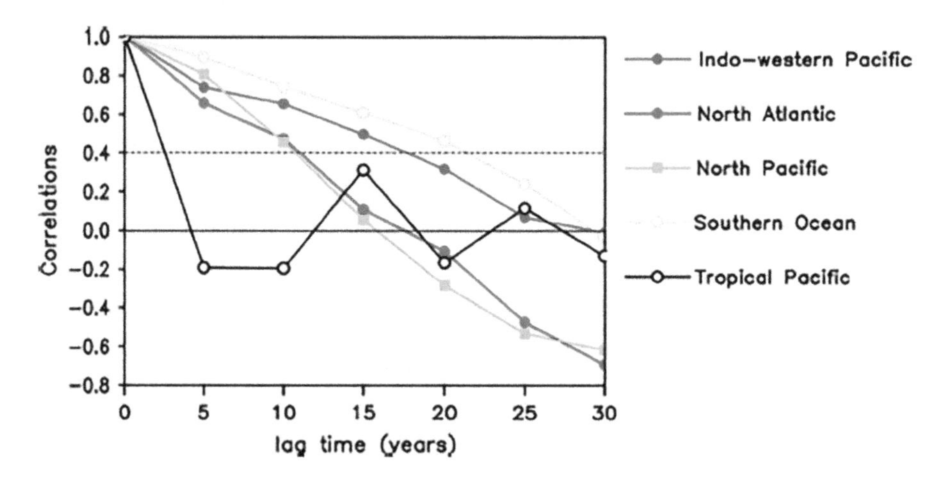

FIGURE 10.8 Autocorrelations of 5-year average SST averaged over the Indo-western Pacific (15°S to 25°N, 40° to 140°E), North Atlantic (20° to 65°N, 90° to 0°W), North Pacific (20° to 60°N, 120°E to 120°W), Southern Ocean (65° to 40°S, 0° to 360°E), and tropical Pacific (5°S to 5°N, 180° to 85°W) as a function of lag time. The horizontal dashed line indicates the 95% significance level. (From: Ding et al., 2016.)

4 years to 12 years as Figure 10.7 shows. The limit is longer than seven years in the North Atlantic Ocean, the North Pacific Ocean, the Southern Oceans, the tropical Indian Ocean, and the tropical western Pacific Ocean (Figure 10.8). The limit is shorter (four to six years) in tropical central-eastern Pacific Ocean. It is very interesting to note that the predictability limit is longer in regions of stronger

decadal variability. Ding et al. (2016) also found that lagged autocorrelations of the same SST data are consistent with the predictability limits shown in Figure 10.8. Predictability limits of the low-pass-filtered PDO and AMO indices were found to be 9 years and 11 years, respectively, as shown by the average error growths in Figure 10.9. The limits of both indices were found to be consistent with lagged autocorrelations. To compare with predictability of low-pass filtered random noise, Ding et al. (2016) applied the NLLE and Monte Carlo methods to random data having the same spectral characteristics as the actual PDO and AMO indices. In these simulations, they found that the predictability limit of low-pass filtered noise was approximately three years, implying that the predictability limits of the actual PDO and AMO indices were due to decadal variability generated by physical processes. Although Ding et al. (2016) found that the predictability limits of the SLP data were generally similar to those of the SST data, the predictability limits of NAM and

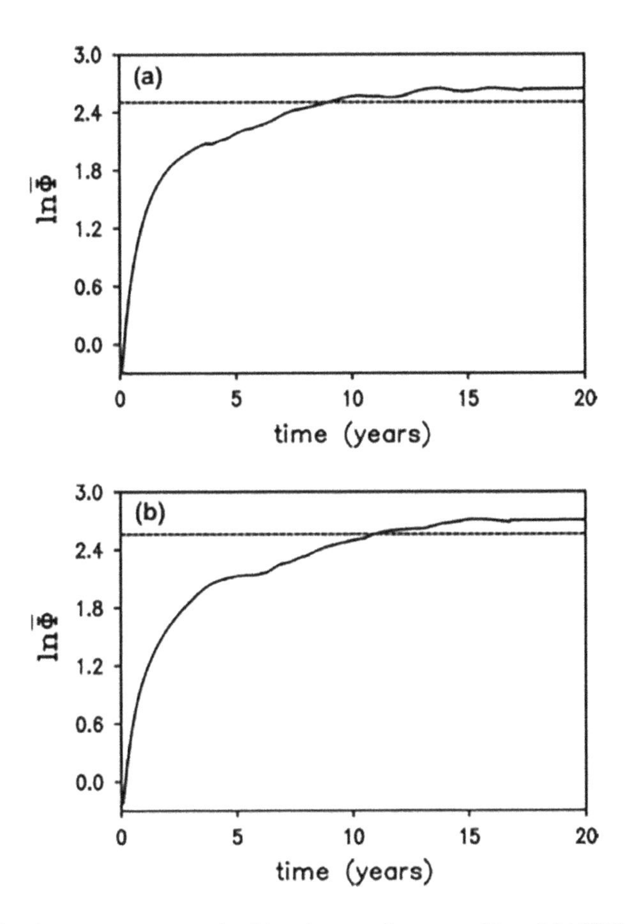

FIGURE 10.9 Average error growth of the nine-year low-pass-filtered (a) PDO and (b) AMO indices, obtained using the NLLE method. The dashed line represents the 95% level of the saturation value obtained by taking the average of the average error growth after 15 years. (From Ding et al., 2016.)

SAM indices were found to be shorter and mainly attributable to low-pass filtering. Thus, the NLLE analyses of predictability limits of SST data and derived indices showed that the empirical predictability limits appear to be at least seven years, perhaps as long as 9 to 11 years, in regions of strong decadal variability. These estimates also showed that the predictability limit of SSTs in the tropical Indo-Pacific Oceans is at least as long as that in extratropical oceans. This has an important potential for decadal predictability of climate on continents because tropical SST variability at decadal timescales is associated with climate variability on continents (see, for example, Mehta (2017) and references therein). It must be emphasized that, as explained in the introduction, potential predictability estimates may only provide an upper limit on the actual predictability and prediction skill.

10.3.1.3 Phase Transition Probabilities

Another approach was evolved from the point of view of users of decadal climate information—farmers, water managers, and other stakeholders and policymakers. Associations between positive and negative DCV phases; and dry/wet epochs, and changes in river flows and crop yields are known reasonably well via analyses of empirical data and via experiments with numerical models of the global atmosphere. Therefore, data and information such as phase (positive or negative) of average anomaly in precipitation and temperature, river flow, drought index, and other quantities over the next 2–10 years can be very useful for management decisions in water and agriculture sectors (Mehta et al., 2013a; Mehta, 2017). A study of the value of decadal climate information to the agriculture sector in the MRB – the largest river basin in the USA and a major "bread basket" of the world – with a water and crop choices model showed that the correct prediction of the PDO and other important DCV phenomena one year in advance can be worth approximately $80 million per year (Fernandez et al., 2016). This study also showed that the correct prediction of even the phase of the DCV phenomena one year in advance can realize $29 million per year. Therefore, accurate predictions of DCV phase transitions sustained for several months to a year or longer can be useful in the adaptation of worldwide agriculture and water resources to DCV-related hydro-meteorological conditions, with important consequences for water and food securities. Since decision processes in these sectors utilize probabilistic information, accurate predictions of DCV phase transition probabilities would be very useful to these sectors. Understanding and prediction of DCV phases is also important for attribution of DCV phase transitions to internal ocean–atmosphere processes or changes in external forcings.

Therefore, Mehta et al. (2019) estimated phase transition probabilities of the PDO, the tropical Atlantic SST gradient (TAG), and the WPWP indices using ERSST data from 1961 to 2010 CE. Annual and three-month average (December–January–February (DJF), March–April–May (MAM), June–July–August (JJA), and September–October–November (SON)) indices calculated from the ERSST data were used to estimate occurrence frequencies of each phase and phase transition probabilities, as a percentage of total number of years, were calculated. If it is assumed that positive and negative phases of a DCV index over a multidecadal period have equal probabilities of occurring, then the average occurrence of each phase would be 50%. The actual phase occurrences in annual data are almost 50%, with small departures

(±6% for the PDO index, ±2% each for the TAG and the WPWP indices) from the expected occurrences. Occurrences of the two phases in three-month averages are almost the same as the occurrences in the annual data. Next, transition probabilities between positive and negative phases of each DCV index were found. Phase transition probabilities for annual data are shown in Figure 10.10. For the PDO phases, there is an overwhelming tendency for same-phase transitions, or persistence, of the PDO from one year to the next. TAG phases are less persistence, and their transition

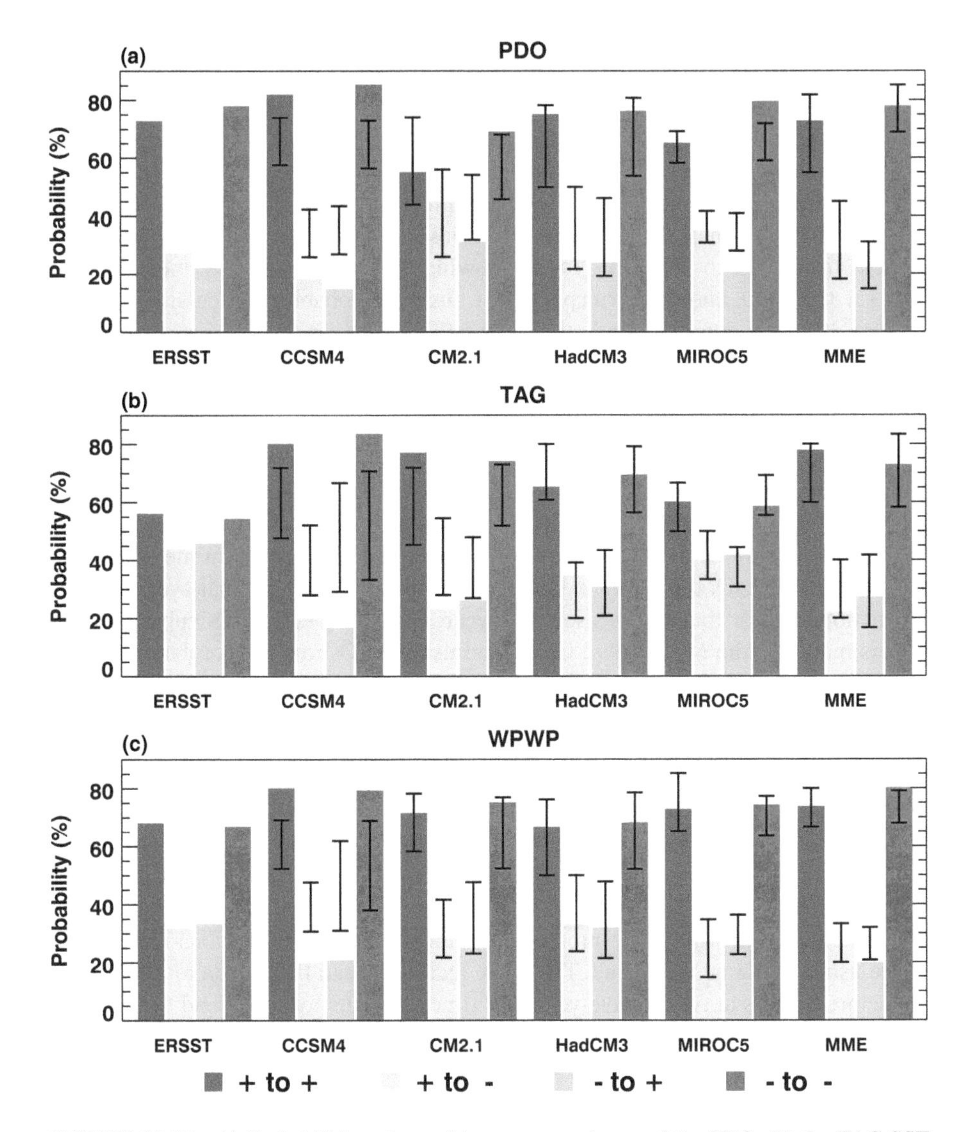

FIGURE 10.10 (a) Probabilities of transitions among phases of the PDO, (b) the TAG SST variability, and (c) the WPWP SST variability from 1961 to 2010 CE in annual ERSST data. From: Mehta et al., 2019, Figure licensed under Creative Commons Attribution (https://creativecommons.org/licenses/by/4.0/).

probabilities are almost equal, although same-phase transitions have higher probabilities. Figure 10.10 also shows that, as for the PDO and TAG phases, same-phase transition probabilities of phases of the WPWP index are much higher compared to the opposite-phase transition probabilities. These phase transition probabilities can be used for actual prediction of phases of the DCV indices and associated other variables as shown in Section 10.3.2.

10.3.2 Empirical Prediction

After establishing a predictability limit with empirical data, the next step is to develop a statistical model and prediction scheme to make actual predictions and compare them with independent data to verify the predictions and estimate the prediction accuracy. Some examples of empirical predictions are given here.

In a prediction scheme based on the DCV phase transition probabilities described in Section 10.3.1, phase transition probabilities are calculated using observed DCV indices from 1951 to 1990 CE. Then, knowing the phase of a DCV index in 1991 CE, say the PDO, and the corresponding transition probability calculated from the 1951 to 1990 CE data, the prediction of the PDO phase in 1992 CE is made. Next, transition probabilities are calculated using observed data from 1951 to 1991 CE, and the PDO phase in 1993 CE is predicted based on the PDO phase in 1992 CE. In this way, the prediction scheme marches forward and makes predictions up to 2018 CE. The phase predictions from 1992 to 2018 CE are then verified against observed phase in the corresponding year. These verifications of prediction accuracy against independent data (not used in calculating phase transition probabilities) are shown in Table 10.1 for annual-average and seasonal-average, one-year lead time phase predictions of the PDO, the TAG, and the WPWP indices. The seasonal-average phase predictions are for the same season next year; for example, the PDO phase prediction is made for the MAM 1992 CE period using PDO transition probabilities calculated with MAM data from 1951 to 1990 CE and the MAM 1991 CE PDO phase. Table 10.1 shows that the highest accuracy (81% to 82%) is for the prediction of the PDO's annual and JJA phases, and is almost the same for positive and negatives phases. The accuracies are smaller in other seasons, with the smallest accuracies in the DJF season. Table 10.1 also shows that positive PDO phase is more predictable than negative phase in MAM and SON seasons. In comparison, there are large variations in prediction accuracies of the TAG index as Table 10.1 shows. In all averaging periods, however, the accuracies of the positive TAG phase prediction are higher than the negative TAG phase prediction. Multiyear phase predictions can also be made using the same technique. The same technique can be applied to make DCV phase predictions at two- to five-year lead times. Results for these lead times, using annual-average data, are also shown in Table 10.1.

Based on these probabilistic DCV phase predictions and composites of anomalous Standardized Precipitation and Evapotranspiration Index (SPEI; Vicente-Serrano et al., 2010) for each combination of phases of the PDO and the TAG indices, Mendoza and Mehta (2020) have generated worldwide dryness–wetness outlooks at one to four seasons and one- to five-year lead times. In these experimental outlooks, dryness and wetness are defined as below- and above-average values of SPEI at each grid point of

TABLE 10.1
Accuracy (% of total number of years) of prediction phases of PDO and TAG variability from 1992 to 2018 CE

Period	PDO		TAG	
	Positive	Negative	Positive	Negative
December–January–February	54	64	94	0
March–April–May	73	67	77	36
June–July–August	81	82	50	47
September–October–November	75	47	75	29
Annual – Year 1	81	82	47	33
Annual – Year 2	63	64	60	50
Annual – Year 3	44	46	53	33
Annual – Year 4	75	73	100	0
Annual – Year 5	75	9	87	25

the 0.5° longitude – 0.5° latitude global grid. Hindcasts of the dryness–wetness outlooks are tested against observed SPEI data to estimate accuracies of outlooks at one season to five-year lead times. The probabilistic DCV phase predictions and dryness–wetness outlooks, along with verified past predictions, are available publicly.[8]

10.4 DYNAMICAL PREDICTABILITY AND PREDICTION

10.4.1 Estimation of Dynamical Predictability

In Section 10.2.1, we saw pioneering research to study weather predictability and what limits weather predictability. We also saw that longer-term sources of predictability such as planetary scale atmospheric waves, and oceanic Rossby and Kelvin waves may provide predictability of monthly-average and seasonal to multiseason climate predictability, respectively. As the need and, consequently, the search for multiyear to decadal and longer-term climate predictability increased, attention turned to using multidecades to century long runs of ESMs to estimate decadal predictability. But, what should be the measures of ESM-based decadal climate predictability and how should they be estimated? Also, as Figure 10.2 shows, prediction becomes a mixed initial- and forced boundary-value problem in the multiyear to decadal lead time range, and then becomes a forced boundary-value problem at longer lead times. We will now see what measures have been devised and what estimates of decadal predictability have been made from a variety of ESMs.

George Boer made pioneering contributions in defining measures of dynamical predictability at multiyear to multidecadal timescales and then in estimating these measures using data from long runs of ESMs. Boer (2000) adopted two approaches in defining measures of predictability – diagnostic and prognostic. In the diagnostic

[8] crces.org/past-present-and-future/the-future/.

approach, long simulations with ESMs are analyzed to estimate predictability of time-averaged variables such as surface air temperature. There can be two types of diagnostic predictability estimates: (1) analog predictability, which estimates the separation rate of observed states which are initially close and where initial differences are assumed to be errors in initial conditions; and (2) potential predictability, which estimates the part of observed or modeled variability which is not noise and which is assumed to arise from physical processes which are potentially predictable. In the prognostic approach, predictability is estimated by the rate of separation of trajectories of time-average variables in ESM forecasts. In this approach also, there can be two types of predictability estimates – perfect model predictability and practical predictability – both of which are derived from numerical and/or statistical models to make forecasts in the presence of error in initial condition and/or model. The error growth in forecasts by perfect or imperfect models gives information on the two types of predictability depending on whether the model is perfect or not.

Using these definitions, conceptual formulations of quantitative estimates, and successively larger/longer/multi-model ensembles of simulations with ESMs, Boer (2000, 2004, 2009, 2011) and Boer and Lambert (2008) estimated worldwide potential predictability of surface air temperature and precipitation. Recognizing the overlap between initial and forced boundary value problems at multiyear to multidecadal lead times, Boer (2009, 2011) also estimated potential predictability due to oceanic initial conditions and that due to increasing GHGs, and overlapping ranges of predictability associated with the two sources. After the initial test of predictability formulations, Boer (2004) applied them to surface air temperature data from 11 ESMs in the World Climate Research Program (WCRP)'s CMIP1 to estimate potential predictability of 5-, 10-, and 25-year averages of surface air temperature. Potential predictability was found predominantly over high-latitude oceans and to a lesser extent over the tropical Atlantic and tropical Pacific Oceans. Potential predictability of surface air temperatures was small over land areas. Interestingly, it was found that potential predictability oscillated in time over tropical oceans compared to extratropical oceans, implying the presence of quasi-oscillatory variability over tropical oceans and inconsistency with the hypothesis of the climate variability being generated solely by simple oceanic damping of atmospheric variability. Boer (2004) interpreted these results as suggesting that potential predictability in these ESM simulations was associated with long timescale oceanic variability in the extratropics and with coupled ocean–atmosphere processes in the tropics.

As the CMIP projects gathered steam, an increasing number of ESMs began to participate in these projects and, consequently, sample sizes for statistical studies of simulated climate and its predictability also increased. Taking advantage of the increasing number of ESMs, Boer and Lambert (2008) analyzed 8000 years of surface air temperature and precipitation data simulated by 21 ESMs in CMIP3. Again, the focus was on estimating potential predictability of 5-, 10-, and 25-year averages of these two model variables. As with results from the much smaller sample size in Boer (2000), Boer and Lambert (2008) found that there was significant potential predictability mainly over the extratropical oceans and predominantly where the ocean surface was connected to deeper ocean, implying that the predictability was due to long timescale memory of sub-surface oceans. They also found that there was a weak predictability of surface air temperature over land in areas bordering the oceans in the Northern Hemisphere, but there was practically no predictability of precipitation. The authors added a caveat that these estimates were those of long-term average predictability over the model runs and that potential predictability could be higher in specific situations.

In parallel with these developments in estimating multiyear to multidecadal predictability, the idea began to strike root that the overlapping nature of initial-value (predictability of the first kind) and forced boundary condition (predictability of the second kind) complexity at these timescales (Figure 10.2) required that there should be predictability estimates due to each individually as well as the two collectively. Collins and Allen (2002) were perhaps the first researchers to begin to explore this complex problem systematically. They argued that predictability of the first kind can be estimated by measuring the ensemble spread at increasing lead time against the spread expected from random sampling of climatological noise in an ensemble of climate model simulations with small perturbations to the initial conditions. They also argued that predictability of the second kind can be estimated by measuring an ensemble's average of data from experiments with changing radiative forcing and comparing the average with some estimate of climatological internal variability. They formulated both predictability measures essentially as ratios of predictable variance to noise (or background) variance (see Collins and Allen (2002) for the mathematics of these formulations). An application of these formulations to nine experiments with the HadCM2 ESM was carried out. Four experiments with different ocean initial conditions but increasing GHGs and sulfate aerosols from 1860 to 1996 CE were conducted. Four other experiments with fixed ocean initial conditions from the previous set and a small radiative perturbation to atmosphere were also conducted. A 200-year control experiment with fixed concentrations of GHGs and sulfate aerosols was also conducted to estimate background internal variability.

Measures of predictability, based on analyses of data from these nine experiments, are shown in Figures 10.11 and 10.12. I, a measure of predictable variance (or signal-to-noise ratio) due to a perfect knowledge of initial conditions, is shown in Figure 10.11. I equals 1 for a perfect forecast and is 0 when the ensemble spread equals climatological spread in magnitude. Figure 10.11 shows that global-average temperature may be predictable at approximately two years' lead time (when $I \leq 0$), North Atlantic SSTs may be predictable up to a decade in advance, Niño3 SSTs may be predictable up to one to two years with seasonal modulation, and Northern Hemisphere land temperature may not be predictable beyond approximately one month as Figure 10.11 shows. β, a measure of predictable variance (or signal-to-noise ratio) due to changing boundary conditions, is shown as a function of time for global-average and Northern Hemisphere land temperatures, and North Atlantic and Niño3 SSTs in Figure 10.12. In response to increasing radiative forcing, global-average temperature signal emerges from noise in approximately one decade and then continues to show warming in Collins and Allen (2002)'s experiments. The Northern Hemisphere land temperature shows a somewhat similar behavior, but with winter being predictable whereas summer is not (perhaps due to snow-albedo feedback) as

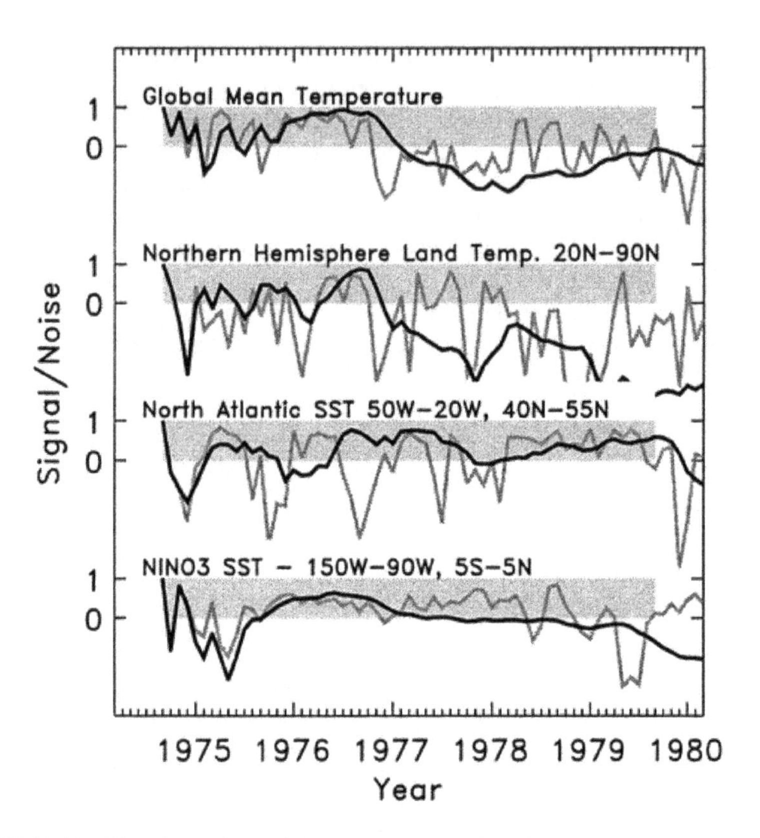

FIGURE 10.11 Signal-to-noise ratio I, as a measure of the initial condition predictability, which is 1 for a perfect forecast and 0 when the ensemble spread is of the same magnitude as the climatological spread (as indicated by the shaded region). Curves are shown for four different temperature indices. The black curve is I computed using linear trends, and the gray curve is that computed by taking simple differences in the monthly average temperature. (From Collins and Allen, 2002 © American Meteorological Society. Used with permission.)

indicated by a net change diagnosis (Collins and Allen, 2002). Interestingly, while these large-area averages show some predictability of the second kind due to an averaging-induced increase in β, there is no such predictability for the North Atlantic or Niño3 SSTs. Collins and Allen (2002) argued that these land-ocean differences in predictability may also be because of smaller heat capacity of land compared to that of oceans, and may also be because much of the excess radiative forcing may drive increased evaporation from oceans rather than increased temperature. Thus, as these estimates of potential predictabilities of the first and second kind show, both initial conditions and changing boundary conditions impact overall predictability, especially at multiyear to multidecadal lead times. This conclusion also implies that ocean initial conditions must be known precisely at any given time for initializing climate prediction models and the evolution of external forcings must be known well in advance for realizing the potential predictability. Both requirements present daunting challenges and impose limits on realizing potential predictability.

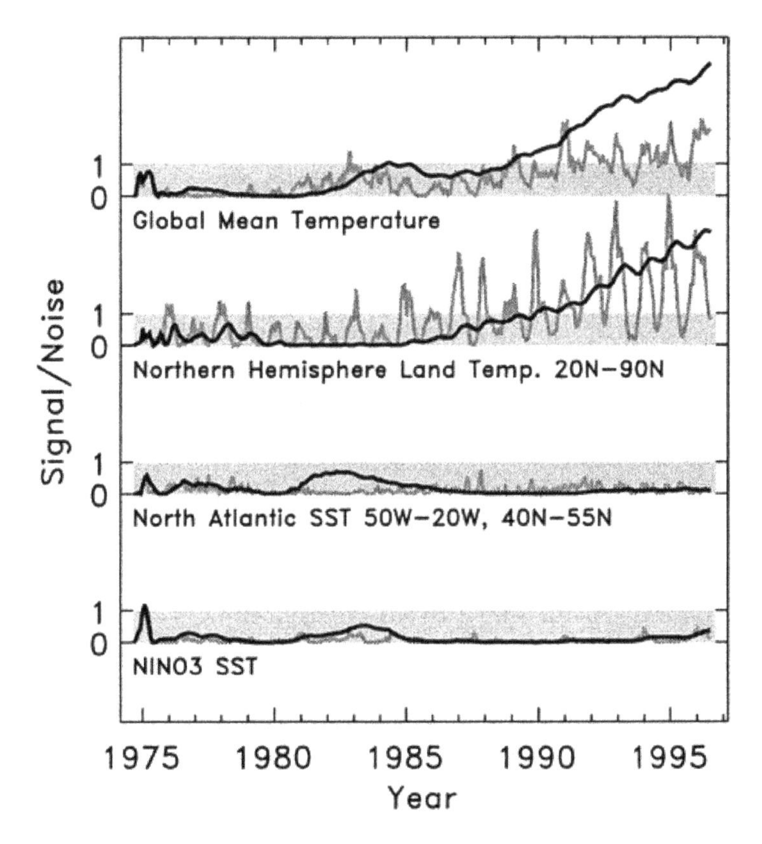

FIGURE 10.12 Signal-to-noise ratio β, a measure of predictability due to changing boundary conditions, which is 0 when there is no significant response and greater than 1 when the response is significantly greater than climatological noise (as indicated by the shaded region). Curves are shown for four different temperature indices as indicated. The black curve is β computed using linear trends over the forecast period, and the gray curve is that computed by taking a simple temperature difference between the initial and final months. (From Collins and Allen, 2002 © American Meteorological Society. Used with permission.)

This line of exploration of the two kinds of multiyear to multidecadal climate predictability attracted an increasing number of researchers as ESMs evolved, computational capabilities of numerous modeling groups increased, and the CMIPs gathered momentum. Also, some new approaches and some modified old approaches were developed to attack these twin peaks of climate predictability. Among the first to develop new approaches were Branstator and Teng (2010) who developed formulations based on relative entropy estimates from information theory (Kleeman, 2002) to quantify the two kinds of predictability in 40-member climate change experiments with the NCAR Community Climate System Model 3 (CCSM3). Two different emission scenarios were used – the Special Report on Emissions Scenarios (SRES) A1B (a mid-range scenario with a balanced emphasis on all energy sources) and Commitment (all GHG emissions stabilized at the 2000 CE level) (Meehl et al., 2006). Using the relative entropy approach, they quantified predictability of the upper 300 m ocean temperature in eight basins – North, tropical, and South Atlantic; North, tropical, and South Pacific; tropical Indian Ocean, including the Bay of Bengal and the Arabian Sea; and southern Indian Ocean. Analyses of the CCSM3 runs showed that information from ocean initial conditions exceeds that from forced response by approximately seven years and the predictability limit of the first kind is reached after approximately 10 years. Then, predictability beyond 10 years becomes a pure boundary condition problem. The evolution of relative entropy averaged over the eight ocean basins, as a measure of the two kinds of predictability, is shown in Figure 10.13 for the A1B and Commitment scenarios. It is clear that evolutions of predictability due to ocean initial conditions are almost identical, but the evolutions of predictability due to the two forced boundary condition scenarios diverge after approximately 10 years, with the stronger A1B scenario increasing the predictability much faster than the Commitment scenario.

Extending his formulation of potential predictability variance fraction to include both initial-value and forced boundary condition contributions, Boer (2011) looked forward to 21st century climate and its potential predictability with surface air temperature and precipitation data from the CMIP3 project. Simulations of 21st century climate with 18 ESMs under the SRES B1 scenario (a relatively low-emission

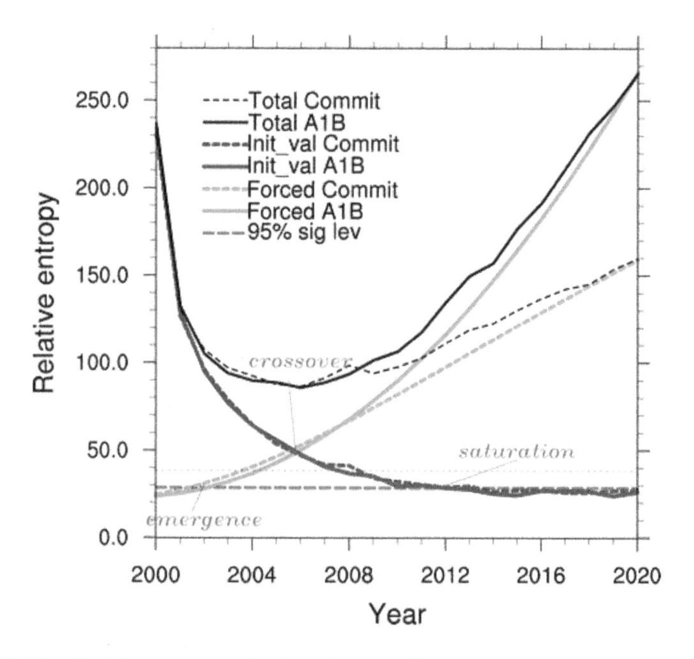

FIGURE 10.13 Relative entropy of T0–300 anomalies relative to climatology of year 1999 (black), relative entropy of the initial-value component (blue), and of the forced component (green) in the A1B (solid) and Commitment (dashed) ensembles. The red dashed line indicates the 95% significance level derived from the control run, and the red dotted line indicates the relative entropy value with 10 bits more information than the 95% significance value. All relative entropy values are sums from eight ocean basins, and each basin is represented by its leading 15 EOFs. (From Branstator and Teng, 2010 © American Meteorological Society. Used with permission.)

scenario) were analyzed to estimate both kinds of potential predictability. The predictability of the second kind due to increases in GHG emissions implicit in the B1 scenario was found to grow as the 21st century progressed and eventually dominated the predictability of the first kind due to initial ocean conditions. The largest potential predictability was for surface air temperature over tropical and mid-latitude oceans, with the tropical Pacific an exception. Potential predictability declined with latitude and was relatively low over mid-to-high latitude land. Potential predictability for precipitation was generally low in the 21st century and was due almost entirely to the forced component and mainly at high latitudes. In estimating the change in potential predictability from one decade to the next, Boer (2011) found that predictabilities of both kinds were important for surface air temperature. However, there was no decade-to-decade potential predictability for precipitation in the 18 CMIP3 models for the B1 scenario.

10.4.2 Dynamical Prediction

We will now see the relatively short history of actual decadal climate predictions, ending with the most recent prediction effort. Before embarking on this journey proceeding at a dizzying pace, however, a detour to clarify differences between climate prediction and climate projection is in order. Dynamical climate prediction is essentially an extension of dynamical weather prediction and follows the same template of components. At the heart of a climate prediction system is the climate model with specified solar radiation at the top of the atmosphere; and atmospheric constituents such as CO_2, various other trace gases, natural and anthropogenic aerosols, and stratospheric ozone. Then, there are the initial data to provide starting values for model variables such as winds, temperature, and pressure in the atmosphere; and temperature, currents, and salinity in the ocean. The initial data are typically based on actual observations at a given time and are generally made independently of each other. The climate model variables, however, are inter-connected via quantitative equations. Thus, there is a discrepancy between relationships among the independently observed initial data and relationships among the model variables. This discrepancy can "shock" the model into unrealistic behavior and so various techniques are devised to balance the initial data for the model and bring their respective interrelationships as close as possible. These techniques, generally labeled data assimilation, are a branch of climate modeling and prediction, and their detailed description is not within the scope of this chapter. The goal of data assimilation systems is to minimize the initial data shock to the climate prediction model by optimizing a balance among the observed data according to the model which is to be initialized. We saw earlier that major sources of decadal climate predictability are ocean heat content variations associated with ocean circulation variations and oceanic Rossby and Kelvin waves. Therefore, it is very important to provide initial ocean data to the climate prediction model which are representative of the ocean state and dynamically balanced (see, for example, Pohlmann et al. (2009)). It is believed that the specification of ocean initial data can be improved by data assimilated in coupled ocean–atmosphere models such that the entire ocean–atmosphere state that is specified would be in a dynamical balance close to that in the climate prediction model (see, for example, Kirtman et al. (2013), Laloyaux et al. (2016), Penny and Hamill (2017), and references therein).

In contrast to climate predictions which are initialized at a particular time, projections are used to estimate impacts on evolutions of time-average climate by hypothesized evolutions of radiative forcings due to GHGs and anthropogenic aerosols. Thus, climate projections are potential, hypothetical trajectories of climate. Also, climate projections do not benefit from initial-value predictability and can only benefit from forced boundary condition predictability to the extent that specified, future evolutions of atmospheric constituents actually occur. Both decadal climate prediction and longer-term climate projection efforts require enormous computational resources which are now available in dozens of climate modeling centers around the world to various extents. It is also obvious that further progress in decadal and longer-term climate prediction will depend critically on faster supercomputers, larger data storage systems, and faster communications within and among computers.

Let us now see a contemporary history of decadal climate predictions. Some fundamental and substantial problems of decadal climate prediction are (Meehl et al. 2009, 2014; Mehta et al., 2011): (1) relatively short time series of instrument-based global ocean observations, especially sub-surface observations, for understanding, model initialization, and comparison with prediction; (2) an insufficient understanding of fundamental physics of DCV; (3) an insufficient theoretical understanding of possible behaviors of geographically varying, complex, and nonlinear dynamical systems with mixed initial and forced boundary values; (4) global climate models displaying less than satisfactory skill in simulating climate in general and DCV in particular; and (5) insufficient guidance from stakeholders and policymakers as to which DCV-related climate, weather, and impacts information would be useful for applications to societal impacts of DCV if predicted. Despite these problems, however, there have been many encouraging decadal prediction studies with ESMs. In these pioneering studies, ESMs were initialized from observed data, and natural and anthropogenic changes in aerosol optical depth (AOD) were prescribed from observations-based estimates (or scenarios). Smith et al. (2007) showed that skillful decadal prediction of global-average temperature may be possible. Keenlyside et al. (2008) and Pohlmann et al. (2009) showed that skillful prediction of decadal, North Atlantic SSTs may be possible. Building on these studies, Yang et al. (2012) found that an inter-hemispheric, multidecadal SST pattern in the Atlantic may be predictable 4 to 10 years in advance.

Almost concurrently with these and other pioneering studies, the WCRP's CMIP5 project was initiated. CMIP5 had an explicit decadal hindcast (retrospective forecast) component. Two sets of core decadal prediction experiments were conducted under CMIP5 (Taylor et al., 2012). The first set was a series of ensembles of 10-year hindcasts starting approximately in 1960, 1970, 1980, 1990, and 2000 CE. The second was a series of ensembles of 30-year hindcasts starting in 1960, 1980, and 2005 CE, the last a combined hindcast–forecast. In both sets, AODs (including those due to volcanic eruptions) and solar radiation were prescribed from past observations. As CMIP5 progressed, there were also decadal hindcast experiments with yearly initializations. These experiments were somewhat idealistic and exploratory, especially in view of near impossibility of predicting volcanic eruptions. Over 30 ESMs from many climate modeling groups around the world participated in CMIP5. Meehl et al. (2014) have described results from CMIP5 and other decadal hindcasting experiments, so only major results pertaining to predictability of indices of decadal SST variability are reviewed here. There have been two types of assessments of prediction skill of the PDO index; one, correlation coefficient between observed and predicted indices or area-average SSTs over several decades, and two, prediction skill of specific warm or cold events. An example of the former type is a skill assessment of decadal hindcasts of the PDO index in five CMIP5 ESMs by Kim et al. (2012) who found that there was a reasonably significant prediction skill for up to five years after prediction initialization, but that this skill was less than that derived from persistence of the PDO index. An example of the latter type of skill assessment is the improved prediction skill of the mid-to-late 1970s CE change in the PDO phase from negative (cool) to positive (warm) in combined initial and boundary value experiments with several CMIP5 and other ESMs by Meehl and Teng (2012, 2014) compared to uninitialized simulations as forced boundary value experiments. As mentioned earlier and described in detail by Meehl et al. (2014), reasonably high skill of area-average North Atlantic SSTs is shown by several ESMs (see, for example, Keenlyside et al. (2008), Pohlmann et al. (2009), van Oldenborgh et al. (2012), Yang et al. (2012), Hazeleger et al. (2013), Ham et al. (2014), and others). Using decadal hindcast data from four CMIP5 ESMs, Mehta et al. (2013b) found that there was significant, but variable, decadal hindcast skill of global- and tropical ocean basin-average SSTs during 1961–2010 CE. The skill varied by averaging region and decade. It was also found that low-latitude volcanic eruptions can be one of the sources of decadal SST hindcast skill when major eruptions occurred. In the four ESMs, decadal hindcast skills of SST anomalies over ocean basin size averaging regions generally improved due to model initialization with observed data.

Mehta et al. (2019) also used CMIP5 data to assess the ability of four ESMs (UK Meteorological Office HadCM3, NCAR CCSM4, GFDL CM2.1, and MIROC5) to hindcast SST indices of the PDO, the TAG variability, and the WPWP variability from 1961 to 2010 CE. Deterministic and probabilistic skill estimates showed predictability of detrended WPWP index to 5 years' lead time and of non-detrended WPWP index to 10 years' lead time. These estimates also showed atypical skill dependence of PDO and TAG indices on lead times, with increasing skill in the middle to end of 10-year hindcasts in three of the four ESMs; MIROC5 hindcasts showed a similar skill dependence on lead time as for a persistence hindcast. It was also found that all ESMs hindcast occurrence frequencies of positive and negative phases of the three DCV indices, and probabilities of phase transitions from one year to the next reasonably well. Major, low-latitude volcanic eruptions were associated with phase transitions of all observed and some of the ensemble-average hindcast indices. All ESMs' WPWP index hindcasts responded correctly to all four eruptions as did three observed PDO phase transitions. Some of the ESMs hindcast correct phase transitions in the absence of eruptions also, implying that initializations with observed data were beneficial in predicting phase transitions.

Encouraged by these and other experimental decadal hindcast results, there are now efforts under way to make operational decadal climate forecasts. An informal decadal forecast exchange began in 2010 CE, facilitated by a Decadal Exchange established in the UK Meteorological Office – Hadley Centre and containing decadal forecasts made by close to ten dynamical climate models and a few empirical models (Smith et al., 2013). As noted by Smith et al. (2013), initialization had little impact beyond four years in most regions, but this informal exchange effort was the first, operational surface air temperature forecast of the 2011–2020 CE decade. Following up with a larger multi-model ensemble (71 ensemble members from seven different ESMs), Smith et al. (2019) estimated decadal prediction skill (average over years two to nine of decadal forecasts) of near-surface temperature, precipitation, and atmospheric circulation. They also assessed the impact of initialization on the decadal forecasts. Decadal hindcasts, initialized every year, from 1960 to 2005 CE were used in this skill assessment; assessment techniques are described by Smith et al. (2019). Figure 10.14 shows total skill and skill due to initialization (both where the correlation coefficients are positive) for worldwide near-surface temperature, precipitation, and SLP. As Figure 10.14a shows, temperature forecast has high skill almost all around the world, except northeast Pacific and parts of the Southern Oceans, which is generally consistent with previous studies. Precipitation forecasts (Figure 10.14c) show moderate skill in the Sahel region of Africa, and across northern Europe and Eurasia. Precipitation forecast skill is lower, but significant, in northern Canada and western Alaska; northern Greenland; northwestern USA; and parts of South America, India, southeast Asia, and the Maritime Continent (MC). SLP forecasts (Figure 10.14e) are significantly skillful in many parts of the world, except in most of Africa, western Indian Ocean, and eastern South Atlantic; the western tropical Pacific-MC region; a wide, east–west swath across central Asia; and parts of the Southern Oceans. Corresponding figures (Figure 10.14b, d, and f) show forecast improvements to various extents due to model initializations. The forecast changes are depicted as the ratio of predicted signal due to initialization and the total predicted signal, so positive ratios denote improvement and negative ratios denote degradation due to initialization (Smith et al., 2019). As Figures 10.14b, d, and f show, there are skill increases as well as decreases in many parts of the world due to initialization.

Building on the relatively 10 to 15-year long history of experimental decadal climate predictions, the World Meteorological Organization (WMO) recently instituted the Near-Term Climate Prediction (NTCP) program as a Grand Challenge to produce skillful and reliable forecasts on annual to decadal timescales (Kushnir et al., 2019) and designated the UK Meteorological Office as the WMO Lead Centre for Annual to Decadal Prediction.[9] NTCP will include both internally generated climate variability and climate variability/change due to forced boundary conditions, using future projections of anthropogenic forcings. Also, building on encouraging results and insights from CMIP5 decadal hindcast experiments, CMIP6 will have much more refined and specific sets of experiments to address decadal predictability, and impacts of solar variability and volcanic eruptions on DCV and decadal climate predictability (Eyring et al., 2016). Specifically, there will be a set of CMIP6 experiments pertaining to decadal climate prediction – the Decadal Climate Prediction Project (DCPP; Boer et al., 2016). Over 100 model combinations are registered to provide data under CMIP6 which should be available over the next one to two years.

[9] www.metoffice.gov.uk/research/climate/seasonal-to-decadal/long-range/wmolc-adcp.

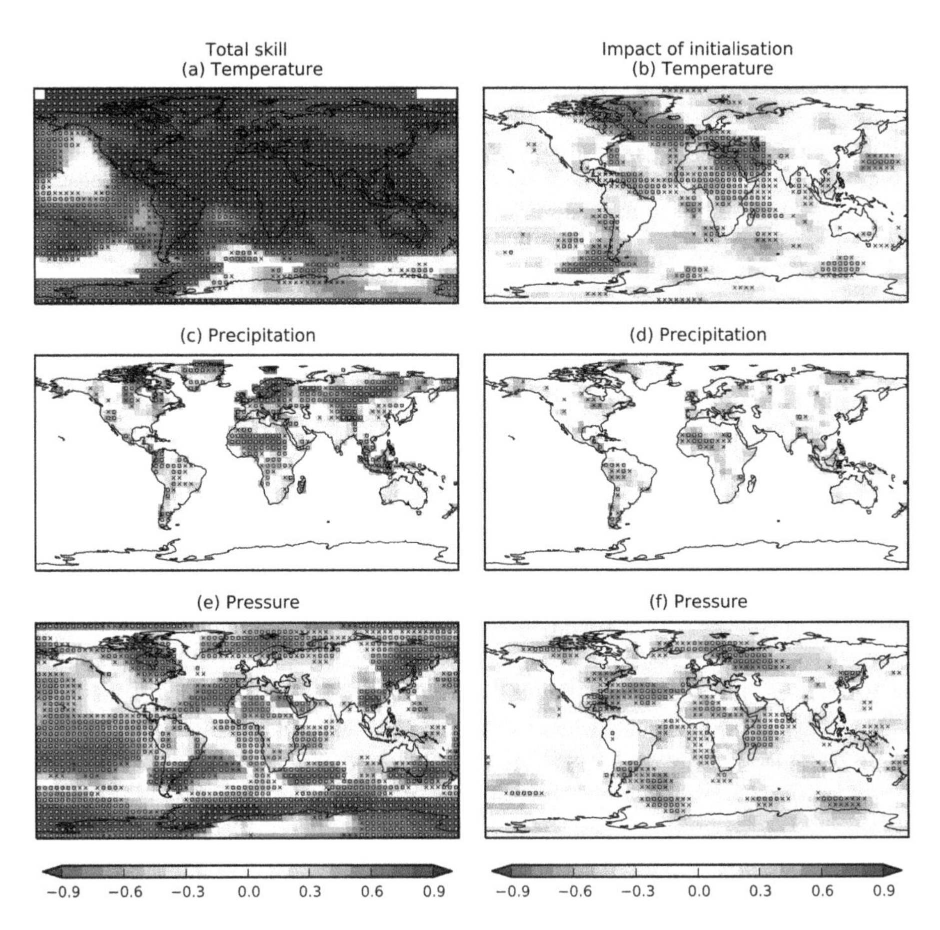

FIGURE 10.14 (a) Correlation coefficients between year 2 to 9 initialized ensemble-average forecasts and observations for near-surface temperature. (b) The impact of initialization computed as the ratio of predicted signal arising from initialization divided by the total predicted signal (where positive/negative values show improved/reduced skill). (c) and (d) As (a) and (b), but for precipitation. (e) and (f) As (a) and (b), but for SLP. Stippling shows where correlations with observations (a, c, e) and of residuals (b, d, f) are significant (crosses and circles show 90% and 95% confidence intervals, respectively). (From Smith et al., 2019, Figure licensed under Creative Commons Attribution http://creativecommons.org/licenses/by/4.0/.)

10.5 SUMMARY AND CONCLUSIONS

The need for decadal climate information, including decadal climate prediction, is increasing as impacts of DCV in water resources, agriculture, fisheries, transportation, and other societal sectors are coming to light. Decadal climate predictions are also important to attribute apparent changes in near-term climate to natural DCV or anthropogenic climate change. In order to make progress in prediction, it is very important to estimate potential decadal predictability using a variety of methods, models, and data; and relate potential decadal predictability to actual climate predictability and prediction. The evolution of numerical weather prediction in the 20th century led to seasonal to multiseasonal climate prediction as initial-value problems in which a set of partial differential equations are integrated forward in time from initial values of primary variables. In potential climate change projections, the equations are integrated forward and the solutions evolve in response to changes in boundary forcings such as atmospheric constituents. In decadal climate prediction, both the potential predictability which can be derived from initial values and the potential predictability from changing boundary forcings are important. The decadal climate predictability assessment and actual prediction efforts are learning to understand effects of both types of physics, and estimate potential predictability and its spatiotemporal dependence. These efforts are handicapped by the availability of quantitatively reliable, observed climate data to the last 100 years at the most, especially over and in the oceans which play a critically important role.

Braving these difficulties, modern-day researchers began to delve into mysteries of decadal climate predictability approximately 20 years ago and have made impressive progress. As the experience and understanding have grown, decadal predictability is being defined in several different ways, and methods are being developed to estimate potential predictability and make actual predictions. Empirical predictability estimates, based largely on observed data, indicate that there may be skill in predicting SSTs at mid- to high latitudes in the Atlantic and the Southern Oceans at two-year to nine-year lead times. These two- to nine-year skill estimates also extend to the PDO index and perhaps to longer lead times for the AMO index. However, it is not very clear that these potential predictability estimates, especially about surface air temperature, consistently exceed persistence skill everywhere in the world. It is also estimated that forced response to changing boundary conditions contributes potential predictability after approximately seven- to ten-year lead times. Dynamical predictability estimates indicate predictability at mid- to high latitudes in the Atlantic and the Southern Oceans where the surface ocean interacts with the deep ocean. Dynamical predictability estimates are dependent on an ESM's ability to simulate observed decadal phenomena, so these estimates may have limited application in real world.

The pace of decadal prediction research and operational decadal prediction is accelerating as described in this chapter. The WCRP's CMIP5 contributed immensely in these efforts and CMIP6 is expected to continue this process. As Kirtman et al. (2013) concluded, "Predictions for averages of temperature, over large regions of the planet and for the global mean, exhibit positive skill when verified against observations for forecast periods up to 10 years". Most recent studies indicate area-dependent, robust skills of SST and surface air temperature, precipitation, and SLP predictions attributable both to initializations with ocean observations and forced boundary conditions. Empirical predictions are also indicating skill not only in predicting the PDO, the TAG, and the AMO indices, but also in using the predicted indices in empirical, probabilistic predictions of societally useful information such as land dryness and wetness. Thus, there is a slow and incremental, but definite, progress in making skillful decadal climate predictions. However, the following areas need sustained attention for improvements in decadal prediction skill: (1) improvements in the ability of ESMs to simulate observed DCV phenomena, (2) development of data assimilation systems based on ESMs and assimilation of soil moisture and vegetation data, (3) increasing ensemble sizes to extract predictable signals, (4) continued increases in computing power, with associated data storage and communication, facilities, and (5) capacity building in climate modeling and prediction.

So, what does the future portend for decadal climate predictions? We can say with high confidence that the rigorous research expected from CMIP6 on decadal hindcasts should lead to substantial improvements in at least our understanding of the sources of decadal climate predictability and its limits, and the respective roles of initialization and forced boundary conditions in these hindcasts. We can say with less confidence that this understanding will lead to more skillful, operational decadal predictions, especially of surface air temperature, precipitation, winds, and other societally useful variables on land. However, fledgling efforts to develop probabilistic decadal predictions of hydro-meteorological variables on land, using SST predictions by ESMs in statistical models to predict surface air temperature, precipitation, and other variables on land may show more skill than either approach alone and so may be more useful in the near term. Despite these efforts by the climate modeling community, we can say with even less confidence that skillful decadal predictions will be used by stakeholder communities in the near future. There are numerous reasons for this low confidence. Among the major reasons are (1) impacts of DCV on water resources, agriculture, transportation, fisheries, and other societally relevant sectors modulated by other regional and local conditions such as local costs of inputs and global demands and prices of products, availability of crop insurance and/or other safety nets, availability of storage and transportation options and prices, laws governing allocations of water resources, and laws governing fishing; (2) a lack of boundary organizations between decadal climate prediction producers and consumers; (3) a general paucity of information about what type of decadal climate information stakeholders need in various societal sectors in various parts of the world; and (4) a lack of programs to bring climate and other scientists together with stakeholders for mutual education and appreciation of each community's needs and difficulties. Also, there are bureaucratic walls between programs such as the World Meteorological Organization's NTCP and Global Framework for Climate Services (GFCS) although both programs profess to work together. Last but perhaps the most important is the need to build human resource capacities in climate modeling and prediction. Despite the accelerating pace of decadal climate prediction research, we should always remember the quotation attributed to Professor Niels Bohr cited at the beginning of this chapter – "Prediction is very difficult, especially if it is about the future".

11 Epilogue

11.1 WHY THIS JOURNEY

Research on decadal climate variability (DCV) and its predictability began several centuries ago with solar variability, indicated by Sunspot numbers, invoked as the major cause. Since then, front-rank science journals have continued to publish research papers on Sunspot–climate associations well into the 21st century Current Era (CE). In the late 1980s CE, following the development of a theory of the El Niño–La Niña events as an internally generated, coupled ocean–atmosphere phenomenon, tentative efforts began to model DCV also as an internally generated, coupled ocean–atmosphere phenomenon. The end of the Cold War in the early 1990s CE enabled public releases of long-term ocean temperature and salinity data that the United States of America (USA), British, Soviet, and other navies held in their archives. These and other data collected by merchant ships since 1850s CE soon became available to researchers for analyses. These analyses revealed major DCV phenomena – decadal variability of the tropical Atlantic sea surface temperature (SST) gradient (TAG for brevity), the Pacific Decadal Oscillation (PDO)/the Interdecadal Pacific Oscillation (IPO), decadal variability of the West Pacific Warm Pool (WPWP) and the East Indian Warm Pool (EIWP), decadal variability of tropical cyclones, and decadal variability of extratropical atmospheric phenomena known as the North Atlantic Oscillation (NAO)/Northern Annular Mode (NAM)/the arctic Oscillation (AO), and the Southern Annular Mode (SAM)/the Antarctic Oscillation (AAO). Also in the 1990s CE, it started to become obvious to some climate scientists that natural decadal–multidecadal climate variability can confound the attribution of changing climate to anthropogenic effects. So, the modern era in DCV research began and now has the added motivation of understanding and predicting impacts on water – food – energy – public health securities and other societal sectors.

In the modern era, thousands of peer-reviewed research papers and many reports have been published on DCV and its predictability. There is much confusion created by labeling variability decadal, interdecadal, or multidecadal without much specificity regarding timescales or oscillation periods of the analyzed or modeled variabilities. Therefore, the purpose of this book is to synthesize knowledge, largely from peer-reviewed publications, of empirically-identified DCV phenomena, causes and mechanisms of these phenomena, and fledgling efforts to predict the future evolution of these phenomena. The main focus of this book is on *decadal* variability, defined as oscillation periods approximately between eight and 20 years. The justification for focusing only on this band of timescales was provided by substantial and significant peaks at these oscillation periods in the Fourier and other spectra of multidecades to multicentury time series of paleoclimate proxy data and instrument-measured data. This focus is necessary to clearly understand causes and mechanisms, and to make progress on decadal climate predictability. This book is also meant to be a useful and important companion to published books, research papers, and reports on anthropogenic climate change and its societal impacts. This chapter, the Epilogue, summarizes highlights of DCV phenomena, their causes and mechanisms, and decadal climate predictability described in the previous chapters. Areas that need more illumination to make further progress conclude this chapter and the book.

After this brief description of the need for this book, highlights of this book, including research areas needing further clarity, are reviewed in Section 11.2. Then, areas needing further illumination are briefly outlined in Section 11.3 and closing remarks are presented in Section 11.4.

11.2 THE STORY SO FAR

In the Prologue (Chapter 1), the stage was set for detailed descriptions and discussions of DCV phenomena and their predictability by an introduction to major Earth System components – the atmosphere, the oceans, the land–vegetation–snow-ice system, and sea ice – and interactions between the atmosphere and the underlying components. Annual cycles and other major features of these components which play or are hypothesized to play important roles in DCV phenomena were described. Since possible effects of solar emissions variability are a major part of the DCV story, an introduction to basic solar physics and the Sun's emissions was provided in Chapter 2. In the following five sub-sections, highlights of observed DCV, its hypothesized mechanisms, interactions between natural DCV and anthropogenic climate change, modulation of tropical cyclones by DCV, and decadal predictability and prediction in Chapters 3 to 10 are summarized. Chapter references are provided in each sub-section.

11.2.1 Observed Decadal Climate Variability

We begin this description with associations between decadal variability of solar emissions and paleoclimate proxy data. Sunspot numbers, total solar and ultraviolet (UV) irradiances, 10.7 cm radio flux, galactic cosmic ray (GCR) flux, and geomagnetic field were included in these analyses (Chapter 2). Paleoclimate proxy data such as tree rings, oxygen and carbon isotopes, and lake sediments in many parts of the world show DCV in all geographic regions from where the data originated and the data also show substantial and significant statistical associations with variability of solar emissions, solar cycle length, or the GCR flux. Some of the paleoclimate proxy data tell tales about climate as far back as 12,000 years ago and, in one case, tree-ring data from approximately 290 million years ago in southeastern Germany show associations with the 11-year solar cycle. Analyses of instrument-measured climate observations, including model-assimilated climate observations, show substantial and significant associations between decadal variability of solar emissions and the GCR flux, and instrument-based climate observations in the last approximately 150 years (Chapter 2).

Focusing on DCV in the Pacific region, we saw that the PDO, the IPO, the South Pacific Decadal Oscillation, and decadal–multidecadal variability of the Niño3.4 SST indices show that they represent essentially the same phenomenon (Chapter 3).

It was also shown that decadal SST variability has the largest variance and significant spectral peaks in the tropical central and western Pacific region. Analyses of instrument-measured and model-assimilated atmosphere and ocean data show that anomalous Hadley and Walker circulations and atmospheric Rossby waves are associated with the tropical Pacific decadal variability, which can influence or even impact extratropical climate. The analyses also show that anomalous thermocline depths and associated sub-surface ocean temperature anomalies, and anomalous surface current and shallow tropical circulations (STCs) in the Pacific Ocean, especially the Southwest Pacific STC, are associated with the decadal SST variability in the tropical central and western Pacific. Analyses of a 350-years long, coral oxygen isotope $\delta^{18}O$ record confirmed the presence of prominent decadal variability in the South Pacific Convergence Zone (SPCZ) region, corroborating the likely role of the STC in this region in tropical Pacific decadal variability. Comparisons of 12 paleoclimate proxy-based records of the IPO index indicated general inconsistencies among most of them, raising doubts about the ability of such records to provide reliable information about the IPO variability over the last several centuries (Chapter 3).

Moving eastward from the Pacific, we addressed the long-running mystery of the cross-equatorial dipole SST pattern in the tropical Atlantic region which undergoes decadal variability (Chapter 4). It was shown by numerous analyses of century-long and longer time series of tropical Atlantic SST data that there is substantial decadal variability with a significant spectral peak at 12 to 14-year timescales. However, the presence or absence of this spectral peak apparently depends on the SST data set, the spectral analysis and significance testing techniques used, and the time span of the analyses. Multidecadal variability appears to be stronger than decadal variability in the first half of the 20th century CE in the tropical Atlantic, and the latter appears to be stronger since then. These analyses also showed that there is nearly-independent decadal SST variability in the tropical North and South Atlantic, and that decadal dipole variability pattern revealed by statistical analyses should be interpreted as variability of the inter-hemispheric SST gradient or the TAG. Thus, from the observational point of view, the dipole controversy is finally laid to rest with the conclusion that there is no dipole mode of decadal variability of the tropical Atlantic ocean–atmosphere system. Empirical analyses also showed substantial evidence of SST anomalies traveling clockwise in the North Atlantic and counter-clockwise in the South Atlantic at decadal timescales. Their natural alignments on both sides of the Equator occasionally create anomalies of the same sign or opposite signs on two sides of the Equator, thus creating either a broad anomaly pattern spanning the entire tropical Atlantic or the bipolar TAG pattern.

Moving further east to the tropical Indian Ocean (Chapter 5), we saw that a century long coral-based $\delta^{18}O$ record showed the presence of a distinct 11 to 14-year oscillation timescale in SST variability. This and other records also showed that the tropical Indian Ocean decadal variability is coherent with variability in the tropical western and central Pacific regions. Instrument-measured data since the 1950s CE confirmed the tropical Indian Ocean decadal SST variability and also showed that the total area of the EIWP undergoes decadal variability. The Walker circulation; and low-level winds, upper-ocean currents and implied heat transport, and surface heat flux in the tropical–subtropical Indian Ocean were shown to be varying in phase with basin-average SSTs. An increase (decrease) in ocean heat transport from the EIWP and a net, area-averaged reduction (increase) in ocean-to-atmosphere surface heat flux were shown to be associated with a warmer (cooler) tropical Indian Ocean in the warm (less warm) or positive (negative) phase of the decadal variability. These observations support the conclusion that the tropical Indian Ocean decadal variability is largely controlled by the tropical Pacific Ocean decadal variability via the Walker circulation, ocean heat transport, and surface heat flux connections between the two basins.

On DCV in the extratropical Northern Hemisphere (NH) region (Chapter 6), we saw that weather observations by Scandinavian missionaries in Greenland in the 18th century CE pioneered attempts to understand and predict variability in what was later called the NAO/NAM phenomenon – an oscillation in atmospheric mass or sea level pressure (SLP) between subpolar and subtropical latitudes, with associated wind variations from the Earth's surface to well over 100 km in the atmosphere. The NAO, the AO, and the NAM appear to represent the same phenomenon. These surface wind and consequent heat flux variations generate a tripole SST anomaly pattern from subpolar seas to subtropical North Atlantic. Frequency spectra of time series of NAO indices formed from SLP data and the tripole SST pattern show several significant peaks between eight and 20-year periods. The decadal surface wind and heat fluxes also interact with sea ice in the NH.

The Southern Hemisphere (SH) counterpart to the NAO/AO/NAM phenomenon is the SAM/AAO phenomenon (Chapter 7). The time series of instrument-measured surface data in extratropical SH are relatively short, longer than 100 years only at a few stations. It was shown that in these data and also in multidecade-long reanalysis of surface and upper air data, decadal co-variability of SLP and SST is clearly evident at approximately 13 to 14-year period. Multiproxy-based and tree-ring-based, multicentury reconstructions of the SAM/AO indices going back to the 16th century CE corroborate the presence of decadal variability at oscillation periods similar to those in the instrument-measured and reanalysis data.

Thus, as described in this brief summary, decadal spectral peaks are ubiquitous in paleoclimate proxy data and in instrument-measured SST, SLP, and other data in the tropical–subtropical oceanic regions, and in the extratropical NH and SH regions. The presence of longer timescale variability is also evident in these data.

11.2.2 Mechanisms of Decadal Climate Variability

Following these descriptions of empirical analyses, it was shown by model simulations (Chapters 2, 3, 6, and 7) that the so-called "top-down" mechanism of solar cycle–climate connection can be initiated by solar UV radiation absorption in the stratosphere. Subsequently, meridional temperature gradient created by the heating due to UV radiation absorption can influence stratospheric circulations and, then, vertically-propagating planetary waves and stratospheric circulations can influence tropospheric winds and SLP in mid- and high latitudes. These wind variations can cause surface heat flux variations which, in turn, can cause SST and upper ocean temperature variations. It was also shown by model simulations that the so-called "bottom-up" mechanism of solar–climate connection can generate DCV by the absorption of visible solar radiation in subtropical oceans, whose subsequent effects on trade winds and unstable ocean–atmosphere interactions can generate DCV (Chapters 2, 3, and 6). As a synergistic culmination of the two mechanisms, it was shown that simulations with coupled ocean–atmosphere models forced by estimates of decadal variability in solar radiation can generate DCV in the tropical–subtropical Pacific region which resembles the observed association between decadal solar emissions variability and the Pacific region's climate.

These top-down and bottom-up mechanisms have also been hypothesized to explain decadal NAO-NAM variability (Chapter 6). Simulations with global coupled ocean–atmosphere models, with the atmosphere extending to 200 km, show that the 11-year solar cycle can act via the stratosphere on the troposphere–ocean coupled system as described above and can act as a pace-maker of the decadal NAO–SST–upper ocean temperature variability. Interactions among the extratropical NH atmosphere, Arctic and North Atlantic Oceans and marginal seas, and sea ice are also hypothesized as possible mechanisms of the decadal NAO variability. Among other mechanisms hypothesized and tested to some extent by model simulations is the possible effect of moderate-to-severe volcanic eruptions, especially at tropical latitudes. Such eruptions act on the stratosphere–troposphere–ocean system in the North Atlantic region to generate multiyear to decadal and longer timescale NAO variations. It is also possible that such eruptions can also influence the SAM variability.

It has been shown with coupled ocean–atmosphere–sea ice models that they can intrinsically generate SAM variability at preferred decadal timescales due to combinations of negative and positive feedbacks among SSTs, meridional ocean heat transport, and winds (Chapter 7) at mid- and high latitudes in the SH. The top-down solar influence hypothesis proposed for decadal variability of the NAO has also been proposed for decadal variability of the SAM. So, it is possible that the 11-year solar cycle may be acting as a pace-maker of decadal SAM variability as in the decadal NAO variability. Another possible candidate for both decadal NAO and SAM variability is intrinsic atmospheric variability in winter due to nonlinear dynamical interactions between the zonally-averaged winds, and stationary and transient atmospheric eddies (Chapter 6).

It was shown that simulations with uncoupled and coupled ocean–atmosphere models identified important roles of STCs and off-equatorial oceanic Rossby waves, and their interactions with the atmosphere in generating DCV in the tropical central and western Pacific region (Chapter 3). The possible role of the 11-year solar cycle in forcing decadal SST variability in the tropical–subtropical Pacific region was also illustrated as described above. For pan-Pacific multidecadal variability, it was shown that – although extratropical processes such as changes in winds around the Aleutian and South Pacific low-pressure systems reaching towards the tropical Pacific, and spectral reddening due to integration of atmospheric higher-frequency forcings by mid-latitude Pacific Ocean appear to be playing important roles – the variability appears to be primarily due to STCs, the Bjerknes feedbacks, and oceanic Rossby waves at tropical–subtropical latitudes. The success of numerous global coupled ocean–atmosphere models in simulating IPO-like multidecadal variability reasonably well was also described. Moderate-to-severe low-latitude volcanic eruptions were shown to perturb surface energy balance over the Pacific Ocean such that they can change amplitudes and even phases of the IPO and the WPWP decadal variability. The two phenomena's recoveries from such eruption-forced changes can take several years which can provide predictability over several seasons to years.

Moving eastward to the tropical Atlantic and decadal variability of the TAG (Chapter 4), idealized coupled ocean–atmosphere models showed in the 1990s CE that damped decadal coupled modes can exist for some ranges of coupling parameters, but that stochastic forcing would have to be invoked to generate such modes in the actual Atlantic climate system and that such forcing would break the TAG SST pattern or weaken it considerably. Simulations with more realistic uncoupled and coupled ocean–atmosphere models suggest that tropical–subtropical surface wind forcing can generate SST and sub-surface oceanic variability, which can result in a weak influence back to the atmosphere via the wind–evaporation–SST feedback. These wind anomalies can also spin up and spin down subtropical gyre circulations in the North and South Atlantic Oceans which can generate SST variability due to anomalous oceanic heat advection. Effects of wind anomalies on one side of the Equator can generate wind anomalies and weak SST anomalies on the other side, but such SST anomalies would not always be strong and have cross-equatorial coherence. Three possible mechanisms of the preferred decadal timescale were identified: (1) a combination of top-down and bottom-up solar effects working in the tropical Atlantic region; (2) influence of the tropical Pacific decadal variability acting on the tropical Atlantic via atmospheric east–west circulation; and (3) influence of the preferred decadal timescale of the NAO and/or the SAM acting on the tropical Atlantic via trade winds, surface heat fluxes, and ocean heat transport by NH and/or SH subtropical gyres.

Moving further east to the tropical Indian Ocean, observations support the conclusion that the tropical Indian Ocean decadal variability is largely controlled by the tropical Pacific Ocean decadal variability via the Walker circulation, ocean heat transport, and surface heat flux connections between the two basins. This conclusion, however, needs to be tested by simulation experiments with appropriate models.

11.2.3 Natural Decadal Climate Variability and Anthropogenic Climate Change

Although it was anticipated among DCV researchers since the 1990s CE that natural decadal–multidecadal climate variability can and will make attribution of observed climate changes difficult if not impossible, the so-called global warming hiatus in the last decade has brought this conundrum to the forefront of public discussions on anthropogenic climate change. We saw that the IPO made a substantial contribution in the global warming hiatus, and will accelerate or slow down future changes in global-mean surface temperature. Such IPO contributions will make verifications of international climate frameworks – such as the Paris Climate Agreement (UNFCCC, 2015) – dependent on reasonably accurate IPO predictions (Chapter 8).

In a review of the nascent field of possible anthropogenic climate change effects on DCV, we saw that changes in vertical ocean stratification due to warming of the upper ocean can influence the speed of planetary-scale oceanic Rossby waves which, in turn, can change the travel time of such waves and periodicities of DCV phenomena in which such Rossby waves play very important roles (Chapter 8).

11.2.4 Tropical Cyclones

We saw that decadal variability of SSTs is ubiquitous at tropical–subtropical latitudes. It is well known that tropical–subtropical SSTs play a very important role in tropical cyclones' (TCs') formation, growth, and tracks. Does the decadal and longer timescale SST variability also cause variability in numbers, intensities, and tracks of TCs? This question was addressed in Chapter 9. We saw that empirical data on TCs going back to at least the 1950s CE show decadal and longer timescale variability in the numbers of intense and very intense TCs in the tropical–subtropical Atlantic, Pacific, and Indian Oceans; and also in the South China Sea, the Bay of Bengal, and the Arabian Sea. Data on landfalling TCs, however, go back to mid- or late 19th century CE. For the Atlantic region, these data show that the TC frequency and the number of intense TC days vary at decadal–multidecadal timescales. Data on landfalling TCs along the US coast and along the Caribbean Sea coast also show such variability over late 19th and entire 20th centuries CE. It was shown that decadal SST variability in the Main Development Region for Atlantic TCs in the North Atlantic off the west coast of Africa (Chapter 4) plays a very important role in the decadal modulation of the frequency of intense Atlantic TCs.

Frequencies of TCs in the western North Pacific and the South China Sea show a physically consistent relationship with the PDO variability. Vertical shear of zonal wind, moist static energy, and lower atmospheric stability are modulated by tropical Pacific SST variability, and the TC frequency is then affected by this modulation. Decadal variations in the Bay of Bengal TC frequency are associated with large-scale atmospheric conditions, possibly modulated by remote effects of the PDO via vertical wind shear, moist static energy, and low-level relative vorticity. It was also shown that TC frequency variability in the Arabian Sea has a significant spectral peak at 10 to 13 years which is the same timescale as that found in tropical Indian Ocean SST variability (Chapter 5). Thus, decadal SST variability in the tropical–subtropical oceans and seas appears to be a major driver of frequencies of intense and very intense TCs.

11.2.5 Decadal Predictability and Prediction

Is DCV predictable with useful skill? Are decadal climate predictions being made? These questions were addressed in Chapter 10. Potential decadal predictability is being estimated using a variety of methods, models, and data; and is also being related to actual climate predictability and prediction. In estimating decadal climate predictability, both the potential predictability which can be derived from evolving DCV phenomena and the potential predictability from changing boundary conditions such as atmospheric constituents and solar radiation are important. The decadal climate predictability assessment and actual prediction efforts are learning to understand effects of both types of sources of potential predictability, and estimate potential predictability and its spatio-temporal dependence. These efforts are handicapped by the availability of quantitatively reliable, observed climate data to the last 100 years at the most, especially over and in the oceans which play a critically important role. Despite these and other difficulties, there is a substantial progress in estimating potential decadal predictability and understanding sources of such predictability. As the experience and understanding have grown, decadal predictability is being defined in several different ways, and methods are being developed to estimate potential predictability and make actual predictions.

Empirical predictability estimates, based on observed data, indicate that there may be skill in predicting decadal and longer timescale evolution of SSTs at mid- to high latitudes in the Atlantic and the Southern Oceans at two- to nine-year lead times. These two- to nine-year significant skill estimates also extend to the PDO. However, it is not very clear that these potential predictability estimates consistently exceed skill due to persistence since DCV phenomena, by definition, persist for many years. It is also estimated that forced response to changing boundary conditions contributes potential predictability after approximately seven- to ten-year lead times.

Dynamical predictability estimates indicate predictability at mid- to high latitudes in the Atlantic and the Southern Oceans where the surface ocean interacts with the deep ocean. Dynamical predictability estimates are dependent on an Earth System Model (ESM)'s ability to simulate observed decadal phenomena, so these estimates may have limited application in real world since ESMs are not yet able to simulate spatial patterns and temporal evolution of DCV phenomena very accurately. However, deterministic and probabilistic skill estimates from CMIP5 decadal hindcast experiments indicate that the evolutions of the PDO and the WPWP SST variabilities may be predictable for at least three to five years. A warming trend in WPWP SSTs increases the lead time of significant skill for up to a decade.

Most recent decadal prediction studies with ESMs indicate area-dependent, robust skills of SST and surface air temperature, precipitation, and SLP predictions attributable to both initializations with ocean observations and forced boundary conditions. Empirical predictions are indicating skill not only in predicting the PDO and TAG indices, but also in using the predicted indices in empirical, probabilistic predictions of societally useful information such as land dryness and wetness. Thus, there is a slow and incremental, but definite, progress in making skillful decadal climate predictions. Although the collective scientific community experience of seasonal to interannual climate prediction appears to be shortening the learning period about decadal climate prediction, we may be many years or even decades away from useful, skillful, and trustworthy decadal climate predictions. But, what appeared to be impossible even a decade ago seems to be possible in some near or not-very-distant future. Therefore, the state of decadal predictability research should be looked upon as a partially filled glass rather than a largely empty glass.

11.3 NEED FOR FURTHER ILLUMINATION

Despite the amazing progress in the last 30 years, there are still outstanding problems and areas of ignorance which need further illumination.

1. *Influence of the solar cycle:*

   The spectral peak at 11 to 14-year oscillation periods in instrument-measured SSTs and other data, and in paleoclimate proxy data appears to be a robust feature of tropical and extratropical climate variability because it is evident in many independent analyses of independent data types spanning overlapping time periods. It is also becoming noticeable in simulations with global coupled ocean–atmosphere models both without and with some measure of the 11-year solar cycle included among the boundary conditions. It appears from global coupled model simulations that there are subtle but potentially insightful differences between decadal variability generated by free-running models and that generated by solar cycle-forced simulations, such as the extent of the stratosphere's involvement in troposphere–ocean variability; and the roles of evaporation, clouds, and trade winds in tropical–subtropical decadal variability. Can these differences be quantified and their robustness checked via ensembles of model experiments?

Empirical analyses and simulation experiments indicate that the solar activity cycle may be playing a very important role in decadal NAM and SAM variabilities. It is not clear from these studies, however, if they are synchronized. So, can the possible synchronization of decadal variabilities in the NAM and SAM by the solar cycle be studied via empirical analyses and simulations with appropriate global coupled ocean–atmosphere models?

Approximate symmetry with respect to the Equator is a major feature of tropical–subtropical Pacific region DCV. Can the possible synchronization of such symmetry be studied via empirical analyses and appropriate global coupled ocean–atmosphere models?

2. *Ocean–atmosphere coupling and feedbacks*:

   The strength, robustness, and transience of ocean–atmosphere coupling appear to be very important in generating intrinsic DCV and in the coupled ocean–atmosphere–sea ice system's response to the solar activity cycle. Also, it is well known that ocean-to-atmosphere feedback processes are different in tropical–subtropical and mid–high latitudes. Can combined observation and modeling programs be designed to reduce uncertainties in our knowledge of the coupling and feedback processes?

3. *Influence of volcanic eruptions*:

   Moderate-to-severe volcanic eruptions, especially at low latitudes, appear to make substantial impacts on tropical and extratropical DCV phenomena. Can these impacts be unambiguously isolated and quantified in the presence of DCV generated by internal ocean–atmosphere variability and by the solar activity cycle?

4. *Intrinsic atmospheric variability*:

   In the rush to develop and employ ESMs of increasing complexity, intriguing results of idealized experiments with models of the global atmosphere which indicated that linear and nonlinear atmospheric dynamics can generate decadal variability appear to have been largely forgotten by the climate modeling community. This intriguing possibility should be re-visited with state-of-the-art models of the global atmosphere, and their ability to generate decadal variability should be studied in the presence of fixed boundary conditions. Such studies may be particularly applicable to decadal NAO and SAM variabilities.

5. *Local and remote influences on TCs*:

   TCs develop under the influences of SST and other variabilities from multiple sources in local and remote ocean basins. Can such influences be identified and quantified for better understanding and prediction of decadal variability in TC frequencies, intensities, and tracks?

6. *Interactions between natural variability and anthropogenic climate change*:

   While traces of natural DCV can be found in a variety of proxy climate records going back thousands of years, and anthropogenic climate change is relatively recent and its effects are likely to be felt mainly in the future, both of them occur in the same Earth system and it is inconceivable that there is no interaction between them. Results from preliminary studies were described in Chapter 8. Therefore, research on interactions between natural DCV and anthropogenic climate change must be placed on a more prominent level than is presently the case.

7. *Decadal predictability and prediction*:

   It was described in Chapter 2 that the possibility of climate and impacts prediction using Sunspot numbers was a major motivation for early DCV research. Early pioneers of DCV and prediction research understood the importance of prediction to society, especially successful prediction's implications for the economy. In the last ten years or so, experimental decadal climate prediction has again begun to attract the attention of a substantial number of scientists, especially in the USA and Western Europe as described in Chapter 10. The discovery of several DCV phenomena involving the oceans and the need to make near-term climate predictions to test and validate climate change models have driven this new search for decadal climate predictability. Concurrently with these initial decadal climate predictability studies, the World Climate Research Program (WCRP) organized the Coupled Model Intercomparison Project 5 (CMIP5) to assess the ability of the current generation of ESMs used in climate and impacts assessments by the Inter-governmental Panel on Climate Change (IPCC) to simulate and hindcast/forecast decadal climate. Over 30 global climate models participated in the CMIP5 project. These experiments were somewhat idealistic and exploratory, especially in view of the well-known difficulty of predicting volcanic eruptions well in advance. But, decadal hindcast results from CMIP5 have been encouraging. The next in the series of Coupled Model Intercomparison Projects is CMIP6 (Eyring et al., 2016), which is building on CMIP5 and seeks to address a bigger range of scientific questions about climate predictability and its sensitivity to various external processes and initialization techniques.

8. *Multiple timescale interactions*:

   As stated unambiguously throughout this book, the focus is on DCV as defined in the Foreword and Chapter 1. There are occasional descriptions of and references to longer timescale variability, but only if a broader context is required. However, other than the modulation of tropical cyclones by decadal and longer timescale variability (Chapter 9), the modulations of higher frequency phenomena such as the El Niño–Southern Oscillation (ENSO), the Madden–Julian Oscillation, and other intraseasonal variability by DCV are not included in this book for the sake of maintaining the stated focus. It is obvious from published research that these multiple timescale interactions are very important for the understanding and prediction of intraseasonal weather variability and interannual climate variability. This is also very important from the point of view of societal impacts because all climate variability/change impacts are largely felt by the society via changing weather such as changes in seasonal precipitation and temperature patterns, and frequency and/or duration of precipitation events and heat waves. Therefore, a substantial part of future research must include a multiple timescale approach, including the possibility that there are two-way or multi-way interactions between and among timescales.

11.4 THE STORY CONTINUES

We now come to the end of this journey, during which we saw that there has been a quantum leap from statistical Sunspot number–climate variability associations to a much better understanding of possible physics of DCV and tentative efforts to actually make decadal climate predictions in this modern era of DCV research. This progress on a truly historical scale was brought about by the convergence of data, models, and needs to understand and predict DCV. A search on the Web of Science global citation data base shows that the annual number of published research papers with decadal climate variability in title or topic increased from 0 in 1990 CE to 775 in 2019 CE, the annual number of published papers with decadal variability in title or topic increased from 2 in 1990 CE to 1014 in 2019 CE, and the annual number of published research papers with interdecadal climate variability in title or topic increased from 0 in 1990 CE to 164 in 2019 CE as shown in Figure 11.1. The total number of published research papers in from 1990 to 2019 CE was 8264, 10969, and 2187, in each search, respectively.

The modern era in DCV research and the quantum leap as indicated by the numbers of published research papers were stimulated by newly available instrument-measured data and ESMs; also, long paleoclimate records, many of them assembled to see past signatures of climate change, helped in extending DCV records back by several centuries. This quantum progress in the last 30 years was unimaginable in 1990 CE[1]. In this moment of success, we must not forget the scientists and non-scientists who toiled to interpret, justify, and use Sunspot–climate associations over several centuries; they are vindicated to a large extent by the achievements in this modern era of DCV research and prediction. At the least, they showed that DCV was

[1] A personal note: A prescient, anonymous reviewer of my paper in Journal of Climate, based on my Ph.D. dissertation and published in 1992, wrote, "This paper should be published after minor revisions. It is an exceptionally lucid presentation of a pioneering attempt to consider the coupled modes of the global ocean-atmosphere system with periods in the interannual to decadal range. Though it is now virtually unique I expect many related efforts to appear in the next few years." The numbers in Figure 11.1 testify to a very highly skillful prediction made by the anonymous reviewer!

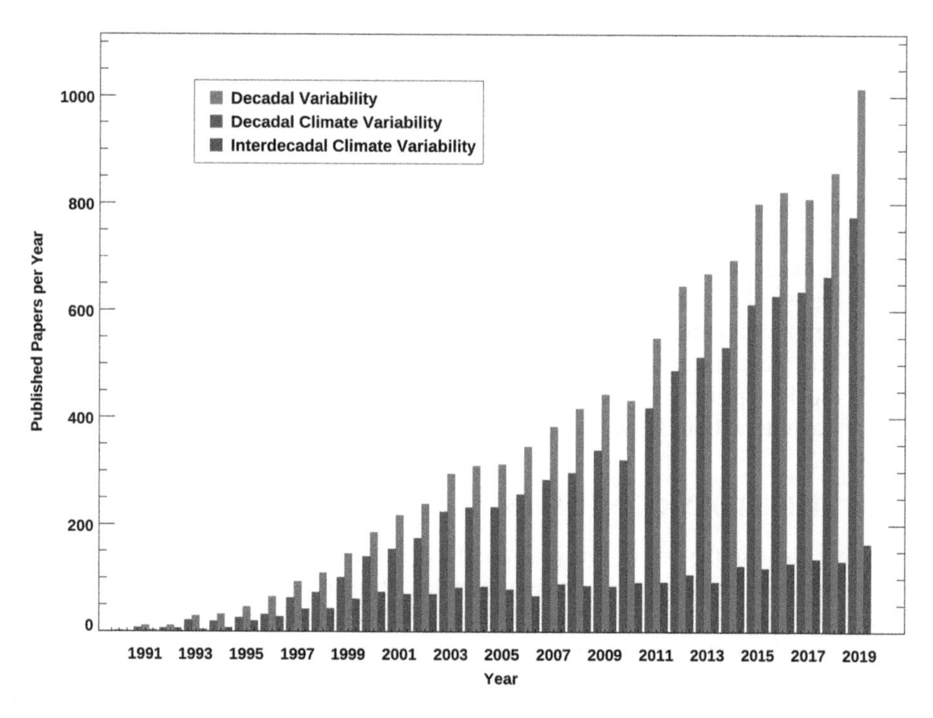

FIGURE 11.1 The annual numbers of published research papers with decadal climate variability in title or topic, decadal variability in title or topic, and interdecadal climate variability in title or topic. (The Web of Science global citation database.)

present and ubiquitous; now, we know that perhaps their belief in the Sun's influence on DCV is an almost certain possibility.

It has been mentioned several times in this book and in Mehta (2017) that it was wrong and unscientific of the climate science community to focus on two ends of the climate spectrum – potential climate change and seasonal to interannual variability – in the last 30 years. The progress described in this book has occurred in spite of this mistake. To a large extent, the climate science community is now paying the price of this mistake in terms of uncertainties in potential global warming projections; and in its inability to provide useful, skillful, and trustworthy climate and societal impacts predictions at multiyear to decadal lead times. It is very important to recognize that there have been attempts in the USA – mainly by the National Science Foundation (NSF), US Department of Agriculture (USDA)'s National Institute for Food and Agriculture (NIFA), National Oceanic and Atmospheric Administration (NOAA), and National Aeronautics and Space Administration (NASA) – to encourage and support research on DCV. These attempts, however, are too feeble or too short-lived to make lasting impacts. Science has progressed so far by the development of hypotheses and theories, and their verification by critical examinations of empirical evidence. We must continue this tradition in climate science and use natural DCV information to prepare for multiyear to multidecadal climate variability, test and validate climate models, correctly attribute real or perceived observed climate anomalies to natural DCV or climate change due to increasing greenhouse gases, and plan and execute adaptation scenarios. However, the society in general and the climate science community in particular must accept the possibility that we may not be able to completely understand causes and mechanisms of DCV phenomena, given the relatively short record of observed data and far from perfect climate models, in the next decade or two. We must also accept that climate is not stationary. The much longer history of intensive research on the interannual ENSO phenomenon teaches us that conclusions about even seasonal to interannual climate variability are not stationary, so we cannot expect complete certainty about the still-evolving field of DCV, at least in the next one to two decades. The increased awareness of DCV and its predictability, as evidenced in the last IPCC AR5 Report, is a welcome change and must be built upon for faster progress. Clearly, substantial efforts should be made in natural DCV research and applications, comparable to the efforts for research on potential climate change and its impacts. Moreover, objective and sound science requires us to pay at least as much attention to natural DCV as to anthropogenic climate change. Therefore, the story of DCV must continue.

Appendix 1: Abbreviations

AA	anthropogenic aerosols
AAO	Antarctic Oscillation
ACC	anthropogenic climate change
AGCM	Atmospheric General Circulation Model
AO	Arctic Oscillation
AOD	aerosol optical depth
AMO	Atlantic Multidecadal Oscillation
AMOC	Atlantic Meridional Overturning Circulation
APT	average predictability time
BAS	British Antarctic Survey
BCE	Before Current Era
CCCma	Canadian Centre for Climate modelling and analysis
CCSM	Community Climate System Model
CE	Current Era
CESM	Community Earth System Model
CFES	Coupled general circulation model For the Earth Simulator
CMIP	Coupled Model Intercomparison Project
CO_2	carbon dioxide
COADS	Comprehensive Ocean–Atmosphere Data Set
DCPP	Decadal Climate Prediction Project
DCV	decadal climate variability
DePreSys	Hadley Centre Decadal Prediction System
DJF	December–January–February
ECHO	coupled European Centre atmospheric GCM (ECHAM) and the Hamburg Primitive Equation Ocean GCM (HOPE)
ECMWF	European Centre for Medium-Range Weather Forecasts
EIWP	East Indian Warm Pool
EN	El Niño
ENIAC	Electronic Numerical Integrator And Computer
ENSO	El Niño–Southern Oscillation
EOF	empirical orthogonal function
ERA	European reanalysis
ERSST	Extended Reconstructed Sea Surface Temperature
ESM	Earth System Model
ET	evapotranspiration
FOAM	Fast Ocean–Atmosphere Model
GCM	General Circulation Model
GCR	galactic cosmic rays
GFCS	Global Framework for Climate Services
GFDL	Geophysical Fluid Dynamics Laboratory
GHG	greenhouse gases

GISST	Global Sea Ice and Sea Surface Temperature
GMST	global-mean surface temperature
GOSTA	Global Ocean Surface Temperature Atlas
HadISST	Hadley Centre Sea Ice and Sea Surface Temperature
HFP	higher frequency period
hPa	hectoPascals
IDPO	Interdecadal–Decadal Pacific Oscillation
IGY	International Geophysical Year
IMD	India Meteorological Department
IOD	Indian Ocean diploe
IPCC	Inter-Governmental Panel on Climate Change
IPO	Interdecadal Pacific Oscillation
IPWP	Indo-Pacific Warm Pool
IPY	International Polar Year
ISCCP	International Satellite Cloud Climatology Project
ITCZ	InterTropical Convergence Zone
JJA	June–July–August
KOE	Kuroshio–Oyashio Extension
LDA	local dynamical analog
LE	Lyapunov Exponent
LFP	low-frequency period
LIM	Linear Inverse Model
LN	La Niña
LSW	Labrador Sea water
MAM	March–April–May
MBT	mechanical bathythermograph
MC	Maritime Continent
MDR	Main Development Region
MIT	Massachusetts Institute of Technology
MME	multi-model ensemble
MODIS	MODerate resolution Imaging Spectroradiometer
MPI	Max Planck Institute
MRB	Missouri River Basin
MSLP	mean sea-level pressure
NAM	Northern Annular Mode
NAO	North Atlantic Oscillation
NASA	National Aeronautics and Space Administration
NCAR	National Center for Atmospheric Research
NCEP	National Centers for Environmental Prediction
NDVI	Normalized Difference Vegetation Index
NH	Northern Hemisphere
NIFA	National Institute of Food and Agriculture
NLLE	Nonlinear Local Lyapunov Exponent
NOAA	National Oceanic and Atmospheric Administration
NPO	North Pacific Oscillation
NSF	National Science Foundation

NTC net tropical cyclone index
NTCP Near-Term Climate Prediction
OCCAM Ocean Circulation and Climate Advanced Model
ORAS Ocean Reanalysis System
PBL planetary boundary layer
PC principal component
PDO Pacific Decadal Oscillation
PDSI Palmer Drought Severity Index
QBO Quasi-Biennial Oscillation
RCP8.5 Representative Concentration Pathway 8.5
REOF rotated empirical orthogonal function
RMSE root-mean-square error
SAM Southern Annular Mode
SAT surface air temperature
SCL Sunspot cycle length
SH Southern Hemisphere
SIC sea ice concentration
SIE sea ice edge
SLA sea-level anomaly
SLP sea level pressure
SO Southern Oscillation
SODA Simple Ocean Data Analysis
SOHO SOlar and Heliospheric Observatory
SON September–October–November
SORCE Solar Radiation and Climate Experiment
SPCZ South Pacific Convergence Zone
SPDO South Pacific Decadal Oscillation
SPEEDY Simplified Parameterizations primitivE-Equation DYnamics
SPEI Standardized Precipitation Evapotranspiration Index
SRES Special Report on Emissions Scenario
SST sea surface temperature
STC shallow tropical circulations or cells
STPI Summer Trans-Polar Index
TAG tropical Atlantic SST gradient
TC tropical cyclone
THC thermohaline circulation
TNA tropical North Atlantic
TPI Trans-Polar Index
TSA tropical South Atlantic
TWP tropical warm pool
UCN ultrafine condensation nuclei
UK United Kingdom
UNFCCC United Nations Framework Convention on Climate Change
US United States
USA United States of America
USDA United States Department of Agriculture

UV ultraviolet
VEI Volcanic Explosivity Index
WCRP World Climate Research Program
WES wind–evaporation–SST
WMO World Meteorological Organization
WPWP West Pacific Warm Pool
XBT expendable bathythermograph

References

Abram, N. J., M. K. Gagan, J. Cole, W. S. Hantoro, and M. Mudelsee, 2008: Recent intensification of tropical climate variability in the Indian Ocean. *Nat. Geosci.*, **1**, 849–853. doi:10.1038/ngeo357.

Allan, R., 2017: Gilbert Walker: A pioneer of modern day climatology. climate.envsci.rutgers.edu/climdyn2017/Gilbert_Walker.pdf.

Alexander, M., 2010: Extratropical air–sea interaction, sea surface temperature variability, and the Pacific decadal oscillation. Climate dynamics: Why does climate vary? *Geophys. Monogr.*, **189**, 123–148.

Alexander, M., and C. Deser, 1995: A mechanism for the recurrence of wintertime midlatitude SST anomalies. *J. Phys. Oceanogr.*, **25**, 122–137.

Alexander, M., and J. D. Scott, 2008: The role of Ekman ocean heat transport in the Northern Hemisphere response to ENSO. *J. Clim.*, **21**, 5688–5707.

Alexander, M., C. Deser, and M. S. Timlin, 1999: The reemergence of SST anomalies in the North Pacific Ocean. *J. Clim.*, **12**, 2419–2431.

Alexander, M. A., M. S. Timlin, and J. D. Scott, 2001: Winter-to-winter recurrence of sea surface temperature, salinity and mixed layer depth anomalies. *Prog. Oceanogr.*, **49**, 41–61.

Alpers, E. A., 2013: *The Indian Ocean in World History.* Oxford University Press, Oxford, MS. 172 p.

Ambaum, M. H. P., B. J. Hoskins, and D. B. Stephenson, 2001: Arctic oscillation or North Atlantic oscillation? *J. Clim.*, **14**, 3495–3507.

Andrews, M. B., J. R. Night, and L. J. Gray, 2015: A simulated lagged response of the North Atlantic Oscillation to the solar cycle over the period 1960–2009. *Environ. Res. Lett.*, **10**, 054022.

Appenzeller, C., T. F. Stocker, and M. Anklin, 1998: North Atlantic oscillation dynamics recorded in Greenland ice cores. *Science*, **282**, 446–449.

Aristarain, A. J., J. Jouzel, and C. Lorius, 1990: A 400 year isotope record of the Antarctic Peninsula climate. *Geophys. Res. Lett.*, **17**, 2369–2372.

Aristarain, A. J., J. Jouzel, and M. Pourchet, 1986: Past Antarctic Peninsula climate (1850–1980) deduced from ice core isotope record. *Clim. Change*, **8**, 69–89.

Armal, S., N. Devineni, and R. Khanbilvardi, 2018: Trends in extreme rainfall frequency in the contiguous United States: Attribution to climate change and climate variability modes. *J. Clim.*, **31**, 369–385.

Ashok, K., W.-L. Chan, T. Motoi, and T. Yamagata, 2004: Decadal variability of the Indian Ocean dipole. *Geophys. Res. Lett.*, **31**, L24207. doi:10.1029/2004GL021345.

Azad, S., T. S. Vignesh, and R. Narasimha, 2010: Periodicities in Indian monsoon rainfall over spectrally homogeneous regions. *Int. J. Climatol.*, **30**, 2289–2298.

Baede, A. P. M., 2015: Annex I. Glossary: IPCC – Intergovernmental Panel on Climate Change. http://www.ipcc.ch/pdf/glossary/ar4-wg1.pdf.

Bajish, C. C., S. Aoki, B. Taguchi, N. Komori, and S.-J. Kim, 2013: Quasi-decadal circumpolar variability of Antarctic sea ice. *SOLA*, **9**, 32–35.

Balachandran, N. K., D. Rind, P. Lonergan, and D. T. Shindell, 1999: Effects of solar cycle variability on the lower stratosphere and the troposphere. *J. Geophys. Res.*, **104**, 27321–27339.

Balling, R. C., and R. S. Cerveny, 2003: Cosmic ray flux impact on clouds? An analysis of radiosonde, cloud cover, and surface temperature records from the United States. *Theoret. Appl. Climatol.*, **75**, 225–231.

Balmaseda, M. A., M. K. Davey, and D. L. T. Anderson, 1995: Decadal and seasonal dependence of ENSO prediction skill. *J. Clim.*, **8**, 2705–2715.
Balmaseda, M. A., K. Mogensen, and A. T. Weaver, 2013: Evaluation of the ECMWF ocean reanalysis system ORAS4. *Quart. J. Roy. Meteorol. Soc.*, **139**, 1132–1161.
Barlow, L. H., J. W. C. White, R. G. Barry, J. C. Rogers, and P. M. Grootes, 1993: The North Atlantic Oscillation signature in deuterium and deuterium excess signals in the Greenland ice sheet Project 2 ice core. *Geophys. Res. Lett.*, **20**, 2901–2904.
Barnett, T. P., 1983: Interaction of the monsoon and Pacific trade wind system at interannual time scales. Part I: The equatorial zone. *Mon. Wea. Rev.*, **111**, 756–773.
Barnett, T. P., 1985: Variations in near-global sea level pressure. *J. Atmos. Sci.*, **42**, 478–501.
Barnston, A. G., and R. E. Livezey, 1987: Classification, seasonality, and persistence of low frequency atmospheric circulation patterns. *Mon. Wea. Rev.*, **115**, 1083–1126.
Battisti, D. S., and A. C. Hirst, 1989: Interannual variability in a tropical atmosphere–ocean model: Influence of the basic state, ocean geometry and nonlinearity. J. *Atmos. Sci.*, **46**, 1687–1712.
Bengtsson, L., 1984: Computer requirements for atmospheric modelling. Workshop on Using Multiprocessors in Meteorological Models, European Centre for Medium-Range Weather Forecasts, December 1984. www.ecmwf.int/en/learning/workshops-and-seminars/past- workshops/1984-mulitprocessors-in-meteorological-models.
Berndtsson, R., C. Uvo, M. Matsumoto, K. Jinno, A. Kawamura, S. G. Xu, and J. Olsson, 2001: Solar-climatic relationship and implications for hydrology. *Nordic Hydrol.*, **32**, 65–84.
Bennett, A. J., and R. G. Harrison, 2008: Surface measurement system for the atmospheric electrical vertical conduction current density, with displacement current correction. *J. Atmos. Sol. Terr. Phys.*, **70**, 1373–1381. doi:10.1016/j.jastp.2008.04.014.
Bhalme, H. N., and D. A. Mooley, 1981: Cyclic fluctuations in the flood area and relationship with the double (Hale) sunspot cycle. *J. Appl. Meteor.*, **20**, 1041–1048.
Biondi, F., A. Gershunov, and D. R. Cayan, 2001: North Pacific decadal climate variability since 1661. *J. Clim.*, **14**, 5–10.
Bjerknes, V., 1904: Das Problem der Wettervorhersage, betrachtet von Standpunkt der Mechanik und Physik. *Meteorol. Z.*, **21**, 1–7.
Bjerknes, J., 1966: A possible response of the atmospheric Hadley circulation to equatorial anomalies of ocean temperature. *Tellus*, **18**, 820–828.
Bjerknes, J., 1969: Atmospheric teleconnections from the equatorial Pacific. *Mon. Wea. Rev.*, **97**, 163–172.
Black, D. E., M. A. Abahazi, R. C. Thunell, A. Kaplan, E. J. Tappa, and L. C. Peterson, 2007: An 8-century tropical Atlantic SST record from the Cariaco Basin: Baseline variability, twentieth-century warming, and Atlantic hurricane frequency. *Paleoceanography*, **22**, PA4204. doi:10.1029/2007PA001427.
Black, D. E., L. C. Peterson, J. T. Overpeck, A. Kaplan, M. N. Evans, and M. Kashgarian, 1999: Eight centuries of Atlantic Ocean atmosphere variability. *Science*, **286**, 1709–1713.
Black, D. E., R. C. Thunell, A. Kaplan, L. C. Peterson, and E. J. Tappa, 2004: A 2000-year record of Caribbean and tropical North Atlantic hydrographic variability. *Paleoceanography*, **19**, PA2022.
Blasing, T. J., and D. Duvick, 1984: Reconstruction of precipitation history in North American corn belt using tree rings. *Nature*, **307**, 143–145.
Boberg, F., and H. Lundstedt, 2002: Solar wind variations related to fluctuations of the North Atlantic Oscillation. *Geophys. Res. Lett.*, **29**, 1718. doi:10.1029/2002GL014903.
Boeing, G., 2016: Visual analysis of nonlinear dynamical systems: Chaos, fractals, self- similarity and the limits of prediction. *Systems*, **4**, 37.
Boer, G. J., 2000: A study of atmosphere–ocean predictability on long time scales. *Clim. Dyn.*, **16**, 469–477.

Boer, G. J., 2004: Long time-scale potential predictability in an ensemble of coupled climate models. *Clim. Dyn.*, **23**, 29–44.

Boer, G. J., 2009: Changes in interannual variability and decadal potential predictability under global warming. *J. Clim.*, **22**, 3098–3109.

Boer, G. J., 2011: Decadal potential predictability of twenty-first century climate. *Clim. Dyn.*, **36**, 1119–1133.

Boer, G. J., and S. J. Lambert, 2008: Multi-model decadal potential predictability of precipitation and temperature. *Geophys. Res. Lett.*, **35**, L05706. doi:10.1029/2008GL033234.

Boer, G. J., et al., 2016: The decadal climate prediction project (DCPP) contribution to CMIP6. *Geosci. Model Dev.*, **9**, 3751–3777.

Boivin, N., A. Crowther, M. Prendergast, and D. Q. Fuller, 2014: Indian Ocean food globalisation and Africa. *Afr. Archaeol. Rev.*, **31**, 547–581.

Bottomley, M., C. K. Folland, J. Hsiung, R. E. Newell, and D. E. Parker, 1990: *Global Ocean Surface Temperature Atlas (GOSTA).* Her Majesty's Stationery Office, London.

Bradley, J., 1728: An account of a new discovered motion of the fixed stars. *Philos. Trans. R. Soc. London B*, **35**, 637–661. doi:10.1098/rstl.1727.0064.

Bradley, R. S., and P. D. Jones (eds.), 1992: *Climate since AD 1500.* Routledge, London, 679 p. ISBN 0-415-07593-9.

Branstator, G., and H. Teng, 2010: Two limits of initial-value decadal predictability in a CGCM. *J. Clim.*, **23**, 6292–6311.

Bravais, A., 1846: Analyse mathématique sur les probabilités des erreurs de situation d'un point (Mathematical analysis of the probabilities of errors in a point's location). *Mémoires Presents Par Divers Savants à l'Académie des Sciences de l'Institut de France. Sciences Mathématiques et Physiques*, **9**, 255–332.

Bretherton, C. S., and D. S. Battisti, 2000: An interpretation of the results from atmospheric general circulation models forced by the time history of the observed sea surface temperature distribution. *Geophys. Res. Lett.*, **27**, 767–770.

Breugem, W.-P., W. Hazeleger, and R. J. Haarsma, 2007: Mechanisms of northern tropical Atlantic variability and response to CO_2 doubling. *J. Clim.*, **20**, 2691–2705.

Bronniman, S., T. Ewen, T. Griesser, and R. Jenne, 2006: Multidecadal signal of solar variability in the upper troposphere during the 20th century. *Space Sci. Rev.*, **125**, 305–317.

Brown, G. M., and P. W. Leftwich Jr., 1982: A compilation of eastern and central North Pacific tropical cyclone data. NOAA Tech. Memo. NWS NHC 16, Miami, FL, 15 pp.

Burke, H. K., and A. A. Few, 1978: Direct measurements of the atmospheric conduction current. *J. Geophys. Res.*, **83**, 3093–3098. doi:10.1029/JC083iC06p03093.

Butler, C. J., 1994: Maximum and minimum temperatures at Armagh-observatory, 1844–1992, and the length of the sunspot cycle. *Solar Phys.*, **152**, 35–42.

Camuffo, D., 2001: Lunar influences on climate. *Earth Moon Planet*, **85**, 99–113.

Cane, M. A., S. E. Zebiak, and S. C. Dolan, 1986: Experimental forecasts of El Niño. *Nature*, **321**, 827–832.

Capotondi, A., and M. A. Alexander, 2001: Rossby waves in the tropical North Pacific and their role in decadal thermocline variability. *J. Phys. Oceanogr.*, **31**, 3496–3515.

Capotondi, A., M. A. Alexander, and C. Deser, 2003: Why are there Rossby wave maxima in the Pacific at 10°S and 13°N? *J. Phys. Oceanogr.*, **33**, 1549–1563.

Capotondi, A., M. A. Alexander, C. Deser, and M. J. McPhaden, 2005: Anatomy and decadal evolution of the Pacific subtropical–tropical cells (STCs). *J. Clim.*, **18**, 3739–3758.

Carrington, R. C., 1863: *Observations of the Spots on the Sun.* Williams and Norgate, London, 1–248 pp.

Carton, J. A., X. Cao, B. S. Giese, and A. M. da Silva, 1996: Decadal and interannual SST variability in the tropical Atlantic. *J. Phys. Oceanogr.*, **26**, 1165–1175.

Carton, J. A., G. Chepurin, X. Cao, and B. S. Giese, 2000: A simple ocean data assimilation analysis of the global upper ocean 1950–95. Part I: Methodology. *J. Phys. Oceanogr.*, **30**, 294–309.

Castagnoli, G. C., G. Bonino, C. Taricco, and S. M. Bernasconi, 2002a: Solar radiation variability in the last 1400 years recorded in the carbon isotope ratio of a Mediterranean sea core. *Adv. Space Res.*, **29**, 1989–1994. Solar Variability and Solar Physics Missions.

Castagnoli, G. C., G. Bonino, and C. Taricco, 2002b: Long term solar-terrestrial records from sediments: Carbon isotopes in planktonic foraminifera during the last millennium. *Adv. Space Res.*, **29**, 1537–1549. International Solar Cycle Study.

Cerveny, R. S., and J. A. Shaffer, 2001: The moon and El Nino. *Geophys. Res. Lett.*, **28**, 25–28.

Cerveny, R. S., and R. C. Balling, 1999: Lunar influence on diurnal temperature range. *Geophys. Res. Lett.*, **26**, 1605–1607.

Chambers, F., 1886: Sunspots and prices of Indian food-grains. *Nature*, **34**, 100–104.

Chan, J. C. L., 2008: Decadal variations of intense typhoon occurrence in the western North Pacific. *Proc. Math. Phys. Eng. Sci.*, **464**, 249–272.

Chang, P., L. Ji, and H. Li, 1997: A decadal climate variation in the tropical Atlantic Ocean from thermodynamic air-sea interactions. *Nature*, **385**, 516–518.

Chang, P., L. Ji, and R. Saravanan, 2001: A hybrid coupled model study of tropical Atlantic variability. *J. Clim.*, **14**, 361–390.

Chang, P., R. Saravanan, L. Ji, and G. C. Hegerl, 2000: The effect of local sea surface temperatures on the atmospheric circulation over the tropical Atlantic sector. *J. Clim.*, **13**, 2195–2216.

Charles, C. D., D. E. Hunter, and R. G. Fairbanks, 1997: Interaction between the ENSO and the Asian Monsoon in a coral record of tropical climate. *Science*, **277**, 925–927.

Charney, J. G., and J. G. De Vore, 1979: Multiple flow equilibria in the atmosphere and blocking. *J. Atmos. Sci.*, **36**, 1205–1216.

Charney, J. G., and J. Shukla, 1981: Predictability of monsoons. In: J. Lighthill and R. P. Pearce, *Monsoon Dynamics*, Cambridge University Press, New York, 735p.

Charney, J. G., R. Fjortoft, and J. von Neumann, 1950: Numerical integration of the barotropic vorticity equation. *Tellus*, **2**, 237–254.

Chaudhuri, S., J. Pal, and S. Guhathakurta, 2015: The influence of galactic cosmic ray on all India annual rainfall and temperature. *Adv. Space Res.*, **55**, 1158–1167.

Chelliah, M., and G. D. Bell, 2004: Tropical multidecadal and interannual climate variability in the NCEP–NCAR reanalysis. *J. Clim.*, **17**, 1777–1803.

Chelton, D. B., R. A. de Szoeke, M. G. Schlax, K. El Naggar, and N. Siwertz, 1998: Geographical variability of the first baroclinic Rossby radius of deformation. *J. Phys. Oceanogr.*, **28**, 433–460.

Chen, J. Y., 1999: The study on the responding of ENSO and severe floods in the Yangtze River to the astronomical factors. *Chinese J. Geophys.-Chinese edition*, **42**, 30–40.

Chen, G., C. Y. Fang, C. Y. Zhang, and Y. Chen, 2004: Observing the coupling effect between warm pool and "rain pool" in the Pacific Ocean. *Remote Sens. Environ.*, **91**, 153–159.

Chiodo, G., J. Oehrlein, L. M. Polvani, J. C. Fyfe, and A. K. Smith, 2019: Insignificant influence of the 11-year solar cycle on the North Atlantic Oscillation. *Nature Geosci.*, **12**, 94–99.

Christoforou, P., and S. Hameed, 1997: Solar cycle and the Pacific 'centers of action'. *Geophys. Res. Lett.*, **24**, 293–296.

Chu, P. S., 2002: Large-scale circulation features associated with decadal variations of tropical cyclone activity over the central North Pacific. *J. Clim.*, **15**, 2678–2689.

Chu, P. S., and J. D. Clark, 1999: Decadal variations of tropical cyclone activity over the central North Pacific. *Bull. Amer. Meteor. Soc.*, **80**, 1875–1881.

Chu, P. S., and J. Wang, 1997: Tropical cyclone occurrences in the vicinity of Hawaii: Are the differences between El Niño and non-El Niño years significant? *J. Clim.*, **10**, 2683–2689.

Clauser, B. E., 2007: The life and labors of Francis Galton: A review of four recent books about the father of behavioral statistics. *J. Edu. Behav. Stat.*, **32**, 440–444.

Clayton, H. H., 1923: *World Weather.* MacMillan Co., New York, 393 p.

Clegg, S. L., and T. M. L. Wigley, 1984: Periodicities in precipitation in north-east China. *Geophys. Res. Lett.*, **11**, 1219–1222.

Clement, A. C., R. Seager, and R. Murtugudde, 2005: Why are there tropical warm pools? *J. Clim.*, **18**, 5294–5311.

Cobb, K. M., C. D. Charles, and D. E. Hunter, 2001: A central tropical Pacific coral demonstrates Pacific, Indian, and Atlantic decadal climate connections. *Geophys. Res. Lett.*, **28**, 2209–2212. doi:10.1029/2001GL012919.

Cohen, T. J., and E. I. Sweetser, 1975: The 'spectra' of the solar cycle and of data for Atlantic tropical cyclones. *Nature*, **256**, 295–296.

Cole, J. E., R. B. Dunbar, T. R. McClanahan, and N. A. Muthiga, 2000: Tropical Pacific forcing of decadal SST variability in the western Indian Ocean over the past two centuries. *Science*, **287**, 617–619. doi:10.1126/science.287.5453.617.

Collins, M., and M. Allen, 2002: Assessing the relative roles of initial and boundary conditions in interannual to decadal climate predictability. *J. Clim.*, **15**, 3104–3109.

Comiso, J. C., R. A. Gersten, L. V. Stock, J. Turner, G. J. Perez, and K. Cho. 2017: Positive trend in the Antarctic Sea ice cover and associated changes in surface temperature. *J. Clim.*, **30**, 2251–2267. doi:10.1175/JCLI-D-16-0408.1.

Cook, E. R., R. D. D'Arrigo, and K. R. Briffa, 1998: The North Atlantic Oscillation and its expression in circum-Atlantic tree-ring chronologies from North America and Europe. *Holocene*, **8**, 9–17.

Cook, E. R., R. D. D'Arrigo, and M. E. Mann, 2001: A well-verified, multi-proxy reconstruction of the winter North Atlantic Oscillation Index since AD 1400. *J. Clim.*, **15**, 1754–1764.

Cook, E. R., D. M. Meko, and C. W. Stockton, 1997: A new assessment of possible solar and lunar forcing of the bidecadal drought rhythm in the western United States. *J. Clim.*, **10**, 1343–1356.

Cooper, M. C., P. E. O'Sullivan, and A. J. Shine, 2000: Climate and solar variability recorded in Holocene laminated sediments - a preliminary assessment. *Quat. Int.*, **68**, 363–371.

Courant, R., K. Friedrichs, and H. Lewy, 1928: Über die partiellen Differentialgleichungen der mathematischen Physik. *Math. Ann.*, **100**, 32–74.

Coward, A. C., and B. A. de Cuevas, 2005: The OCCAM 66 level model: Physics, initial conditions and external forcing. Southampton Oceanography Centre Internal Rep., 99. Southampton Oceanography Centre, Southampton. 21 pp.

Crantz, D., 1765: *The History of Greenland.* www.explorenorth.com/library/history/history_of_greenland-1766.html.

CRC, 1995: *Natural Climate Variability on Decade-to-Century Time Scales.* Climate Research Committee, Board of Atmospheric Sciences and Climate, National Research Council, National Academy Press, Washington. D.C., 615 p.

Cubasch, U., R. Voss, G. C. Hegerl, J. Waszkewitz, and T. C. Crowley, 1997: Simulation with an O-AGCM of the influence of variations of the solar constant on the global climate. *Clim. Dyn.*, **13**, 757–767. doi:10.1007/s003820050196.

Cubasch, U., E. Zorita, F. Kaspar, J. F. Gonzales-Rouco, H. von Storch, and K. Prommel, 2006: Simulation of the role of solar and orbital forcing on climate. *Adv. Space Res.*, **37**, 1629–1634. doi:10.1016/j.asr.2005.04.076.

Cullen, H. M., R. D'Arrigo, E. R. Cook, and M. E. Mann, 2000: Multiproxy reconstructions of the North Atlantic Oscillation. *Paleooceonography*, **16**, 27–39.

Currie, R. G., 1974: Solar cycle signal in surface air temperature. *J. Geophys. Res.*, **79**, 5657–5660.

Currie, R. G., 1976: The spectrum of sea level from 4 to 40 years. *Geophys. J. Roy. Astron. Soc.*, **46**, 513–520.

Currie, R. G., 1979: Distribution of solar cycle signal in surface air temperature over North America. *J. Geophys. Res.*, **84**, 753–761.

Currie, R. G., 1980: Detection of the 11-year sunspot cycle signal in earth rotation. *Geophys. J. Roy. Astronom. Soc.*, **61**, 131–139.

Currie, R. G., 1981: Solar cycle signal in air temperature in North America: Amplitude, gradient, phase and distribution. *J. Atmos. Sci.*, **38**, 808–818.

Currie, R. G., 1984: On bistable phasing of 18.6-year induced flood in India. *Geophys. Res. Lett.*, **11**, 50–53.

Currie, R. G., 1993: Luni-solar 18.6-year and 10–11-year solar-cycle signals in South-African rainfall. *Int. J. Climatol.*, **13**, 237–256.

Currie, R. G., and R. W. Fairbridge, 1985: Periodic 18.6-year and cyclic 11-year induced drought and flood in northeastern China and some global implications. *Quatern. Sci. Rev.*, **4**, 109–134.

Currie, R. G., and D. P. O'Brien, 1988: Periodic 18.6-year and cyclic 10 to 11 year signals in northeastern United States precipitation data. *Int. J. Climatol.*, **8**, 255–281.

Currie, R. G., and D. P. O'Brien, 1990: Deterministic signals in USA precipitation records. *Int. J. Climatol.*, **10**, 795–818.

Currie, R. G., and R. G. Vines, 1996: Evidence for luni-solar M(n) and solar cycle S-c signals in Australian rainfall data. *Int. J. Climatol.*, **16**, 1243–1265.

Currie, R. G., T. Wyatt, and D. P. O'Brien, 1993: Deterministic signals in European fish catches, wine harvests, and sea-level, and further experiments. *Int. J. Climatol.*, **13**, 665–687. doi:10.1002/joc.3370130607.

Curry, R. G., and M. S. McCartney, 1996: Labrador sea water carries northern climate signal south. *Oceanus*, **39**, 24–28.

Czaja, A., P. van der Vaart, and J. Marshall, 2002: A diagnostic study of the role of remote forcing in tropical Atlantic variability. *J. Clim.*, **15**, 3280–3290.

Dai, A., J. C. Fyfe, S. P. Xie, and X. Dai, 2015: Decadal modulation of global surface temperature by internal climate variability. *Nature Clim. Change*, **5**, 555–559.

Damnati, B., and M. Taieb, 1995: Solar and ENSO signatures in laminated deposits from Lake Magadi (Kenya) during the Pleistocene/Holocene transition. *J. African Earth Sci.*, **21**, 373–382.

D'Arrigo, R., and R. Wilson, 2006: On the Asian expression of the PDO. *Int. J. Climatol.*, **26**, 1607–1717.

D'Arrigo, R., R. Villalba, and G. Wiles, 2001: Tree-ring estimates of Pacific decadal climate variability. *Clim. Dyn.*, **18**, 219–224.

D'Arrigo, R., et al., 2006: The reconstructed Indonesian warm pool sea surface temperatures from tree rings and corals: Linkages to Asian monsoon drought and El Niño–Southern Oscillation. *Paleoceanography*, **21**, PA3005. doi:10.1029/2005PA001256.

Da Silva, A., A. C. Young, and S. Levitus, 1994: Atlas of Surface Marine Data, Volume 1: Algorithms and Procedures. US Department of Commerce: Washington, DC. NOAA Atlas NESDIS 6.

Dee, D. P., et al., 2011: The ERA-Interim reanalysis: Configuration and performance of the data assimilation system. *Quart. J. Roy. Meteorol. Soc.*, **137**, 553–597.

Defant, A., 1924: Die Schwankungen der atmosphärischen Zirkulation über dem Nordatlantischen Ozean im 25-jährigen Zeitraum 1881–1905. *Geogr. Ann.*, **6**, 13–41.

DelSole, T., and M. K. Tippett, 2008: Predictable components and singular vectors. *J. Atmos. Sci.*, **65**, 1666–1678.

DelSole, T., and M. K. Tippett, 2009a: Average predictability time. Part I: Theory. *J. Atmos. Sci.*, **66**, 1172–1187.

DelSole, T., and M. K. Tippett, 2009b: Average predictability time. Part II: Seamless diagnosis of predictability on multiple time scales. *J. Atmos. Sci.*, **66**, 1188–1204.

DelSole, T., L. Jia, and M. K. Tippett, 2013: Decadal prediction of observed and simulated sea surface temperatures. *Geophys. Res. Lett.*, **40**, 2773–2778.

Delworth, T., and V. M. Mehta, 1998: Simulated interannual to decadal climate variability of tropical and subtropical Atlantic. *Geophys. Res. Lett.*, **25**, 2825–2828.

Deser, C., 2000: On the teleconnectivity of the "Arctic oscillation." *Geophys. Res. Lett.*, **27**, 779–782.

Deser, C., and M. L. Blackmon, 1993: Surface climate variations over the North Atlantic Ocean during winter: 1900–1989. *J. Clim.*, **6**, 1743–1753.

Deser, C., J. E. Walsh, and M. S. Timlin, 2000: Arctic sea ice variability in the context of recent atmospheric circulation trends. *J. Clim.*, **13**, 617–633.

DeWeaver, E., and S. Nigam, 2000a: Do stationary waves drive the zonal-mean jet anomalies of the northern winter? *J. Clim.*, **13**, 2160–2176.

DeWeaver, E., and S. Nigam, 2000b: Zonal-eddy dynamics of the North Atlantic Oscillation. *J. Clim.*, **13**, 3893–3914.

Ding, R. Q., and J. P. Li, 2007: Nonlinear finite-time Lyapunov exponent and predictability. *Phys. Lett. A*, **364**, 396–400.

Ding, R. Q., J. P. Li, and K. J. Ha, 2008: Trends and interdecadal changes of weather predictability during 1950s–1990s. *J. Geophys. Res.*, **113**, D24112. doi:10.1029/2008JD010404.

Ding, R. Q., J. P. Li, and K. H. Seo, 2010: Predictability of the Madden–Julian oscillation estimated using observational data. *Mon. Wea. Rev.*, **138**, 1004–1013.

Ding, R. Q., J. P. Li, and K. H. Seo, 2011: Estimate of the predictability of boreal summer and winter intraseasonal oscillations from observations. *Mon. Wea. Rev.*, **139**, 2421–2438.

Ding, R. Q., J. P. Li, F. Zheng, J. Feng, and D. Liu, 2016: Estimating the limit of decadal–scale climate predictability using observational data. *Clim. Dyn.*, **46**, 1563–1580.

Dommenget, D., and M. Latif, 2000: Interannual to decadal variability in the tropical Atlantic. *J. Clim.*, **13**, 777–792.

Dommenget, D., and M. Latif, 2002: A cautionary note on the interpretation of EOFs. *J. Clim.*, **15**, 216–225.

Doney, S. C., S. Yeager, G. Danabasoglu, W. G. Large, and J. C. McWilliams, 2003: Modeling global oceanic interannual variability (1958–1997): Simulation design and model-data evaluation. NCAR Technical Report NCAR/TN-4521STR, 48 pp.

Dong, L., and M. J. McPhaden, 2017: The role of external forcing and internal variability in regulating global mean surface temperatures on decadal timescales. *Environ. Res. Lett.*, **12**, 034011. doi:10.1088/1748-9326/aa5dd8.

Eade, R., et al., 2014: Do seasonal-to-decadal climate predictions underestimate the predictability of the real world? *Geophys. Res. Lett.*, **41**, 5620–5628.

Easterling, D. R., and M. F. Wehner, 2009: Is the climate warming or cooling? *Geophys. Res. Lett.*, **36**, L08706. doi:10.1029/2009GL037810.

Elsner, J. B., and T. H. Jagger, 2008: United States and Caribbean tropical cyclone activity related to the solar cycle. *Geophys. Res. Lett.*, **35**, L18705.

Elsner, J. B., and S. P. Kavlakov, 2001: Hurricane intensity changes associated with geomagnetic variation. *Atmos. Sci. Lett.*, **2**, 86–93.

Elsner, J. B., T. H. Jagger, and R. E. Hodges, 2010: Daily tropical cyclone intensity response to solar ultraviolet radiation. *Geophys. Res. Lett.*, **37**, L09701.

Elsner, J. B., T. Jagger, and X.-F. Niu, 2000: Changes in the rates of North Atlantic major hurricane activity during the 20th century. *Geophys. Res. Lett.*, **27**, 1743–1746.

Elsner, J. B., G. S. Lehmiller, and T. B. Kimberlain, 1996: Objective classification of Atlantic basin hurricanes. *J. Clim.*, **9**, 2880–2889.

Emanuel, K. A., 1991: The theory of hurricanes. *Annu. Rev. Fluid. Mech.*, **23**, 179–196.

Enfield, D. B., and D. A. Mayer, 1997: Tropical Atlantic sea surface temperature variability and its relationship to El Niño-Southern Oscillation. *J. Geophys. Res.*, **102**, 929–945.

Enfield, D. B., A. M. Mestas-Nuñez, D. A. Mayer, and L. Cid-Serrano, 1999: How ubiquitous is the dipole relationship in the tropical Atlantic sea surface temperatures? *J. Geophys. Res.*, **104**, 7841–7848.

Enfield, D. B., A. M. Mestas-Nuñez, and P. J. Trimble, 2001: The Atlantic multidecadal oscillation and its relation to rainfall and river flows in the continental U.S. *Geophys. Res. Lett.*, **28**, 2077–2080.

England, M. H., et al., 2014: Recent intensification of wind-driven circulation in the Pacific and the ongoing warming hiatus. *Nature Clim. Change*, **4**, 222–227.

Exner, F. M., 1913: Übermonatliche Witterungsanomalien auf der nördlichen Erdhälfte im Winter. *Sitzungsberichte d. Kaiserl. Akad. der Wissenschaften*, **122**, 1165–1241.

Exner, F. M., 1924: Monatliche Luftdruck- und Temperaturanomalien auf der Erde, Sitzungsberichted. *Kaiserl. Akad. der Wissenschaften*, **133**, 307–408.

Eyring, V., S. Bony, G. A. Meehl, C. A. Senior, B. Stevens, R. J. Stouffer, and K. E. Taylor, 2016: Overview of the coupled model intercomparison project phase 6 (CMIP6) experimental design and organization. *Geosci. Model Dev.*, **9**, 1937–1958.

Fang, C., L. Wu, and X. Zhang, 2014: The impact of global warming on the Pacific decadal oscillation and the possible mechanism. *Adv. Atmos. Sci.*, **31**, 118–130.

Farneti, R., F. Molteni, and F. Kucharski, 2014: Pacific interdecadal variability driven by tropical-extratropical interactions. *Clim. Dyn.*, **42**, 3337–3355.

Fasullo, J., and P. Webster, 1999: Warm pool SST variability in relation to the surface energy balance. *J. Clim.*, **12**, 1292–1305.

Feldstein, S. B., and C. Franzke, 2006: Are the North Atlantic Oscillation and the northern annular mode distinguishable? *J. Atmos. Sci.*, **63**, 2915–2930.

Fels, S. B., J. D. Mahlman, M. D. Schwarzkopf, and R. W. Sinclair, 1980: Stratospheric sensitivity to perturbations in ozone and carbon dioxide: Radiative and dynamical responses. *J. Atmos. Sci.*, **37**, 2265–2297.

Fernandez, M., P. Huang, B. McCarl, and V. M. Mehta, 2016: Value of decadal climate variability information for agriculture in the Missouri River Basin. *Clim. Change*, **139**, 517–533.

Ferrel, W., 1858: The influence of the Earth's rotation upon the relative motion of bodies near its surface. *Astron. J.*, **109V**, 97–100.

Fogt, R. L., J. Perlwitz, A. J. Monaghan, D. H. Bromwich, J. M. Jones, and G. J. Marshall, 2009: Historical SAM variability. Part II: Twentieth century variability and trends from reconstructions, observations, and the IPCC AR4 models. *J. Clim.*, **22**, 5346–5365. doi:10.1175/2009JCLI2786.1.

Folland, C. K., and D. E. Parker, 1995: Correction of instrumental biases in historical sea surface temperature data. *Q. J. Roy. Meteor. Soc.*, **121**, 319–367.

Folland, C. K., J. A. Owen, M. N. Ward, and A. W. Colman, 1991: Prediction of seasonal rainfall in the Sahel region using empirical and dynamical methods. *J. Forecast.*, **10**, 21–56.

Folland, C. K., D. E. Parker, A. W. Colman, and R. Washington, 1998: Largescale modes of ocean surface temperature since the late nineteenth century. Hadley Centre, UK meteorological office. *Clim. Res. Tech. Note CRTN*, **81**, 45 pp.

Folland, C. K., T. N. Palmer, and D. E. Parker, 1986: Sahel rainfall and worldwide sea temperatures. *Nature*, **320**, 602–606.

Folland, C. K., J. A. Renwick, M. J. Salinger, and A. B. Mullan, 2002: Relative influences of the interdecadal Pacific Oscillation and ENSO on the South Pacific convergence zone. *Geophys. Res. Lett.*, **29**, 1643. doi:10.1029/2001GL014201.

Foltz, G. R., and M. J. McPhaden, 2006: The role of oceanic heat advection in the evolution of tropical North and South Atlantic SST anomalies. *J. Clim.*, **19**, 6122–6138.

Fontaine, B., S. Janicot, and P. Roucou, 1999: Coupled ocean-atmosphere surface variability and its climate impacts in the tropical Atlantic region. *Clim. Dyn.*, **15**, 451–473.

Frank, N., and S. A. Husain, 1971: The deadliest tropical cyclone in history? *Bull. Amer. Meteorol. Soc.*, **52**, 438–444.

Frankignoul, C., and K. Hasselmann, 1977: Stochastic climate models. Part II: Application to sea-surface temperature anomalies and thermocline variability. *Tellus*, **29**, 289–305. doi:10.1111/j.2153-3490.1977.tb00740.x.

Frankignoul, C., and E. Kestenare, 2002: The surface heat flux feedback. Part 1: Estimates from observations in the Atlantic and the North Pacific. *Clim. Dyn.*, **19**, 633–647.

Frankignoul, C., and E. Kestenare, 2005: Air–Sea interactions in the tropical Atlantic: A view based on lagged rotated maximum covariance analysis. *J. Clim.*, **18**, 3874–3890.

Frankignoul, C., and R. W. Reynolds, 1983: Testing a dynamicalmodel formidlatitude sea surface temperature anomalies. *J. Phys. Oceanogr.*, **13**, 1131–1145. doi:10.1175/1520-0485(1983)013,1131:TADMFM.2.0.CO;2.

Friis-Christensen, E., 1993: Solar-activity variations and global temperature. *Energy*, **18**, 1273–1284.

Friis-Christensen, E., and K. Lassen, 1991: Length of the solar-cycle - an indicator of solar-activity closely associated with climate. *Science*, **254**, 698–700.

Fu, R., A. D. Del Genio, and W. B. Rossow, 1994: Influence of ocean surface conditions on atmospheric vertical thermodynamic structure and deep convection. *J. Clim.*, **7**, 1092–1108.

Fukumori, I., 2006: What is data assimilation really solving, and how is the calculation actually done? In: E. P. Chasignet and J. Verron (eds.) *Ocean Weather Forecasting: An Integrated View of Oceanography*. Springer, Dordrecht, pp. 317–339.

Fyfe, J. C., et al., 2016: Making sense of the early-2000s warming slowdown. *Nature Clim. Change*, **6**, 224–228.

Galton, F., 1888: Co-relations and their measurement, chiefly from anthropometric data. *Proc. Roy. Soc. London*, **45**, 135–145.

Garcia, R., L. Gimeno, E. Hernandez, R. Prieto, and P. Ribera, 2000: Reconstructing the North Atlantic atmospheric circulation in the 16th, 17th and 18th centuries from historical sources. *Clim. Res.*, **14**, 147–151.

Garnett, R., N. Nirupama, C. E. Haque, and T. S. Murty, 2006: Correlates of Canadian prairie summer rainfall: Implications for crop yields. *Clim. Res.*, **32**, 25–33.

Gedalof, Z., and D. J. Smith, 2001: Interdecadal climate variability and regime-scale shifts in Pacific North America. *Geophys. Res. Lett.*, **28**, 1515–1518.

Gershunov, A., and T. P. Barnett, 1998: Interdecadal modulation of ENSO teleconnections. *Bull. Amer. Meteor. Soc.*, **79**, 2715–2725.

Giese, B., S. Urizar, and N. Fuckar, 2002: Southern Hemisphere origins of the 1976 climate shift. *Geophys. Res. Lett.*, **29**, 1014. doi:10.1029/2001GL013268.

Gillett, N. P., T. D. Kell, and P. D. Jones, 2006: Regional climate impacts of the southern annular mode. *Geophys. Res. Lett.*, **33**, L23704. doi:10.1029/2006GL027721.

Gleisner, H., and P. Thejll, 2003: Patterns of tropospheric response to solar variability. *Geophys. Res. Lett.*, **30**. doi:10.1029/2003GL017129.

Gleisner, H., P. Thejll, M. Stendel, E. Kaas, and B. Machenhauer, 2005: Solar signals in tropospheric re-analysis data: Comparing NCEP/NCAR and ERA40. *J. Atmos. Solar-Terrestr. Phys.*, **67**, 785–791.

Glueck, M. F., and C. W. Stockton, 2001: Reconstruction of the North Atlantic Oscillation. *Int. J. Climatol.*, **21**, 1453–1465.

Godfrey, J. S., and E. J. Lindstrom, 1989: The heat budget of the equatorial western Pacific surface mixed layer. *J. Geophys. Res.*, **94**, 8007–8017.

Goldernberg, S. B., C. W. Landsea, A. M. Mestas-Nuñez, and W. M. Gray, 2001: The recent increase in Atlantic hurricane activity: Causes and implications. *Science*, **293**, 474–479.

Gong, D., and S. Wang, 1998: Antarctic oscillation: concept and applications. *Chin. Sci. Bull.*, **43**, 734–738.

Gong, D., and S. Wang, 1999: Definition of Antarctic Oscillation index. *Geophys. Res. Lett.*, **26**, 459–462.

Goni, G. J., and J. A. Trinanes, 2003: Ocean thermal structure monitoring could aid in the intensity forecast of tropical cyclones. *EOS Trans. Amer. Geophys. Union*, **84**, 577–578.

Goodkin, N. F., K. A. Hughen, S. C. Doney, and W. B. Curry, 2008: Increased multidecadal variability of the North Atlantic Oscillation since 1781. *Nature Geosci.*, **1**, 844–848.

Graham, N. E., and T. P. Barnett, 1987: Observations of sea surface temperature and convection over tropical oceans. *Science*, **238**, 657–659.

Gray, W. M., 1990: Strong association between West African rainfall and U.S. landfall of intense hurricanes. *Science*, **249**, 1251–1256.

Gray, W. M., C. W. Landsea, P. W. Mielke Jr., and K. J. Berry, 1992: Predicting Atlantic seasonal hurricane activity 6–11 months in advance. *Weather Forecast.*, **7**, 440–455.

Gray, W. M., C. W. Landsea, P. W. Mielke Jr., and K. J. Berry, 1994: Predicting Atlantic basin seasonal tropical cyclone activity by 1 June. *Weather Forecast.*, **9**, 103–115.

Gray, L. J., et al., 2010: Solar influences on climate. *Rev. Geophys.*, **48**, RG4001. doi:10.1029/2009RG000282.

Grötzner, A., M. Latif, and T. P. Barnett, 1998: A decadal climate cycle in the North Atlantic Ocean as simulated by the ECHO coupled GCM. *J. Clim.*, **11**, 831–847.

Hadley, G., 1735: The cause of the general trade-wind. *Philos. T. R. Soc. Lond.*, **29**, 58–62.

Haigh, J. D., 1994: The role of stratospheric ozone in modulating the solar radiative forcing of climate. *Nature*, **370**, 544–546. doi:10.1038/370544a0.

Haigh, J. D., 1996: The impact of solar variability on climate. *Science*, **272**, 981–984.

Haigh, J. D., 2003: The effects of solar variability on the Earth's climate. *Philos. Trans. R. Soc. Lon. Ser. A*, **361**, 95–111.

Haigh, J. D., and M. Blackburn, 2006: Solar influences on dynamical coupling between the stratosphere and troposphere. *Space Sci. Rev.*, **125**, 331–344. doi:10.1007/s11214-006-9067-0.

Haigh, J. D., M. Blackburn, and R. Day, 2005: The response of tropospheric circulation to perturbations in lower-stratospheric temperature. *J. Clim.*, **18**, 3672–3685.

Hale, G. E., 1908: On the probable existence of a magnetic field in sun-spots. *Astrophys. J.*, **28**, 315.

Hale, G. E., and S. B. Nicholson, 1925: The law of sun-spot polarity. *Astrophys. J.*, **62**, 270–300.

Hall, A., and M. Visbeck, 2002: Synchronous variability in the Southern Hemisphere atmosphere, sea ice, and ocean resulting from the annular mode. *J. Clim.*, **15**, 3043–3057.

Halley, E., 1686: An historical account of the trade-winds and monsoons observable in the seas between and near the tropicks, with an attempt to assign the physical cause of said Winds. *Philos. T. R. Soc. Lond.*, **16**, 153–168.

Ham, Y.-G., M. M. Rienecker, M. J. Suarez, Y. Vikhliaev, B. Zhao, J. Marshak, G. Vernieres, and S. D. Schubert, 2014: Decadal prediction skill in the GEOS-5 forecast system. *Clim. Dyn.*, **42**, 1–20.

Han, W. Q., G. A. Meehl, A. X. Hu, J. Zheng, J. Kenigson, J. Vialard, B. Rajagopalan, and Yanto, 2017. Decadal variability of the Indian and Pacific Walker cells since the 1960s: Do they covary on decadal time scales? *J. Clim.*, **30**, 8447–8468.

Han, W. Q., D. Stammer, G. A. Meehl, A. Hu, F. Sienz, and L. Zhang, 2018: Multi-decadal trend and decadal variability of the regional sea level over the Indian Ocean since the 1960s: Roles of climate modes and external forcing. *Climate*, **6**, 51. doi:10.3390/cli6020051.

Han, W., et al., 2014: Indian Ocean decadal variability: A review. *Bull. Amer. Meteorol. Soc.*, **95**, 1679–1703.

Hansen, D. V., and H. F. Bezdek, 1996: On the nature of decadal anomalies in North Atlantic sea surface temperature. *J. Geophys. Res.*, **101**, 8749–8758.

Hardy, D. M., and J. J. Walton, 1978: Principal components analysis of vector wind measurements. *J. Appl. Meteor.*, **17**, 1153–1162.

Harper, K., L. W. Uccellini, E. Kalnay, K. Carey, and L. Morone, 2007: 50th anniversary of operational numerical weather prediction. *Bull. Amer. Meteorol. Soc.*, **88**, 639–650.

Harrison, R. G., and W. J. Ingram, 2005: Air-earth current measurements at Kew, London, 1909–1979. *Atmos. Res.*, **76**, 49–64. doi:10.1016/j.atmosres.2004.11.022.

Harrison, R. G., and I. Usoskin, 2010: Solar modulation in surface atmospheric electricity. *J. Atmos. Sol. Terr. Phys.*, **72**, 176–182. doi:10.1016/j.jastp.2009.11.006.

Hartmann, D. L., 1995: A PV view of zonal flow vacillation. *J. Atmos. Sci.*, **52**, 2561–2576.

Hartmann, D. L., and F. Lo, 1998: Wave-driven zonal flow vacillation in the southern hemisphere. *J. Atmos. Sci.*, **55**, 1303–1315.

Hasselmann, K., 1976: Stochastic climate models. Part I. Theory. *Tellus*, **28**, 473–485. doi:10.1111/j.2153-3490.1976.tb00696.x.

Hastenrath, S., 1990: Decadal-scale changes of the circulation in the tropical Atlantic sector associated with Sahel drought. *Int. J. Climatol.*, **10**, 459–472.

Hastenrath, S., 1991: *Climate Dynamics of the Tropics.* Kluwer Academic Publishers, Dordrecht, 488 pp.

Hazeleger, W., et al., 2013: Predicting multiyear North Atlantic Ocean variability. J. *Geophys. Res.*, **118**, 1087–1098. doi:10.1002/jgrc.20117.

He, X. G., Z. J. Chen, W. Chen, X. M. Shao, H. He, and Y. Sun, 2007: Solar activity, global surface air temperature anomaly and pacific decadal oscillation recorded in urban tree rings. *Ann. For. Sci.*, **64**, 743–756.

Held, I. M., and B. J. Soden, 2006: Robust responses of the hydrological cycle to global warming. *J. Clim.*, **19**, 5686–5699.

Helfrich, S. R., D. McNamara, B. H. Ramsay, T. Baldwin, and T. Kasheta. 2007: Enhancements to, and forthcoming developments in the interactive multisensor snow and ice mapping system (IMS). *Hydrol. Process.*, **21**, 1576–1586. doi:10.1002/hyp.6720.

Henley, B. J., 2017: Pacific decadal climate variability: Indices, patterns and tropical- extratropical interactions. *Global Planet. Change*, **155**, 42–55.

Henley, B. J., and A. D. King, 2017: Trajectories toward the 1.5°C Paris target: Modulation by the Interdecadal Pacific Oscillation. *Geophys. Res. Lett.*, **44**, 4256–4262.

Henley, B. J., J. Gergis, D. J. Karoly, S. Power, J. Kennedy, C. K. Folland, 2015: A Tripole Index for the Interdecadal Pacific Oscillation. *Clim. Dyn.*, **45**, 3077–3090.

Henley, B. J., M. A. Thyer, G. Kuczera, and S. W. Franks, 2011: Climate-informed stochastic hydrological modeling: incorporating decadal-scale variability using paleo data. *Water Resour. Res.*, **47**, W11509.

Henley, B. J., et al., 2017: Spatial and temporal agreement in climate model simulations of the Interdecadal Pacific Oscillation. *Environ. Res. Lett.*, **12**, 044011. doi:10.1088/1748-9326/aa5cc8.

Herschel, W., 1801: Observations tending to investigate the nature of the Sun, in order to find the causes or symptoms of its variable emission of light and heat; with remarks on the use that may possibly be drawn from solar observations. *Phil. Trans. R. Soc. London*, **91**, 265–318.

Hiremath K. M., 2006: Influence of solar activity on the rainfall over India. *International Living With A Star Workshop*, Goa, India, 19–24 February 2006.

Hiremath K. M., and P. I. Mandi, 2004: Influence of the solar activity on the Indian Monsoon rainfall. *New Astron.*, **9**, 651–662.

Ho, C. H., J. J. Baik, J. H. Kim, and D. Y. Gong, 2004: Interdecadal changes in summertime typhoon tracks. *J. Clim.*, **17**, 1767–1776.

Ho, M., A. S. Kiem, and D. C. Verdon-Kidd, 2012: The southern annular mode: A comparison of indices. *Hydrol. Earth Syst. Sci.*, **16**, 967–982.

Hodges, R. E., and J. B. Elsner, 2011: Evidence linking solar variability with US hurricanes. *Int. J. Climatol.*, **31**, 1897–1907.

Horel, J. D., 1981: A rotated principal component analysis of the interannual variability of the northern hemisphere 500 mb height field. *Mon. Wea. Rev.*, **109**, 2080–2092.

Horel, J. D., 1984: Complex principal component analysis: Theory and examples. *J. Clim. Appl. Meteor.*, **23**, 1660–1673.

Houghton, R. W., and Y. M. Tourre, 1992: Characteristics of low frequency sea surface temperature fluctuations in the tropical Atlantic. *J. Clim.*, **5**, 765–771.

Hoyt, D. V., and K. H. Schatten, 1998: *The Role of the Sun in Climate Change.* Oxford University Press, New York. 275 pp.

Hu, Z.-Z., and B. Huang, 2006: Physical processes associated with the tropical Atlantic SST meridional gradient. *J. Clim.*, **19**, 5500–5518.

Huang, B., and Z.-Z. Hu, 2007: Cloud-SST feedback in southeastern tropical Atlantic anomalous events. *J. Geophys. Res.*, **112**, C03015.doi:10.1029/2006JC003626.

Huang, B., and V. M. Mehta, 2004: The response of the Indo-Pacific warm pool to interannual variations in net atmospheric freshwater. *J. Geophys. Res.*, **109**, C06022. doi:10.1029/2003JC002114.

Huang, B., and V. M. Mehta, 2005: The response of the Pacific and Atlantic Oceans to interannual variations in net atmospheric freshwater. *J. Geophys. Res. Oceans*, **110**, C08008. doi:10/1029/2004JC002830.

Huang, B., and V. M. Mehta, 2010: Influences of freshwater from major rivers on global ocean circulation and temperatures in the MIT ocean general circulation model. *Adv. Atmos. Sci.*, **27**, 455–468.

Huang, B., and J. Shukla, 1997: Characteristics of the interannual and decadal variability in a general circulation model of the tropical Atlantic Ocean. *J. Phys. Oceanogr.*, **27**, 1693–1712.

Huang, B., V. F. Banzon, E. Freeman, J. Lawrimore, W. Liu, T. C. Peterson, T. M. Smith, P. W. Thorne, S. D. Woodruff, and H.-M. Zhang, 2015: Extended reconstructed sea surface temperature version 4 (ERSST.v4). Part I: Upgrades and intercomparisons. *J. Clim.*, **28**, 911–930.

Huang, B., P. W. Thorne, V. F. Banzon, T. Boyer, G. Chepurin, J. H. Lawrimore, M. J. Menne, T. M. Smith, R. S. Vose, and H.-M. Zhang, 2017: NOAA extended reconstructed sea surface temperature (ERSST), Version 5. NOAA National Centers for Environmental Information. doi:10.7289/V5T72FNM.

Huete, A., Didan, K., Miura, T., Rodriguez, E. P., Gao, X., Ferreira, L. G., 2002: Overview of the radiometric and biophysical performance of the MODIS vegetation indices. *Remote Sens. Environ.*, **83**, 195–213.

Hurrell, J. W., 1995: Decadal trends in the North Atlantic Oscillation: Regional temperatures and precipitation. *Science*, **269**, 676–679.

Ineson S., A. A. Scaife, J. R. Knight, J. C. Manners, N. J. Dunstone, L. J. Gray, and J. D. Haigh, 2011: Solar forcing of winter climate variability in the Northern Hemisphere. *Nature Geosci.*, **4**, 753–757.

Jacob, R. L., 1997: Low frequency variability in a simulated atmosphere–ocean system. Ph.D. dissertation, Department of Atmospheric and Oceanic Sciences, University of Wisconsin- Madison, 155 pp.

James, I. N., and P. M. James, 1989: Ultra-low-frequency variability in a simple atmospheric circulation model. *Nature*, **342**, 53–55.

Jevons, W. S., 1879: Sun-spots and commercial crises. *Nature*, **8**, 588–590.

Jones, J. M., and M. Widmann, 2003: Instrument- and tree-ringbased estimates of the Antarctic Oscillation. *J. Clim.*, **16**, 3511–3524.

Jorgensen, T. S., A. W. Hansen, 2000. Comments on variation of cosmic ray flux relationships. *J. Atmos. Solar Terrestr. Phys.*, **62**, 73–77.

Joshi, M. K., and A. C. Pandey, 2011: Trend and spectral analysis of rainfall over India during 1901–2000. *J. Geophys. Res.-Atmos.*, **116**, D6. doi:10.1029/2010JD014966.

Joyce, T. M., C. Frankignoul, J. Yang, and H. E. Phillips, 2004: Ocean response and feedback to the SST dipole in the tropical Atlantic. *J. Phys. Oceanogr.*, **34**, 2525–2540.

Juan, Z., Z. Juan, and Y. B. Han, 2005: Determination of precipitation cycle in Beijing area and comparison with solar activity cycle. *Earth Moon Planets*, **97**, 69–78.

Kaiser, H. F., 1958: The varimax criterion for analytic rotation in factor analysis. *Psychometrika*, **23**, 187–200.

Kalnay, E., et al., 1996: The NCEP/NCAR 40-year reanalysis project. *Bull. Am. Met. Soc.*, **77**, 437–531.w

Kaplan, A., M. A. Cane, Y. Kushnir, A. C. Clement, M. B. Blumenthal, and B. Rajagopalan, 1998: Analysis of global sea surface temperatures 1856–1991. *J. Geophys. Res.*, **103**, 18567–18589.

Kaplan, A., Y. Kushnir, M. Cane, and B. Blumenthal, 1997: Reduced space optimal analysis for historical datasets: 136 years of Atlantic sea surface temperatures. *J. Geophys. Res.*, **102**, 27835–27860.

Karoly, D. J., 1990: The role of transient eddies in low-frequency zonal variations of the Southern Hemisphere circulation. *Tellus*, **42A**, 41–50.

Karoly, D. J, P. Hope, and P. D. Jones, 1996: Decadal variations for the southern hemisphere circulation. *Int. J. Climatol.*, **16**, 723–738.

Kawamura, R., 1994: A rotated EOF analysis of global sea surface temperature variability with interannual and interdecadal scales. *J. Phys. Oceanogr.*, **24**, 707–715.

Keenlyside, N., M. Latif, J. Jungclaus, L. Kornblueh, and E. Roeckner, 2008: Advancing decadal-scale climate prediction in the North Atlantic sector. *Nature*, **453**, 84–88.

Kernthaler, S. C., R. Toumi, J. D. Haigh, 1999. Some doubts concerning a link between cosmic ray fluxes and global cloudiness. *Geophys. Res. Lett.*, **26**, 863–865.

Kidson, J. W., 1975: Eigenvector analysis of monthly mean surface data. *Mon. Wea. Rev.*, **103**, 177–186.

Kidson, J. W., 1988: Indices of the Southern Hemisphere zonal wind. *J. Clim.*, **1**, 183–194.

Kim, J.-H., K.-B. Kim, and H.-Y. Chang, 2017: Solar influence on tropical cyclone in western North Pacific Ocean. *J. Astron. Space Sci.*, **34**, 257–270.

Kim, H.-M., P. J. Webster, and J. A. Curry, 2012: Evaluation of short-term climate change prediction in multi-model CMIP5 decadal hindcasts. *Geophys. Res. Lett.*, **39**, L10701. doi:10.1029/2012GL051644.

King, M. P. and F. Kucharski, 2006: Observed decadal connections between the tropical oceans and the North Atlantic Oscillation in the 20th Century. *J. Clim.*, **19**, 1032–1041.

Kirtman, B. P., 1997: Oceanic Rossby wave dynamics and the ENSO period in a coupled model. *J. Clim.*, **10**, 1690–1704.

Kirtman, B., et al., 2013: Near-term climate change: projections and predictability. In: T. F. Qin, G.-K. Plattner, M. Tignor, S. K. Allen, J. Boschung, A. Nauels, Y. Xia, V. Bex, and P. M. Midgley (eds.) *Climate Change 2013: The Physical Science Basis. Contribution of Working group I to the Fifth Assessment Report of the Intergovernmental Panel on Climate Change.* Cambridge University Press, Cambridge, UK.

Kleeman, R., 2002: Measuring dynamical prediction utility using relative entropy. *J. Atmos. Sci.*, **59**, 2057–2072.

Kleeman, R., J. McCreary, and B. Klinger, 1999: A mechanism for generating ENSO decadal variability. *Geophys. Res. Lett.*, **26**, 1743–1746.

Klinger, B., J. McCreary, and R. Kleeman, 2002: The relationship between oscillating subtropical wind stress and equatorial temperature. *J. Phys. Oceanogr.*, **32**, 1507–1521.

Klotzbach, P. J., 2006: Trends in global tropical cyclone activity over the past twenty years (1986–2005). *Geophys. Res. Lett.*, **33**, L10805. doi:10.1029/2006GL025881.

Klotzbach, P. J., and C. W. Landsea, 2015: Extremely intense hurricanes: Revisiting Webster et al. (2005) after 10 years. *J. Clim.*, **28**, 7621–7629.

Knight, J. R., R. J. Allan, C. K. Folland, M. Vellinga, and M. E. Mann, 2005: A signature of persistent natural thermohaline circulation cycles in observed climate. *Geophys. Res. Lett.*, **32**, L20708. doi:10.1029/2005GL024233.

Knutson, T. R., and S. Manabe, 1998: Model assessment of decadal variability and trends in the tropical Pacific Ocean. *J. Clim.*, **11**, 2273–2296.

Kodera, K., 2002: Solar cycle modulation of the North Atlantic Oscillation: Implications in the spatial structure of the NAO. *Geophys. Res. Lett.*, **29**, 1218. doi:10.1029/2001GL014557.

Kodera, K., 2003: Solar influence on the spatial structure of the NAO during the winter 1900–1999. *Geophys. Res. Lett.*, **30**. doi:10.1029/2002GL016584.

Kodera, K., and Y. Kuroda, 2002: Dynamical response to the solar cycle: Winter stratopause and lower stratosphere. *J. Geophys. Res.*, **107**, 4749.

Kodera, K., and Y. Kuroda, 2005: A possible mechanism of solar modulation of the spatial structure of the North Atlantic Oscillation. *J. Geophys. Res.*, **110**, D02111.

Komori, N., A. Kuwano-Yoshida, T. Enomoto, H. Sasaki, and W. Ohfuchi, 2008: High- resolution simulation of the global coupled atmosphere-ocean system: Description and preliminary outcomes of CFES (CGCM for the Earth Simulator). In: K. Hamilton and W. Ohfuchi (eds.), *High Resolution Numerical Modelling of the Atmosphere and Ocean*, Springer, New York, pp. 241–260.

Komori, N., K. Takahashi, K. Komine, T. Motoi, X. Zhang, and G. Sagawa, 2005: Description of sea-ice component of coupled ocean-sea-ice model for the Earth Simulator (OIFES). *J. Earth Simul.*, **4**, 31–45.

Kosaka, Y., and S. P. Xie, 2013: Recent global-warming hiatus tied to equatorial Pacific surface cooling. *Nature*, **501**, 403–407.

Kossin, J. P., T. L. Olander, and K. R. Knapp, 2013: Trend analysis with a new global record of tropical cyclone intensity. *J. Clim.*, **26**, 9960–9976.

Kravchenko, V. O., O. M. Evtushevsky, A. V. Grytsai, and G. P. Milinevsky, 2011: Decadal variability of winter temperatures in the Antarctic Peninsula region. *Antarct. Sci.*, **23**, 614–622.

Krishnamurti, T. N., S.-H. Chu, and W. Iglesias, 1986: On the sea level pressure of the southern oscillation. *Arch. Met. Geoph. Biocl. A.*, **34**, 385. doi:10.1007/BF02257768.

Kristjánsson, J. E., A. Staple, J. Kristiansen, and E. Kaas, 2002: A new look at possible connections between solar activity, clouds and climate. *Geophys. Res. Lett.*, **29**, 2107. doi:10.1029/2002GL015646.

Kubota, H., and J. C. L. Chan, 2009: Interdecadal variability of tropical cyclone landfall in the Philippines from 1902 to 2005. *Geophys. Res. Lett.*, **36**, L12802. doi:10.1029/2009GL038108.

Kucharski, F., F. Molteni, and A. Bracco, 2005: A western tropical Pacific relation to the NAO. *Bull. Amer. Meteorol. Soc.*, **86**, 1418–1419.

Kucharski, F., F. Molteni, and A. Bracco, 2006: Decadal interactions between the western tropical Pacific and the North Atlantic oscillation. *Clim. Dyn.*, **26**, 79–91.

Kuleshov, Y., R. Fawcett, L. Qi, B. Trewin, D. Jones, J. McBride, and H. Ramsay, 2010: Trends in tropical cyclones in the South Indian Ocean and the South Pacific Ocean. *J. Geophys. Res.*, **115**, D01101. doi:10.1029/2009JD012372.

Kuroda, Y., 2018: On the origin of the solar cycle modulation of the southern annular mode. *J. Geophys. Res. Atmos.*, **123**, 1959–1969.

Kuroda, Y., and M. Deushi, 2016: Influence of the solar cycle on the polar-night jet oscillation in the southern hemisphere. *J. Geophys. Res. Atmos.*, **121**, 11,575–11,589.

Kuroda, Y., and K. Kodera, 2005: Solar cycle modulation of the southern annular mode. *Geophys. Res. Lett.*, **32**, L13802. doi:10.1029/2005GL022516.

Kuroda, Y., and K. Shibata, 2006: Simulation of solar-cycle modulation of the Southern Annular Mode using a chemistry-climate model. *Geophys. Res. Lett.*, **33**, L05703. doi:10.1029/2005GL025095.

Kuroda, Y., M. Deushi, and K. Shibata, 2007: Role of solar activity in the troposphere stratosphere coupling in the Southern Hemisphere winter. *Geophys. Res. Lett.*, **34**, L21704. doi:10.1029/2007GL030983.

Kushnir, Y., et al., 2019: Towards operational predictions of the near-term climate. *Nature Clim. Change*, **9**, 94–101.

Küttel, M., E. Xoplaki, D. Gallego, J. Luterbacher, R. Garcia-Herrera, R. Allan, M. Barriendos, P. D. Jones, D. Wheeler, and H. Wanner, 2010: The importance of ship log data: Reconstructing North Atlantic, European and Mediterranean sea level pressure fields back to 1750. *Clim. Dyn.*, **34**, 115–128.

Kutzbach, J. E., 1970: Large-scale features of monthly mean northern hemisphere anomaly maps of sea-level pressure. *Mon. Wea. Rev.*, **98**, 708–716.

Kuwano-Yoshida, A., T. Enomoto, and W. Ohfuchi, 2010: An improved PDF cloud scheme for climate simulations. *Quart. J. Roy. Meteorol. Soc.*, **136**, 1583–1597.

Kwok, R., and J. C. Comiso, 2002: Spatial patterns of variability in Antarctic surface temperature: Connections to the Southern Hemisphere Annular Mode and the Southern Oscillation. *Geophys. Res. Lett.*, **29**, 1705. doi:10.1029/2002GL015415.

Labitzke, K., 1987: Sunspots, the QBO and the stratospheric temperature in the north polar region. *Geophys. Res. Lett.*, **14**, 535–537. doi:10.1029/GL014i005p00535.

Labitzke, K., and H. van Loon, 1995: Connections between the troposphere and stratosphere on a decadal scale. *Tellus Ser. A*, **47**, 275–286. doi:10.1034/j.1600-0870.1995.t01-1-00008.x.

Laloyaux, P., M. Balmaseda, D. Dee, K. Mogensen, and P. Janssen, 2016: A coupled data assimilation system for climate reanalysis. *Q. J. Roy. Meteorol. Soc.*, **142**, 65–78.

Landsea, C. W., 1993: A climatology of intense (or major) Atlantic hurricanes. *Mon. Wea. Rev.*, **121**, 1703–1713.

Landsea, C. W., B. A. Harper, K. Hoarau, and J. A. Knaff, 2006: Can we detect trends in extreme tropical cyclones? *Science*, **313**, 452–453.

Landsea, C. W., R. A. Pielke Jr., A. M. Mestas-Nuñez, and J. A. Knaff, 1999: Atlantic basin hurricanes: Indices of climatic changes. *Clim. Change*, **42**, 89–129.

Lassen, K., and E. Friis-Christensen, 1995: Variability of the solar-cycle length during the past 5 centuries and the apparent association with terrestrial climate. *J. Atmos. Terrestr. Phys.*, **57**, 835–845.

Latif, M., and T. P. Barnett, 1994: Causes of decadal climate variability over the North Pacific and North America. *Science*, **266**, 634–637.

Latif, M., T. Stockdale, J. O. Wolff, G. Burgers, E. Maier-Reimer, M. M. Junge, K. Arpe, and L. Bengtsson, 1994: Climatology and variability in the ECHO coupled GCM. *Tellus*, **46A**, 351–366.

Lau, N. C., 1981: A diagnostic study of recurrent meteorological anomalies appearing in a fifteen-year simulation with a GFDL general circulation model. *Mon. Wea. Rev.*, **109**, 2287–2311.

Lean, J., 1991: Variations in the Suns radiative output. *Rev. Geophys.*, **29**, 505–535. doi:10.1029/91RG01895.

Lean, J., G. Rottman, J. Harder, and G. Kopp, 2005: SORCE contributions to new understanding of global change and solar variability. *Sol. Phys.*, **230**, 27– 53.

Lee, H., and A. K. Smith, 2003: Simulation of the combined effects of solar cycle, quasi-biennial oscillation, and volcanic forcing on stratospheric ozone changes in recent decades. *J. Geophys. Res.*, **108**, 4049. doi:10.1029/2001JD001503.

Lee, J. N., D. T. Shindell, and S. Hameed, 2009: The influence of solar forcing on tropical circulation. *J. Clim.*, **22**, 5870–5885.

Lefebvre, W., H. Goosse, R. Timmermann, and T. Fichefet, 2004: Influence of the southern annular mode on the sea ice–ocean system. *J. Geophys. Res.*, **109**, C09005. doi:10.1029/2004JC002403.

Li, R. C. Y., and W. Zhao, 2018: Revisiting the intraseasonal, interannual and interdecadal variability of tropical cyclones in the western North Pacific. *Atmos. Oce. Sci. Lett.*, **11**, 198–208.

Lim, G. H., Y. C. Suh, and B. M. Kim, 2006: On the origin of the tropical Atlantic decadal oscillation based on the analysis of the ICOADS. *Q. J. Roy. Meteorol. Soc.*, **132**, 1139–1152.

Limpasuvan, V., and D. L. Hartmann, 1999: Eddies and the annular modes of climate variability. *Geophys. Res. Lett.*, **26**, 3133–3136.

Limpasuvan, V., and D. L. Hartmann, 2000: Wave-maintained annular modes of climate variability. *J. Clim.*, **13**, 4414–4429.

Lindstrom, E., R. Lukas, R. Fine, E. Firing, S. Godfrey, G. Meyers, and M. Tsuchiya, 1987: The western equatorial Pacific Ocean circulation study. *Nature*, **330**, 533–537.

Linkin, M. E., and S. Nigam, 2008: The North Pacific oscillation–West Pacific teleconnection pattern: Mature-phase structure and winter impacts. *J. Clim.*, **21**, 1979–1997.

Linsley, B. K., P. Zhang, A. Kaplan, S. S. Howe, and G. M. Wellington, 2008: Interdecadal-decadal climate variability from multicoral oxygen isotope records in the South Pacific Convergence Zone region since 1650 AD. *Paleoceanography*, **23**, PA2219.

Liu, Z., 1994: A simple model of the mass exchange between the subtropical and tropical ocean. *J. Phys. Oceanogr.*, **24**, 1153–1165.

Liu, Z., 2012: Dynamics of interdecadal climate variability: A historical perspective. *J. Clim.*, **25**, 1963–1995.

Liu, Z., L. Wu, R. Gallimore, and R. Jacob, 2002: Search for the origins of Pacific decadal climate variability. *Geophys. Res. Lett.*, **29**, 1404. doi:10.1029/2001GL013735.

Lohmann, K., and M. Latif, 2005: Tropical Pacific decadal variability and the subtropical–tropical Cells. *J. Clim.*, **18**, 5163–5178.

Lorenz, E. N., 1951: Seasonal and irregular variations of the Northern Hemisphere sea-level pressure profile. *J. Meteor.*, **8**, 52–59.

Lorenz, E. N., 1963: Deterministic nonperiodic flow. *J. Atmos. Sci.*, **20**, 130–141.

Lorenz, E. N., 1969: The predictability of a flow which possesses many scales of motion. *Tellus*, **21**, 289–307.

Love, J. J., 2013: On the insignificance of Herschel's sunspot correlation. *Geophys. Res. Lett.*, **40**, 4171–4176. doi:10.1002/grl.50846.

Lu, P., and J. P. McCreary Jr., 1995: Influence of the ITCZ on the flow of thermocline water from the subtropical to the equatorial Pacific Ocean. *J. Phys. Oceanogr.*, **25**, 3076–3088.

Ludlum, D. M., 1989: *Early American Hurricanes 1492–1870.* Lancaster Press, Inc., Lancaster, PA, 198 pp.

Luebbecke, J. F., C. W. Boening, and A. Biastoch, 2008: Variability in the subtropical–tropical cells and its effect on nearsurface temperature of the equatorial Pacific: A model study. *Ocean Sci.*, **4**, 73–88.

Lukas, R., and E. Lindstrom, 1991: The mixed layer of the western equatorial Pacific ocean. *J. Geophys. Res.*, **96**, 3343–3357.

Luo, J.-J., and T. Yamagata, 2001: Long-term El Niño-Southern oscillation (ENSO)-like variation with special emphasis on the South Pacific. *J. Geophys. Res.*, **106**, 22211–22227.

Luo, Y., L. Rothstein, and R. H. Zhang, 2009: Response of Pacific subtropical-tropical thermocline water pathways and transports to global warming. *Geophys. Res. Lett.*, **36**, L04601. doi:10.1029/2008GL036705.

Luterbacher, J., C. Schmutz, D. Gyalistras, E. Xoplaki, and H. Wanner, 1999: Reconstruction of monthly NAO and EU indices back to AD 1675. *Geophys. Res. Lett.*, **26**, 2745–2748.

Luterbacher, J., E. Xoplaki, D. Dietrich, P. D. Jones, T. D. Davies, D. Portis, H. Gonzalez-Rouco, H. von Storch, D. Gyalistras, C. Casty, and H. Wanner, 2002a: Extending NAO reconstructions back to 1500. *Atmos. Sci. Lett.* doi:10.1006/asle.2001.0044.

Luterbacher, J., E. Xoplaki, D. Dietrich, R. Rickli, J. Jacobeit, C. Beck, D. Gyalistras, C. Schmutz, and H. Wanner, 2002b: Reconstruction of sea level pressure fields over the eastern North Atlantic and Europe back to 1500. *Clim. Dyn.*, **18**, 545–561.

Luthardt, L., and R. Rößler, 2017: Fossil forest reveals sunspot activity in the early Permian. *Geology*, **45**, 279–282.

Lyatkher, V. M., 2000: Solar cycle length stochastic association with Caspian Sea level. *Geophys. Res. Lett.*, **27**, 3727–3730.

Lynch, P., 2004: Richardson's forecast: What went wrong? *Symposium on the 50th anniversary of Operational Numerical Weather Prediction*, University of Maryland, College Park, MD, U.S.A., 14–17 June, 2004.

Lynch, P., 2006: *The Emergence of Numerical Weather Prediction: Richardson's Dream.* Cambridge University Press, Cambridge. 290 p. ISBN 978-0-521-85729-1.

MacDonald, G. M., and R. A. Case, 2005: Variations in the Pacific decadal oscillation over the past millennium. *Geophys. Res. Lett.*, **32**, L08703.

Maddison, A., 2003: *The World Economy: Historical Statistics.* OECD Publishing, Paris. 274 p.

Manabe, S., R. J. Stouffer, M. Spelman, and K. Bryan, 1991: Transient responses of a coupled ocean–atmosphere model to gradual changes of atmospheric CO_2. Part I: Annual-mean response. *J. Clim.*, **4**, 785–817.

Mann, M. E. and J. Lees, 1996: Robust estimation of background noise and signal detection in climatic time series. *Clim. Change*, **33**, 409–445.

Mann, M. E., et al., 2009: Global signatures and dynamical origins of the little ice age and medieval climate anomaly. *Science*, **326**, 1256–1260.

Mantua, N. J., S. R. Hare, Y. Zhang, J. M. Wallace, and R. C. Francis, 1997: A Pacific decadal climate oscillation with impacts on salmon. *Bull. Amer. Meteorol. Soc.*, **78**, 1069–1079.

Maravilla, D., B. Mendoza, E. Jauregui, and A. Lara, 2004: The main periodicities in the minimum extreme temperature in northern Mexico and their relation with solar variability. *Adv. Space Res.*, **34**, 365–369.

Markham, C. G., and D. R. McLain, 1977: Sea surface temperature related to rain in Ceara, Northeastern Brazil. *Nature*, **265**, 320–323.

Markson, R., and M. Muir, 1980: Solar wind control of the Earth's electric field. *Science*, **208**, 979–990. doi:10.1126/science.208.4447.979.

Marsh, N., and H. Svensmark, 2003: Galactic cosmic ray and El Niño Southern Oscillation trends in international satellite cloud climatology project d2 low-cloud properties. *J. Geophys. Res.*, **108**, 4195. doi:10.1029/2001JD001264.

Marsh, N., and H. Svensmark, 2004: Comment on "Solar influences on cosmic rays and cloud formation: A reassessment" by Bomin Sun and Raymond S. Bradley. *J. Geophys. Res.*, **109**, D14205. doi:10.1029/2003JD004063.

Marshall, G. J., 2003: Trends in the southern annular Mode from observations and reanalyses. *J. Clim.*, **16**, 4134–4143. doi:10.1175/1520-0442(2003)016<4134:TITSAM>2.0.CO;2.

Marshall, G. J., 2007: Half-century seasonal relationships between the southern annular mode and Antarctic temperature. *Int. J. Climatol.*, **27**, 373–383.

Marshall, J., Y. Kushnir, D. Battisti, P. Chang, A. Czaja, R. Dickson, J. Hurrell, M. McCartney, R. Saravanan, and M. Visbeck, 2001: North Atlantic climate variability: Phenomena, impacts and mechanisms. *Int. J. Climatol.*, **21**, 1863–1898.

Marshall, G. J., D. W. J. Thompson, and M. R. van den Broeke, 2017: The signature of southern hemisphere atmospheric circulation patterns in antarctic precipitation. *Geophys. Res. Lett.*, **44**, 11580–11589.

Matsuura, T., and S. Iizuka, 2000: Zonal migration of the Pacific warm-pool tongue during El Niño events. *J. Phys. Oceanogr.*, **30**, 1582–1600.

Matthes, K., Y. Kuroda, K. Kodera, and U. Langematz, 2006: Transfer of the solar signal from the stratosphere to the troposphere: Northern winter. *J. Geophys. Res.*, **111**, D06108. doi:10.1029/2005JD006283.

McCreary Jr., J. P., 1983: A model of tropical ocean atmosphere interaction. *Mon. Wea. Rev.*, **111**, 370–387.
McCreary Jr., J. P., and P. Lu, 1994: Interaction between the subtropical and equatorial ocean circulations: The subtropical cell. *J. Phys. Oceanogr.*, **24**, 466–497.
McGregor, S., A. Timmermann, and O. Timm, 2010: A unified proxy for ENSO and PDO variability since 1650. *Clim. Past*, **6**, 1–17.
McKinnell, S. M., and W. R. Crawford, 2007: The 18.6-year lunar nodal cycle and surface temperature variability in the northeast Pacific. *J. Geophys. Res. Oceans*, **112**, C02002.
McRae, N., 1999: *John Von Neumann: The Scientific Genius Who Pioneered the Modern Computer, Game Theory, Nuclear Deterrence, and Much More.* Second Edition, American Mathematical Society, Providence. 406 p.
Meehl, G. A., and J. M. Arblaster, 2009: A lagged warm event like response to peaks in solar forcing in the Pacific region. *J. Clim.*, **22**, 3647–3660.
Meehl, G. A., and A. Hu, 2006: Megadroughts in the Indian monsoon region and southwest North America and a mechanism for associated multi-decadal Pacific sea surface temperature anomalies. *J. Clim.*, **19**, 1605–1623.
Meehl, G. A., and H. Teng, 2012: Case studies for initialized decadal hindcasts and predictions for the Pacific region. *Geophys. Res. Lett.*, **39**, L22705. doi:10.1029/2012G L0534 23.
Meehl, G. A., and H. Teng, 2014: CMIP5 multi-model initialized decadal hindcasts for the mid-1970s shift and early-2000s hiatus and predictions for 2016–2035. *Geophys Res Lett.* doi:10.1002/2014G L059256.
Meehl, G. A., J. M. Arblaster, G. Branstator, and H. van Loon, 2008: A coupled air-sea response mechanism to solar forcing in the Pacific region. *J. Clim.*, **21**, 2883–2897.
Meehl, G. A., J. M. Arblaster, K. Matthes, F. Sassi, and H. van Loon, 2009: Amplifying the Pacific climate system response to a small 11 year solar cycle forcing. *Science*, **325**, 1114–1118.
Meehl, G. A., J. M. Arblaster, J. T. Fasullo, A. Hu, and K. E. Trenberth, 2011: Model-based evidence of deep-ocean heat uptake during surface-temperature hiatus periods. *Nature Clim. Change*, **1**, 360–364.
Meehl, G. A., C. Covey, T. Delworth, M. Latif, B. McAvaney, J. F. B. Mitchell, R. J. Stouffer, and K. E. Taylor, 2007: The WCRP CMIP3 multimodel dataset: A new era in climate change research. *Bull. Amer. Meteor. Soc.*, **88**, 1383–1394.
Meehl, G. A., L. Goddard, J. Murphy, R. J. Stouffer, G. Boer, G. Danabasoglu, K. Dixon, M. A. Giorgetta, A. Greene, E. Hawkins, G. Hegerl, D. Karoly, N. Keenlyside, M. Kimoto, B. Kirtman, A. Navarra, R. Pulwarty, D. Smith, D. Stammer, and T. Stockdale, 2009: Decadal Prediction: Can it be skillful? *Bull. Am. Meteorol. Soc.*, **90**, 1467. doi:10.1175/2009BAMS2778.1.
Meehl, G. A., A. Hu, J. M. Arblaster, J. Fasullo, and K. E. Trenberth, 2013: Externally forced and internally generated decadal climate variability associated with the Interdecadal Pacific Oscillation. *J. Clim.*, **26**, 7298–7310.
Meehl, G. A., W. M. Washington, T. M. L. Wigley, J. M. Arblaster, and A. Dai, 2003: Solar and greenhouse gas forcing and climate response in the 20th century. *J. Clim.*, **16**, 426–444.
Meehl, G. A., et al., 2006: Climate change projections for the twenty-first century and climate change commitment in the CCSM3. *J. Clim.*, **19**, 2597–2626.
Meehl, G. A., et al., 2014: Decadal climate prediction: An update from the trenches. *Bull. Am. Meteorol. Soc.*, 243–267.
Mehta, V. M., 1991: Meridional oscillations in an idealized ocean-atmosphere system, Part I: Uncoupled modes. *Clim. Dyn.*, **6**, 49–65.
Mehta, V. M., 1992: Meridionally-propagating interannual to interdecadal variability in a linear ocean-atmosphere model. *J. Clim.*, **5**, 330–342.

Mehta, V. M., 1998: Variability of the tropical ocean surface temperatures at decadal-multidecadal timescales. Part I: The Atlantic Ocean. *J. Clim.*, **11**, 2351–2375.

Mehta V. M., 2017: *Natural Decadal Climate Variability: Societal Impacts.* CRC Press, Taylor & Francis Group, Boca Raton, FL, 326 p.

Mehta, V. M., and M. Coughlan, 1998: Summary of the proceedings of the JCESS-CLIVAR workshop on decadal climate variability. *Bull. Am. Meteorol. Soc.*, **79**, 301–303.

Mehta, V. M., and T. Delworth, 1995: Decadal variability of the tropical Atlantic Ocean surface temperature in shipboard measurements and in a global ocean-atmosphere model. *J. Clim.*, **8**, 172–190.

Mehta, V. M., and K.-M. Lau, 1997: Influence of solar irradiance on the Indian monsoon-ENSO relationship at decadal-multidecadal time scales. *Geophys. Res. Lett.*, **24**, 159–162.

Mehta, V. M., Y. Kushnir, J. Lean, D. Legler, R. Lukas, A. Proshutinsky, N. Rosenberg, H. von Storch, P. Schopf, and W. White, 2006: The CRCES workshop on decadal climate variability. *Bull. Am. Meteorol. Soc.*, **87**, 1223–1225.

Mehta, V. M., C. L. Knutson, N. J. Rosenberg, J. R. Olsen, N. A. Wall, T. K. Bernadt, and M. J. Hayes, 2013a: Decadal climate information needs of stakeholders for decision support in water and agriculture production sectors: a case study in the Missouri River Basin. *Weather Clim. Soc.*, **5**, 27–42.

Mehta, V. M., G. Meehl, L. Goddard, J. Knight, A. Kumar, M. Latif, T. Lee, A. Rosati, D. Stammer, 2011: Decadal climate predictability and prediction: where are we? *Bull. Am. Meteorol. Soc.*, **92**, 637–640.

Mehta, V. M., K. Mendoza, and H. Wang, 2019: Predictability of phases and magnitudes of natural decadal climate variability phenomena in CMIP5 experiments with the UKMO HadCM3, GFDL-CM2.1, NCAR-CCSM4, and MIROC5 global earth system models. *Clim. Dyn.*, **52**, 3255–3275.

Mehta, V. M., M. Suarez, J. Manganello, and T. Delworth, 2000a: Oceanic influence on the North Atlantic Oscillation and associated Northern Hemisphere climate variations: 1959–1993. *Geophys. Res. Lett.*, **27**, 121–124.

Mehta, V. M., E. Lindstrom, A. Busalacchi, T. Delworth, C. Deser, L.-L. Fu, J. Hansen, G. Lagerloef, K.-M. Lau, S. Levitus, G. Meehl, G. Mitchum, J. Susskind, W. White, 2000b: Summary of the proceedings of the NASA workshop on decadal climate variability. *Bull. Am. Meteorol. Soc.*, **81**, 2983–2985.

Mehta, V. M., H. Wang, and K. Mendoza, 2013b: Decadal predictability of tropical basin-average and global-average sea-surface temperatures in CMIP5 experiments with the HadCM3, GFDL-CM2.1, NCAR-CCSM4, and MIROC5 global earth system models. *Geophys. Res. Lett.* doi:10.1002/grl.50236.

Mehta, V. M., H. Wang, and K. Mendoza, 2018: Simulations of three natural decadal climate variability phenomena in CMIP5 experiments with the UKMO-HadCM3, GFDL-CM2.1, NCAR-CCSM4, and MIROC5 global earth system models. *Clim. Dyn.*, **51**, 1559–1584.

Meinardus, W., 1898: Der Zusammenhang des Winterklimas in Mittel- und Nordwest-Europa mit dem Golfstrom. *Z. d. Ges. f. Erdkunde in Berlin*, **23**, 183–200.

Meldrum, C., 1872: On a periodicity in the frequency of cyclones in the Indian Ocean south of the equator. *Nature*, **6**, 357–358.

Meldrum, C., 1885: On a supposed periodicity of the cyclones in the Indian Ocean south of the equator. *Nature*, **32**, 613–614.

Mendoza, B., A. Lara, D. Maravilla, and E. Jauregui, 2001: Temperature variability in central Mexico and its possible association to solar activity. *J. Atmos. Solar-Terrestr. Phys.*, **63**, 1891–1900.

Mendoza, K., and V. M. Mehta, 2020: Seasonal to multiyear, worldwide dryness-wetness outlooks based on probabilistic prediction of decadal climate variability indices. *Geophys. Res. Lett.* (In preparation).

Meneghini, B., I. Simmonds, and I. N. Smith, 2007: Association between Australian rainfall and the southern annular mode. *Int. J. Climatol.*, **27**, 109–121, 2007.

Meredith, M. P., and A. M. Hogg, 2006: Circumpolar response of Southern Ocean eddy activity to a change in the southern annular mode. *Geophys. Res. Lett.*, **33**, L16608. doi:10.1029/2006GL026499.

Merryfield, W., and G. Boer, 2005: Variability of upper Pacific Ocean overturning in a coupled climate model. *J. Clim.*, **18**, 666–683.

Milana, J. P., and S. Lopez, 1998: Solar cycles recorded in Carboniferous glacimarine rhythmites (Westem Argentina): relationships between climate and sedimentary environment. *Palaeogeogr. Palaeoclimatol. Palaeoecol.*, **144**, 37–63.

Miller, A. J., D. R. Cayan, T. P. Barnett, N. E. Graham, and J. M. Oberhuber, 1994: Interdecadal variability of the Pacific Ocean: Model response to observed heat flux and wind stress anomalies. *Clim. Dyn.*, **9**, 287–302. doi:10.1007/BF00204744.

Mitford, W., 1808: The history of Greece. *Edinb. Rev.*, **12**, 478ff. books.google.com/books?id=szEeAQAAIAAJ&pg=PA480#v=onepage&q&f=false.

Molteni, F., R. Buizza, T. N. Palmer, and T. Petroliagis, 1996: The ECMWF ensemble prediction system: methodology and validation. *Q. J. Roy. Meteorol. Soc.*, **122**, 73–119.

Mosley-Thompson, E., 1992: Paleoenvironmental conditions in Antarctica since A.D. 1500. In: R. S. Bradley, and P. D. Jones (eds.), *Climate since A.D. 1500*. Routledge, London, pp. 572–591.

Mosley-Thompson, E., 1996: Holocene climate changes recorded in an East Antarctica ice core. In: P. D. Jones, R. Bradley, and J. Jouzel (eds.), *Climatic Variations and Forcing Mechanisms of the Last 2000 Years*. NATO ASI Series, Vol I41. Springer, Heidelberg, pp. 263–279.

Müller, W. A., C. Frankignoul, and N. Chouaib, 2008: Observed decadal tropical Pacific-North Atlantic teleconnections. *Geophys. Res. Lett.*, **35**, L24810. doi:10.1029/2008GL035901.

Munoz, A., J. Ojeda, and B. Sanchez-Valverde, 2002: Sunspot-like and ENSO/NAO-like periodicities in lacustrine laminated sediments of the Pliocene Villarroya Basin (La Rioja, Spain). *J. Paleolimnol.*, **27**, 453–463.

Murphy, A. H., 1988: Skill scores based on the mean square error and their relationships to the correlation coefficient. *Mon. Wea. Rev.*, **116**, 2417–2424.

Murphy, J. O., 1990: Australian tree ring chronologies a proxy data for solar variability. *Astronom. Soc. Austral.*, **8**, 292–297.

Murphy, J. O., 1991: The downturn in solar activity during solar cycles 5 and 6. *Astronom. Soc. Austral.*, **9**, 330–331.

Mysak, L. A., and S. A. Venegas, 1998: Decadal climate oscillations in the Arctic: A new feedback loop for atmosphere–ice–ocean interactions. *Geophys. Res. Lett.*, **25**, 3607–3610.

Mysak, L. A., D. K. Manak, and R. F. Marsden, 1990: Sea–ice anomalies observed in the Greenland and Labrador Seas during 1901–1984 and their relation to an interdecadal Arctic climate cycle. *Clim. Dyn.*, **5**, 111–133.

Narasimha, R., and S. Bhattacharyya, 2010: A wavelet cross-spectral analysis of solar–ENSO– rainfall connections in the Indian monsoons. *Appl. Comput. Harmonic Anal.*, **28**, 285–295.

Neale, R., and J. Slingo, 2003: The maritime continent and its role in the global climate: A GCM study. *J. Clim.*, **16**, 834–848.

Newell, N. E., R. E. Newell, J. Hsiung, and W. Zhongxiang, 1989: Global marine temperature variation and the solar magnetic cycle. *Geophys. Res. Lett.*, **16**, 311–314.

Newhall, C. G., and S. Self, 1982: The volcanic explosivity index (VEI): an estimate of explosive magnitude for historical volcanism. *J. Geophys. Res.*, **87**, 1231–1238.

Newman, M., 2007: Interannual to decadal predictability of tropical and North Pacific sea surface temperatures. *J. Clim.*, **20**, 2333–2356.

Newman, M., 2013: An empirical benchmark for decadal forecasts of global surface temperature anomalies. *J. Clim.*, **26**, 5260–5269.
Newman, M., et al., 2016: The Pacific decadal oscillation, revisited. *J. Clim.*, **29**, 4399–4427.
Ng, E. K. W., and J. C. L. Chan, 2012: Interannual variations of tropical cyclone activity over the north Indian Ocean. *Int. J. Climatol.*, **32**, 819–830.
Njau, E. C., 2003: Solar cycles reflected in the English climate. *Nuovo Cimento Della Societa Italiana Di Fisica C-Geophys. Space Phys.*, **26**, 23–37.
Nobre, P., and J. Shukla, 1996: Variations of sea surface temperature, wind stress, and rainfall over the tropical Atlantic and South America. *J. Clim.*, **9**, 2464–2479.
Nonaka, M., S.-P. Xie, and J. McCreary, 2002: Decadal variations in the subtropical cells and equatorial Pacific SST. *Geophys. Res. Lett.*, **29**, 1116. doi:10.1029/2001GL013717.
O'Brien, D. P., and R. G. Currie, 1993: Observations of the 18.6-year cycle of air-pressure and a theoretical-model to explain certain aspects of this signal. *Clim. Dyn.*, **8**, 287–298.
Olson, R., 2005: A Critical Juncture Analysis, 1964–2003. Final Report. The Office of U.S. Foreign Disaster Assistance (OFDA) of the United States Agency for International Development (USAID). www.usaid.gov/our_work/humanitarian_assistance/disaster_assistance/publications/ofda _cjanalysis_02_21–2005.pdf
Otterå, O. H., M. Bentsen, H. Drange, and L. Suo, 2010: External forcing as a metronome for Atlantic multidecadal variability. *Nature Geosci.*, **3**, 688–694.
Palmer, T., 2008: Edward Norton Lorenz. *Phys. Today*, **61**, 81–82.
Palmer, T., 2018: The ECMWF ensemble prediction system: Looking back (more than) 25 years and projecting forward 25 years. *Q. J. Roy. Meteorol. Soc.* doi:10.1002/qj.3383.
Palmer, T. N., and Z. Sun, 1985: A modeling and observational study of the relationship between sea surface temperature in the northwest Atlantic and the atmospheric general circulation. *Q. J. Roy. Meteor. Soc.*, **111**, 947–975.
Parker, D. E., 1994: Effects of changing exposure of thermometers at land stations. *Int. J. Climatol.*, **14**, 1–31.
Parker, D. E., C. K. Folland, M. Jackson, 1995a: Marine surface temperature: Observed variations and data requirements. *Clim. Change*, **31**, 559–600.
Parker, D. E., C. K. Folland, A. Bevan, M. N. Ward, M. Jackson, and F. Maskell, 1995b: Marine surface data for analysis of climate fluctuations on interannual to century timescales. In: D. G. Martinson et al. (eds.), *Natural Climate Variability on Decadal to Century Time Scales*, National Academy Press, Washington DC, pp. 241–250.
Pattanaik, D. R., 2005: Variability of oceanic and atmospheric conditions during active and inactive periods of storms over the Indian region. *Int. J. Climatol.*, **25**, 1523–1530.
Patwardhan, S. K., and H. N. Bhalme, 2001: A study of cyclonic disturbances over India and the adjacent oceans. *Int. J. Climatol.*, **21**, 527–534.
Paulsen, D. E., H. C. Li, and T. L. Ku, 2003: Climate variability in central China over the last 1270 years revealed by high-resolution stalagmite records. *Quat. Sci. Rev.*, **22**, 691–701.
Peel, D. A., 1992: Ice core evidence from the Antarctic Peninsula region. In: R. S. Bradley, and P. D. Jones (eds.), *Climate Since A.D. 1500*, Routledge, London, pp. 549–571.
Peel, D. A., R. Mulvaney, E. C. Pasteur, and C. Chenery, 1996: Climate changes in the Atlantic sector of Antarctica over the past 500 years from ice-cores and other evidence. In: P. D. Jones, R. Bradley, and J. Jouzel (eds.), *Climatic Variations and Forcing Mechanisms of the Last 2000 Years*. NATO ASI Series, Vol. 41, Springer, Heidelberg, pp. 243–262.
Penny, S. G., and T. M. Hamill, 2017: Coupled data assimilation for integrated earth system analysis and prediction. *Bull. Am. Meteorol. Soc.*, **98**, ES169–ES172.
Penland, C., and P. D. Sardeshmukh, 1995: The optimal growth of tropical sea surface temperature anomalies. *J. Clim.*, **8**, 1999–2024.
Pettersson, O., 1896: Über die Beziehungen zwischen hydrographischen und meteorologischen Phänomenen. *Meteorol. Z.*, **13**, 285–321.

Phillips, N., 1951: A simple three-dimensional model for the study of large-scale extratropical flow patterns. *J. Meteorol.*, **8**, 381–394.

Phillips, N., 1956: The general circulation of the atmosphere: A numerical experiment. *Q. J. Roy. Meteorol. Soc.*, **82**, 123–164.

Pierce, D. W., 2001: Distinguishing coupled ocean–atmosphere interactions from background noise in the North Pacific. *Prog. Oceanogr.*, **49**, 331–352. doi:10.1016/S0079-6611(01)00029-5.

Pierce, J. R., and P. J. Adams, 2009: Can cosmic rays affect cloud condensation nuclei by altering new particle formation rates? *Geophys. Res. Lett.*, **36**, L09820. doi:10.1029/2009GL037946.

Pingree, D., 1998: Legacies in Astronomy and Celestial Omens. In: S. Dalley (eds.), *The Legacy of Mesopotamia*, Oxford University Press, Oxford, pp. 125–137.

Pinto, J. G., and C. C. Raible, 2012: Past and recent changes in the north Atlantic oscillation. *WIREs Clim. Change*, **3**, 79–90.

Pittock, A. B., 1973: Global meridional interactions in the stratosphere and troposphere. *Q. J. Roy. Meteorol. Soc.*, **99**, 424–437.

Pittock, A. B., 1980: Patterns of climatic variations in Argentina and Chile – I. Precipitation, 1932–1960. *Mon. Wea. Rev.*, **108**, 1347–1361.

Poey, M. A., 1873: Sur les rapports entre les taches solaires et les ourages des Antilles de l'Atlantique-nord et de l'Ocean Indian sud. *Compt. Rend.*, **77**, 1223–1226.

Pohlmann, H., J. H. Jungclaus, A. Kohl, D. Stammer, and J. Marotzke, 2009: Initializing decadal climate predictions with the GECCO oceanic synthesis: Effects on the North Atlantic. *J. Clim.*, **22**, 3926–3938.

Power, S., T. Casey, C. Folland, A. Colman, and V. M. Mehta, 1999: Interdecadal modulation of the impact of ENSO on Australia. *Clim. Dyn.*, **15**, 319–324.

Poynting, J. H., 1884: A comparison of the fluctuations in the price of wheat and in the cotton and silk imports into Great Britain. *J. Stat. Soc. Lond.*, **47**, 34–74.

Proctor, R. A., 1880: Sun-spots and financial panics. *Scribner's Mon.*, **20**, 170–178.

Proctor, C. J., J. Baker, W. L. Barnes, and M. A. Gilmore, 2000: A thousand year speleothem proxy record of North Atlantic climate from Scotland. *Clim. Dyn.*, **16**, 815–820.

Pustil'nik, L. A., and G. Yom Din, 2004a: Influence of solar activity on the state of the wheat market in medieval Europe. *Solar Phys.*, **223**, 335–356.

Pustil'nik, L. A., and G. Yom Din, 2004b: Space climate manifestations in Earth prices – from medieval England up to modern U.S.A. *Solar Phys.*, **224**, 473–481.

Pustil'nik, L. A., and G. Yom Din, 2009: Possible space weather influence on the Earth wheat markets. *Sun Geosphere*, **4**, 35–41.

Pustil'nik, L. A., and G. Yom Din, 2013: On possible influence of space weather on agricultural markets: Necessary conditions and probable scenarios. *Astrophys. Bull.*, **68**, 107–124.

Rahmstorf, S., 2003: The concept of the thermohaline circulation. *Nature*, **421**, 699.

Rajagopalan, B., Y. Kushnir, and Y. M. Tourre, 1998: Observed decadal midlatitude and tropical Atlantic climate variability. *Geophys. Res. Lett.*, **25**, 3967–3970.

Rajeevan, M., J. Srinivasan, K. Niranjan Kumar, C. Gnanaseelan, and M. M. Ali, 2013: On the epochal variation of intensity of tropical cyclones in the Arabian Sea. *Atmos. Sci. Lett.*, **14**, 249–255.

Ramage, C. S., 1968: Role of a tropical "maritime continent" in the atmospheric circulation. *Mon. Wea. Rev.*, **96**, 365–370.

Ramanathan, V., and W. Collins, 1991: Thermodynamic regulation of ocean warming by cirrus clouds deduced from observations of the 1987 El Niño. *Nature*, **351**, 27–32.

Ramanathan, V., B. Subasilar, G. J. Zhang, W. Conant, R. D. Cess, J. T. Kiehi, H. Grassi, and L. Shi, 1995: Warm pool heat budget and shortwave cloud forcing: A missing physics? *Science*, **267**, 499–503.

Randall, C. E., V. L. Harvey, C. S. Singleton, S. M. Bailey, P. F. Bernath, M. Codrescu, H. Nakajima, and J. M. Russell III, 2007: Energetic particle precipitation effects on the southern hemisphere stratosphere in 1992–2005. *J. Geophys. Res.*, **112**, D08308. doi:10.1029/2006JD007696.

Randel, W. J., et al., 2009: An update of observed stratospheric temperature trends. *J. Geophys. Res.*, **114**, D02107. doi:10.1029/2008JD010421.

Rashid, H. A., and I. Simmonds, 2004: Eddy-zonal flow interactions associated with the Southern Hemisphere annular mode: results from NCEP-DOE reanalysis and a quasi-linear model. *J. Atmos. Sci.*, **61**, 873–888.

Rashid, H. A., and I. Simmonds, 2005: Southern Hemisphere annular mode variability and the role of optimal nonmodal growth. *J. Atmos. Sci.*, **62**, 1947–1961.

Rasmusson, E. M., P. A. Arkin, W. Y. Chen, and J. B. Jalickee, 1981: Biennial variations in surface temperature over the United States as revealed by singular decomposition. *Mon. Wea. Rev.*, **109**, 181–192.

Raspopov, O. M., V. A. Dergachev, T. Kolström, 2004: Periodicity of climate conditions and solar variability from dendrochronological and other palaeoclimatic data. *Paleogeogr. Paleoclimatol. Palaeoecol.*, **209**, 127–139.

Raspopov, O. M., O. I. Shumilov, E. A. Kasatkina, E. Turunen, M. Lindholm, T. Kolström, 2001: The nonlinear character of the effect of solar activity on climatic processes. *Geomag. Aeron.*, **41**, 407–412.

Rayner, N. A., E. B. Horton, D. E. Parker, C. K. Folland, and R. B. Hackett, 1996: Version 2.2 of the global sea-ice and sea surface temperature data set, 1903–1994. UKMO Climate Research Tech. Note CRTN74, 46 pp. (Available from National Meteorological Library, Meteorological Office, London Rd., Bracknell, Berkshire RG12 2SY, United Kingdom.)

Rayner, N. A., D. E. Parker, E. B. Horton, C. K. Folland, L. V. Alexander, D. P. Rowell, E. C. Kent, and A. Kaplan, 2003: Global analyses of sea surface temperature, sea ice, and night marine air temperature since the late nineteenth century. *J. Geophys. Res.*, **108**, 4407. doi:10.1029/2002JD002670.

Reid, G. C., 1991: Solar total irradiance variations and the global sea surface temperature record. *J. Geophys. Res.*, **96**, 2835–2844.

Reid, G. C., 1999: Solar variability and its implications for the human environment. *J. Atmos. Sol.-Terres. Phys.*, **61**, 3–14.

Renard, B., X. Sun, and M. Lang, 2013: Bayesian methods for nonstationary extreme value analysis. In: A. AghaKouchak, et al. (eds.), *Water Science and Technology Library, Extremes in a Changing Climate*, Vol. 65, Springer, pp. 39–95. doi:10.1007/978-94-007-4479-0_3.

Reuter, F., 1936: Die synoptische Darstellung der ½ jährigen Drukwelle. *Veröff. Geophys. Inst. Univ. Leipzig*, **7**, 257–295.

Rhodes, L. A., and B. A. McCarl, 2020: The value of ocean decadal climate variability information to United States agriculture. *Atmosphere*, **11**, 318. doi:10.3390/atmos11040318.

Richardson, L. F., 1922: *Weather Prediction by Numerical Processes*. Cambridge University Press, Boston, MA. 258 p. ISBN 9780511618291.

Richman, M. B., 1986: Rotation of principal components. *J. Climatol.*, **6**, 293–335.

Rigozo, N. R., D. J. R. Nordemann, E. Echer, and L. E. A. Vieira, 2004: Search for solar periodicities in tree-ring widths from Concordia (SC, Brazil). *Pure Appl. Geophys.*, **161**, 221–233.

Rigozo, N. R., D. J. R. Nordemann, H. E. da Silva, M. P. Souza Echer, and E. Echer, 2007: Solar and climate signal records in tree ring width from Chile (AD 1587–1994). *Planet. Space Sci.*, **55**, 158–164.

Robinson, W. A., 1991: The dynamics of the zonal index in a simple model of the atmosphere. *Tellus*, **43A**, 295–305.

Rodell, M., P. R. Houser, U. Jambor, J. Gottschalck, K. Mitchell, C. Meng, K. Arsenault, B. Cosgrove, J. Radakovich, M. Bosilovich, J. K. Entin, J. P. Walker, D. Lohmann, and D. Toll, 2004: The global land data assimilation system. *Bull. Amer. Meteor. Soc.*, **85**, 381–394. doi:10.1175/BAMS-85-3-381.

Rodrigo, F. S., D. Pozo-Vázquez, M. J. Esteban-Parra, and Y. Castro-Diez, 2001: A reconstruction of the winter North Atlantic Oscillation Index back to AD 1501 using documentary data in southern Spain. *J. Geophys. Res.*, **106**, 14805–14818.

Rodwell, M. J., D. P. Rowell, and C. K. Folland, 1999: Oceanic forcing of the wintertime North Atlantic Oscillation and European climate. *Nature*, **398**, 320–333.

Rogers, J. R., and H. van Loon, 1982: Spatial variability of sea level pressure and 500 mb height anomalies over the Southern Hemisphere. *Mon. Wea. Rev.*, **110**, 1375–1392.

Rosenberg, N. J., V. M. Mehta, J. R. Olsen, H. von Storch, R. G. Varady, M. J. Hayes, and D. Wilhite, 2007: Societal adaptation to decadal climate variability in the United States: CRCES workshop on societal impacts of decadal climate variability in the United States. *Eos Trans. Am. Geophys. Union*, **88**, 444.

Rossby, C.-G., 1939: Relations between variations in the intensity of the zonal circulation of the atmosphere and the displacements of the semipermanent centers of action. *J. Mar. Res.*, **3**, 38–55.

Roy, I., and M. Collins, 2015: On identifying the role of Sun and the El Niño southern oscillation on Indian summer monsoon rainfall, *Atmos. Sci. Lett.*, **16**, 162–169.

Ruiz-Barradas, A., J. A. Carton, and S. Nigam, 2000: Structure of interannual-to-decadal climate variability in the tropical Atlantic sector. *J. Clim.*, **13**, 3285–3297.

Russell, B., 1950: *Unpopular Essays*. George Allen & Unwin Ltd., London. todayinsci.com/R/Russell_Bertrand/RussellBertrand-Arithmetic-Quotations.htm.

Saenko, O. A., 2006: Influence of global warming on baroclinic Rossby radius in the ocean: A model intercomparison. *J. Clim.*, **19**, 1354–1360.

Saji, N. H., B. N. Goswami, P. N. Vinayachandran, and T. Yamagata, 1999: A dipole mode in the tropical Indian Ocean. *Nature*, **401**, 360–363.

Sardeshmukh, P. D., and B. J. Hoskins, 1988: The generation of global rotational flow by steady idealized tropical divergence. *J. Atmos. Sci.*, **45**, 1228–1251.

Scaife, A. A., J. Austin, N. Butchart, S. Pawson, M. Keil, J. Nash, and I. N. James, 2000: Seasonal and interannual variability of the stratosphere diagnosed from UKMO TOVS analyses. *Q. J. Roy. Meteorol. Soc.*, **126**, 2585–2604. doi:10.1002/qj.49712656812.

Scaife, A. A., S. Ineson, J. R. Knight, L. Gray, K. Kodera, and D. M. Smith, 2013: A mechanism for lagged North Atlantic climate response to solar variability. *Geophys. Res. Lett.*, **40**, 434–439.

Scaife, A. A., et al., 2014: Skillful long-range prediction of European and North American winters. *Geophys. Res. Lett.*, **41**, 2514–2519.

Schlesinger, M. E., and N. Ramankutty, 1992: Implications for global warming of intercycle solar irradiance variations. *Nature*, **360**, 330–333.

Schmutz C., J. Luterbacher, D. Gyalistras, E. Xoplaki, and H. Wanner, 2000: Can we trust proxy based NAO index reconstructions? *Geophys. Res. Lett.*, **27**, 1135–1138.

Schneider, N., T. Barnett, M. Latif, and T. Stockdale, 1996: Warm pool physics in a coupled GCM. *J. Clim.*, **9**, 219–239.

Schneider, U., A. Becker, P. Finger, A. Meyer-Christoffer, B. Rudolf, and M. Ziese, 2011: GPCC Full Data Reanalysis Version 6.0 at 1.0: Monthly Land-Surface Precipitation from Rain-Gauges built on GTS-based and Historic Data. doi:10.5676/DWD_GPCC/FD_M_V7_100

Schneider, D. P., E. J. Steig, and J. C. Comiso, 2004: Recent climate variability in Antarctica from satellite-derived temperature data. *J. Clim.*, **17**, 1569–1583.

Schoenefeldt, R., and F. A. Schott, 2006: Decadal variability of the Indian Ocean cross- equatorial exchange in SODA. *Geophys. Res. Lett.*, **33**, L08602. doi:10.1029/2006GL025891.

Schopf, P. S., and M. J. Suarez, 1988: Vacillations in a coupled ocean-atmosphere model. *J. Atmos. Sci.*, **45**, 549–566.

Schopf, P. S., and M. J. Suarez, 1990: Ocean wave dynamics and the timescale of ENSO. *J. Phys. Oceanogr.*, **20**, 629–645.

Schott, F. A., M. Dengler, and R. Schoenefeldt, 2002: The shallow overturning circulation of the Indian Ocean. *Prog. Oceanogr.*, **52**, 57–103.

Schott, F. A., J. P. McCreary, and G. C. Johnson, 2004: Shallow overturning circulation of the tropical-subtropical oceans. Earth's climate: The ocean-atmosphere interaction. *Geophys. Monogr. Amer. Geophys. Union*, **147**, 261–304.

Schwabe, H., 1843: Die Sonne. *Astron. Nachr.*, **20**, 495.

Schwabe, H., 1844: Sonnen-Beobachtungen im Jahre 1843. *Astron. Nachr.*, **21**, 233–236.

Schwerdtfeger, W., 1960: The seasonal variation of the strength of the southern circumpolar vortex. *Mon. Wea. Rev.*, **88**, 203–208.

Schwerdtfeger, W., 1967: Annual and semi-annual changes of atmospheric mass over Antarctica. *J. Geophys. Res.*, **72**, 3543–3547.

Schwerdtfeger, W., and F. Prohaska, 1955: Análisis de la marcha anual de la presión y sus relaciones con la circulación atmosferica en Sudamérica austral y la Antártida. *METEOROS*, **5**, 223–237.

Schwerdtfeger, W., and F. Prohaska, 1956: Der Jahresgang des Luftdrucks auf der Erde und seine halbjährige Komponente. *Meteor. Rundsch.*, **9**, 33–43.

Screen, J. A., N. P. Gillett, D. P. Stevens, G. J. Marshall, and H. K. Roscoe, 2009: The role of eddies in the Southern Ocean temperature response to the southern annular mode. *J. Clim.*, **22**, 806–818.

Seager, R., Y. Kushnir, P. Chang, N. Naik, J. Miller, and W. Hazeleger, 2001: Looking for the role of the ocean in tropical Atlantic decadal climate variability. *J. Clim.*, **14**, 638– 655.

Seip, K. L., Ø. GrØn, and H. Wang, 2019: The North Atlantic Oscillations: Cycle Times for the NAO, the AMO and the AMOC. *Climate*, **7**, 43. doi:10.3390/cli7030043.

Seleshi, Y., G. R. Demaree, and J. W. Delleur, 1994: Sunspot numbers as a possible indicator of annual rainfall at Addis-Ababa, Ethiopia. *Int. J. Climatol.*, **14**, 911–923.

Semazzi, F. H. M., V. M. Mehta, and Y. C. Sud, 1988: An investigation of the relationship between sub-Saharan rainfall and global sea surface temperatures. *Atmos. Ocean*, **26**, 118–138.

Sen Gupta, A. S., and M. H. England, 2006: Coupled ocean–atmosphere–ice response to variations in the southern annular mode. *J. Clim.*, **19**, 4457–4486.

Servain, J., 1991: Simple climate indices for the tropical Atlantic Ocean and some applications. *J. Geophys. Res.*, **96**, 15,137–15,146.

Servain, J., I. Wainer, J. P. McCreary, and A. Dessier, 1999: Relationship between the equatorial and meridional modes of climatic variability in the tropical Atlantic. *Geophys. Res. Lett.*, **26**, 485–488.

Servain, J., M. Seva, S. Lukas, and G. Rougier, 1987: Climatic atlas of the tropical Atlantic wind stress and sea surface temperature: 1980–1984. *Ocean Air Interact.*, **1**, 109–182.

Shakun, J. D., and J. Shaman, 2009: Tropical origins of North and South Pacific decadal variability. *Geophys. Res. Lett.*, **36**, L19711.

Shaw, N., 1928: *Manual of Meteorology II*. Cambridge University Press, Cambridge, UK.

Shen, C. M., W. C. Wang, W. Gong, and Z. Hao, 2006: A Pacific Decadal Oscillation record since 1470 AD reconstructed from proxy data of summer rainfall over eastern China. *Geophys. Res. Lett.*, **33**, L03702.

Shermatov, E., B. Nurtayev, U. Muhamedgalieva, et al., 2004: Analysis of water resources variability of the Caspian and Aral sea basins on the basis of solar activity. *J. Mar. Syst.*, **47**, 137–142.

Shibata, K., M. Deushi, T. Sekiyama, and H. Yoshimura, 2005: Development of an MRI chemical transport model for the study of the global climate. *Pap. Meteorol. Geophys.*, **55**, 75–119.

Shibata, K., et al., 1999: A simulation of troposphere, stratosphere and mesosphere with an MRI/JMA98 GCM, *Pap. Geophys. Meteorol.*, **50**, 15–53.
Shindell, D., D. Rind, N. Balachandran, J. Lean, and P. Lonergan, 1999: Solar cycle variability, ozone and climate. *Science*, **284**, 305–308.
Shiotani, M., 1990: Low-frequency variations of the zonal mean state of the Southern Hemisphere troposphere. *J. Meteor. Soc. Jpn.*, **68**, 461–471.
Shukla, J., 1981: Dynamical predictability of monthly means. *J. Atmos. Sci.*, **38**, 2547–2572.
Shukla, J., 1996: Predictability in the midst of chaos: A scientific basis for climate forecasting. *Science*, **282**, 728–731.
Simmons, A. J., and J. K. Gibson, 2000: The ERA-40 project plan, ERA-40 Proj. Rep. Ser. 1, 63 pp., Eur. Cent. for Medium Range Weather Forecasts, Reading, UK.
Simpson, R. H., 1974: The hurricane disaster potential scale. *Weatherwise*, **27**, 169–186.
Sinclair, A. R. E., J. M. Gosline, G. Holdsworth, C. J. Krebs, S. Boutin, J. N. M. Smith, R. Boonstra, and M. Dale, 1993: Can the solar-cycle and climate synchronize the Snowshoe Hare cycle in Canada - Evidence from tree rings and ice cores. *Am. Nat.*, **141**, 173–198.
Singh, O. P., T. M. Ali Khan, and M. S. Rahman, 2000: Changes in the frequency of tropical cyclones over the north Indian Ocean. *Meteorol. Atmos. Phys.*, **75**, 11–20.
Slonosky, V. C., L. A. Mysak, and J. Derome, 1997: Linking Arctic sea-ice and atmospheric circulation anomalies on interannual and decadal timescales. *Atmos. Ocean*, **35**, 333–366.
Smith, A. K., and K. Matthes, 2008: Decadal-scale periodicities in the stratosphere associated with the solar cycle and the QBO. *J. Geophys. Res.*, **113**, D05311. doi:10.1029/2007JD009051.
Smith, D. M., S. Cusack, A. Colman, A. Folland, G. Harris, and J. Murphy, 2007: Improved surface temperature prediction for the coming decade from a global circulation model. *Science*, **317**, 796–799.
Smith, D. M., et al., 2013: Real-time multi-model decadal climate predictions. *Clim. Dyn.*, **41**, 2875–2888.
Smith, D. M., et al., 2019: Robust skill of decadal climate predictions. *NPJ Clim. Atmos. Sci.*, **13**. doi:10.1038/s41612-019-0071-y.
Smith, T. M., R. W. Reynolds, R. E. Livezey, and D. C. Stokes, 1996: Reconstruction of historical sea surface temperatures using empirical orthogonal functions. *J. Clim.*, **9**, 1403–1420.
Solomon, A., J. McCreary, R. Kleeman, and B. Klinger, 2003: Interannual and decadal variability in an intermediate coupled model of the Pacific region. *J. Clim.*, **16**, 383–405.
Soon, W., S. Baliunas, E. S. Posmentier, et al., 2000: Variations of solar coronal hole area and terrestrial lower tropospheric air temperature from 1979 to mid-1998: astronomical forcings of change in earth's climate? *New Astron.*, **4**, 563–579.
Soukharev, B. E., and L. L. Hood, 2006: Solar cycle variation of stratospheric ozone: Multiple regression analysis of long-term satellite data sets and comparisons with models. *J. Geophys. Res.*, **111**, D20314. doi:10.1029/2006JD007107.
Stager, J. C., A. Ruzmaikin, D. Conway, et al., 2007: Sunspots, El Nino, and the levels of Lake Victoria, East Africa. *J. Geophys. Res. Atmos.*, **112**, D15106.
Stager, J. C., D. Ryves, B. F. Cumming, et al., 2005: Solar variability and the levels of Lake Victoria, East Africa, during the last millenium. *J. Paleolimn.*, **33**, 243–251.
Stenchikov, G., K. Hamilton, R. J. Stouffer, A. Robock, V. Ramaswamy, B. Santer, and H.-F. Graf, 2006: Climate impacts of volcanic eruptions in the IPCC AR4 climate models. *J. Geophys. Res.*, **111**, D07107.
Stenchikov, G., A. Robock, V. Ramaswamy, M. D. Schwarzkopf, K. Hamilton, and S. Ramachandran, 2002: Arctic oscillation response to the 1991 Mount Pinatubo eruption: Effects of volcanic aerosols and ozone depletion. *J. Geophys. Res.*, **107**, 4803. doi:10.1029/ 2002JD002090.

Steves, B., and A. J. Maciejewski (eds.) 2001: *The Restless Universe Applications of Gravitational N-Body Dynamics to Planetary Stellar and Galactic Systems*. CRC Press, Boca Raton, FL, 648 p. ISBN 0750308222.

Straus, D. M., and J. Shukla, 2002: Does ENSO force the PNA? *J. Clim.*, **15**, 2340–2358.

Suarez, M. J., and P. S. Schopf, 1988: A delayed action oscillator for ENSO. *J. Atmos. Sci.*, **45**, 3283–3287.

Subbarayappa, B. V., 1989: Indian astronomy: An historical perspective. In: S. K. Biswas, D. C. V. Mallik, and C. V. Vishveshwara, (eds.), *Cosmic Perspectives*, Cambridge University Press, New York, pp. 25–40.

Sun, B., and R. S. Bradley, 2002: Solar influences on cosmic rays and cloud formation: A reassessment. *J. Geophys. Res.*, **107**, 4211. doi:10.1029/2001JD000560.

Sun, B., and R. S. Bradley, 2004: Reply to comment by N. D. Marsh and H. Svensmark on "Solar influences on cosmic rays and cloud formation: A reassessment". *J. Geophys. Res.*, **109**, D14206. doi:10.1029/2003JD004479.

Sutton, R. T., and M. R. Allen, 1997: Decadal predictability of North Atlantic sea surface temperature and climate. *Nature*, **388**, 563–567.

Sutton, R. T., S. P. Jewson, and D. P. Rowell, 2000: The elements of climate variability in the tropical Atlantic region. *J. Clim.*, **13**, 3261–3284.

Svensmark, H., and E. Friis-Christensen, 1997: Variations of cosmic ray flux and global cloud coverage—A missing link in solar-climate relationships. *J. Atmos. Terr. Phys.*, **59**, 1225–1232. doi:10.1016/S1364-6826(97)00001-1.

Svensmark, H., 1998: Influence of cosmic rays on earth's climate. *Phys. Rev. Lett.*, **81**, 5027.

Szeredi, I., and D. J. Karoly, 1987: Horizontal structure of monthly fluctuations of the Southern Hemisphere troposphere from station data. *Austr. Meteor. Mag.*, **35**, 119–129.

Taylor, K. E., R. J. Stouffer, and G. A. Meehl, 2012: An overview of CMIP5 and the experiment design. *Bull. Amer. Meteorol. Soc.*, **93**, 485–498.

Thiéblemont, R., K. Matthes, N.-E. Omrani, K. Kodera, and F. Hansen, 2015: Solar forcing synchronizes decadal North Atlantic climate variability. *Nat. Commun.* doi:10.1038/ncomms9268.

Thompson, D. W. J., and J. M. Wallace, 1998: The Arctic oscillation signature in the wintertime geopotential height and temperature fields. *Geophys. Res. Lett.*, **25**, 1297–1300.

Thompson, D. W. J., and J. M. Wallace, 2000: Annular modes in the extratropical circulation. Part I: Month-to-month variability. *J. Clim.*, **13**, 1000–1016.

Thompson, L. G., D. A. Peel, E. Mosley-Thompson, R. Mulvaney, J. Dai, P. N. Lin, M. E. Davis, and C. F. Raymond, 1994: Climate since AD 1510 on Dyer Plateau, Antarctic Peninsula: evidence for recent climatic change. *Ann. Glaciol.*, **20**, 420–426.

Thresher, R. E., 2002: Solar correlates of Southern Hemisphere mid-latitude climate variability. *Int. J. Climatol.*, **22**, 901–915.

Tian, B., G. Zhang, and V. Ramanathan, 2001: Heat balance in the Pacific warm pool atmosphere during TOGA COARE and CEPEX. *J. Clim.*, **14**, 1881–1893.

Timlin M. S., M. A. Alexander, and C. Deser, 2002: On the reemergence of North Atlantic SST anomalies. *J. Clim.*, **15**, 2707–2712.

Tinsley, B. A., and Dean, G. W., 1991: Apparent tropospheric response to MeV-GeV particle flux variations: a connection via electrofreezing of supercooled water in high level clouds? *J. Geophys. Res.*, **96**, 22283–22296.

Toth, Z., and E. Kalnay, 1997: Ensemble Forecasting at NCEP and the breeding method. *Mon. Wea. Rev.*, **125**, 3297–3319.

Tourre, Y. M., B. Rajagopalan, Y. Kushnir, M. Barlow, and W. B. White, 2001: Patterns of coherent decadal and interdecadal climate signals in the Pacific Basin during the 20th Century. *Geophys. Res. Lett.*, **28**, 2069.

Tozuka, T., J. Luo, S. Masson, and T. Yamagata, 2007: Decadal modulations of the Indian Ocean dipole in the SINTEX-F1 coupled GCM. *J. Clim.*, **20**, 2881–2894.

Treguier, A., J. L. Sommer, J. Molines, and B. de Cuevas, 2010: Response of the Southern Ocean to the southern annular mode: Interannual variability and multidecadal trend. *J. Phys. Oceanogr.*, **40**, 1659–1668.

Trenberth, K. E., 1976: Fluctuations and trends in indices of the southern hemisphere circulation. *Q. J. Roy. Meteorol. Soc.*, **102**, 65–75.

Trenberth, K. E., and D. A. Paolino, 1981: Characteristic patterns of variability of sea level pressure in the Northern Hemisphere. *Mon. Wea. Rev.*, **109**, 1169–1189.

UNFCCC Conference of the Parties, 2015: Adoption of the Paris Agreement. FCCC/CP/2015/10/Add.1, pp. 1–32, Paris.

Uppala, S. M., et al., 2005: The ERA-40 re-analysis. *Q. J. Roy. Meteorol. Soc.*, **131**, 2961–3012.

Urban, F. E., J. E. Cole, and J. T. Overpeck, 2000: Influence of mean climate change on climate variability from a 155-year tropical Pacific coral record. *Nature*, **407**, 989–993.

Vance, T. R., J. L. Roberts, C. T. Plummer, A. S. Kiem, T. D. van Ommen, 2015: Interdecadal Pacific variability and eastern Australian megadroughts over the last millennium. *Geophys. Res. Lett.*, **42**, 129–137.

van Loon, H., 1967: The half yearly oscillations in middle and high southern latitudes and the coreless winter. *J. Atmos. Sci.*, **24**, 472–486.

van Loon, H., 1971: On the interaction between Antarctica and middle latitudes. In: *Research in the Antarctic*, L. O. Quam (ed.), American Association for the Advancement of Science, Washington, D.C., pp. 447–487.

van Loon, H., and G. A. Meehl, 2012: The Indian summer monsoon during peaks in the 11 year sunspot cycle. *Geophys. Res. Lett.*, **39**, L13701.

van Loon, H., and J. C. Rogers, 1978: The seesaw in winter temperatures between Greenland and Northern Europe. I, General description. *Mon. Wea. Rev.*, **106**, 296–310.

van Loon, H., G. A. Meehl, and J. M. Arblaster, 2004: A decadal solar effect in the tropics in July–August. *J. Atmos. Sol. Terr. Phys.*, **66**, 1767–1778. doi:10.1016/j.jastp.2004.06.003.

van Loon, H., G. A. Meehl, and D. Shea, 2007: The effect of the decadal solar oscillation in the Pacific troposphere in northern winter. *J. Geophys. Res.*, **112**, D02108. doi:10.1029/2006JD007378.

van Oldenborgh, G., F. Doblas Reyes, B. Wouters, and W. Hazeleger, 2012: Decadal prediction skill in a multi-model ensemble. *Clim. Dyn.*, **38**, 1263–1280.

Vecchi, G. A., and B. J. Soden, 2007: Global warming and the weakening of the tropical circulation. *J. Clim.*, **20**, 4316–4340.

Vecchi, G. A., B. J. Soden, A. T. Wittenberg, I. M. Held, A. Leetmaa, and M. J. Harrison, 2006: Weakening of tropical Pacific atmospheric circulation due to anthropogenic forcing. *Nature*, **441**, 73–76.

Verdon, D. C., and S. W. Franks, 2006: Long-term behaviour of ENSO: interactions with the PDO over the past 400 years inferred from paleoclimate records. *Geophys. Res. Lett.*, **33**, L06712.

Veretenenko, S. V., V. A. Dergachev, and P. B. Dmitriyev, 2005: Long-term variations of the surface pressure in the North Atlantic and possible association with solar activity and galactic cosmic rays. *Adv. Space Res.*, **35**, 484–490.

Vicente-Serrano, S. M., S. Beguería, and J. I. Lopez-Moreno, 2010: A multiscalar drought index sensitive to global warming: The standardized precipitation evapotranspiration index. *J. Clim.*, **23**, 1696–1718.

Villalba, R., E. R. Cook, R. D. D'Arrigo, G. C. Jacoby, P. D. Jones, M. J. Salinger, J. Palmer, 1997: Sea-level pressure variability around Antarctica since A.D. 1750 inferred from subantarctic tree-ring records. *Clim. Dyn.*, **13**, 375–390.

Visbeck, M., 2009: A station-based southern annular mode index from 1884 to 2005. *J. Clim.*, **22**, 940–950. doi:10.1175/2008JCLI2260.1.

Visbeck M., and A. Hall, 2004: Comments on "Synchronous variability in the Southern Hemisphere atmosphere, sea ice, and ocean resulting from the annular mode" – Reply. *J. Clim.*, **17**, 2255–2258.

Vitart, F., and J. L. Anderson, 2001: Sensitivity of Atlantic tropical storm frequency to ENSO and interdecadal variability of SSTs in an ensemble of AGCM integrations. *J. Clim.*, **14**, 533–545.

Vowinckel, E., 1955: Southern Hemisphere weather map analysis: Five year mean pressures. *Notos*, **4**, 17–50 and 204–216.

Wahl, E., 1942: Untersuchungen über den jährlichen Luftdruckgang. *Veröff. Geophys. Inst. Univ. Berlin*, **4**, 3–71.

Waliser, D. E., and N. E. Graham, 1993: Convective cloud systems and warm pool SSTs: Coupled interactions and self-regulation. *J. Geophys. Res.*, **98**, 12881–12893.

Walker, G. T., 1909: Correlation in seasonal variation of climate. *Mem. Ind. Met. Dept.*, **20**, 122.

Walker, G. T., 1924: Correlation in seasonal variation of weather, IX. *Mem. Ind. Met. Dept.*, **25**, 275–332.

Walker, G. T., 1928: World weather. *Q. J. Roy. Meteorol. Soc.*, **54**, 79–87.

Walker, G. T., and E. W. Bliss, 1932: World weather V. Mem. *Roy. Meteor. Soc.*, **4**, 53–84.

Walker, J. M., 1997: Pen portrait of Sir Gilbert Walker, CSI, MA, ScD, FRS. *Weather*, **52**, 217–220. www.rmets.org/sites/default/files/papers/walkergt.pdf.

Wallace, J. M., 2000: North Atlantic Oscillation/annular mode: two paradigms—one phenomenon. *Q. J. Roy. Meteorol. Soc.*, **126**, 791–805.

Wallace, J. M., and D. S. Gutzler, 1981: Teleconnections in the geopotential height field during the northern hemisphere winter. *Mon. Wea. Rev.*, **109**, 784–812.

Wallace, J. M., C. Smith, and Q. Jiang, 1990: Spatial patterns of atmospheric–ocean interaction in the northern winter. *J. Clim.*, **3**, 990–998.

Wang, B., and X. Xie, 1998: Coupled modes of the warm pool climate system. Part I: The role of air–sea interaction in maintaining Madden–Julian oscillation. *J. Clim.*, **11**, 2116–2135.

Wang, D., and M. A. Cane, 2011: Pacific shallow meridional overturning circulation in a warming climate. *J. Clim.*, **24**, 6424–6439.

Wang, H., and R. Fu, 2000: Winter monthly mean atmospheric anomalies over the North Pacific and North America associated with El Niño SSTs. *J. Clim.*, **13**, 3435–3447.

Wang, H., and V. M. Mehta, 2008: Decadal variability of the Indo-Pacific warm pool and its association with atmospheric and oceanic variability in the NCEP–NCAR and SODA reanalyses. *J. Clim.*, **21**, 5545–5565.

Wang, L., R. Huang, and R. Wu, 2013: Interdecadal variability in tropical cyclone frequency over the South China Sea and its association with the Indian Ocean sea surface temperature. *Geophys. Res. Lett.*, **40**, 768–771.

Wang, H., A. Kumar, and W. Wang, 2013: Characteristics of subsurface ocean response to ENSO assessed from simulations with the NCEP climate forecast system. *J. Clim.*, **26**, 8065–8083.

Wang, Y.-H., G. Magnusdottir, H. Stern, X. Tian, and Y. Yu, 2012: Decadal variability of the NAO: Introducing an augmented NAO index. *Geophys. Res. Lett.*, **39**, L21702. doi:10.1029/2012GL053413.

Wanner, H., S. Bronnimann, C. Casty, D. Gyalistras, J. Luterbacher, C. Schmutz, D. B. Stephenson, and E. Xoplaki, 2001: North Atlantic oscillation. Concepts and studies. *Surv. Geophys.*, **22**, 321–382.

Ward, M. N., and C. K. Folland, 1991: Prediction of seasonal rainfall in the north Nordeste of Brazil using eigenvectors of sea-surface temperature. *Int. J. Climatol.*, **11**, 711–743.

Webster, P. J., and R. Lukas, 1992: TOGA-COARE: The coupled ocean-atmosphere response experiment. *Bull. Amer. Meteorol. Soc.*, **73**, 1377–1416.

Webster, P. J., G. J. Holland, J. A. Curry, and H.-R. Chang, 2005: Changes in tropical cyclone number, duration, and intensity in a warming environment. *Science*, **309**, 1844–1846.

Weeks, W. F., 2010: *On Sea Ice*. University of Alaska Press, Fairbanks. 664 pp.

Weng, H. Y., 2005: The influence of the 11 yr solar cycle on the interannual-centennial climate variability. *J. Atmos. Solar Terrestr. Phys.*, **67**, 793–805.

White, J. W. C., D. Gorodetzky, E. R. Cook, and L. K. Barlow, 1996: Frequency analysis of an annually resolved, 700 year paleoclimate record from the GISP2 ice core, In: R. S. Bradley, et al. (eds.), *Climate Variations and Forcing Mechanisms of the Last 2000 Years*, Springer-Verlag, New York, pp. 193–213.

White, W. B., and Z. Liu, 2008a: Resonant response of the quasi-decadal oscillation to the 11-yr period signal in the Sun's irradiance. *J. Geophys. Res.*, **113**, C01002. doi:10.1029/2006JC004057.

White, W. B., and Z. Liu, 2008b: Non-linear alignment of El Niño to the 11-yr solar cycle. *Geophys. Res. Lett.*, **35**, L19607. doi:10.1029/2008GL034831.

White, W. B., and Y. M. Tourre, 2003: Global SST/SLP waves during the 20th century. *Geophys. Res. Lett.*, **30**. doi:10.1029/2003GL017055.

Wilks, D. S., 1995: *Statistical Methods in the Atmospheric Sciences.* Academic Press, San Diego, CA. p. 467.

Wilks, D. S., 2011: *Statistical Methods in the Atmospheric Sciences.* Academic Press, San Diego, CA. 704 p. ISBN 9780123850225.

Wilson, C. T. R., 1906: On the measurement of the earth-air current and on the origin of atmospheric electricity. *Proc. Cambridge Philos. Soc.*, **13**, 363–382.

Wilson, R. M., 1998: Evidence for solar-cycle forcing and secular variation in the Armagh Observatory temperature record (1844–1992). *J. Geophys. Res. Atmos.*, **103**, 11159–11171.

Wolfram, S., 2002: *A New Kind of Science*. Wolfram Media. p. 998. ISBN 978-1579550080.

Woodruff, S. D., R. J. Slutz, R. L. Jenne, and P. M. Steurer, 1987: A comprehensive ocean-atmosphere data set. *Bull. Amer. Meteorol. Soc.*, **68**, 1239–1250.

Wu, M.-C., K.-H. Yeung, and W.-L. Chang, 2006: Trends in western North Pacific tropical cyclone intensity. *Eos Trans. Amer. Geophys. Union*, **87**, 537–538.

Xie, S.-P., 1999: A dynamic ocean–atmosphere model of the tropical Atlantic decadal variability. *J. Clim.*, **12**, 64–70.

Xie, S.-P., and S. G. H. Philander, 1994: A coupled ocean-atmosphere model of relevance to the ITCZ in the eastern Pacific. *Tellus*, **46A**, 340–350.

Xie, S.-P., and Y. Tanimoto, 1998: A pan-Atlantic decadal climate oscillation. *Geophys. Res. Lett.*, **25**, 2185–2188.

Xie, S.-P., H. Annamalai, F. Schott, and J. P. McCreary Jr., 2002: Origin and predictability of South Indian Ocean climate variability. *J. Clim.*, **15**, 864–874.

Xue, Y., M. A. Balmaseda, T. Boyer, N. Ferry, S. Good, I. Ishikawa, A. Kumar, R. Rienecker, A. J. Rosati, and Y. Y. Yin, 2012: A comparative analysis of upper ocean heat content variability from an ensemble of operational ocean reanalyses. *J. Clim.*, **25**, 6905–6929.

Yang, H., Z. Liu, and H. Wang, 2004: Influence of extratropical thermal and wind forcings on equatorial thermocline in an ocean GCM. *J. Phys. Oceanogr.*, **34**, 174–187.

Yang, J. Y., 1999: A linkage between decadal climate variations in the labrador sea and the tropical Atlantic Ocean. *Geophys. Res. Lett.*, **26**, 1023–1026.

Yang, X., et al., 2012: A predictable AMO-like pattern in the GFDL fully coupled ensemble initialization and decadal forecasting system. *J. Clim.*, **26**, 650–661.

Yeo, S.-R., and K.-Y. Kim, 2015: Decadal changes in the Southern Hemisphere sea surface temperature in association with El Niño–southern oscillation and southern annular mode. *Clim. Dyn.*, **45**, 3227–3242.

Yndestad, H., 1999: Earth nutation influence on system dynamics of Northeast Arctic cod. *ICES J. Mar. Sci.*, **56**, 652–657. doi:10.1006/jmsc.1999.0491.

Yndestad, H., 2003: The code of the long term biomass cycles in the Barents Sea. *ICES J. Mar. Sci.*, **60**, 1251–1264.
Yndestad, H., 2006: The influence of the lunar nodal cycle on Arctic climate. *ICES J. Mar. Sci.*, **63**, 401–420.
Yu, J.-Y., and D. L. Hartmann, 1993: Zonal flow vacillation and eddy forcing in a simple GCM of the atmosphere. *J. Atmos. Sci.*, **50**, 3244–3259.
Yuan, X., and E. Yonekura, 2011: Decadal variability in the Southern hemisphere. *J. Geophys. Res.*, **116**, D19115. doi:10.1029/2011JD015673.
Yuan, X., and C. Li, 2008: Climate modes in southern high latitudes and their impacts on Antarctic sea ice. *J. Geophys. Res.*, **113**, C06S91. doi:10.1029/2006JC004067.
Yumoto, M., and T. Matsuura, 2001: Interdecadal variability of tropical cyclone activity in the western North Pacific. *J. Meteorol. Soc. Jpn.*, **79**, 23–35.
Zanchettin, D., C. Timmreck, O. Bothe, S. J. Lorenz, G. Hegerl, H.-F. Graf, J. Luterbacher, and J. H. Jungclaus, 2013: Delayed winter warming: A robust decadal response to strong tropical volcanic eruptions? *Geophys. Res. Lett.*, **40**, 204–209.
Zanchettin, D., C. Timmreck, H.-F. Graf, A. Rubino, S. Lorenz, K. Lohmann, K. Krüger, and J. H. Jungclaus, 2012: Bi-decadal variability excited in the coupled ocean–atmosphere system by strong tropical volcanic eruptions. *Clim. Dyn.*, **39**, 419–444.
Zebiak, S. E., and M. A. Cane, 1987: A model El Niño-Southern Oscillation. *Mon. Wea. Rev.*, **115**, 2262–2278.
Zebiak, S. E., 1984: Tropical atmosphere-ocean interaction and the El Niño/Southern Oscillation phenomenon. Ph.D. thesis, M.I.T., 261 pp.
Zhang, L., and C. Wang, 2013: Multidecadal North Atlantic sea surface temperature and Atlantic meridional overturning circulation variability in CMIP5 historical simulations. *J. Geophys. Res.-Oceans*, **118**, 5772–5791.
Zhang, L., and T. L. Delworth, 2016: Simulated response of the Pacific decadal oscillation to climate change. *J. Clim.*, **29**, 5999–6018.
Zhang, R., and T. L. Delworth, 2005: Simulated tropical response to a substantial weakening of the Atlantic thermohaline circulation. *J. Clim.*, **18**, 1853–1860.
Zhang, R., and T. L. Delworth, 2006: Impact of Atlantic multidecadal oscillations on India/Sahel rainfall and Atlantic hurricanes. *Geophys. Res. Lett.*, **33**, L17712. doi:10.1029/2006GL026267.
Zhang, Y., J. M. Wallace, and D. S. Battisti, 1997: ENSO-like interdecadal variability: 1900–93. *J. Clim.*, **10**, 1004–1020.
Zhang, Z.-Y., D.-Y. Gong, X.-Z. He, Y.-N. Lei, and S.-H. Feng, 2010: Statistical reconstruction of the Antarctic oscillation index based on multiple proxies. *Atmos. Oceanic Sci. Lett.*, **3**, 283–287.
Zhou, K. Q., and C. J. Butler, 1998: A statistical study of the relationship between the solar cycle length and tree-ring index values. *J. Atmos. Solar Terrestr. Phys.*, **60**, 1711–1718.
Zorita, E., V. Kharin, and H. von Storch, 1992: The atmospheric circulation and sea-surface temperature in the North Atlantic area in winter: Their interaction and relevance for Iberian precipitation. *J. Clim.*, **5**, 1097–1108.

Index